Quantitative Genetics
WITH SPECIAL REFERENCE TO
Plant AND Animal Breeding

Quantitative Genetics

WITH SPECIAL REFERENCE TO

Plant AND Animal Breeding

RALPH E. COMSTOCK

IOWA STATE UNIVERSITY PRESS / AMES

To

L. M. Winters,

Gertrude Cox,

and

H. F. "Cotton" Robinson

Ralph E. Comstock, professor emeritus, Department of Genetics, and Biological Sciences Department, University of Minnesota, St. Paul, received his BS, MS, and PhD degrees from the College of Agriculture, Forestry, and Home Economics, University of Minnesota, St. Paul. Dr. Comstock spent his professional career doing research in and teaching animal breeding, applied statistics, and quantitative genetics. He was also a statistical consultant to researchers in agricultural science. Dr. Comstock has published extensively in scientific journals such as *Journal of Animal Science, Agronomy Journal, Crop Science, Genetics, Theoretical and Applied Genetics,* and *Bionetrics* and in *Proceedings* of various national and international conferences.

Publication was made possible by grants from **Pig Improvement Group Ltd.** and **Pioneer Hi-bred International.**

First edition, 1996

Library of Congress Cataloging-in-Publication Data

Comstock, Ralph E.
Quantitative genetics with special reference to plant and animal breeding / Ralph E. Comstock.—1st ed.
p. cm.
Includes bibliographical references (p.) and index.
ISBN 0-8138-2011-1 (acid-free paper)
1. Plant breeding. 2. Livestock—Breeding. 3. Crops—Genetics. 4. Livestock—Genetics. 5. Quantitative genetics. I. Title.
S494.C66 1996
631.5'23—dc20 94-48902

Contents

Preface

The content of this book reflects in many ways the background and experiences that shaped my enduring special interests. I was raised on small farms in southern Minnesota. In the period between high school graduation and enrollment at the University of Minnesota I was an apprentice on our home farm, and during those four years I was a 4H Club member with swine and maize projects. My 4H experience kindled a beginning awareness of (a) the contributions of research relating to the husbandry of agricultural species and (b) opportunity inequalities associated with economic stratification. The former motivated the choice of agriculture as my major field at the university.

Undergraduate and graduate courses that most excited me were Principles of Genetics, Animal Breeding, Principles of Economics, Biometry, and Applied Statistics. It was my good fortune to be influenced by both L. M. Winters and Jay Lush—two of the early leaders in Animal Breeding. I worked for and later with Professor Winters during the 1934-43 years, first as an abstracter of literature pertaining to animal breeding and genetics and as a caretaker of research project animals and later with responsibility for records and analysis of data from his extensive swine breeding investigations. My contacts with Professor Lush derived from the Minnesota and Iowa participations in the U.S.D.A. Regional Swine Breeding Laboratory in the years when I worked as an assistant to Professor Winters. I well remember critiques by Dr. Lush of drafts of my earliest literature contributions.

A later era very significant for me were the ten years, 1947-57, in which as a member of the Department of Experimental Statistics of the North Carolina State University I was charged by Gertrude Cox with statistical consultation in the area of genetics. This wonderful opportunity provided the open door through which I learned about plant breeding methods and problems from H.F. "Cotton" Robinson, Walton Gregory, Paul Harvey, E.B. Morrow, and T.J. Mann among others. It was during those years that I was senior author of the papers that described Design III for investigation of levels of dominance and reciprocal recurrent selection as a system of choice if overdominance was found to be an important feature of the genetics of any economically important quantitative trait. It was also during those years that I began teaching "Statistical Concepts in Genetics," the embryo from which this book emerged.

It cannot go without mention that my world view was greatly expanded by fifteen months in Puerto Rico as Animal Husbandry head at La Agricultura Experimental Estacion, Rio Piedras. It was there that I began to develop an appreciation of people and problems outside of the U.S.A.—an awareness that since has been amplified by contacts with graduate students such as Abdel Rahman Khalil of Egypt, Belum Reddy and Madho Singh from India, and Hiroshi Fujimaki from Japan.

The focus in this book is on a theory basis for optimizing the decisions involved in the structuring of programs for improvement of genetic materials employed, or to be employed, in agriculture. The theory in this book is limited to diploid and functionally diploid species and constitutes an application of Mendelian genetics. As in all treatments of quantitative genetics the strict validity of what is presented hinges on various simplifying assumptions. An effort is made throughout to keep readers aware that this is so and why it is unavoidable. The possible consequences relative to the central themes of the book are examined in Chapter XVI, and the need for experimental data adequate for validity tests of theory is emphasized. I argue in section 1.5 that theory is essential relative to optimum decisions by breeders. Section 1.8.1 provides an explanation of the symbol system for variances and covariances used throughout this book.

Because change in genotypes and their values results from changes in allele frequencies, theory is derived in terms of the expectations of change per unit of time in the frequencies of individual alleles as a consequence of the mode and intensity of selection practiced and the magnitude of "drift"—the random variation around those expectations to be anticipated as a consequence of the effective size of selection populations. I have attempted to elucidate the theory presented in sufficient detail and at a level of mathematical sophistication such that it can be understood by as many interested students as possible. This attempt was motivated by memories of problems I encountered when first attempting to follow the theoretical developments of the great pioneers of quantitative genetics and my wishes to promote the utilization of theory and to facilitate evaluation, identification of possible errors, and improvements where significant errors are discovered.

There is no formal index. Using the extensive Contents, the reader is directed to the major theories and concepts. Equations that are frequently referred to are numbered and often cross-referenced for ease of location for review or comparison.

The audience for which the book has been written includes all students of quantitative genetics including graduate and post-doctoral trainees and the professionals of plant or animal breeding in public institutions or private enterprises. My overall purpose has been to increase the contribution of quantitative genetics to the welfare of society.

Quantitative Genetics
WITH SPECIAL REFERENCE TO
Plant AND Animal Breeding

1 Introduction

• 1.1. Frame of Reference

The number of people in the world is increasing each day. In the absence of major catastrophe, e.g., a nuclear holocaust or successive crop failures over extensive geographical areas, it is judged that world population will increase from 5.3 billion in 1990 to seven billion or more in the year 2010 (Brown et al, 1990). How soon zero population growth will be achieved is problematical, but as a rough guide for the immediate future, and despite current agricultural surpluses in various parts of the world, we must accept the necessity for a 30 percent increase in the production of human food by 2010 A.D. if dramatic increases in malnutrion and famine are to be avoided. This is the basic reality to which at least three notes should be added.

First, to maintain the present balance between people and food is not enough if progress toward a humane, civilized world is to be continued. There are large numbers of people now living miserable lives on the edge of starvation, hundreds of thousands of children whose future fulfillment is being destroyed by malnutrition. If this is to be a better world for all, food production must increase by a larger proportion than population unless compensating adjustments are made in the food now consumed by those of us at the top of the heap.

Second, to the degree possible, the geographical distribution of the production of basic foods must follow that of population. Otherwise the difficulty of getting food to people will remain or become very difficult. Transportation alone poses very grave problems in a potentially energy-short world, but beyond this are the very troublesome barriers of trade balances among nations. I make a point of the distribution requirement in food production because it emphasizes the imperative of extending the genetic adaptations of the great food and feed crops to the greatest possible array of physical and climatic cónditions.

Third, environments in the world must be expected to change (doubtless are changing) in ways that will affect the productivity of presently superior genetic segments of our agricultural species. Topsoils are being lost by erosion. Ground water reserves are being decreased by irrigation practices. Temperature increases (the greenhouse effect) resulting from increasing concentrations of carbon dioxide and other gases in the upper atmosphere are predicted. Secondary impacts of fertilizers, pesticides, and herbicides may militate against continued use of these products at present levels.

It is against this general background that quantitative genetic issues in breeding will be treated herein. Genetic improvement is by no means the only avenue to greater food production, but its possible contribution is substantial and maximum exploitation of the genetic potential of economic species in now urgent. To make genetic improvements is no longer enough. We need now to become very serious about making the greatest possible genetic improvement within the time frame dictated by the world population growth curve and aspire to the synthesis, in each of the important agricultural species, of the best possible genotypes, for each significant production area, that are potential in the allele resoures of those species. This will be the dominant frame of reference of this book (which will be structured in terms of the decisions required in design of breeding programs aimed at the synthesis of "best possible genotypes").

• 1.2. Quantitative Genetics

This branch of genetics is concerned with quantitative traits; e.g., annual milk production in cattle, grain yield of maize, growth rates of swine or trees, intelligence in humans, etc. The difference between quantitative and qualitative traits resides in the relative magnitude of allele substitution effects. If the effect of substituting one allele for another is large relative to total phenotypic variation, the trait is qualitative . If such substitution effects are small relative to total phenotypic variation, the trait is quantitative. If allele substitution effects are small it is because the trait is affected by numerous genes and/or because a substantial portion of the total variation of the trait is nongenetic in origin. Thus the concepts, multiple factors and heritability, that are unique to quantitative genetics are inherent in the nature of quantitative characters. It is worth noting that our criterion embraces as quantitative traits both those (such as grain yield and growth rate) of which variation is continuous through the range between extremes and those (such as number of locules in certain fruits, number of toes or survival under stress) for which phenotypic variation is discontinuous.

The principal objectives of quantitative genetics are to measure the present or potential genetic variation of quantitative traits and to explain that variation in terms of the information that has been provided by basic research in genetics and cytology. Relevance to plant and animal breeding flows from the fact that the majority of the value traits of our agricultural species are quantitative and consequently most objectives of the breeder pose problems that can only be understood through quantitative genetics.

Population genetics is concerned with genetic structure (allele frequencies, their differences among populations, and the forces that affect them)---quantitative genetics focuses on phenotype. The overlap between them is obvious in origin since the genetic variation of traits must be interpreted in terms of genetic structure of the populations in which variation is measured. Focus on the trait calls for attention to all the genes that affect the trait whereas much of population genetics deals with the gene as its unit. But, clearly, quantitative and population genetics are not separable. The term, quantitative genetics, is used in the title of this text to include all aspects of population genetics required for understanding the variation of quantitative traits

• 1.3. Breeding, Eugenics, and Selection

The term, breeding, is used herein to encompass animal breeding, plant breeding, and eugenics. It denotes programs and procedures aimed at the genetic modification of phenotype in breeds, varieties, lines, hybrids, or populations of economic (or potentially economic) species of plants and animals or of humans. The term, eugenics, is reserved for breeding aimed at promoting genetic change in human populations, an activity area where goals and problems are substantially different from those of breeding programs focused on one or another of the economic species.

The directional force in all breeding programs is selection. Genetic change can occur randomly and a part of the change that does occur in any finite population is inevitably random, but only selection, based on the phenotype of individuals or the averge phenotype of families, can impart direction to genetic change. These are the truly basic facts of breeding: (1) selection is the only force that can give direction to genetic change and (2) phenotype is the only basis for effective selection.[1] The latter fact reflects the underlying truth that the values of genotypes must be inferred from their developmental impacts on phenotype. It should, but might not for all readers, go without saying that phenotype refers here to all of the physical attributes of an organism ranging from things visible or that can be measured externally to molecular or physiological attributes of which measurement or characterization may only be possible using sophisticated instruments and technics.

• 1.4. Total Value and Value Traits

Every trait, qualitative or quantitative, no matter the means by which it is discerned or measured, is a component of the total phenotype of an individual. On the other hand, the total value of phenotype relative to the needs or purposes of agriculture or society can always be defined in terms of a comparatively small subset of traits.[2] As an example, consider the corn plant. How the total value of corn (maize) is properly defined depends on uses made of the components of the plant. If only the grain is utilized and the stalks, leaves, and husks are viewed and treated as waste (as is true for a major portion of the corn grown in the U.S.), total value can be identified with harvestable grain yield and the quality attributes (protein content, oil content, amino acid distribution in the protein, color, etc.) of the grain. These value traits are the consequence of the effects and interactions of a myriad of morphological (root system, ear height, etc.), biochemical (photosynthesis, reaction rates,etc.), and physiological (nutrient transport, disease resistance, etc.) aspects of phenotype, but they serve to integrate the contributions to total value of all the traits on which they themselves depend.

While the number of traits that make direct contributions to total value, and must therefore appear as variables in any satisfactory definition of it, is small relative to the very numerous variable aspects of phenotype, it is not this that I wish to emphasize most particularly. Instead I think it crucial to note (1) that to know total value will invariably require the measurement of several traits, say from 3 to 15, and (2) that because the aspects of phenotype that contribute to variation in those value traits are many, the total number of relevant genes must be large.

Previously, (Comstock, 1977), I have used "phenotypic value" in reference to what here is identified as total value. Change to the latter appeared desirable because of the wide-spread use of phenotypic value relative to physical measurements of individual traits as by Falconer (1981) and Comstock and Moll (1963).

1.5. The Critical Role of Theory

Every practicing breeder knows that there is a vast potential for variation in breeding programs and is plagued by questions of how best to proceed. The phenotypic measures on which selection is based, the relative emphasis placed on the several measures, the nature of units among which selection is practiced, whether and how inbreeding is used, the selection intensity, the size of program (number of individuals or families from which selections are made) and the number and type of environments in which phenotypic evaluations are made represent only a portion of the potential for variation in programs. Thus, if we contemplate identification of optimum procedure patterns through experimentation, it is obvious that a very large number of comparisons would be required. Moreover, by the nature of the breeding enterprise each replication of each variant investigated would be relatively expensive (thousands of dollars per year at minimum). Finally, longtime experiments are required for observation of long-term consequences. This means (1) a proportionate increase in cost that would be involved, but even more to the point, (2) waiting many years for answers needed immediately, if programs are to be mounted for synthesis of "best possible genotypes" within the critical time frame imposed by the continuing expansion of the human population.

It has reasonably been proposed that this cost and time dilemma could be partially resolved by conducting the required experiments using pilot species in which generation time is short and individuals are very small. However, (1)unit costs are high with even the most favorable species, (2) there is inevitable uncertainty involved when data from one species is employed as the basis for decisions regarding procedure to be used in another, and (3) the laboratories now available would not accommodate more than a small portion of required experiments.

All this makes it apparent that most of the many significant questions concerning the effects of potential variations in breeding programs or procedures cannot possibly be answered empirically (by actual experiments). The design decisions of breeding must be based largely on theory (or theory supplemented by intuition and some knowledge of what seems to have worked well in the past). What can be, and to a limited extent has been, obtained from experience with either pilot or economic species is partial information concerning concordance between observations and theory.

1.6. General Purposes

This book is devoted primarily to (a) discussion of major design decisions required of the breeder, (b) description of a body of theory for comparison of breeding systems and variations of breeding systems, (c) identification and discussion of information required for application of the theory, of the extent to which such information is available and of ways (or the need for ways) in which it can be obtained, (d) directions in breeding that are indicated by the theory and evidence available, and finally (e) some discussion of policy (national and international) implications.

1.7. Contents of Appendix 1

Probability and statistical concepts involved in the theory presented are with few exceptions limited to those covered in any of the familiar texts on applied statistics. However, for the convenience of readers, some aspects of particular importance relative to material in following chapters are reviewed in this appendix. Presentations in the main text will assume familiarity by the reader with the concepts, terminology and symbol systems outlined therein.

1.8. Some Aspects of Book Organization

Appendices are employed more extensively than in many other textbooks. This was done to enable inclusion of material (a) required by some but not all readers or(b) likely to interest some readers but not others, without undue clutter in the presentation of primary themes.

Principal chapter sections are identified by two numbers. The first is the number of the chapter and the second is the serial number of the section within its chapter, e.g., the third main section of Chapter V is section 5.3. Subsections are identified by three numbers and sub-subsections by four. Thus section 6.3.2 refers to the second subsection of the third main section of Chapter VI.

Both equations and tables are identified in the same way as principal chapter sections. Thus the eleventh table or numbered equation in Chapter XIV is labeled table (or equation) 14.11. Some equations are presented again, for the reader's convenience, in a later chapter or a later section of the same chapter Then two numbers are shown. The first, enclosed in parentheses, is its original number. With few exceptions, only equations referred to later in the text are numbered. Exceptions were unintended. References to equations are usually by number alone. To avoid confusion the number is always preceeded by "section" or "table" in reference to sections or tables.

Symbols, adequately explained or defined, can save many words and simplify many statements. As in most books of this general kind, they are used extensively herein. A small number are discussed below, others are reviewed in Appendix 1 or defined at the point of first use. Once defined, they are used repeatedly for the same purpose, without redefinition whenever that appears reasonable. An effort has been made to avoid using any particular symbol for more than one purpose. However, there are exceptions. For example, the letter k has been assigned one special meaning (in Chapter V) when used by itself but different meanings when used as a subscript. As another example the large case letter Z is used from time to time as a "stand in space saver"for some more complicated expression needing frequent repetition within the discussion of a particular issue. However, care is taken to identify its meaning within the context of each discussion in which it is employed. In general, because different authors have, for reasons of their own, employed symbol sets that differ in some respects from those used by others, no effort has been made to conform to symbols that have been employed elsewhere in the quantitative genetic literature. Instead, the symbolism employed here is reasonably consistent with that of previous publications of which I have been the author or a joint author.

1.8.1. Symbol System for Variances and Covariances

The symbol system employed for variances and covariance is different from that in most, if not all, other books and merits immediate description. With few exceptions only population values (the parameters) are considered or discussed. The greek letter σ is employed, as by many authors, in the identification of variance, standard deviation, or covariance. However, instead of using subscripts to identify variables involved, the appropriate symbol or pair of symbols is enclosed in parentheses following the primary symbol. In the case of covariance symbols for the two variables are always separated by a comma. For example, in the system employed $\sigma^2(W)$ stands for the population variance of the variable symbolized by the letter W, and $\sigma(X_1,X_2)$ stands for the covariance between the variables X_1 and X_2. The greek letter β is employed to indicate regression coefficients. For example, $\beta(W,Y)$ stands for the regression of W on Y. Note that in the case of regression coefficients the symbol for the dependent variable will always appear first within the parentheses.

2 Genetic Background

This is a stage-setting chapter in which (a) the genetic mechanism assumed as the basis for theory to be presented in later chapters and (b) selected elements and ideas from population, quantitative, and molecular genetics, are reviewed.

2.1. Genetic Model—The Assumptions

Because deduction proceeds from premises and because theory to be presented herein is deductive theory, statements concerning genetic assumptions are in order at the outset.

The crucial assumption made throughout (as in the whole body of standard quantitative genetic theory) is that mitosis, meiosis, gametogenesis and fertilization follow patterns described as normal in texts devoted to cytology and genetics. For example, barring specific notice to the contrary meiotic drive is assumed not to be present as a force affecting gametic ratios. This is probably justified; as Hartl (1977) stated, "...meiotic drive is apparently an unusual phenomenon. Mendel's law of segregation is still safe". However, meiotic drive serves well to emphasize the dependence of deductive theory on premises because there can be little doubt that the utility of theory to be elaborated would be gravely affected if, in most populations, the segregations of numerous genes were much disturbed by meiotic drive.

The genetic phenomena obviously relevant to quantitative genetics include the linear arrrangement of genes in chromosomes and the consequent genetic linkage, dominance, pleiotropy , epistasis, mutation, multiple alleles, sex linkage, cytoplasmic inheritance, ploidy, and genotype-environment interaction.

This same list would have been presented at any time within the two to three decades prior to the flood of information from molecular genetics during the last 45 years. This does not mean, however, that our new knowledge of such basic things as the structure, replication, and regulation of genes is not relevant to quantitative genetics. On the contrary, this newer knowledge contributes immeasurably to our insights concerning the components of the general model for quantitative genetics. When the gene is viewed as a strand of DNA the types of mutation that are possible and the vast potential for multiple alleles can be perceived. In like manner, our growing knowledge of developmental genetics and of gene regulation and the mechanisms by which they are, or may be, accomplished is the source of new insights (realized and potential) concerning the mechanisms of epistasis. However, it is the fact of replication not the mechanism, the fact and frequencies of mutations not the physical or chemical nature of changes involved, and the individual and joint effects of genes on phenotypes not the mechanisms by which those effects are mediated that are relevant to the understanding required for predicting population changes in quantitative traits to be expected in breeding programs. Thus the lack of recent change in the premises of quantitative genetics is easily rationalized though complete justification can hardly be guaranteed.

Regarding the genetic models actually employed for the purposes of this book, the reader will learn that all the quantitative theory presented rests on selected simplifying assumptions concerning phenomena listed at the beginning of this section. This is not unusual. There is no quantitative theory in quantitative or population genetics that has had utility and does not rest on simplifying assumptions. The general procedure herein is to employ the simplifications required to make the mathematics (1) manageable and (2) productive in the sense that approximations derived are simple enough in form to be comprehended and used. This involved judgements concerning the effective background of the audience for which the book is written and was affected by my own limitations. The obvious risk is that theory which is rigorous, or nearly so, relative to a partially unrealistic model will be seriously in error relative to real world situations.

Kempthorne (1988) in his Overview emphasized (1) that the mathematical deductions of which available quantitative genetic theory is comprised are all premised on simplified, partially unrealistic models and (2) that this is inevitable because mathematics cannot deal with the complexity of a model completely faithful to the biology of quantitative traits. Though he speaks almost scornfully of our present theory he states, "It is the case that any area of science develops from interplay of theory and observations ·····. It is critically the case that without some theory, which may be appallingly naive ····· one has no idea about what to observe or about what (controlled) experiment to do", and later, "It is true, of course, that we have a rather simple and naive theory that enables us to make some comparisons of various schemes of selection." I agree with those statements and assert that he thereby has provided ample justification for attempting an orderly and explanatory presentation of available theory concerning the issues deemed to be of particular significance. The theory that can be produced mathematically, despite simplifying assumptions required, serves at minimum as an aid to effective designs of experiments and breeding programs. Further, it can be argued (I would) that the benchmarks provided by such theory will always be an essential ingredient of credible interpretations of data from programs involving living organisms or from massive explorations via modern computors (as proposed by Kempthorne).

2.1.1. Summary Relative to Models and Assumptions Employed

1. All theory presented is concerned with diploid, or functionally diploid, organisms. This is not a major limitation since the animal species are diploid and the majority of important agricultural plants are either diploid or allopolyploid.

2. Sex-linked genes are ignored. Of course there are none in monoecious species and it is my judgment that in most of the important dioecious species only a small proportion of genes affecting phenotypic value are sex-linked. Poultry and, among the important pilot organisms, Drosophila are apparent exceptions.

3. Most of the quantitative theory presented assumes no epistasis and no multiple alleles. However, some interrelations between epistasis and theory presented are discussed in the context of problems addressed in this book.

4. Linkage equilibrium is assumed in part of the work but because that assumption is not tenable in various important situations, effects of linkage (in the absence of multiple alleles and epistasis) are quite thoroughly investigated.

5. Mutation is ignored except in the context of expectations over long time periods.

6. Barring specific notice to the contrary, restrictive assumptions relative to dominance, pleiotropy or genotype-environment interaction are not made.

Some evidence concerning magnitudes of errors that result from simplifying assumptions is presented. Care is taken throughout to draw attention to all assumptions and to identify those considered most critical to the utility of derived theory.

2.2. Populations

A population is a group of things, either finite or infinite in number, of a specified class. "Things" is used here in a very general sense. They may have physical substance as do (1) cows on U.S. farms or (2) 40-watt light bulbs, or lack physical substance as do (1) the cattle that could be produced by crossing Shorthorn bulls with Hereford cows or (2) prices paid for radios in New York City. There are innumerable ways in which a "class of things" can be specified; e.g.

(a) Cattle in general

(b) Cattle in Indiana at a specified time

(c) Cattle of the Jersey breed at a specified time

(d) Cattle less than 12 months of age at a specified time

(e) Cattle in the F_2 generation of a cross between random animals of the Jersey and Ayrshire breeds

In the case of (a) the only limitation placed on "things" belonging to the population is that they be cattle. In the other four cases, besides being cattle they must meet one more specification which has to do with geographical location (case b), their ancestry (cases c and e) or their age (case d). Inherent in the concept of population is the concept of relationship. The members of a population have one or more common attributes that distinguish them from nonmembers. It is these common attributes that give the population unity and reality.

The difference between finite and infinite populations is, of course, that the latter have no limit to their size, which means that they can have only conceptual reality. The automobiles owned at a given time by the citizens of Boston constitute a finite population whereas the plants that could be produced in the F_2 generation of a cross between two inbred lines of corn constitute an infinite population. Obviously the number actually produced must be finite but conceptually there is no limit to the number that could be produced; i.e., the population of which the members actually produced are only a sample.

In the context of plant and animal breeding it is worth distinguishing betweeen what can be called static and evolving genetic populations. The static genetic population has a relatively constant base from which new samples of the same infinite population can be generated whenever desired.[1] In the case of a double cross corn hybrdid, the four homozygous parent lines are genetically stable (except for the slow change from unrecognized mutation) so that each time the double cross is made a new sample from the same population is generated. The static nature of a strawberry variety hinges on the asexual mode of reproduction; each new plant has a genotype that is an exact copy (except for rare somatic mutations) of the original variety genotype and new plants of the variety can be generated whenever desired and in whatever number desired. It should be noted, as illustrated by our two examples, that the members of static genetic populations can be genetically uniform or genetically heterogeneous.

The evolving population differs from the static by reason of sexual reproduction by individuals having heterozygous genotypes. Genetic change due to selection, drift, or both is characteristic of the evolving population.[2] It is worth noting that at each point in time any evolving population is a conceptually infinite population but as time passes the chance to draw a new sample from the same population is lost. Consecutive samples constitute the ever changing base (by reason of selection and drift) from which newer samples must be drawn.

2.2.1. Random Mating

The terms, random mating and random mating population, are widely used and will be encountered frequently in the pages that follow. The random mating concept is important in population genetics because the attendant analytical theory is simpler. It is important in breeding by reason of results expected from it.

Definitions of random mating vary (and often lack rigor). Herein the term will be employed quite literally to mean that each member of each mating pair is a random individual from among all those in the population that have reached and not passed reproductive age. In breeding program populations, individuals that have been rejected, rather than selected, for use as parents are, for the purposes of our definition, considered to have been eliminated from the population. In the world of animals a mating pair includes the female inseminated and the male by which she is inseminated (whether naturally or artificially). More than one zygote may result from the mating. In plants the mating must be identified with the fertilization event and consists of the plant that produced the egg cell and the plant that produced the pollen grain.

Note that absolutely random mating is only possible in monoecious organisms. In the case of dioecious species random mating will be understood to mean that each female has the same chance (1/F) of being the female member of each mating pair and each male has the same chance (1/M) of being the male member. Here F and M represent the numbers of reproductive age females and males, respectively, in the population.

Probably the greatest deviation from random mating is seen in the many species where self-fertilization is the rule and cross-fertilization the exception and where no close approximation to random breeding is posssible (except by human manipulative intervention). Lesser deviations from random mating result from geographical proximity and from assortative mating (for example, there is moderate positive correlation between human mates in both stature and I.Q.). In the context of breeding programs geographical proximity is not as important as in natural populations where it has obvious implications for intrapopulation genetic gradients.

• 2.2.2. Migration

The boundaries between genetic populations are often not distinct and absolute. In nature this is frequently the consequence of geography. Plant populations centered in neighboring river basins may encounter each other at low density on the dividing ridges. In the same way valleys may constitute imperfect barriers between populations within alpine species. In the world of domestic species wind transport of pollen can blur the boundaries between open pollinated varieties of corn and individual zeal can lead to violation of the artificial boundaries between breeds of livestock. The imperfect separation of genetic populations is recognized in population genetics by the notion of migration. Quantitatively, migration has to do with the fraction of the total gametic input to a population that comes from any other population. Considering all the swine at any point in time that are officially registered in the Hampshire herd book, what proportion of their immediate ancestry was other than animals registered as Hampshires? That fraction would be the rate of migration into the Hampshire population of swine.

• 2.3. Allele Frequency

Allele frequencies are attributes of populations, groups or individuals. The frequency of an allele in a population or group is its number in that population or group as a fraction of the total number of loci at which it or another allele of the same gene is present. Consider a finite population of a diploid organism consisting of 100 individuals. Suppose that three alleles of a specific gene are found in the population and let them be symbolized as B_1, B_2, and B_3. Suppose further that the numbers of genotypes at the locus of this gene are as follows:

Genotype	Number
B_1B_1	8
B_1B_2	17
B_1B_3	23
B_2B_2	16
B_2B_3	30
B_3B_3	6
	100

The total number of loci for the B gene is 2 x 100 = 200 and the allele frequencies are as follows:

B_1 $[(8 \times 2) + 17 + 23]/200 = 0.28$

B_2 $[(16 \times 2) + 17 + 30]/200 = 0.395$

B_3 $[(6 \times 2) + 23 + 30]/200 = 0.325$

The total of these frequencies is 1.0, which illustrates the elementary fact that in all cases the sum of the frequencies of the alleles of each gene must add to one. In this book I will consistently use Q to symbolize allele frequency in an individual and q to symbolize allele frequency in a population. Note that the average over the individuals of a population of the individual allele frequencies is equal to the allele frequency for the population. In the above example Q_1, the frequency of B_1, is

1.0 in each of the 8 B_1B_1 genotypes,

0.5 in each of the 17 B_1B_2 and the 23 B_1B_3 genotypes, and

0.0 in each of the B_2B_2, B_2B_3, and B_3B_3 genotypes.

The average of Q_1 over all of the 100 individuals is $[8(1) + (17 + 23)(0.5)]/100 = .28 = q_1$ which is the B_1 allele frequency of the population.

The determination or estimation of population allele frequencies is an important basic enterprise of population genetics. The process is most direct when genotypes can be identified by visible phenotypic effects and dominance is not complete so that all genotypes can be distinguished on the basis of phenotype. Consider coat color in Shorthorn cattle. There is an allele pair that controls the red-white contrast. Animals homozygous for the R allele are red, those homozygous for the other member of the allele pair (W) are white and most of the heterozygotes are roan. It follows that the frequencies of R and W can be directly estimated from coat color proportions in an adequate sample from the breed. The obvious necessary qualification is that for all we know there may be two or more variations (allelic forms) of R that are not distinguishable on the basis of phenotype and that the same might be the case for W. Variant forms of R or W might explain the fact that a small proportion of heterozygous animals are not roan.

There are other instances of alleles with invariable visible effects on the phenotype that are not completely dominant so that both homozygous and heterozygous genotypes can be separately identified but the total of these amounts to only a very small fraction of all the genes in any organism.

Another category of genes for which allele frequency estimation is straightforward are those sex-linked genes of which alleles have visibly different phenotypic effects. Incomplete dominance is not a requirement in these cases because in any event the necessary data (counts) can be obtained from individuals of the heterogametic sex.

In still other cases, more or less elaborate laboratory tests enable identification of what might be the complete set of genotypes possible at a specific locus. Probably the best known example is that involving genes responsible for blood antigen differences. The oldest example is the well-known A-B-O group of human blood phenotypes (Bernstein, 1925). Extensive information on the frequencies of the A, B, and O alleles in numerous human populations has accumulated in the literature. The point made earlier, that what we identify as a single allele on the basis of a specific kind of information; e.g., the coat colors in Shorthorns that are distinguishable by the human eye, may really be a subgroup of alleles, has been demonstrated by the A-B-O data. More sophisticated tests have revealed variants of the A allele (Race and Sanger, 1962). We now have information on blood antigen allele frequencies for a rather large number of genes in humans and for a number of genes in several animal species (for examples of data available see Lerner and Libby, 1976).

Highly significant for population genetics has been establishment of the fact that proteins are primary gene products, that a very large class of allele differences are reflected in the amino acid sequences of the protein coded by the specific gene, and the development of electrophoresis technics for distinguishing some variant forms (but not all) of any particular protein. Very substantial additions to our knowledge of allele frequencies have resulted from application of these electrophoretic technics. Pioneer work in this area was reported by Lewontin and Hubby (1966). Scores of important contributions in this category have been published since 1966 and many were reviewed by Lewontin (1974).

In the case of autosomal genes with visible phenotypic effects but with one allele completely dominant to the other, the observed frequency of the recessive phenotype provides a basis for estimation of the frequencies of the two alleles. In the next section we will see that, in the absence of selection, the probability of the recessive phenotype in a random mating population is the square of the frequency of the recessive allele. Thus the square root of the frequency of the recessive phenotype is an estimate of the frequency of the recessive allele. An example is provided by coat color in Angus cattle. The normal color is black but there is a recessive allele that induces red color when homozygous. Lush (1937) stated that one animal in 100-200 is red, that is, the frequency of the recessive phenotype is .005 - .01 and the frequency of the red allele is estimated to be .07 - .10. This means that the frequency of the black allele is 0.9 - 0.93.

Despite the information and information sources now available, questions about allele frequencies remain among the critical problems of population genetics. For example, while allele frequency information has been or can be obtained on many genes using electrophoretic technics we are confronted by such issues as these:

(1) Are the genes for which information is obtained a random sample from the whole gene set of an organism?

(2) What portion of the genes for which information is obtainable have a significant bearing on fitness or, in the case of economic species, on any one of the important quantitative traits of the organism?

The significance of allele frequencies in population biology can hardly be overemphasized. To the extent that phenotype is genetically determined, it is determined by the alleles (the forms) of the various genes that are present in the individual. It follows that the array of phenotypes found in a population must be a strong function of the allele frequencies of that population and that the causes of allele frequency changes must be particularly relevant to the goals of plant and animal breeding. Those causes are widely recognized to be selection, mutation, migration and chance.

2.4. Genotype Frequencies in Random Mating Diploid Populations

A random mating population was defined above as one in which each member of every mating pair is a random individual from among those in the population that are of reproductive age. It follows (1) that the allele contents of the gametes involved in any fertilization are mutually independent and (2) that, if there is no meiotic drive or gamete selection, the probability that a specific allele will be present in any effective gamete is the frequency of that allele in the reproductive age portion of the population.

2.4.1. Single-locus Genotypes

Consider any autosomal gene and let any two alleles of the gene be symbolized as B_k and B_m and the corresponding allele frequencies in potential parents (the reproductive age portion of the population) as q_k and q_m. The probability that a random gamete carries B_k is q_k and the probability that uniting gametes both carry B_k is $q_k \times q_k = q_k^2$. If B_m is any allele of the gene other than B_k, the probability that the male member of a pair of uniting gametes carries B_k and the female member B_m is $q_k q_m$ and the probability that the male member carries B_m and the female member B_k is also $q_k q_m$. (The assumption here has been that allele frequencies are the same in males and females though this may not always be the case.) The implications of all this are as follows.

$$\text{Probability of the } B_kB_k \text{ genotype} = P(B_kB_k) = q_k^2 \qquad 2.1$$

$$\text{Probability of the } B_kB_m \text{ genotype} = P(B_kB_m) = 2q_kq_m \qquad 2.1$$

These are known as the Hardy-Weinberg probabilities (Hardy, 1908; Weinberg, 1908). The reader should take careful note of the fact that in deducing these probabilities no assumption was made concerning the relative frequencies of genotypes among potential parents. The HW (Hardy-Weinberg) probabilities of single locus genotypes in offspring depend only on "random mating".

The special case of two alleles is most frequently discussed and will be a part of the models employed as the basis for most of the theory presented in this book. When two alleles are assumed the symbols employed to represent them will be B and b, B being used to symbolize the more favorable of the two alleles.[3] The frequency of the allele designated as B will be symbolized as q and, since the frequencies of the two alleles must add to 1.0, the frequency of the b allele can then be symbolized as 1 - q. Locus will frequently be specified by a subscript. Then B_i will represent the more favorable of two alleles of the i-th locus gene. Using these symbols in application of equations 2.1 to the two allele case we obtain the following probabilities for the three possible i-th locus genotypes in a random mating population.

Genotype	Probability
B_iB_i	q_i^2
B_ib_i	$2q_i(1 - q_i)$
b_ib_i	$(1 - q_i)^2$

The sum of the probabilities is 1.0 as it should be when all of the genotypes are properly accounted for.

$$q^2 + 2q(1 - q) + (1 - q)^2 = [q + (1 - q)]^2 = 1.0$$

• 2.4.2. **Probabilities vs. Frequencies**

Consider the F_2 generation of a cross between Angus and Red Dane cattle. Because the black-red coat color contrast is determined by a single gene of which the allele for black is dominant to the allele for red, the probabilities that a random F_2 individual (that had black F_1 parents) will be black or red are 3/4 and 1/4, respectively. Given a sample, say 200, of F_2 animals actually produced, it could turn out that 158 were black and 42 red. These (158 and 42) would then be the absolute frequencies of black and red animals. Dividing these numbers by n (the number in the sample) we obtain the relative frequencies (.79 and .21) of black and red. These are similar, but not identical, to the probabilities, .75 and .25, of black and red animals. Probabilities can be and frequently are referred to as expected relative frequencies. These and actual relative frequencies in finite samples will never be exactly the same except by coincidence. On the other hand, the sampling deviations of relative frequencies from the corresponding true probabilities become small as sample size increases. This is reflected in a quantitative way by the standard error of an observed relative frequency, $[p(1 - p)/n]^{1/2}$, where p is the probability of the event in question and n is sample size. Applied to our illustration, we find that the standard error of the observed relative frequency of either black or red animals is $[(.75)(.25)/200]^{1/2} = .0306$. Had the sample included only 50 individuals the standard error would have been twice as large (.0612); if it had been of size 5,000 the standard error would have been reduced to .0061.

• 2.4.3. Two-locus Genotypes

There are two situations to be considered. These are known as "linkage equilibrium" and "linkage disequilibrium". Linkage equilibrium is the special situation in which the probabilities of gametic genotypes are products of the population frequencies of the alleles present in the gametes, as they are when the allele at the i-th locus is an event independent of the allele present at the j-th locus. Then, for example, the probability of a gamete containing the second allele of the i-th gene and the fourth allele of the j-th gene would be $q_{i2}q_{j4}$. Assuming only two alleles per gene, the probabilities of two locus gametes, given linkage equilibrium, would be as follows.

Gamete Genotype	Probability
B_iB_j	q_iq_j
B_ib_j	$q_i(1 - q_j)$
b_iB_j	$(1 - q_i)q_j$
b_ib_j	$(1 - q_i)(1 - q_j)$

Then if gametes were combined in random pairs, as they would be in a random mating population, the gamete pairs, their probabilities, and the resulting genotypes would be as follows:

Gamete Pairs	Probability	Resulting Genotype
B_iB_j B_iB_j	$q_i^2 q_j^2$	$B_iB_iB_jB_j$
B_iB_j B_ib_j	$2q_i^2q_j(1 - q_j)$	$B_iB_iB_jb_j$
B_ib_j B_ib_j	$q_i^2(1 - q_j)^2$	$B_iB_ib_jb_j$
B_iB_j b_iB_j	$2q_i(1 - q_i)q_j^2$	$B_ib_iB_jB_j$
B_iB_j b_ib_j	$2q_iq_j(1 - q_i)(1 - q_j)$	$B_ib_iB_jb_j$
B_ib_j b_iB_j	$2q_iq_j(1 - q_i)(1 - q_j)$	$B_ib_iB_jb_j$
B_ib_j b_ib_j	$2q_i(1 - q_i)(1 - q_j)^2$	$B_ib_ib_jb_j$
b_iB_j b_iB_j	$(1 - q_i)^2q_j^2$	$b_ib_iB_jB_j$
b_iB_j b_ib_j	$2(1 - q_i)^2q_j(1 - q_j)$	$b_ib_iB_jb_j$
b_ib_i b_jb_j	$(1 - q_i)^2(1 - q_j)^2$	$b_ib_ib_jb_j$

When the uniting gametes are genetically dissimilar the probability is twice the product of the gamete frequencies to account for both the case where the male gamete is of the first kind and the female gamete of the second kind and the reverse case where the male gamete is of the second kind and the female gamete is of the first kind. With the exception of the fifth and sixth pairs of gemetes listed each pair yields a different genotype and the probability of the pair is the probability of the resulting genotype. Once the fifth and sixth probabilities have been summed to give $4q_iq_j(1-q_i)(1-q_j)$ as the probability of the $B_ib_iB_jb_j$ genotype we find (1) that the probabilities of the single locus genotypes are their Hardy-Weinberg probabilities, in accord with Section 2.4.1, and (2) that the probability of each two locus genotype is the product of the probabilities of the single locus genotypes of which it is composed. The first of these facts is demonstrated by identifying the two locus genotypes that contain a specific single locus genotype and summing their probabilities to obtain the probability of the the single locus genotype in question. For example, B_iB_i is found in the first three of the ten genotypes listed and the sum of the corresponding probabilities is q_i^2. When this is done for each of the i-th locus genotypes the following expected results are obtained.

Genotype	Probability
B_iB_i	$q_i^2q_j^2 + 2q_i^2q_j(1-q_j) + q_i^2(1-q_j)^2 = q_i^2$
B_ib_i	$2q_i(1-q_i)q_j^2 + 4q_iq_j(1-q_i)(1-q_j) + 2q_i(1-q_i)(1-q_j)^2 = 2q_i(1-q_i)$
b_ib_i	$(1-q_i)^2q_j^2 + 2(1-q_i)^2q_j(1-q_j) + (1-q_i)^2(1-q_j)^2 = (1-q_i)^2$

The second of the facts noted above is clear from inspection of the two locus genotype probabilities. When the probabilities of multilocus genotypes are the products of the Hardy-Weinberg probabilities of the component single locus genotypes, they are sometimes identified as extended Hardy-Weinberg probabilities.

It will now be demonstrated that H-W probabilities at individual loci are not dependent upon the linkage equilibrium state. Consider the cross of two homozygous lines with attention to two loci, on the same chromosome, at which the two lines are homozygous for different alleles. Suppose line 1 has the two locus genotype $B_1B_1b_2b_2$. Then F_1 individuals must all be repulsion double heterozygotes, B_1b_2/b_1B_2, and the allele frequencies are $q_1 = q_2 = 0.5$. The two locus genotypes of gametes produced by the F_1 are listed below along with their probabilities and the products of F_1 frequencies of the alleles contained in the genotypes.

Gametic Genotypes	Probabilities	Products of Allele Frequencies
B_1B_2	$r/2$	1/4
B_1b_2	$(1-r)/2$	1/4
b_1B_2	$(1-r)/2$	1/4
b_1b_2	$r/2$	1/4

Here r is the probability of recombination between the two loci. The condition for linkage equilibrium; i.e., gamete probabilities equal to the product of frequencies of the alleles contained in the gamete genotyupe, is obviously not met unless there is independent assortment so that $r = 1/2$. Despite this the probabilities of the BB, Bb and bb genotypes in the F_2 generation are $q^2 = 1/4$, $2q(1 - q) = 1/2$ and $(1 - q)^2 = 1/4$, respectively, at both loci. Going on to the probabilities of two locus genotypes in the F_2 we find, for example, that the probability of the $b_1b_1b_2b_2$ genotype is $(r/2)^2$ instead of $(1 - q_1)^2(1 - q_2)^2 = 1/16$ and that of the probabilities of two locus genotypes in the F_2 we find, for example, that the probability of the $b_1b_1b_2b_2$ genotype is $(r/2)^2$ instead of $(1 - q_1)^2(1 - q_2)^2 - 1/16$ and that of the $b_1b_1B_2b_2$ genotype is $2[(1 - r)/2][r/2] = r(1 - r)/2$ instead of $2(1 - q_1)^2q_2(1 - q_2) = 1/8$. But, given independent assortment so that $r = 1/2$, $(r/2)^2 = 1/16$ and $r(1 - r)/2 = 1/8$.

The above example suggests that independent assortment might insure linkage equilibrium in a random breeding population. That this is not the case will be demonstrated by another example. Consider a finite population of a monoecious organism such as maize. For the sake of an easily manageable example we will assume only five plants with two locus genotypes as listed below. The probabilities of the four two locus gametic genotypes (assuming independent assortment) are also listed for each of the five individuals.

Individual	Genotype	Gametic Genotype Probabilities			
		B_1B_2	B_1b_2	b_1B_2	b_1b_2
1	$B_1b_1B_2b_2$	1/4	1/4	1/4	1/4
2	$b_1b_1B_2b_2$	0	0	1/2	1/2
3	$B_1b_1B_2B_2$	1/2	0	1/2	0
4	$b_1b_1B_2B_2$	0	0	1	0
5	$B_1B_1b_2b_2$	0	1	0	0
	Averages	0.15	0.25	0.45	0.15

It is clear by inspection of the genotypes that $q_1 = 0.4$ and $q_2 = 0.6$. The probabilities of gametic genotypes in random gametes (provided by the averages of the above table) are listed below along with products of the frequencies of the alleles present in the genotype.

Gametic Genotype	Probability	Product of Allele Frequencies
B_1B_2	0.15	0.24
B_1b_2	0.25	0.16
b_1B_2	0.45	0.36
b_1b_2	0.15	0.24

The differences between actual probabilities and the allele frequency products represent deviations from linkage equilibrium. Here the absence of linkage equilibrium is not the result of linkage (i.e., absence of independent assortment). It results instead because the genotype frequencies of the parent population are not themselves in the linkage equilibrium state; they could not be in this example because the parents are fewer in number than the possible two locus genotypes, but it is true in general that the relative frequencies of genotypes will only be equal to the genotype probabilities by coincidence when the population is finite.

In summary, (1) in a random mating population the probabilities of single locus genotypes are always the H-W probabilities given by the 2.1 equations but (2) the multilocus genotype probabilities of a random breeding population in a particular generation may (usually will) deviate from the extended Hardy-Weinberg probabilities; i.e., from the linkage equilibrium state, as the result of deviation of genotype frequencies from the linkage equilibrium state in the preceding generation. Linkage equilibrium and deviations from it will be considered in greater detail in later chapters.

2.5. Inbreeding and Its Effects on Expected Frequencies of Genotypes

Inbreeding, the mating of related individuals, and crossbreeding are the forms of nonrandom mating that will be of most frequent concern in this book. The effects of inbreeding on genotype frequencies have been examined and reported by many geneticists and mathematicians. (See Kempthorne, 1957; Crow and Kimura, 1970; or Wright, 1969, for discussions of inbreeding and references to much of the important literature.)

The basic significant effect of inbreeding is that the probability of homozygous genotypes is greater and the probability of heterozygous genotypes is less, at every locus, in offspring from matings between relatives, i.e. in inbred individuals, than in offspring from matings between unrelated individuals. The <u>coefficient of inbreeding</u>, defined by Wright (1921) as the "correlation between uniting gametes", quantifies this effect in unselected diploid organisms and is very widely used as a measure of the intensity of inbreeding. Let f be the coefficient of inbreeding of an individual and consider the probabilities in that individual of the genotypes involving any two alleles, B_k and B_m, of any gene. In general, those probabilities are

$$P(B_kB_k) = (1 - f)q_k^2 + fq_k = q_k^2 + fq_k(1 - q_k) \qquad 2.2$$

and $$P(B_kB_m) = 2q_kq_m(1 - f) \qquad 2.2$$

where the q's represent allele frequencies in the random mating population of which all noninbred ancestors of the inbred individuals are assumed to have been members.

When f = zero, i.e., when there is no inbreeding, equations 2.2 reduce to equations 2.1. Note that f measures the expected decrease in frequency of every single locus heterozygous genotype and hence the expected decrease in overall heterozygosity. Algebraically,

$$(1 - f)H_0 = H \quad \text{or} \quad f = (H_0 - H)/H_0 \qquad 2.3$$

where H_0 is the amount of heterozygosity expected in the absence of any inbreeding and H is the expected amount of heterozygosity in individuals with coefficient of inbreeding equal to f.

As the simplest application of equations 2.2, consider a gene with two allelic forms and a population in which all individuals are equally inbred. The expected frequencies of genotypes are as follows:

Genotype	Expected Frequency
BB	$q^2 + fq(1 - q)$
Bb	$2q(1 - q)(1 - f)$
bb	$(1 - q)^2 + fq(1 - q)$

Note that the increase in homozygosity over that expected when mating is random is equally divided between the two homozygous genotypes. The total increase in expected frequency of homozygous genotypes is of course equal to the decrease in expected frequency of the heterozygous genotype.

The expected frequencies of genotypes in a population in which all individuals are equally inbred are shown below for the case of a gene with three alleles (B_1, B_2, and B_3).

Genotype	Expected Frequency
B_1B_1	$q_1^2 + fq_1(1 - q_1) = q_1^2 + fq_1q_2 + fq_1q_3$
B_1B_2	$2q_1q_2(1 - f) = 2q_1q_2 - 2fq_1q_2$
B_1B_3	$2q_1q_3(1 - f) = 2q_1q_3 - 2fq_1q_3$
B_2B_2	$q_2^2 + fq_2(1 - q_2) = q_2^2 + fq_1q_2 + fq_2q_3$
B_2B_3	$2q_2q_3(1 - f) = 2q_2q_3 - 2fq_2q_3$
B_3B_3	$q_3^2 + fq_3(1 - q_3) = q_3^2 + fq_1q_3 + fq_2q_3$

Note that the inbreeding reduction in the expected frequency of B_1B_2 is balanced by equal increases in the expected frequencies of the B_1B_1 and B_2B_2 homozygotes. In like manner, the decrease in frequency of B_1B_3 is compensated by equal increases in the frequencies of the B_1B_1 and B_3B_3 homozygotes and the decrease in frequency of B_2B_3 by equal increases in the frequencies of the B_2B_2 and B_3B_3 genotypes.

We have seen that in random mating populations, multilocus genotype probabilities are the products of the probabilities of the constituent single locus genotypes when the parent generation is in linkage equilibrium. For this to be true in the case of inbred populations requires (1) absence of linkage among loci and (2) uniform amounts of inbreeding among individuals is demonstrated by Crow and Kimura (1970).

2.5.1. Computing the Coefficient of Inbreeding

The systems of inbreeding that will be referred to most frequently in what follows are (1) self-fertilization and (2) full-sib mating. The former is the most extreme form of inbreeding and the latter is one of the two most extreme forms that are possible with dioecious organisms.

In the case of self-fertilization, loci that are homozygous in the parent must also be homozygous (for the allele carried by the parent) in the offspring while loci heterozygous in the parent will segregate in the familiar 1:2:1 ratio so that the expected amount of heterozygosity in offspring is one-half of that in the parent. Thus the amounts of heterozygosity expected after successive generations of self-fertilization are 1/2, (1/2)(1/2) = $(1/2)^2$, (1/2)$(1/2)^2$ = $(1/2)^3$, ...$(1/2)^n$ of the amount expected in noninbred individuals. It follows that the successive values of f are 1/2, 3/4, 7/8······1 - $(1/2)^n$ since $f_n = (H_0 - H_n)/H_0$ (see equation 2.3). These are given by the formula

$$f_n = (1/2)(1 + f_{n-1}) \qquad 2.4$$

where f_n = the coefficient of inbreeding after n generations of inbreeding and

f_{n-1} = the coefficient of inbreeding after n - 1 generations of inbreeding.

This recurrence equation was given by Wright (1921). In the same paper, he derived the following recurrence equation for f resulting from full-sib mating.

$$f_n = (1/4)(1 + 2f_{n-1} + f_{n-2}) \qquad 2.5$$

Numerical values of f for the first nine generations of self-fertilization and full-sib mating (obtained using equations 2.4 and 2.5) are listed below.

	Values of f	
n	Self-fertilization	Full-sib Mating
1	.50	.25
2	.75	.375
3	.075	.500
4	.9375	.5938
5	.9688	.6719
6	.9844	.7344
7	.9922	.7851
8	.9961	.8262
9	.9980	.8594

Wright (1922) provided a general formula from which the coefficient of inbreeding can be computed using the information supplied by any pedigree. That expression, a derivation of it, and instructions for using it are contained in Appendix 2.1.

• 2.6. Temporal Change in Allele Frequencies

Four causes of change, from one point in time to another, in allele frequencies are recognized (Wright, 1955). They are migration, mutation, selection, and chance. The symbol, Δq, has been widely used, and will used herein, to represent the change, during a specified time interval, in the frequency of a specified allele in a specified population. If q_t is the frequency of an allele at time t and $q_{(t+1)}$ its frequency one unit of time later is

$$\Delta q = q_{(t+1)} - q_t. \qquad 2.6$$

The successive generations of some populations are distinct and nonoverlapping. In those cases time is conveniently identified with generations. This is often appropriate when annual plants are being considered. When the generations of a population do overlap, as is usually the case in cattle populations, standard time units such as weeks or years must be employed. In all that follows, discrete, nonoverlapping generations will be assumed unless the opposite is specified.

The total change in the frequency of an allele can be viewed as the sum of changes due to the four causes. Symbolically,

$$\Delta q = \Delta_M + \Delta_m + \Delta_s + \Delta_c$$

where Δ_M is change in q due to migration,

Δ_m is change in q due to mutation,

Δ_s is change in q due to selection, and

Δ_c is change in q due to chance.

A few of the elementary connections between allele frequency changes and selection, mutation, and chance, respectively, will be reviewed here. Selection and chance will be discussed in much greater detail in later chapters. Migration needs no more attention for the purposes of this book than it received in Section 2.2. Unless otherwise stated, migration into breeding program populations will be assumed absent.

• 2.6.1. Selection

Selection implies differential reproduction by the members of a population. Variation in numbers of offspring is a natural phenomenon in wild populations but is, for the most part, imposed by the breeder in programs for the improvement of domestic species populations. The selection process is described as natural selection in the first case, as artificial selection in the second. If reproductive variation is correlated with genotype, the selection involved will usually cause change in allele frequencies.

As the basis for derivation of what must be the most widely known symbolic statements (equations 2.13, 2.14 and 2.15) regarding allele frequency changes due to selection, consider a monoecious population and any single gene of which only two alleles are present in that population. At the outset we will assume the population large enough so that all sampling errors are trivial and can therefore be ignored. The possible genotypes, their frequencies and their reproduction rates (average number of offspring per individual) will be symbolized as follows:

Genotype	Relative Frequency	Reproduction Rate
BB	P	W_2
Bb	2R	W_1
bb	S	W_0

where P + 2R + S = 1.0, P + R = q and the W's are assumed constant through time. For our present purposes, the quantities in this table are defined with reference to the single cell zygote stage of the population. For example, P is the fraction of fertilized eggs in which the diploid nucleus contains the BB genotype and W_2 is the average number of offspring, counted at the single cell (fertilized egg) stage, per BB individual at the single cell stage in the preceding generation. Obviously, the W's will reflect survival to reproductive age, components of reproduction (such as number of seeds produced in the case of plants or fecundity in animals) and in breeding program populations, any phenotypic trait that influences whether an individual is selected or rejected as one of the parents of the next generation. Since one-half of each individual must be credited to its male (or pollen) parent and one-half to its female (or seed) parent, the number of offspring per individual is equal to one-half the number of effective gametes per individual; an effective gamete being defined as one involved in the formation of a fertilized egg.

Let the average reproduction rate for the population be $\bar{W}$. This average is sometimes identified as the <u>absolute fitness</u> of the population. It is apparent that

$$\bar{W} = PW_2 + 2RW_1 + SW_0, \qquad 2.7$$

that $\bar{W} = 1.0$ is required if population size is to be maintained, that $\bar{W} > 1.0$ means an expanding population and that $\bar{W} < 1.0$ means a population decreasing in size.

If the total number of individuals in the parent generation is N, the total number of B locus genes transmitted to offspring will be $2N\bar{W}$ and if the B and b alleles have equal probability of transmission by Bb parents (no meiotic drive or gametic selection), the number of B and b alleles transmitted will be $2PNW_2 + 2RNW_1$ and $2RNW_1 + 2SNW_0$, respectively, if mutation is ignored, as it will be until the following subsection. It follows that the frequency of the B allele in the offspring generation would be

$$2N(PW_2 + RW_1)/2N\bar{W} = (PW_2 + RW_1)/\bar{W} \qquad 2.8$$

We note that N cancels out; i.e., that it is the relative rather than the absolute frequencies of parent genotypes that are relevant. Note also that each W_i could be multiplied (or divided) by any constant without changing the absolute magnitude of the allele frequency given by 2.8. This means that allele frequency in the offspring depends on the relative (not the absolute) reproductive rates of the genotypes. Remembering that $P + R = q$ and $R + S = 1 - q$, we find from 2.6, 2.7, and 2.8 that

$$\Delta_s = (PW_2 + RW_1)/(PW_2 + 2RW_1 + SW_0) - (P + R)$$

$$= [(1 - q)(PW_2 + RW_1) - q(RW_1 + SW_0)]/\bar{W} \qquad 2.9$$

In the case of autosomal genes in dioecious species the reproductive rates (W's) are unlikely to be the same, may be quite different, in males and females. Nevertheless, if W_i ($i = 0, 1,$ or 2) is defined as the average of its values in the two sexes, i.e., $W_i = (W_{im} + W_{if})/2$, equation 2.9 applies as in monoecious species if (a) the probabilities of participation in fertilizations by B carrying gametes and b carrying gametes are not affected by the sex chromosome content of uniting gametes and (b) the frequency of the X-chromosome in effective male gametes (sperm or pollen) is 0.5 (see Appendix 2.2).

The value of Δq for a population in which genotypes are present in HW frequencies is obtained by substituting q^2, $2q(1 - q)$, and $(1 - q)^2$ for P, 2R, and S, respectively, in 2.9. The result is

$$\Delta_s = q(1-q)[q(W_2-W_1) + (1-q)(W_1-W_0)]/\bar{W} \qquad 2.10$$

in which

$$\bar{W} = q^2W_2 + 2q(1-q)W_1 + (1-q)^2W_0$$

As it should, this expression says $\Delta_s = 0$ in all cases where the gene is (1) without effect on reproduction so that $W_2 = W_1 = W_0$ or (2) fixed so that $q = 1.0$ or 0.0. When the gene is segregating ($0.0 < q < 1.0$) and has an effect on reproduction, we see that (1) if $W_2 \geq W_1 \geq W_0$, Δ_s is positive and (2) if $W_0 \geq W_1 \geq W_2$, Δ_s is negative.

If either of these situations persists in successive generations of random mating, fixation (of one allele or the other) is promoted by the selection. It is also apparent from 2.10 that $\Delta q = 0$ when

$$Z = q(W_2 - W_1) + (1-q)(W_1 - W_0) = 0$$

Solving this for q we obtain the special value of q (which will be symbolized as q_e) that makes $Z = 0$. The result is

$$q_e = (W_1 - W_0) / [(W_1 - W_2) + (W_1 - W_0)] \qquad 2.11$$

The relevant (admissible) values (>0.0 and <1.0) of q_e require (1) W_1 greater than both W_2 and W_0 or (2) W_1 less than both W_2 and W_0; i.e., the heterozygous genotype must be either superior or inferior to <u>both</u> homozygous genotypes in its effect on reproduction. To understand what happens when $q \neq q_e$, we must examine the algebraic sign of Z because it determines whether Δ_s is positive or negative. By rearrangement

$$Z = q(W_2 - W_1) + (1-q)(W_1 - W_0) = (W_1 - W_0) + q(W_2 - 2W_1 + W_0)$$

which indicates that the regression of Z on q is positive when W_1 is less than both W_2 and W_0 (heterozygote inferior to homozygotes) but negative when W_1 is greater than both W_2 and W_0 (heterozygote superior to homozygotes). Remembering (1) that $Z = 0$ when $q = q_e$ and (2) that the algebraic signs of Z and Δ_s are always the same, we deduce the following summary for a random mating population.

W_1	q	Δ_s
> W_2 and W_0	< q_e	positive
	> q_e	negative
< W_2 and W_0	< q_e	negative
	> q_e	positive

It is apparent from this summary that when the reproduction rate is higher for the heterozygote than for either homozygous genotype, the effect of selection is always to push q toward q_e, its equilibrium value; the equilibrium is, for that reason, said to be <u>stable</u>. This contrasts with what happens when the reproduction rate of the heterozygous genotype is less than those of the other genotypes. In the latter case, we note that whenever $q \neq q_e$ the effect of selection is to push q still further from q_e in the same direction as its initial deviation from q_e. To summarize now in another way, when the heterozygote is superior to the homozygotes, segregation tends to be maintained by selection[4] but when the heterozygote is inferior, the ultimate consequence of successive generations of selection is fixation of one or the other allele. Which of the the alleles is fixed is not a function of which homozygous genotype is best but depends instead on whether q is larger or smaller than q_e at the outset. The first of these two circumstances (heterozygote superior) is referred to by the terms overdominance and single locus heterosis; the second is the simplest example that can be offered of the multiple peaks of excellence concept elaborated by Wright (1935 and numerous later papers).

2.6.1.1. ALTERNATIVE SYMBOLISM

So far we have employed as many reproduction parameters (W's) as the number of genotypes considered. At the same time we have noted that Δ_s is a function of the relative, not the absolute, reproductive (fitness) values of the genotypes. If each of the W's is multiplied (or divided) by the same quantity, a set of relative reproduction values is obtained. If each is divided by a specific one of the W's, one of the relative values obtained is 1.0. The primary advantage is that the number of parameters is reduced from three to two or, in the special case of the completely additive gene, to one. Three commonly used sets of symbolic relative values are shown below.

	Relative Fitness (reproduction rate)		
Genotype	General	Additive Gene	Overdominance
BB	1	1	1-v
Bb	1-hs	1-s/2	1
bb	1-s	1-s	1-w

Expressions for Δ_s in terms of these symbols are easily obtained by substitution in 2.9 or 2.10. We will assume HW frequencies of genotypes and substitute in 2.10. Using the first set of fitness values shown above, we obtain

$$\Delta_s = q(1-q)[(1-q)s + (2q-1)hs] / [1-2q(1-q)hs - (1-q)^2 s] \qquad 2.12$$

Perhaps the most interesting values of h are 1, 1/2, and 0.0. These confer fitness values as shown below.

	Relative Fitness		
Genotype	h = 1	h = 1/2	h = 0
BB	1	1	1
Bb	1-s	1-s/2	1
bb	1-s	1-s	1-s

Note that h=1 is appropriate when the more favorable allele is recessive, h=1/2 when the heterozygote is intermediate (additive gene) and h=0 when the more favorable allele is dominant. When these values of h are substituted in 2.12, the following expressions are obtained.

Situation	Δ_s=	
Additive gene (h=1/2)	$(1/2)sq(1-q)/[1 - (1-q)s]$	2.13
Better allele dominant (h=0)	$sq(1-q)^2/[1 - (1-q)^2 s]$	2.14
Better allele recessive (h=1)	$sq^2(1-q)/[1 - (1-q)(1+q)s]$	2.15

When s is small the denominators of these expressions approach 1.0 and are often ignored. In the case of heterotic (overdominant) genes, q_e is more interesting than Δ_s. Substitution of 1-v, 1, and 1-w for W_2, W_1, and W_0, respectively, in 2.11 yields

$$q_e = w/(v+w) \qquad 2.16$$

Equations 2.13, 2.14, and 2.15, with or without denominators, and equation 2.16 are presented in most Principles of Genetics textbooks.

- 2.6.2. **Mutations**

The mutations that change allele frequencies are changes in gene structure that transform one allele of the gene into another. Point mutations are changes in the DNA, e.g., substitutions of one nucleotide pair for another or the deletion or addition of one or more nucleotide pairs. They occur in somatic cells and in germ line cells but only those that occur in germ line cells have to do with temporal change in the allele frequencies of populations. Each gene present in an effective gamete of a specific generation traces back through a unique cell line to a specific ancestral gene in an effective gamete of the preceeding generation. The probability that any allele in any effective gamete of generation t differs by virtue of a mutation from its ancestral allele in an effective gamete of generation t-1 is the mutation rate for the gene locus of that allele. Mutation rates differ from locus to locus (i.e., gene to gene) and can be altered by mutagens. However, in the absence of mutagenic treatment mutation rates will be viewed as constant through time but potentially variable from locus to locus. Now consider a specific allele, say the j-th allele of the i-th gene, to be symbolized as B. Let the rate of mutation from B to any other allele of the same gene be m and the rate at which all other alleles mutate to B be m'. Then, given that q is the frequency of B, the one generation change in q in an infinite population arising from mutation would be

$$\Delta_m = -mq + m'(1-q) \qquad 2.17$$

Because any mutation of B contributes to m but only very specific mutations in alleles of B can cause a "not-B to B" mutation and hence contribute to m', the latter must certainly be much smaller than m. Therefore, for our purpose, 2.17 can be rewritten

$$\Delta_m \approx -mq \qquad 2.18$$

Because mutation rates (m's) are known to be low (usually 10^{-4} or less) unless mutagens are used, allele frequency changes due to mutation are necessarily small except over very long time intervals. However, while mutation changes in favorable alleles are usually deleterious, it must be recognized that the result may be an allele superior to the hitherto best allele of the gene. Progressive evoulution could not otherwise have occurred.

- 2.6.2.1. DELETERIOUS MUTATIONS

When the product of mutation is inferior the forces of selection and recurrent mutation have opposing effects on the frequency of B (the original allele) and interest focuses on the equilibrium consequence. The most interesting case is that of dominant favorable allleles. From 2.18 and 2.14 it is apparent that

$$\Delta_m + \Delta_s = -mq + sq(1-q)^2/[1-(1-q)^2 s]$$

The equilibrium value (q_e) of q can be determined by setting

$$-mq + sq(1-q)^2/[1-(1-q)^2s] = 0$$

and solving for q. The result obtained is

$$q_e = 1- [m/s(1+m)]^{1/2}$$

For practical purposes this can be written

$$q_e = 1 - (m/s)^{1/2} \quad 2.19$$

since m is very small relative to 1.0. In the extreme case where the recessive allele of B is lethal, s=1 and 2.19 reduces to

$$q_e = 1-m^{1/2} \quad 2.20$$

If s were as small as 0.01, the value of q_e would be $1-(100m)^{1/2}$ so, assuming 10^{-6} to 10^{-4} as the range of m, the range of q_e, if s values are assumed always to be ≥0.01, would be 0.9 to 0.999. Dominant alleles with frequencies in this range will be shown, in later chapters, to be very important relative to some aspects of quantitative genetics.

In the case of additive genes the process described above yields

$$q_e = 1-2m/s(1+2m) \approx 1-2m/s \quad 2.21$$

as the equilibrium value of q. That this should be much higher than the equilibrium value for dominant favorable alleles was to be expected since comparison of 2.13 and 2.14 indicates that given the same value of s, Δ_s is much greater for additive genes when q is large. For the same ranges of m and s as are employed in the case of the dominant favorable allele the range for q_e in this case is found to be from .98 to .999998.

Comparing 2.13 and 2.15, we see that Δ_s is larger for additive genes than for those of which the better allele is recessive. It follows that q_e will be larger than in the case of additive genes and, hence (assuming s substantially larger than m), that as q_e is approached both q and q^2 approach 1.0 and can be set equal to 1.0 with no more than trivial effect on the end result. The solution then obtained is

$$q_e \approx 1-m/s \quad 2.22$$

and accordingly equilibrium frequencies of the more favorable allele are still closer to 1.0 than in the case of additive genes.

• 2.6.2.2. MUTANT ALLELES SUPERIOR TO THE ONES FROM WHICH THEY AROSE

These become the "best" alleles of their respective genes. As such the effects of selection and drift on their frequencies are considered in detail in up-coming sections and chapters.

• 2.6.3. Chance and Random Drift

Random variation from each of the following three sources deserve recognition and attention:

(1) The actual individuals in any generation constitute a sample from the infinite population that conceptually might have been produced from the same set of parents. Given variation in allele frequencies among the individuals of that infinite population, sampling variance in the allele frequencies of the actual finite population are inevitable.

(2) Mutations are probability events and consequently the number of alleles of any specific gene, among the 2N in the effective gametes of any generation, that arose by mutation during the preceding unit of generation time is subject to random variation.

(3) Variations in Δq result also from variation in the force of selection. This is apparent from equations 2.13 - 2.15 where s quantifies the force (degree) of selection which is potentially variable as the consequence of variation, from generation to generation, in the environments to which any population may be exposed.

Let d_s be the deviation of Δ_s from its expected value, d_m be the deviation of Δ_m from its expected value and d_c be the sampling deviation of allele frequency in the actual individuals of any generation from that of the infinite population from which the actual individuals are a sample. Then random effects are recognized in the following statement.

$$\begin{aligned}\Delta q &= E(\Delta_s) + E(\Delta_m) + d_s + d_m + d_c \\ &= E(\Delta_s) + E(\Delta_m) + d \end{aligned} \qquad 2.23$$

where $d = d_s + d_m + d_c$. It is important to note that the expectations of each of the d's is zero and hence that

$$E(\Delta q) = E(\Delta_s) + E(\Delta_m). \qquad 2.24$$

Because d_s, d_m, and d_c are uncorrelated quantities

$$\sigma^2(\Delta q) = \sigma^2(d_s) + \sigma^2(d_m) + \sigma^2(d_c) = \sigma^2(d). \qquad 2.25$$

The significant practical question posed by 2.23 is whether Δq will be dominated by its random portion, d, or by its expectation, $E(\Delta_s) + E(\Delta_m)$. We perceive intuitively that the answer will hinge on the relative magnitudes of $\sigma(d)$and $E(\Delta q)$. At the extreme represented when a population is "very large" [so that $\sigma^2(d_c) \approx 0.0$]and $E(\Delta q)$ is consistently greater than zero, we anticipate that the equilibrium frequency given by 2.20, 2.21 , or 2.22 will be the approximate limit of change in q; i.e., that $E(\Sigma\Delta q) \approx q_e - q_0$ where q_0 is initial allele frequency. In smaller populations, on the other hand, our intuition tells us that if $\sigma(d)$ is large relative to $E(\Delta q)$, the allele in question might in the end (1) be lost from the population or (2) be fixed (pending another mutation) in the population and that the probabilities of these outcomes will depend in part on the magnitude of q_0. The first of these alternatives means change in allele frequency from q_0 to 0.0; the second means change from q_0 to 1.0. Their probabilities will be identified as the <u>probability of loss</u> and the <u>probability of fixation</u>, respectively. The breeder wants to fix favorable alleles and would be disturbed if their probabilities of loss were known to be high.

When drift over t successive generations is considered,

$$q_t - q_0 = \Sigma(\Delta q) = \Sigma E(\Delta q) + \Sigma d$$

(where summation is over the t generations)

and, because the d's of successive generations are random and consequently uncorrelated,

$$\sigma^2(q_t - q_0) \approx \Sigma_t\, \sigma^2(d) \qquad 2.26$$

The approximation indicated reflects the fact that $E(\Delta q)$ is a function of q and hence in generations after the very first varies, slightly at least, in response to variation in the sum of d's of preceding generations.

At this point, it will be well to identify the variance being discussed with special care. Let us focus on the frequency of a specific allele of a specific gene and imagine a single infinite population at time zero in which the frequency of that allele is q_0. Suppose, further, that a very large number of random samples are taken from that population to initiate an equal number of finite populations (all the same size and all exposed to similar selection and mutation pressures) that are maintained for t generations at the end of which they would vary with respect to frequency of our identified allele. The variance among populations in that allele frequency would then be the $\sigma^2(q_t - q_0)$ of 2.26. While this variance tends to get larger each generation, it is obviously limited by the value reached when the allele has been lost (q=0) or become fixed (q=1.0) in every population. The population distribution at that time can be summarized as follows:

State of population	Frequency	q_t	$q_t - q_0$
allele fixed	P	1	$1-q_0$
allele lost	1-P	0	$-q_0$
averages		P	$P - q_0$

where P is the fraction of thc population in which the allele became fixed. Working from this table it is easy to show that the limiting variance is

$$\sigma^2(q_t - q_0) = \sigma^2(q_t) = P(1-P)$$

It is worth noting that for any set of conditions (population size, mutation rate and selection pressures) such that the allele considered will eventually be lost or fixed, P and 1-P are by definition the probabilities of fixation and loss, respectively.

• 2.6.3.1. VARIANCE FROM THE SAMPLING OF INDIVIDUALS, $\sigma^2(d_c)$

Consider first a random mating monoecious population in which (1) there is no selection, (2) there are N individuals that reach reproductive age and function as parents, and (3) only one offspring is produced per mating. Following Crow and Kimura (1970) such a population will be referred to in the remainder of this text as an ideal population. The absence of selection means that genotypes do not differ with regard to the proportion of single cell zygotes that survive and develop into individuals of reproductive age. It also means that each reproductive age individual has the same chance of supplying any gamete that contributes to a genotype of the following generation. Consequences of these things, when we focus on a single gene of which just two alleles (B and b) are assumed present in the population, are (1) that the probabilities of gamete genotypes are the frequencies of the alleles in the N potential parents in accord with the following table,

Gamete Genotype	Probability
B	q
b	1-q

and (2) that each gamete genotype is an independent event (this would not be so if more than one offspring were possible per mating). It follows that the presence of B in effective gametes is a binomial event, that the 2N gametes through which genes are transmitted to the N parents of the following generation are a random sample from the corresponding binomial distribution, that the proportion of those effective gametes that carry B is the frequency of B in the N parents of the following generation, and that the sampling variance of that allele frequency will be

$$\sigma^2(d_c) = q_{st}(1-q_{st})/2N \qquad 2.27$$

Here q_{st} is the frequency of the allele in the parents of generation t.

Alternatively, $\sigma^2(d_c)$ could have been obtained by considering the distribution of single locus genotypes in the infinite population of individuals from which the N parents of generation t+1 are a random sample. This distribution (subsection 2.4.1) would be as follows.

Genotype	Frequency	Q
BB	q_{st}^2	1
Bb	$2q_{st}(1-q_{st})$	1/2
bb	$(1-q_{st})^2$	0

From this table, we find that the variance of Q among indivduals is $\sigma^2(Q) = q_{st}(1-q_{st})/2$ from which it follows that

$$\sigma^2(d_c) = q_{st}(1-q_{st})/2N \qquad 2.27$$

because the N parents of generation t+1 are a random sample.

Probably no real population satisfies all the specifications outlined as the basis for the result given by 2.27. Animal and some plant species are not monoecious and, in general, mating is not completely random, selection is not absent and the number of offspring per mating is often greater than one. However, the ideal population is important because it is used as a reference point in the definition of effective population size (Section 2.7).

We will now consider autosomal genes in dioecious populations, assuming as above (1) random mating, (2) no selection, and (3) only one offspring per mating. Let M and F be the numbers of male and female parents in each generation and let q_{stm} and q_{stf} be allele frequencies in the generation t male and female parents. Because male and female parents contribute equal numbers of genes to the following generation, the allele frequency for all parents of the generation is properly identified as

$$q_{st} = (q_{stm} + q_{stf})/2 \qquad 2.28$$

In order to ease notation slightly, the remainder of this subsection will be phrased in terms of generation 1 and 2 instead of generations t and t+1. The 2M genes (per locus) transmitted from generation 1 parents to the M individuals that become male parents in generation 2 consist of two random samples, one from male parents, the other from female parents. The number in each of these samples must be M. Let the allele frequencies in these samples be q_{mm} and q_{fm}, respectively. Then

$$q_{s2m} = (q_{mm}+q_{fm})/2$$
$$\sigma^2(q_{mm}) = q_{s1m}(1-q_{s1m})/M$$
$$\sigma^2(q_{fm}) = q_{s1f}(1-q_{s1f})/M$$

and
$$\sigma^2(q_{s2m}) = [\sigma^2(q_{mm}) + \sigma^2(q_{fm})]/4$$
$$= [q_{s1m}(1-q_{s1m}) + q_{s1f}(1-q_{s1f})]/4M$$

In the same way, it is easily shown that

$$\sigma^2(q_{s2f}) = [q_{s1m}(1-q_{s1m}) + q_{s1f}(1-q_{s1f})]/4F$$

Finally, since $q_{s2} = (q_{s2m} + q_{s2f})/2$

$$\sigma^2(q_{s2}) = \sigma^2(d_c) = [\sigma^2(q_{s2m}) + \sigma^2(q_{s2f})]/4$$

$$=[q_{s1m}(1-q_{s1m}) + q_{s1f}(1-q_{s1f})][1/16M + 1/16F] \qquad 2.29$$

Recognizing that q_{s1m} and q_{s1f} should rarely be much different and recalling 2.28, it appears reasonable, for approximation, to set $q_{s1m} = q_{s1f} = q_{s1}$. Then, substituting in 2.29, we obtain

$$\sigma^2(d_c) \approx q_{s1}(1-q_{s1})[(1/8M) + (1/8F)]$$

$$= q_{s1}(1-q_{s1})[(M+F)/8MF] \qquad 2.30$$

If the numbers of male and female parents are equal, M=F=N/2 and 2.30 reduces to

$$\sigma^2(d_c) = q_{s1}(1-q_{s1})/2N$$

the value given by 2.27.

2.6.3.2. VARIANCE DUE TO THE RANDOM NATURE OF MUTATION, $\sigma^2(d_m)$

When mutation is being considered it is hardly realistic to assume only two alleles. Hence in this subsection we shall focus on a single allele (B) without specifying how many other alleles of the same gene are present. In a monoecious population with N parents per generation, 2N copies of each gene are transmitted from the parents of any one generation to the individuals that will become the parents of the next. Of those 2N a fraction , say p, would consist of B alleles or alleles (to be symbolized B_m) that had mutated from B during the preceding unit of generation time. The probabilities of the B and B_m alleles in random gametes would be $q_s(1-m)$ and $q_s m$, respectively, and within the subgroup (2Np in number) consisting of only these two kinds of alleles, their probabilities would be 1 - m and m and their distribution binomial. It follows that the variance in the frequency of B associated with mutation would be

$$\sigma^2(d_m) = m(1-m)/2Np$$

When this is compared to the variance from the sampling of individuals given by 2.27, we obtain

$$\sigma^2(d_m)/\sigma^2(d_c) = m(1-m)/pq_s(1-q_s) \qquad 2.31$$

Because p will never be much different from q_s and since m is always small enough so that 1 - m can be replaced by 1.0, 2.31 can be rewritten

$$\sigma^2(d_m)/\sigma^2(d_c) \approx m/(q_s)^2(1-q_s)$$

Because m is very small, say 10^{-5}, it is obvious that $\sigma^2(d_m)$ is trivial relative to $\sigma^2(d_c)$ unless q_s is very small or very large.

2.6.3.3. SUMMARY

Accepting that $\sigma^2(d_m)$, the variance in Δq that results because mutation is a random event, can be assumed trivial, leaves sampling among individuals (gametes or zygotes) and temporal variation in the force of selection as the potentially significant sources of random drift in breeding program populations. Variance in the force of selection will be discussed in greater detail after more background has been provided.

2.7. Effective Population Size

In Section 2.6.3 the sampling variance of Δq was shown to be

$$\sigma^2(d_c) = q_s(1-q_s)/2N \qquad 2.27$$

in the case of ideal populations. The effective size (N_e) of any population is defined as the number required to make 2.27 correct for that population when N_e is substituted for N.[5] Clearly, N_e = N in the case of ideal populations. To determine N_e for any population that does not qualify as "ideal", one derives $\sigma^2(d_c)$ as a function of actual paramers of the population in question, sets it equal to $q_s(1-q_s)/2N_e$ and solves for N_e. For example, consider the population of subsection 2.6.3 that deviated from the specifications for an ideal population only by being dioecious (bisexual) instead of monoecious. The approximate sampling variance of Δq for such populations is stated by equation 2.30. Setting it equal to $q_s(1-q_s)/2N_e$ we have

$$q_s(1-q_s)(M+F)/8MF = q_s(1-q_s)/2N_e$$

from which we obtain

$$N_e = 4MF/(M+F)$$

This expression, well known and widely used by animal breeders, is given by various authors; e.g., Falconer (1981) and Crow and Kimura (1970).

Clearly, there is a variety of ways in which populations can deviate from the ideal population. For example, mating may not be random or number of offspring per mating may be greater than one (both of these are common in the world of breeding). A series of special cases encountered in breeding programs is treated in Appendix 11.1.

2.8. Probability of Fixation

The probabilities of fixation and loss were defined and very briefly discussed in subsection 2.6.3. Continuing the discussion, we will first consider what is expected when there is no selection (W_2=W_1=W_0) and no mutation. Then Δ_s = 0.0, Δ_m = 0.0, and $\Delta q = d_c$. The latter, d_c, is a sampling deviation wlith expectation zero. Thus in infinite populations Δq would always be zero, i.e., q would not change through time. In finite populations, on the other hand, d_c would almost always be nonzero, positive in some generations, negative in others and variable in absolute magnitude from generation to generation. As a result q would gradually "drift" away from q_0, its original value, though the accumulated deviation of q from q_0 would sometimes increase, sometimes decrease, instead of increasing (in absolute magnitude) each generation. We sense, intuitively, that the "drift"of q would take it eventually to 1.0 or 0.0, a fact documented by Wright (1931). Because populations being considered are ones in which Δ_s = Δ_m = 0, change in q, when averaged over an infinite number of finite populations, would necessarily be zero. Consequently, the average (overall populations) frequency of the allele when q = 1.0 or 0.0 in each population, which would be the proportion of populations in which q = 1.0, must be equal to q_0. Stated differently, the probability of fixation is q_0, the original frequency of the allele, and the probability of loss is $1-q_0$ when there is no selection or mutation. The argument for this conclusion will now be stated in a more mathematical way. Because $\Delta q = d_c$ when selecton and mutation are absent and $E(d_c) = 0.0$,

$$E[\Sigma(\Delta q)] = E[\Sigma d_c] = 0.0$$

in which summation is over generations until q = 1 or 0 in each population. Also

$$\sigma^2[\Sigma(\Delta q)] = \sigma^2(\Sigma d_c) = \Sigma[\sigma^2(d_c)] \qquad 2.32$$

because each d_c is independent relative to every other d_c. Because variances are always positive the sum of variances given by 2.32 must increase from one generation to the next until its limit, P(1-P), given in subsection 2.6.3 is reached. At that limit, P is both the average of q over all populations and the proportion of populations in which q = 1.0. Because it is the average of q over all populations

$$P = E(q) = q_0 + E[\Sigma(\Delta q)] = q_0$$

and because it is the proportion of populations in which q = 1.0 (when q = 1.0 or 0.0 in all populations), it is apparent that the probability of fixation is q_0. This result, which will be intuitively obvious or appealing to many readers, is important as the justification for the frequent assumption that the probability of a single locus homozygous genotype in pure lines extracted without selection from any random mating population is equal to the frequency of the allele in that population at the time inbreeding (of the lines) was initiated. Symbolically,

$$P(B_jB_j) = q_{j0}$$

where $P(B_jB_j)$ is the probability of a pure line homozygous for the j-th allele. It is worth noting that when all populations are homozygous, f (the coefficient of inbreeding) = 1.0 and from equation 2.2 when f = 1.0 and there has been no selection

$$P(B_jB_j) = q_{j0}^2 + fq_{j0}(1 - q_{j0}) = q_{j0}$$

In the real world, some selection and some mutation must always be assumed. However, it is easily shown that the effect of mutation on probability of fixation must ordinarily be trivial. It is clear that the effect of mutation is to lower the probability of fixation of any allele originally present in a population. This is a reflection of the fact that it is highly unlikely that any random change in the DNA that constitutes an allele of a gene will be the source of another allele already present in a finite population. Since $\Delta_m = -mq$ (equation 2.18), it is apparent that in the absence of selection

$$\bar{q} = q_0 - \Sigma_t m\bar{q}_t$$

where $\bar{q}_t$ is average (over all populations) frequency of the allele in the t-th generation and $\bar{q} = \bar{q}_t$ when q = 1.0 or 0.0 in every population. The reduction, $\Sigma_t m\bar{q}_t$, in $\bar{q}$ caused by mutation would be greater in larger populations because more generations are required before homozygosity is reached when populations are larger. Since $\bar{q}_t = q_0$ initially but becomes progressively smaller as the result of mutation, $\Sigma_t mq_0 = tmq_0$ can be used as an upper limit to $\Sigma_t m\bar{q}_t$. Consider populations large enough so that 400 generations would be required for homozygosity to be reached in all populations. Then, given $m = 10^{-5}$, we would have $tmq_0 = 0.004q_0$ and P = P(fixation) > $0.996q_0$ which is very little less than q_0. In general

$$P = P(\text{fixation}) > q_0(1-tm)$$

and, except for genes more mutable than ordinarily expected or very large populations, the effect of mutation on probability of fixation is essentially zero and, in the absence of selection, P(fixation) can, for practical purposes, be equated to q_0, the initial frequency of the allele.

For the practical purposes of breeding, the issue of special importance is the amount by which the probabilities of fixation of useful alleles are increased by selection. As noted in subsection 2.6.3, these will increase, when other things are equal, as $\sigma(d)$ decreases. However, unlike the case when selection is absent, no generalization that provides the basis for quantitative statements concerning probability of fixation is intuitively apparent.

Kimura (1957) considered the special case of a gene with only two alleles, relative fitnesses of the single locus genotypes constant through time, any level of dominance except overdominance and effective population size constant through time. He derived the following which has been shown to be a very close approximation

$$P(\text{fixation}) = \int_0^{q_0} e^{-2N_e sx[2h + x(1-2h)]}dx \div \int_0^1 e^{-2N_e sx[2h + x(1-2h)]}dx \qquad 2.33$$

in which N_e, q_0, s and h have meanings assigned earlier in this chapter. The following two facts are worth noting. First, when s = 0; i.e., when there is no selection, 2.33 reduces to P(fixation) = q in accord with the demonstration earlier in this section. Second, P(fixation) is a function of the product, $N_e s$, rather than the separate values of N_e and s. For example, the answer provided by 2.33 is the same for $N_e = 20$ and s = 0.1 as for $N_e = 5$ and s = 0.4 or for $N_e = 50$ and s = 0.04. Unfortunately, the integrals of 2.33 do not reduce (except when h = 1/2) to explicit algebraic forms and, consequently, numerical evaluation of P(fixation) on the basis of 2.33 is much easier, given h = 1/2, than when any other value of h is assumed. When h = 1/2, 2.33 reduces to the following explicit expression

$$P(\text{fixation}) = [1-e^{-2N_e sq_0}]/[1-e^{-2N_e s}] \qquad 2.34$$

Equations 2.33 and 2.34 have special value for the breeder because they provide a base for judgment concerning adequate population sizes in breeding programs. Applications will be discussed in detail in later chapters. Information will also be provided (see section 11.2) concerning the nature and magnitude of biases that are encountered in probabilities provided by equations 2.33 or 2.34 when one or another of Kimura's initial assumptions is invalid.

The case of overdominant genes is reasonably discussed in terms of two alleles because selection for a superior heterozygote will insure relatively high frequencies of the alleles involved in that heterozygote and hence low frequencies of any other alleles of the gene in question. Assuming two alleles, let the one that is better in homozygous state be symbolized as B and the other as b. Selection would act to keep the frequency of B close to the equilibrium value, q_e, given by equations 2.11 and 2.16 and in doing so would reduce the probability that fixation (of one allele or the other) would occur within any specified period of time. Despite this fact, it is apparent that fixation should ultimately be expected in any finite population. Given random mating, the probability that all of the 2T gametes involved in the T fertilizations of any generation will carry the B allele is $(q_s)^{2T}$ and the probability that all of the 2T gametes will carry the b allele is $(1-q_s)^{2T}$. Thus the probability that fixation will occur at any point in generation time will be

$$(q_{st})^{2T} + (1-q_{st})^{2T},$$

the probability that fixation will not occur at that point in time will be

$$P = 1 - (q_{st})^{2T} - (1-q_{st})^{2T},$$

and the probability that fixation will not occur during m successive generations is

$$P_1P_2P_3...P_m \approx \pi_i P_i. \qquad 2.35$$

This statement is approximate because q_{st} would vary from one generation to the next. Since each of the P's is less than 1.0, their product becomes small as m becomes large; i.e., the probability that fixation will not occur within m generations becomes progressively smaller and the probability that fixation will occur within those m generations becomes progressively larger as m becomes large. Robertson (1962) considered the probability of fixation of the alleles of overdominant genes but, for reasons that will be developed later, the quantitative solution for overdominant genes is less significant relative to breeding procedure than the results produced by Kimura for nonoverdominant genes.

• 2.9. Epistasis

When the effect on genotype value (Y) of any specific change in genotype at any single locus is conditioned by genotype present at any other locus or loci, epistasis is said to be present; i.e., epistasis denotes interaction among genotypes at different loci. The term itself is a variation of epistacy, the word used by Fisher (1918) to indicate gene interaction. The meaning of epistasis is most easily illustrated in terms of genotype values assumed to depend on only two genes, each with only two alleles. Let the various genotype values be symbolized by Y's with subscript's as shown in Figure 2.1. The first subscript to Y indicates the numer of B's in the locus 1 genotype and the second subscript has the same meaning with respect to locus 2, e.g., Y_{12} is the value of the $B_1b_1B_2B_2$ genotype. If

$$Y_{22} - Y_{21} = Y_{12} - Y_{11} = Y_{02} - Y_{01}$$

and

$$Y_{21} - Y_{20} = Y_{11} - Y_{10} = Y_{01} - Y_{00}$$

Genotypes at locus 1	Genotypes at Locus 2 B_2B_2	B_2b_2	b_2b_2
B_1B_1	Y_{22}	Y_{21}	Y_{20}
B_1b_1	Y_{12}	Y_{11}	Y_{10}
b_1b_1	Y_{02}	Y_{01}	Y_{00}

Figure 2.1. Values of two-locus genotypes.

there is no epistasis (no interaction betweeen genotypes of the two loci). When the above equalities exist, it will also be true that

$$Y_{22} - Y_{12} = Y_{21} - Y_{11} = Y_{20} - Y_{10}$$

and

$$Y_{12} - Y_{02} = Y_{11} - Y_{01} = Y_{10} - Y_{00}$$

Any deviation from the indicated equalities constitutes (is evidence of) epistasis. The 3 x 3 number sets in Figure 2.2 are presented to illustrate a few of the possible relationships among Y values given absence or presence of epistasis. Within each set the arrangement of numbers corresponds to the arrangement of Y's in Figure 2.1 and hence, for example, the second number in the first row is Y_{21} and the third in row three is Y_{00} in all cases. Sets (a), (b), and (d) illustrate no epistasis but different levels of dominance. The absence of epistasis is attested by existence of all of the equalities of 2.35. Set (a) illustrates complete additivity---no epistasis and no dominance. Set (b) reflects complete dominance and (d) portrays overdominance at both

(a)			(b)			(c)			(d)		
4	3	2	2	2	1	2	2	0	4	5	3
3	2	1	2	2	1	2	2	0	6	7	5
2	1	0	1	1	0	0	0	0	2	3	1

(e)			(f)			(g)		
6	6	3	1.792	1.792	1.099	8	4	0
4	4	2	1.386	1.386	.693	4	3	2
2	2	1	.693	.693	.000	0	2	4

Figure 2.2. Example sets of two-locus genotype values.

loci. Epistasis is reflected by sets (c), (e), and (g) in each of which a portion of the equalities of 2.35 is violated. For example, in each of these $Y_{11} - Y_{10} \neq Y_{01} - Y_{00}$. Set (c) depicts the classical case of complementary genes; an effect is produced when the favorable dominant allele is present (homozygous or heterozygous) at both loci but not when it is absent at one or both of the loci. The considerable, but not absolute, similarity between sets (b) and (c) is worth noting. The type of epistasis exemplified in set (c) is inherent in all sets of genes that code the enzymes responsible for promoting any single chain of biochemical reactions. Set (e) depicts multiplicative effects of genes (plus dominance at locus 2). Rows 1 and 2 are multiples of row 3 and columns 1 and 2 are both two times column 3. The epistasis involved is of the special kind that can be elimated by appropriate transformation of the measurement scale, in this case by substituting logarithms of the original scale numbers. Set (f) consists of the natural logarithms of the numbers of set (e) and the reader can easily verify that the equalities that characterize absence of epistasis are all exhibited.

2.9.1. Categories of Epistasis

In the context of breeding programs aimed at syntheses of "best possible" genotypes, it is important (Sections 3.6.4 and 7.3) to distinguish two classes of epistasis, simple and multiple peak. In the case of simple epistasis there are no reversals in rank order values of single locus genotypes as a consequence of change in genotype at any one or more other loci. Sets (c) and (e) in Figure 2.2 depict simple epistasis. In set (c) we note that B_1b_1 is better than b_1b_1 in the presence of B_2B_2 or B_2b_2 but equal to b_1b_1 in the presence of b_2b_2. However, the rank order of B_1b_1 and b_1b_1 is not reversed and hence the defining condition for simple epistasis is not violated as it is in set (g) which depicts multiple peak epistasis.

The "peaks" of multiple peak epistasis are the values of genotypes that are superior to any "adjacent genotype", an adjacent genotype being one obtainable by a single allele substitution in the "peak genotype". Note, for example, that the $B_1B_1B_2B_2$ genotype of set (g) is superior in value to both of its adjacent genotypes, $B_1B_1B_2b_2$ and $B_1b_1B_2B_2$, and that the same is true for the other peak genotype ($b_1b_1b_2b_2$). The term, multiple peak, is a variation of terms such as selective peaks and adaptive peaks used as early as 1932 by Sewall Wright in discussion of implications of epistasis.

2.9.2. Symbolic Representation of Genotypic Value

In much of quantitative genetics, genotypic value is visualized and depicted as a linear function of gene effects. The following representation (model) is used extensively in this book.

$$Y = \alpha + y_1 + y_2 + \ldots\ldots y_i + \ldots\ldots y_n = \alpha + \Sigma_i \, y_i \qquad 2.36$$

Where α = the value of the genotype that is homozygous for the least favorable allele of each gene,

n = number of genes, and

y_i is a quantity that depends on genotype at the i-th locus.

In the case of the genotype homozygous for the least favorable allele at each locus, we have, by the definition of α, that $Y = \alpha$ and hence that Σy_i =zero. To conform with that condition, and because it is convenient, the value of y when the least favorable allele is homozygous at the locus in question is always taken to be 0.0. This defines y_i as the increase in Y that results when another i-th locus genotype is substituted for the least favorable homozygote.

For illustration, imagine a trait affected by only two genes and that genotypic values are those of set (d) in Figure 2.2. In accord with above definitions and the Y's of set (d), $\alpha = 1$, and y values are as follows.

Locus 1		Locus 2	
Genotype	y	Genotype	y
BB	2	BB	1
Bb	4	Bb	2
bb	0	bb	0

As examples, $Y = \alpha + \Sigma y_i$ becomes 1+4+2 = 7 in the case of $B_1b_1B_2b_2$ genotye and 1+2+2=5 in the case of the $B_1B_1B_2b_2$ genotype.

The model doesn't work when there is epistasis because then, by the definition of epistasis, the effects on Y of genotype substitutions at any single locus are not independent of genotype at other loci. For illustration, examine the Y's of set (g) in Figure 2.2. The change in Y resulting from the B_1B_1 for B_1b_1 substitution is 4, 1, or -2 depending on genotype present at locus 2. Clearly, equation 2.36 implies no epistasis; i.e., when 2.36 is employed, absence of epistasis is being assumed. When the genetic model includes epistasis, an appropriate linear function description of genotype value must include at least one more term to reflect interaction effects.

It is worth noting that equation 2.36 places no restrictions on number of genes or on the magnitude of α. The number of genes, n, may be viewed as the number that affects the aspect of phenotype being considered or as the total number of genes in the entire genome. The latter view would include the notion that for some loci (perhaps many) all y values would be zero; i.e., that at some of the n loci change in genotype present would have no impact on one or more aspects of phenotype.

• 2.10. Composition of Variance among Genotypes

Qualitatively, the value (Y) of a total genotype is a function of the average effects of the alleles of which it is composed, dominance (interaction of the alleles in single locus genotypes), and epistasis. It is no surprise then that the genotype variance, σ_Y^2, of a population is recognized as the sum of three parts,

$$\sigma_Y^2 = \sigma_g^2 + \sigma_d^2 + \sigma_i^2$$

where

σ_g^2 = additive genetic variance (the portion that reflects the average effects of alleles),

σ_d^2 = dominance variance (the portion arising from dominance relations between alleles of the same gene), and

σ_i^2 = epistatic variance (the residual after variance from average effects of alleles and dominance has been accounted for).

2.11. Inbred Lines

To speak of inbred lines implies the existence of lines that are not inbred and raises questions concerning what is meant by "line" in relation to other terms used in the designation of populations or subpopulations and what exactly is meant by "inbreeding". Clearly, some definitions and distinctions are required.

2.11.1. Terms and Definitions

Inbreeding was loosely defined in section 2.5 as the mating of related individuals without emphasizing that degree of relationship depends on the point of reference. Whether individuals are "related" depends on whether they have one or more common ancestors. Thus for them to be completely unrelated would require that they be members of an infinite population and while real populations can be very large they cannot be infinite and at the same time be accommodated in finite space. It follows that all individuals of any real population must be related (though in large populations the degree of relationship can be very, very small) and hence, by the above definition, that some amount of inbreeding, small though it might be, would be inherent in all matings. Crow and Kimura (1970) say "Inbreeding occurs when mates are more closely related than they would be if they had been chosen at random from the population". Their definition, called the broad scientific definition by Lush (1937), will suffice in relation to the content of this book. It recognizes that a reference point for relationship is required but not that random mates may be more closely related in one population than in another. The practical implication is that inter-population comparisons in terms of the coefficient of inbreeding (Section 2.5) should be interpreted with caution. Consider the fact that though mating may be random within species in which the number of sexually mature individuals is very small, such species are considered to be "endangered" by the inescapable inbreeding within them.

Closed population: This is one within which membership is circumscribed relative to the species or any other population to which it belongs. Familiar examples in the world of breeding are the "pure breeds" of livestock. To illustrate, purebred Holsteins and purebred Jerseys belong to the same species but as populations are circumscribed by the requirement that their ancestry, as far back as the origin of their breed, shall not include animals not themselves recognized (in the official herdbook) as members of their breed. Animals with ancestry belonging largely, but not exclusively, to a single breed are referred to as "grade" members of the breed. Populations that are closed relative to the breed or variety to which they belong can of course be established at will.

Line: Within this book only closed populations will be referred to as lines. However, closed populations commonly identified as breeds or varieties will usually be refered to as such rather than as lines.

Inbred line: This term denotes a line in which effective population size is or has been small so that its average level of inbreeding is greater than required by the definition of line or closed population.

Clearly, the distinctions between "closed populations", "lines", and "inbred lines" are fuzzy because all three are by definition closed population. In general, however, effective population size is largest in what are referred to as closed populations and smallest in inbred lines. As a consequence the inbreeding and rates of inbreeding are usually greatest in what will be referred to as "inbred lines" and least in "closed populations".

Pure lines: These are populations in which all individuals have the same completely homozygous genotype. The pure line state is the end result approached by any closed population but the time required increases as the effective population size of the population increases. The most rapid approach is obtained when $N_e = 1$, i.e., in lines developed using self-fertilization.

2.11.2. Probabilities of Single Locus Genotypes

Random mating is not obligatory within closed populations but it is the common practice and will be assumed for our purposes. Given random mating, the within line probabilities of single locus genotypes are the HW frequencies given by equations 2.1 (in dioecious organisms this is exactly correct only when the allele frequencies in male and female parents are identical). They are, of course, functions in each generation of the allele frequencies in the parents of that generation and hence will vary from one generation to the next. Note that despite the inbreeding that exists in a closed population the probabilities in question are not those given by equations 2.2. The latter pertain to the expected frequencies of genotypes among all members of the entire collection of closed populations that might be generated in the same way from a common source population (the q of the equations being that of the source population and the f being the common coefficient of inbreeding of all individuals in each of the closed populations).[6]

In pure lines the only genotypes are homozygotes so the probabilities of interest are the probabilities of the two or more possible homozygotes in any single pure line. These, as noted in sections 2.6.3 and 2.8, are the probabilities of fixation of the two or more alleles present in the noninbred population from which the pure line originated. Given no selection, and ignoring any effect of mutation, the probabilities of fixation of the two or more alleles are their frequencies in the noninbred source population. Thus, for example, in pure lines extracted without selection from a common source population the probability of the B_jB_j genotype is

$$P(B_jB_j) = q_i$$

where q_j = the frequency of the B_j allele in the source population.

2.12. Inbreeding Depression and Heterosis

Vast amounts of data have been recorded concerning the effects of inbreeding and crossbreeding and their general consequences are well known to breeders and geneticists. On average, inbred individuals are smaller, weaker and produce fewer seeds or less offspring than their non-inbred counterparts. This reduction in level of performance is commonly called inbreeding depression. In contrast, the average crossbred individual is stronger, grows faster and is more productive. This superior performance is known as hybrid vigor or heterosis. The performance advantage of plant hybrids (particularly maize and sorghum) and crossbred livestock have contributed enormously to agricultural production in the developed countries.

Outbreeding, the opposite of inbreeding, results when mates are less closely related than the average within the reference population to which they both belong. Crossbreeding is the form of outbreeding that involves the mating of individuals from different closed populations (lines, breeds, or varieties). Just as inbreeding increases the probability of homozygosity at each locus, crossbreeding increases the probability of heterozygosity.

It is readily apparent, as outlined by Bruce (1910) and by Jones (1917), that both inbreeding depression and heterosis can be consequences of dominance of favorable alleles. In general, if the contribution of a heterozygous genotype (B_jB_k) to total genotypic value is greater than the average for the corresponding homozygous genotypes (B_kB_k and B_jB_j), the effect of inbreeding, which increases homozygosity at the expense of heterozygosity, will be to decrease average genotype value and the effect of crossbreeding, which increases heterozygosity and decreases homozygosity, will be to increase average genotype value. For various reasons it will be useful to present the logic of these generalizations symbolically and in greater detail.

Consider two segregating populations (C and D) that belong to the same species. Assuming no epistasis, the value of genotype with respect to any single trait can be expressed symbolically as

$$Y = \alpha + y_1 + y_2 + \ldots y_i + \ldots y_n = \alpha + \Sigma_i y_i \qquad 2.36$$

(see section 2.9.2). The corresponding expression for the average of Y within a population is

$$\ddot{Y} = \alpha + \ddot{y}_1 + \ddot{y}_2 + \ldots \ddot{y}_i + \ldots \ddot{y}_n = \alpha + \Sigma_i \ddot{y}_i \qquad 2.37$$

The subscript in these expressions identifies the gene or locus with which the y-value is associated. Assuming two alleles per gene, let the y-value of single locus genotypes be symbolized as follows

Genotype	$y_{.}$
BB	2u
Bb	u + au
bb	0.0

For example, if the genotypes of an individual at loci 1, 3, and 7 were B_1b_1, b_3b_3, and B_7B_7, the associated y's would be $y_1 = u_1 + a_1u_1$, $y_3 = 0.0$, and $y_7 = 2u_7$. Assuming B to be the better allele (and that measurement of the trait is such that higher values are desirable), u is a positive quantity. On the other hand a, which reflects the dominance relation between the alleles, can be either positive or negative (a> 0.0 indicates dominance of the more favorable allele, that the effect of the Bb genotype is greater than the average of the effects of the BB and bb genotypes, and $a < 0.0$ indicates dominance of the less favorable allele, that the effect of Bb is less than the average of the effects of BB and bb). Finally, let p_i and q_i be the frequencies of the better allele of the i-th gene in populations C and D, respectively.

When matings are at random within a population, the probabilities (expected frequencies) of genotypes are the HW frequencies of equations 2.1. It follows that the expectation of y_i (the population mean of y_i) for non-inbred individuals of population D is

$$\ddot{y}_{iDn} = q_i^2(2u_i) + 2q_i(1-q_i)(u_i+a_iu_i) + (1-q_i)^2(0.0)$$

$$= 2q_iu_i + 2q_i(1-q_i)a_iu_i$$

Here the second subscript (D) indicates the population and the third (n) indicates non-inbred individuals. Summation over loci yields

$$\ddot{Y}_{Dn} = \alpha + \Sigma_i 2q_iu_i + \Sigma_i 2q_i(1-q_i)a_iu_i \qquad 2.38$$

as the average of Y for non-inbred individuals of population D.

Now consider inbred individuals of population D and assume the same inbreeding coefficient (f) for all. The probabilities of genotypes (from section 2.5) are as follows

<u>Genotype</u>	<u>Probability</u>
BB	$q^2 + fq(1-q)$
Bb	$2q(1-q)(1-f)$
bb	$(1-q)^2 + fq(1-q)$

Using these, the expectation of y_i for inbred individuals is found to be

$$\ddot{y}_{iDI} = [q_i^2 + fq_i(1-q_i)]2u_i + 2q_i(1-q_i)(1-f)(u_i + a_iu_i)$$

$$= 2q_iu_i + 2q_i(1-q_i)(1-f)a_iu_i.$$

Summation over loci yields in this case

$$\ddot{Y}_{DI} = \alpha + \Sigma_i 2q_iu_i + \Sigma_i 2q_i(1-q_i)(1-f)a_iu_i \qquad 2.39$$

as the expected average of Y for inbred individuals of population D.

By (a) subtraction of 2.38 from 2.39 and (b) dividing the difference by f, the <u>inbreeding effect</u> (I) per unit of inbreeding is found to be

$$I = [\ddot{Y}_{DI} - \ddot{Y}_{Dn}]/f = -\Sigma_i 2q_i(1-q_i)a_iu_i \qquad 2.40$$

The same result is obtained by taking the derivative of $\ddot{Y}_{DI}$ with respect to f.

$$I = d\ddot{Y}_{DI}/df = -\Sigma_i 2q_i(1-q_i)a_iu_i$$

Obviously, the expected effect of inbreeding for population C would differ only because of differences in allele frequencies and would be $-\Sigma_i 2q_i(1-q_i)a_iu_i$.

If random members of populations C and D were mated the probabilities of B and b would be p and 1-p in gametes from the population C parent and q and 1-q in gametes from the population D parent. It follows that the probabilities of single locus genotypes in the crossbred offspring of such matings would be as follows.

Genotype	Probability
BB	pq
Bb	$p(1-q) + q(1-p)$
bb	$(1-p)(1-q)$

Given these probabilities, the expectation of y_i in the crossbred offspring would be

$$\ddot{y}_i = p_iq_i(2u_i) + [p_i(1-q_i) + q_i(1-p_i)](u_i+a_iu_i)$$

$$= (p_i+q_i)u_i + [p_i(1-q_i) + q_i(1-p_i]a_iu_i$$

Note that this reduces to 2.37 if $q_i = p_i$. Substitution in 2.37 yields

$$\ddot{Y}_{CD} = \alpha + \Sigma_i(p_i + q_i)u_i + \Sigma_i[p_i(1-q_i)u_i + q_i(1-p_i)]a_iu_i \qquad 2.41$$

as the expectation of Y for the C x D crossbred individuals, i.e., the CxD cross. Note that if $p = q$ at all loci, 2.41 becomes2.38, as it logically should because in that event C and D would be identical populations and to mate random members from the two would be equivalent to mating two random members of either.

To evaluate the effect of inbreeding the expectation of Y for inbreds was compared to E(Y) for non-inbreds of the same population. What is the appropriate base from which to measure hybrid vigor? Because the parent populations may (and usually will) differ in average genotypic value and since they contribute equally to the genotypes of the crossbred offspring, the mean of the parent population averages (for non-inbred individuals) has intuitive appeal. From 2.38 it is apparent that

$$\ddot{Y}_{Cn} = \alpha + \Sigma_i 2p_iu_i + \Sigma_i 2p_i(1-p_i)a_iu_i \qquad 2.42$$

Then from 2.38 and 2.42

$$[\ddot{Y}_{Cn} + \ddot{Y}_{Dn}]/2 = \alpha + \Sigma_i(p_i+q_i)u_i + \Sigma_i[p_i(1-p_i) + q_i(1-q_i)]a_iu_i \qquad 2.43$$

Taking the difference between 2.41 and 2.43 as hybrid vigor (H), we obtain

$$H = \ddot{Y}_{CD} - [\ddot{Y}_{Cn} + \ddot{Y}_{Dn}]/2 = \Sigma_i[p_i(1-q_i) + q_i(1-p_i) - p_i(1-p_i) - q_i(1-q_i)]a_iu_i$$

$$= \Sigma_i[p_i^2 - 2p_iq_i + q_i^2]a_iu_i = \Sigma_i(p_i-q_i)^2a_iu_i \qquad 2.44$$

It is natural as well as very much to the point to inquire whether the heterosis expected in the F_1 cross of two populations is likely to be retained in the F_2 (the population obtained from matings of random F_1 individuals) and, if not, how much retention can be expected. Note first that in absence of selection expected allele frequencies in F_1 parents (of the F_2 generation) are averages of those in the parent populations, C and D, and second that because matings to produce the F_2 are between random members of the F_1 Hardy-Weinberg frequencies of single locus genotypes are to be expected in the F_2. It follows that E(Y) of the F_2 is obtained by substituting $(p+q)/2$ for q in 2.38. We obtain

$$\ddot{Y}_{F2} = \alpha + \Sigma_i(p_i + q_i)u_i + \Sigma_i[(p_i + q_i)(2-p_i-q_i)/2]a_iu_i$$

and subtracting this from 2.41, the expectation of Y in the case of the F_1 cross, yields

$$\ddot{Y}_{CD} - E(Y)_{F2} = \Sigma_i[p_i(1-q_i) + q_i(1-p_i) - (p_i+q_i)(2-p_i-q_i)/2]a_iu_i$$

$$= \Sigma_i(p_i-q_i)^2a_iu_i/2 \qquad 2.45$$

The following facts, apparent from 2.40, 2.44, and 2.45 merit special attention.

(1) Both inbreeding depresssion and hybrid vigor can be explained in terms of dominance ($a>0.0$) of favorable alleles. Variation in a, including negative as well as positive values, must be assumed but if the distribution of a-values and allele frequencies is such that both $\Sigma_i q_i(1-q_i)a_i u_i$ and $\Sigma_i(p_i-q_i)^2 a_i u_i$ are positive quantities both inbreeding depression and hybrid vigor are expected. It should, of course, be obvious that though the general consequences of inbreeding and crossbreeding are explainable as effects of dominance, epistasis may also be involved.

(2) Assuming that dominance (of favorable alleles) accounts for a major portion of the observed hybrid vigor and effects of inbreeding and that the distributions of a-values are variable from trait to trait, there should be a strong trait to trait correlation between amounts of inbreeding depression and hybrid vigor exhibited. This is consistent with the extensive data concerning these phenomena. Traits that exhibit both are often referred to as heterotic traits.

(3) Both inbreeding depression and hybrid vigor are functions of allele frequencies. Hence, to some extent the former should vary among populations and the latter among pairs of populations.

(4) In the absence of epistasis, the expected effect of inbreeding is a linear function of f, the coefficient of inbreeding.[7] This provides a theoretical basis for testing whether epistasis is involved in the action of genes affecting heterotic traits.

(5) The amount of hybrid vigor depends on allele frequency differences between the parent populations; in the absence of epistasis the contribution from each single locus is $(p_i-q_i)^2 a_i u_i$. This is consistent with the principle enunciated by early corn breeders (see Hayes and Immer, 1942) to the effect that the best maize hybrids will ordinarily be obtained using lines derived from "genetically divergent" sources.

(6) Finally, when we compare 2.44 and 2.45 we find that when epistasis is absent one-half of the hybrid vigor exhibited by an F_1 cross is expected to be retained in the F_2 generation of the cross, i.e., the F_2 mean of any quantitative trait is expected to be about midway between its mean in the F_1 and the average of its means in the two parent populations.

If later generations (F_3, F_4, etc.) are generated using random matings, their mean genotype values are expected to be the same as in the F_2, i.e., half the hybrid vigor of the F_1 would be retained in generations beyond the F_2, as well as in the F_2.

• 2.13. Genotype, Environment, and Phenotype

The following are basic concepts in quantitative genetics: (1) quantitative trait phenotypes are influenced by both genotype and environment and (2) the relative effects of genotypes on phenotype are not constant from one environment to another. The obvious corollary is that the relative effects of environment on phenotype are not constant from one genotype to another.

The terminology (value of genotype, effect of environment, genotype-environment interaction, etc.) and the symbolic descriptions (mathematical models) that have evolved in discussion of these concepts and their implications have become standard components of the language of quantitative genetics. This section will define terms and introduce models and symbols that will be employed in later chapters.

It should be noted that genotype, as the word is used in this section, refers to the total genotype (the entire collection of alleles present in an individual organism). This contrasts with the single locus and two locus genotypes discussed earlier in this chapter. Breeders and quantitative geneticists are committed to the notion that genotypes vary in "value". In fact the breeding enterprise revolves around continuing efforts to develop, discover or identify genotypes or populations of genotypes that are "superior". Since genotypes and collections of genotypes (breeds, varieties, or cultivars) are evaluated in terms of performance (phenotype) it is clear that the value of a genotype rests in the phenotypic expressions that it evokes. However, these depend also on the environment and the relative effects of genotype are variable from one environment to another. Thus, <u>the values of genotypes depend on the kind of environments in which individuals possessing those genotypes will be grown</u>. For example, a maize hybrid that is superior in (adapted to) the environments of southern Minnesota is likely to be inferior (have lower value) in the environments of Missouri. This is by no means a new idea; it is instead one well appreciated by breeders and farmers generally.

It should also be emphasized, though it will be obvious to many, that the values of genotypes depend on the aspect of phenotype employed as the criterion of value. For example, the breeds of cattle rank differently when evaluated in terms of milk production than when evaluated in terms of beef production, and a cotton genotype with high value relative to lint length might rank much lower by the criterion of lint per seed.

The word, environment, is used in two ways. We speak on the one hand of the environment of an individual organism as opposed to that of another living at the same time and in almost the same place and, on the other hand, of the environment identified with a defined location, a period of time, and specified conditions of culture or husbandry. Specificity is absolute in the first case but not in the second. To achieve clarity herein I will employ these three terms: microenvironment, macroenvironment and population of environments. The exact total environment of any single plant or animal will be called a <u>microenvironment</u>. However, microenvironments will be perceived as existing in the absence of the organisms that might grow and develop in them. The general environment associated with a location, a time and one or more other specifics will be referred to as a <u>macroenvironment</u>. The microenvironments that are possible within the specifications of any macroenvironment constitute a population of microenvironments and such a population is in fact the macroenvironment. However, environment populations can, and usually will, be delineated to include the microenvironments of a number of macroenvironments.

The complexity of environment deserves special emphasis. It is made up of all the things, other than genotype, that affect the development of an individual animal or plant. In the case of a plant, for example, the microenvironment includes physical and chemical attributes of the soil in which the plant grows; climatic variables (temperature, humidity, amount and quality of solar radiation, etc.); and the number and kinds of biological organisms (pathogens, insects, etc.) to which the plant is exposed. The compexity is multiplied by the infinite variety of patterns in time (relative to the life-span of plants) that is exhibited by most of the components of environment. For practical purposes the potential number of different single plant (or animal) environments is infinite, even within a very restricted area, and the probability that two plants in the same field at the same time have had precisely the same microenvironment is infinitesimal.

Consider now a population of n genotypes and another of m microenvionments and let n and m be very large so that for practical purposes both populations can be assumed infinite. Next imagine that an individual (plant or animal) of each genotype is (could be) grown in each microenvironment to provide a phenotypic expression for each genotype in each of the microenvironments. Let X_{kp} be the phenotype, characterized numerically, associated with the k-th genotype and the p-th microenvironment. The mn phenotypic values, X's, comprise a population that can usefully be visualized in terms of a two-way table (see Figure 2.3). In accord with the principle that values of genotypes are reflected by phenotypes, the value of the k-th genotype relative to the environment population in question is defined as

$$Y_k = \Sigma_p X_{kp}/m \qquad 2.46$$

	Genotype					
Environment	1	2	3	k	n	Mean
1	X_{11}	X_{21}	X_{31}	X_{k1}	X_{n1}	E_1
2	X_{12}	X_{22}	X_{32}	X_{k2}	X_{n2}	E_2
3	X_{13}	X_{23}	X_{33}	X_{k3}	X_{n3}	E_3
.	.	.	.	.	.	.
.	.	.	.	.	.	.
.	.	.	.	.	.	.
p	X_{1p}	X_{2p}	X_{3p}	$X_{kp} = \ddot{X} + z_k + e_p + (ze)_{kp}$	X_{np}	$E_p = \ddot{X} + e_p$
.	.	.	.	.	.	.
.	.	.	.	.	.	.
m	X_{1m}	X_{2m}	X_{3m}	X_{km}	X_{nm}	E_m
Mean	Y_1	Y_2	Y_3	Y_k	Y_n	$\ddot{X}$

Figure 2.3. Tabular representation of a population of phenotype values and the effects into which phenotype value is divided.

The effect of the k-th genotype is defined as the deviation of its value from the average value ($\ddot{Y}$) of the population of genotypes.

$$z_k = Y_k - \ddot{Y}$$

from which we see that

$$Y_k = \ddot{Y} + z_k$$

In like manner, the value and effect of the p-th environment are defined as

$$E_p = \Sigma_k X_{kp}/n \qquad 2.47$$

and

$$e_p = E_p - \ddot{E}$$

respectively. Actually, $\ddot{Y} = \ddot{E} = \ddot{X} = \Sigma_p \Sigma_k X_{pk}/nm$

If there were no genotype-environment interactions; i.e., if the difference between phenotypes evoked by any pair of genotypes in the same microenvironment was the same in all microenvironments, X_{kp} would always equal $\ddot{X} + z_k + e_p$. Since differences between X_{kp} and $\ddot{E} + z_k + e_p$ result from genotype-environment interaction, the effect of interaction between the k-th genotype and the p-th environment is logically defined as

$$(ze)_{kp} = X_{kp} - \ddot{X} - z_k - e_p$$

Rearranging we obtain one of the most familiar mathematical models (symbolic descriptions) for phenotype

$$X_{kp} = \ddot{X} + z_k + e_p + (ze)_{kp} \qquad 2.48$$

Equation 2.46 reemphasizes the fact that the value of a genotype is not an inherent absolute quality of that genotype. Instead, value, the average of phenotypes evoked by a genotype, depends on the population of environments over which the average is taken. In like manner, the value of an environment, equation 2.47, depends on the genotype population with respect to which value is measured. For example, the environments of Iowa are far better when evaluated by the performance of Iowa maize hybrids than when evaluated by performance of Texas hybrids.

Equation 2.48 does not recognize macroenvironments but can easily be extended to do so. The population of microenvironments pertinent to any genotype population of plant or animal breeding is typically delineated in terms of space, time and husbandry practices; i.e., the breeder is alway interested in value of genotypes with respect to microenvironments identified by geographical area, a period of time extending into the future, and husbandry norms. Within those limits a particular macroenvironment may be identified by more specific boundary conditions and would be comprised of the microenvironments potential within those closer bounds. The effect, f_j, of such a macroenvironment is logically defined as the average of the effects of its constituent microenvironments and the effect of interaction of the k-th genotype with the j-th macroenvironment, $(fz)_{jk}$, as the average of the effects of interaction of the k-th genotype with the microenvironments of the macroenvironment. Accordingly,

$$f_j = \Sigma_p e_{jp}/m_j$$

and

$$(fz)_{jk} = \Sigma_p (ze)_{jkp}/m_j$$

where m_j = number of microenvironments in the j-th macroenvironment. Equation 2.48 can now be rewritten as follows

$$X_{jkp} = \ddot{X} + z_k + f_j + (fz)_{jk} + e_{jp} - f_j + (ze)_{jkp} - (fz)_{jk}$$
$$= \ddot{X} + z_k + f_j + (fz)_{jk} + e_{jkp} \qquad 2.49$$

where X_{jkp} is the value of phenotype for the k-th genotype in the p-th microenvironment of the j-th macroenvironment, and

$$e_{jkp} = (e_{jp} - f_j) + [(ze)_{jkp} - (fz)_{jk}]$$

In words, e_{jkp} is made up of (a) the deviation of the effect of the p-th microenvironment of the j-th macroenvironment from the mean of such effects for the macroenvironment and (b) the deviation of the effect of interaction between the k-th genotype and the p-th microenvironment from the mean of such effects over all microenvironments of the j-th macroenvironment.

Because populations of genotypes usually have a family structure, it is very frequently desirable to include effects associated with families in our mathematical model for phenotype. Let $\bar{z}_i$ be the effect of the i-th family of genotypes and $(f\bar{z})_{ij}$ the interaction between the i-th family and the j-th macroenvironment. They are logically defined as

$$\bar{z}_i = \Sigma_k z_{ik}/n_i$$
$$(f\bar{z})_{ij} = \Sigma_k (fz)_{ijk}/n_i \qquad 2.50$$

where n_i = number of genotypes in the i-th family. We can now rewrite 2.49 as follows.

$$X_{ijkp} = \ddot{X} + \bar{z}_i + f_j + (f\bar{z})_{ij} + e_{ijkp} \qquad 2.51$$

where X_{ijkp} is the value of the phenotype for the k-th genotype of the i-th family in the p-th microenvironment of the j-th macroenvironment and

$$e_{ijkp} = (e_{jp} - f_j) + [(ze)_{ijkp} - (fz)_{ijk}]$$
$$+ (z_{ik} - \bar{z}_i) + [(fz)_{ijk} - (f\bar{z})_{ij}$$
$$= (e_{jp} - f_j) + (z_{ik} - \bar{z}_i) + (ze)_{ijkp} - (f\bar{z})_{ij}] \qquad 2.52$$

Equations 2.48, 2.49. and 2.51 or modifications appropriate to special situations provide the basis for expressing the phenotypic variance among individuals and the variance of family phenotypic means in terms of components that are estimable from analyses of variance of phenotypic measurements. It will be shown in later chapters that the understanding of phenotypic variances provided by such expressions is specifically relevant to an important aspect of optimum design in many breeding programs.

A very early example, possibly the first, of the sort of mathematical model for phenotype discussed in this section was provided by Sprague and Federer (1951). The actual models and symbols presented here follow those set out by Comstock and Moll (1963) in their detailed treatment of genotype-environment interaction relative to problems in quantitative genetics.

3 Breeding Programs

The within generation (or cycle, see section 3.3) selection of breeding programs may all be done at a single point in time. That is single stage selection. On the other hand, when the significant performance data are obtained at different times selection decisions may, for the sake of reducing program costs, be made sequentially as more and more information becomes available, i.e., by multi-stage selection. Consider selection among males in the case of dairy cattle. Many are eliminated at the outset on the basis of records of production by their mothers and/or sisters while selection among those remaining is delayed until progeny performance data are available. As another example, adequate information on traits of wheat such as rust resistance and plant height can be obtained using relatively few plants per line but sufficient data on grain yield requires comparisons in larger plots at more than one location in more than one year. Total cost is obviously reduced if the lines on which yield data are obtained are limited to those already found to be satisfactory with respect to traits that can be effectively evaluated using relatively few plants.

In single as well as multi-stage selection the breeder is usually faced with selection on the basis of more than one measure of performance, e.g., growth rate, backfat thickness, and litter size in the case of swine. Hazel and Lush (1942) and Young (1961) discussed this problem and concluded that the most efficient use of multiple measurements is obtained by combining them properly in a single numerical index. Smith (1936) and Hazel (1943) were the first of many to discuss the optimum structure of such indices. Baker (1986) provides an exhaustive review of literature concerned with selection indices, presents derivations and related logic, and discusses their applications in plant breeding.

Unless stated otherwise, as in section 9.6.2 where selection among lines of maize in successive generations of inbreeding by self-fertilization is considered, this book will be dealing with single-stage selection based on one measure (a single trait or an index based on two or more measurements).

• 3.1. Terminology for Description

The following terms will be employed in specification of the significant details of breeding programs.

Selection unit. In many cases the selection practiced is among individuals; in other cases it is among some kind of family groups (half-sib families, full-sib families, pure lines, etc.). The units among which selection is practiced will be identified as the selection units.

Selections. Those specific units that are chosen to be used as parents (or the source of parents) of the next generation (or cycle) will be termed selections.

Criterion unit. Selection is sometimes, but not always, based on performance of the selection units themselves. For example, individuals may be selected on the basis of their own performance or, alternatively, on the performance of relatives. The units that provide the data used as the basis for selection will be identified as the criterion units. Thus when selection is among individuals but is based on progeny performance, the criterion units are the progenies though individuals are the selection units. Note also that more than one criterion unit may be involved as when selection among individuals is based on performance of both themselves and of parents, sibs, or progeny.

Selection criterion. The variable on which selection is based will be called the selection criterion. It may be a single phenotypic measure (grain yield, growth rate, plant height, number of eggs, etc.) or any function of any number of phenotypic measures. It will be designated in all that follows by the letter X and may be visualized as

$$X = F(X_1, X_2 \ldots X_m)$$

where F is to be read "a function of" and $X_1, X_2 \ldots X_m$ are $\underline{m}$ measures of phenotype. In practice, X is almost always a linear function of its component measures. Whether it is a single phenotypic measure or is compounded from two or more separate measures, it can logically be considered a single measure.

It is well worth noting that specification of the individual measures contributing to X (the selection criterion) specifies as well the criterion unit(s). If, in a corn breeding program, selection among progenies produced by self-fertilization of noninbred plants is based on grain yield (X_1), plant height (X_2), and lodging; i.e., failure of plants to remain erect (X_3), of those same progenies, the latter are both the selection units and the criterion units. If, on the other hand, selection among those same inbred progenies were based on one or more measures of performance of related progenies produced by crossing the noninbred parents to a "tester" population, the latter set of progenies are obviously the criterion units. Finally, if X_1 were grain yield of the "testcross" progenies and X_2 a measure of lodging of the inbred progenies both the inbred progenies (the selection units) and the "testcross" progenies would be criterion units.

Selection population. The population in which selection is practiced will be identified as the selection population and, for brevity, will often be referred to as the S-population. When the term, *genetic population*, is employed it will have reference to all the selection units that might have been produced, i.e., an infinite population of individuals or families.

Genetic sample. This term will be used to designate the sample of selection units from the genetic population (the S-population) that is actually observed and from which selections are chosen.

Target population. The prime objective of a breeding program is often improvement of the S-population itself, in which case, the S-population and the target population are the same. On the other hand, the primary purpose is frequently the improvement of a population other than, but related to the S-population. The program may, for example, be aimed at improvement of the cross between the S and some other population. The population that is the prime target for improvement by the selection effort will be designated the target population, or frequently the T-population.

Response criterion. The responses to (effects of) selection can be seen or measured in any phenotypic trait or combination of traits. The responses of greatest concern are those flowing from genotypic change in the target population. The phenotypic measure by which response to selection is to be characterized will be called the response criterion.

When the response criterion and the selection criterion are the same, the response observed is called *direct response*. Alternatively, when the selection criterion and response criterion are different, the observed change in the response criteion is referred to as *correlated response*. In practical breeding programs, the critical response criterion is total value and hence a function of all value traits. Since selection is rarely, if ever, based on total value itself, *correlated response is in general more crucial than direct response in the real worlds of plant and animal breeding.*

Single (versus two or more) population program. All breeding programs in which the selection population is also the target population are single population programs. Examples are the ordinary programs for the improvement of any of the breeds of livestock. On the other hand, when the target population is some type of cross; e.g., a single cross, three-way cross or rotation cross of different populations or of lines derived from different populations, the selection program may involve two or more distinct populations and would then be designated as a two (or three or more) population program.

Target population of environments (TPE). By virtue of genotype-environment interactions, the rank order values of genotypes are functions of the population of environment concerned and can be much different for different environment populations. Breeders are well aware of this (section 2.12) and, consequently, their programs for genetic improvement are directed toward performance in a particuar population of environments, the *target population of environments* that includes only a fraction of those in which the species in question is, or conceivably could be, grown.

The TPE is doubtless identified more specifically in the breeder's mind than is usual in what is recorded concerning program plans or results. A complete specification would necesarily deal with both geographical area snd numerous aspects of husbandry.

• 3.2. The Breeder's Goal—Ideal Genotypes

It goes without saying that there would be nothing better that breeding (applied agricultural genetics) could do than to supply immediately the best possible genotype(s) of each economic species for every environment in which the species is an appropriate choice for production purposes. Because this cannot be done immediately and because of continuing and partly unpredictble changes in environment, it represents a goal that can be approached but never completely realized. In practice the breeder is acutely aware of the relatively near future and of the rewards of genetic improvement within short time periods, facts reflected in coverage by the accumulated literature of plant and animal breeding. Nevertheless, the implicit and increasingly urgent long range goal of breeding is to produce what were earlier referred to as "best possible genotypes" but will now be called ideal genotypes.

What is meant by ideal genotypes can be illuminated by discussing their structure in terms of genes, alleles, and single locus genotypes. In doing so, let us first assume no genotype environment interaction, no epistasis, no overdominance, and that the dominance of one allele to another is in no case complete. In that situation one specific allele of each gene that affects total value would be the best (most useful) allele, the ideal genotype would be homozygous for the best allele of each relevant gene and there would be only one ideal genotype.[1] Of course none of our four assumptions can be treated as valid, so we must proceed to consider the situation as one after another of them is abandoned.

Let us begin by dropping our assumptions concerning dominance. If the best allele is completelly dominant to one or more others in the case of one or more genes, there will be not one ideal genotype that exceeds all others in value but, instead, a group with values that are maximum and equal. Among these one would be homozygous for all "best" alleles while the others would be heterozygous at one or more loci. If overdominance is present at one or more loci, each ideal genotype would be heterozygous at each such locus. It is worth noting and emphasizing that an ideal genotype that is heterozygous at one or more loci would have to be reproduced asexually or as the hybrid of two pure lines if segregation is to be avoided.

Given genotype environment interaction, any genotype that is ideal for one environment may or may not be ideal for another. We can, however, imagine a population of environments sufficiently homogeneous so that genotypes ideal for one of them are ideal as well for all others in that population of environments. The corollary for the breeder is that ideal genotypes are likely to be different from one environment population to another. It has, of course, long been recognized that varieties or breeds that are superior in one region of the world may be, and often are, inferior in another.

Nothing is changed by the presence of simple epistasis; one allele of each gene (or, given overdominance, one single locus genotype) is still the most useful over the whole range of genotypes at other loci. However, multiple peak epistasis means that what is the most useful allele or genotype at any one locus may vary depending on genotype at other loci. Accordingly, it is conceivable that genotypes having different alleles at two or more loci could both be ideal; i.e., have equal and maximum value relative to a specific population of environments. As perhaps the simplest example, $A_1A_1B_2B_2$ and $A_2A_2B_1B_1$ might be equal in value and superior to all of the other possible genotypes for the A and B genes.

3.3. Recurrent Selection

Recurrent selection is a cyclic process. In each cycle selection units are produced, selection criterion data are obtained and, as the final step, selections are identified. Recurrent selection programs are those in which (1) there are successive cycles and (2) the selections of the k-th cycle are employed as the parents or gene source of cycle k+1.

Recurrent selection is a natural mode of procedure, almost automatic, in long term, open-ended breeding programs. If selection is effective, selections will on average be superior to previously available material and hence the natural choice as parent stock for the next sample of selection units. Thus recurrent selection is common, in one form or another, in both animal and plant breeding and examples of long term programs in which recurrent selection is completely absent are hard to find. In general the issue is not whether recurrent selection should be employed but rather how to balance between recurrent and nonrecurrent selection in work with species in which the latter is feasible and in one way or another attractive.

The contrast between the two modes can be exemplified in terms of the classical inbred-hybrid system of corn breeding. The development of a sample of inbred lines plus the various steps leading to the selection of one or more specific hybrids constitutes a cycle of the program. If the lines of the next cycle are extracted from the same source material as those of the preceding cycle, the selection is nonrecurrent. But if superior hybrids or their inbred line parents are used in the synthesis of a new source from which lines of the next cycle are developed, the selection is recurrent. As another example, consider a naturally self-fertilizing species such as barley or rice in which normally the cultivars are pure lines. The double cross of four superior cultivars might be used as the source of new pure lines (selection units) to be evaluated as potential selections. If the program were continued by always returning to the same source of selection units, the selection involved would be nonrecurrent but if a double cross of four selections (or any other population synthesized from selections)were employed as the source of the next sample of selection units, the mode of selection would be recurrent.

In nonrecurrent selection, successive samples from the same genotype population are screened for superior selection units. It is obvious that this process will always be subject to diminishing returns, and that after a very few cycles (two or more) the chance of finding substantially better selection units than those already identified will become very small unless the nature of environments in which high performance is desired has changed in some significant way. Changes in environment populations do occur and do cause changes in the relative values of genotypes (see section 2.12). Possible causes for such changes come readily to mind. New variants of disease pathogens are a well-known source of change in genotypic values. If fertilizer prices (relative to grain prices) increase, reduced fertilizer usage will result and environments will be changed. Many other changes in husbandry or cultural practices are possible that would shift genotype values. When genotype values change, the genotypes selected in the past may obviously become inferior relative to others present in the same genetic population.

• 3.4. The Logic of Recurrent Selection

It was noted in section 2.4 that the probability of a genotype in a random mating population at linkage equilibrium is a function of allele frequencies. More specifically, it is the product of the population frequencies of the alleles contained in the genotype in question. Thus, in diploid organisms,

$$P(G) = \pi_i q_{im} q_{if} \qquad 3.1$$

where G symbolizes a specific genotype, q_{im} is the population frequency of the i-th locus allele received via the male parent gamete, q_{if} is the population frequency of the i-th locus allele received via the female parent gamete, and π_i specifies the product of all quantities identified by the subscript i (i = 1,2 ..... n). It is apparent that this probability will be smaller than the smallest of the q's (unless all q's save one are equal to 1.0) and that it is likely to be very small if the number of significant genes that are segregating is even moderately large.

This is crucial because if the probability of an ideal genotype is extremely small, the probability that one or more individuals possessing that genotype will be present in a sample may also be very small. In that event the ideal genotype is not likely to be obtained by nonrecurrent selection. On the other hand, the number of useful alleles possessed by selections should usually be greater than in the population from which they were chosen. Therefore, the frequencies of useful alleles are expected to be larger in successive cycles of recurrent selection. After sufficient elevation of those frequencies the probability of an ideal genotype will be increased enough so that the probability of its presence in a sample of moderate size will be large. To summarize, the logic of recurrent selection is to achieve in successive cycles what would almost certainly never be achieved by nonrecurrent selection.

For illustration, suppose that the value of genotype in a particular species is affected by n genes, that a random breeding population is at linkage equilibrium, that the frequency of the best allele is q at all loci and that the best allele is in no case completely dominant to any other allele of the same gene. Then the ideal genotype would have to be homozygous for the best allele at each of the n relevant loci and its probability, from 3.1, would be

$$P = \pi_i q_i^2 = q_i^{2n}$$

It follows that (a) the probability of an individual with other than the ideal genotype would be (1 - P), (b) the probability that each of m random genotypes would be other than ideal would be $(1 - P)^m$, and (c) the probability that one or more individuals in a random sample of size m would possess the ideal genotype would be

$$P_m = 1 - (1 - P)^m = 1 - (1 - q^{2n})^m \quad 3.2$$

Table 3.1 shows some sample values of P and Table 3.2 shows values of P_m for various combinations of P and m. The two tables are most usefully employed together. Consider three examples. With 10 segregating genes the probability of the ideal genotype increases from 0.000037 to 0.0115 as q increases from 0.6 to 0.8. Moving to Table 3.2 we find that the corresponding increase in the probability that there will be one or more ideal genotypes in a sample of 200 is from less than 0.01 to more than 0.87. As a second example, note from Table 3.1 that, with q = 0.75, P decreases from 0.0032 when there are 10 segregating genes to 0.00001 when there are 20. Table 3.2 shows the corresponding decrease in the probability that there will be one or more ideal genotypes in a sample of 400 to be from more than 0.699 to 0.004. Note also, using first Table 3.1 and then Table 3.2, that with n = 40 as q increases from

Table 3.1. The probability (P) of an ideal genotype associated with various values of n, the number of significant segregating genes, and q, the frequency of best alleles

n	0.6	0.7	0.75	0.8	0.85	0.9	0.95
10	.000037	.000800	.00317	.01153			
14		.000046	.00032	.00193			
16		.000011	.00010	.00079	.00550		
20			.00001	.00013	.00150		
40					.0000023	.000220	.01650

Note: Assumptions are H-W frequencies of genotypes, linkage equilibrium, equal frequencies of all best alleles, no overdominance, and no best allele that is completely dominant.

Table 3.2. The probability (P_m) of one or more ideal genotypes in a sample of size m associated with various values of P and m

P	m = 50	m = 100	m = 200	m = 400
.00001	.0005	.0010	.0020	.0040
.00005	.0025	.0050	.0100	.0198
.0001	.0050	.0100	.0198	.0392
.0005	.025	.049	.095	.181
.001	.049	.095	.181	.330
.003	.139	.260	.452	.699
.005	.22	.39	.63	.87
.01	.39	.63	.87	.98

0.85 (or any lesser value) to 0.95 the probability of one or more ideal genotypes in a sample of 100 increases from less than 0.001 to >0.63 The difference between q = 0.85 and q = 0.95 appears relatively small but we will find (see section 6.3 and Table 6.3) that when there are as many as 40 significant segregating genes, this apparently small increase cannot often be expected in less than 10-20 cycles of what we call mass selection.

For various reasons the frequencies of best alleles would almost certainly never be equal. However, calculations such as those made to produce Table 3.1 are much easier when that assumption is made and the significant principles are in no way exaggerated or distorted as a result.

• 3.5. Complete Dominance and the Utility of Systems for Exact Reproduction of Genotypes

As discussed in Section 3.2, in the absence of overdominance and complete dominance the single best genotype at any locus is the one homozygous for the best allele. However, if the best allele is completely dominant to other alleles of the same gene, some or all of the genotypes heterozygous for the best allele are as good as the one homozygous for that allele. It follows that if the best allele is completely dominant at one or more of the significant segregating loci, there will be more than the one, completely homozygous, ideal genotype that exists in the absence of any complete dominance. To illustrate how this affects the probability that any random genotype will be ideal, we will make assumptions employed in Section 3.4 except the one concerned with complete dominance. Instead we will assume that at x of the n significant segregating loci the best allele is completely dominant to all other alleles of the same gene. At any of those x loci the probability of a genotype homozygous or heterozygous for the best allele would be $1 - (1-q)^2$, the probability that any random genotype will be one of the possible ideal genotypes would be

$$P = q^{2(n - x)}[1 - (1 - q)^2]^x$$

and the probability that one or more individuals in a random sample of size m would possess an ideal genotype would be

$$P_m = 1 - (1 - P)^m$$

as given by equation 3.2. For the special case in which x = n/2, Table 3.3 lists values of P for comparison with those in Table 3.1.

As anticipated the probabilities in Table 3.3 are substantially higher than those for the same values of n and q in Table 3.1. Consider, for example, the specific case in which n = 40 and q = 0.9. The probability listed in Table 3.1 is 0.00022 whereas that in Table 3.3 is 0.01209. The first of these is the probability of the genotype that is homozygous for the best allele of all 40 significant genes. The second is the probability of an ideal genotype that is homozygous for the best allele at all twenty of the loci where the best allele is not conpletely dominant to all others and homozygous or heterozygous for the best allele at the twenty loci where the best allele is completely dominant. It is apparent that 0.01209 - 0.00022 =0.01187 is the probability of an ideal genotype that is heterozygous for the best allele at one or more of the loci at which the best allele is completely dominant. The significance of 0.01209 versus 0.00022 as the

Table 3.3. The probability (P) of an ideal genotype associated with various values of n, the number of significant segregating loci, and q, the frequency of best alleles, when the best allele is completely dominant to all others at ½ of the n loci

n	0.5	0.6	0.7	0.8	0.85	0.9
10	.00023	.00253	.01763	.08755	...	...
14	...	.00023	.00351	.03305	...	...
16	...	.00007	.00156	.02031	...	...
20	...	...	.00031	.00767	.03087	...
40	...	...	...	.00006	.00095	.01209

Note: Assumptions are H-W frequencies of genotypes, linkage equilibrium, equal frequencies of all best alleles, and no overdominance.

probability of an ideal genotype is emphasized by reference to Table 3.2. Consider the probability (P_m) that a random sample of 200 genotypes will include one or more ideal genotypes. Given P = 0.00022 we find that P_m is about 0.04 but when P is 0.012, P_m is greater than 0.87. In the former case it is quite unlikely that an ideal genotype will be found in a sample of 200 but in the latter case it is almost certain that there will be one or more in a sample of that size.

The full value of selecting a genotype that is ideal but heterozygous with respect to one or more significant genes can be realized only if the genotype is reproducible. The systems by which unlimited reproduction of specific partially heterozygous genotypes is possible are (1) asexual reproduction and (2) single crosses between pure lines. Clearly these have special value (or may have) when the best allele is completely dominant at a portion of the significant segregating loci of a species.

The special significance of the reproducibility of genotypes is emphasized by the generally inferior performance of sexual as compared to asexual offspring in the case of the cultivars of species such as apples, oranges, irish potatoes, or sugar cane in which asexual reproduction is possible and is the normal commercial process.

• 3.6. Special Issues Relating to Potential Success via Recurrent Selection

For the purpose of this subsection, "success" will mean the development of an ideal genotype that can be reproduced for production purposes. The basic condition for success is that the alleles required in the ideal genotype must have positive selection coefficients in enough cycles of selection so that through the course of recurrent selection the expected change in their frequencies will be positive. The five related issues to be discussed here in a preliminary way are overdominance, the selection criterion, epistasis, population size, and change in the values of genotypes.

• 3.6.1. Overdominance

Given overdominance in the action of some portion of the significant genes, an ideal genotype would be heterozygous at all overdominant loci and reproduction of the ideal genotype would require asexual reproduction or the hybridization of appropriate homozygous lines. In species where one of these is a practical possibility, the remaining concern is the probability of an ideal genotype that can be achieved through recurrent selection.

In a single population program, the genotype favored by selection at overdominant loci would be the superior heterozygote. The obvious result is that selection would work to maintain segregation at such loci and intuition suggests that the frequencies of the alleles contained in the superior heterozygote would approach and be maintained at or near equilibrium values by recurrent selection.

Let those equilibrium frequencies be symbolized as q_e and $(1 - q_e)$. Then the probability of the superior (heterozygous) genotype in the selection population would be near $2q_e(1 - q_e)$ of which the maximum value is 0.5. Assuming the same value of q_e at each of x overdominance loci, that the frequencies of best alleles at other loci have reached 1.0, and linkage equilibrium, the probability of the ideal genotype would be

$$P = [2q_e(1 - q_e)]^x = Z^x$$

in which $Z = 2q_e(1-q_e)$.

From Table 3.2 we see that if this quantity is 0.005 or greater, the probabilities of one or more ideal genotypes in samples of 200 and 400 are 0.63 and 0.87, respectively. The largest values of x for which $P > 0.005$ are shown below for selected values of q_e.

q_e	Largest x
0.5	7
0.6	7
0.7	6
0.8	4
0.9	3

Realistically, we should go at least one step further and determine the probability (P) of a genotype that is heterozygous at all or at all save one of x overdominance loci. Based on binomial distribution probabilities, it is apparent that in this case

$$P = Z^x + xZ^{x-1}(1-Z) \qquad 3.3$$

where, as before, $Z = 2q_e(1-q_e)$.

Using 3.3, we find that largest values of x for which $P > 0.005$ are as follows.

q_e	Largest x
0.5	11
0.6	10
0.7	9
0.8	7
0.9	4

The significance of these results is that overdominance at a small number of loci does not seriously limit what can be accomplished via single population recurrent selection if heterozygous genotypes can be replicated at will for commercial purposes. We have noted (section 3.5) that this can be done by asexual reproduction or by the use of single crosses between pure lines.

When the number of overdominance loci is large (greater, say, than 12) or when asexual reproduction is not possible and the development of numerous homozygous lines is not feasible, a different procedure is required. Hull (1945) and Comstock et al (1949) proposed recurrent selection systems in which the target population is the cross of two populations and the selection procedure is such that, at overdominance loci, the expected eventual result is that one population will be homozygous for one of the required alleles and the other population homozygous for the other required allele. These systems are described in detail in Chapter IX. The essential facts to be noted here are (1) a limited amount of overdominance is not a barrier to success, as here defined, via single population recurrent selection in certain species and (2) in other species or if the number of overdominant loci is large, success may be possible via recurrent selection in appropriately designed two population systems.

3.6.2. The Selection Criterion

The key to effective recurrent selection is the increase through generation time in frequencies of favorable alleles. The responsible element of the process is the selection criterion, the measure by which selections are distinguished. If the presence of a specific allele affects phenotype in such a way that the average value of the selection criterion is greater than it would otherwise be, that specific allele is favored by the selection process, the expected change in its frequency as the result of a cycle of selection is positive and the expected total change in its frequency increases with the number of recurrent selection cycles.

If the frequencies of all best alleles are to be increased during the course of recurrent selection, an obvious basic requirement of the program is that the presence of each favorable (best) allele must be reflected by a positive increment in the selection criterion. For this to be the case, the selection criterion must vary in appropriate degrees and directions with each aspect of the phenotype that has a bearing on total value (see section 12.4). Inherent in this requirement is the very special problem, for plant and animal breeders, of knowing in complete detail what the optimum phenotype in a species will be in future years.

- 3.6.3. **Population Size**

The actual, in contrast to the expected, change per cycle in the frequency of an allele includes a random increment. This random increment is the source of the "random drift" of allele frequencies that was recognized and discussed in the preceding chapter. Because it will be negative about as often as positive and, in any specific cycle of selection, may be larger in absolute magnitude than the expected change in frequency, it is possible for the total actual change in frequency of a best allele to be negative instead of positive and possible, also, in finite selection populations for a useful allele to be lost instead of being increased in frequency and ultimately fixed in the population.

All real populations must be finite, though they may be very large, and in section 2.7 we noted that the amplitude of random drift varies inversely with "effective population size". It is apparent that effective size can, at least theoretically, be great enough to make random drift trivial relative to expected changes in the frequencies of favorable alleles due to selection and it follows that the size required, in recurrent selection programs, to trivialize the effects of drift relative to those of selection is a major issue in breeding (see Chapters XI and XIV).

- 3.6.4. **Epistasis**

Two broad classes of epistasis were identified in Chapter II. In the case of simple epistasis, the rank order values of the alleles of any particular gene are not conditioned by the frequencies of the alleles of other genes and no interference with success via recurrent selection arises from the epistasis. But in the case of multiple peak epistasis, the rank order values of a set of alleles are contingent on the frequencies of the alleles of one or more other genes so that, depending on the matrix of allele frequencies in the foundation population, the allele frequency changes fostered by recurrent selection may be such that the limits approached are associated with a lesser peak (a genotype that is superior to many others but is not "ideal"). Briefly, until we know more concerning the significance of multiple peak epistasis relative to the genetic variation within species, we must recognize that this form of epistasis may in some instances interfere with the synthesis of an ideal genotype by recurrent selection.

- 3.6.5. **Change in the Values of Genotypes**

The relative values of genotypes may change over time (a) because the way total value is affected by different aspects of phenotype changes or (b) because the target population of environments changes. Consider the following examples. (1) Before the advent of the now familiar substitutes for lard in food preparation the optimum fat content in swine carcasses was much higher than today. (2) The development of synthetic fibres resulted inevitably in reduced demand for wool with the result that the production cost and quality of lamb carcasses have become more significant relative to wool yield as components of total value in sheep. (3) Maize hybrids respond differentially to increases (or decreases) in level of fertilization and the levels employed by maize producers have changed in the past and, for one reason or another, are likely to change in the future.

4 Comparisons of Breeding Procedures

It is emphasized in Chapter 1 that the cost of resolving all questions concerning the values of procedures in breeding programs by experimentation would be prohibitive. The number of questions is great, many of the required experiments would need to be very large and a significant portion of the questions could have different answers in different species. Moreover, though answers are needed now, most experiments designed to yield any one or more of the needed answers would require long periods of time.

The obvious alternative to complete dependence on experimentation is the development of a theory base for comparison of procedural variations. Given that the development of ideal genotypes requires recurrent selection, it is clearly appropriate that primary attention be focused on theory for comparing the numerous ways, and variations of them, in which recurrent selection can be practiced. The basic genetic questions in evaluation of a breeding program proposal are:

(1) What are the expected response rates; i.e., the rates of genetic changes that would be promoted by the selection process of the program?

(2) When total value is considered, are the amounts of overdominance or hybrid vigor sufficient to make a two (or more) population program a substantially better choice than a single population program?

(3) For any best allele, what is the probability that, in the long run, its frequency will be increased rather than decreased? Or, alternatively, what effective population size would be required to insure progressive increases in the frequencies of most best alleles so that the probability of near ideal genotypes would eventually be large?

The overriding practical question concerns operational cost (a) per unit of genetic improvement per unit of time and (b) per unit of total genetic improvement when the response limit of the program will have been reached.

The basic genetic questions will receive prior and detailed attention in what follows. The cost question is nongenetic but cannot be ignored in a treatment of the implications of quantitative genetic theory for plant and animal breeding practice. Costs will be considered in general ways and suggestions for minimizing cost per unit of response will be made, based on theory to be outlined.

Finally, it should be noted that judgments concerning whether the breeding programs identified as superior (individually or as a class) should or should not be conducted belong in the social policy area and depend in the final analysis on the successive increments of total value likely to be achieved in relation to values placed on the products of alternative uses of human effort and natural resources. Because I think it proper for scientists to consider the "contributions to the common good" that would result from applications of their own science, I will set forth personal views on this subject in a later chapter.

The literature relative to theory for comparison of breeding programs has been accumulating through the last seven decades. The roots of the present extensive literature are in such well-known publications as those by Fisher (1918 and 1930), Wright (1921, 1931, and 1935) and Haldane (1924 - 1932). The most familiar theoretical statement concerning the genetics of breeding programs; i.e., that direct response to selection among individuals (as the selection units) in recurrent selection is equal to the product of the selection differential and the heritability of the trait, appeared (without assignment of origin) in Lush (1937). Its intrinsic origin appears to have been Fisher's Fundamental Theorem of Natural Selection first set out in The Genetical Theory of Natural Selection (1930). The first theoretical comparison of breeding systems was, to my knowledge, that by Lush (1947) followed closely by Comstock et al (1949). However, these were preceded by important papers by Smith (1936), Hazel and Lush (1942), and Hazel (1943) dealing with variations in the selection criterion. The concept of effective population size was originated by Wright (1931) and the germinal papers dealing with the probability of fixation of favorable alleles and response limits were by Kimura (1957) and Robertson (1960), respectively.

5 Responses to Artificial Selection

Knowledge of changes to be expected in the populations of breeding programs is ability to predict; i.e., statements of expectations constitute predictions and the expectations for different breeding programs provide the theory base for the comparative evaluation of programs. This chapter focuses on two quantities, the expected changes per cycle of selection in (1) the frequency of an allele and (2) the average value of the genotypes of a population. None of the expectations presented is completely general in its validity; instead each depends on some assumptions concerning variables that affect it.

This chapter is devoted to selection based directly on measures of phenotype. The role and potential of marker assisted selection is discussed in Chapter XVI in the light of extended background.

• 5.1. Change in Average Value of Genotypes—A Preliminary Form

Consider the value of a genotype relative to a specific population of environments and let

Y_i = the value of the i-th genotype for that environment population and

f_i = the frequency of the i-th genotype in a target population.

Then if we assume that Y_i is a constant,[1] the average in the target population of genotypes is

$$\ddot{Y} = \Sigma_i f_i Y_i. \qquad 5.1$$

The assumption that the Y's are constants implies that they are not frequency-dependent, that in every case the value of a genotype does not vary either with its own frequency or with the frequency of any other genotype or genotypes. It cannot be claimed that this is a completely valid assumption (consider the value of multiline varieties in crop production; see Borlaug, 1953, and Parlevliet, 1979). However, it is an assumption very widely employed in quantitative genetics, often without specific recognition, and will be employed throughout this book except where otherwise indicated.

Now consider the change in $\ddot{Y}$ resulting from one cycle of selection and let this change be symbolized as

$$(\Delta\ddot{Y})_k = (\ddot{Y})_{k+1} - (\ddot{Y})_k$$

Where $(\ddot{Y})_{k+1}$ is the average value of genotypes in the target population after the k-th cycle of selection. Given constant values of the individual Y's, it is apparent, from 5.1, that a change in $\ddot{Y}$, ($\Delta\ddot{Y}\neq 0.0$), can result only from changes in the frequencies of genotypes. These we know (Chapter II) to be functions of allele frequencies in parent populations, degrees of linkage disequilibrium and inbreeding, and random sampling. However, the purpose and prime effect of selection is change in allele frequencies and consistently positive changes in the average genotype value expected in a T-population can only be those associated with changes in allele frequencies. Therefore, the impact of selection on $\ddot{Y}$ is reasonably viewed as the sum of the effects of changes in the frequencies of individual alleles.

The frequency, in the S-population, of the j-th allele at the i-th locus will be symbolized by q_{ij}. The change in $\ddot{Y}$ per unit change in q_{ij} is, for small changes in q_{ij}, closely approximated by the partial derivative, $\partial\ddot{Y}/\partial q_{ij}$. Let Δq_{ij} be the change in q_{ij} from one cycle of selection to the next. Then the consequent change in $\ddot{Y}$ is closely approximated by $(\Delta q_{ij})\partial\ddot{Y}/\partial q_{ij}$, the product of the change in frequency of the allele and the change in $\ddot{Y}$ per unit change in the frequency of the allele. Summing these products over all alleles of all genes provides an approximation to the total change in $\ddot{Y}$ resulting from all changes in allele frequencies of the S-population from one cycle to the next in a breeding program. Thus

$$\Delta\ddot{Y} = \Sigma_i\Sigma_j(\Delta q_{ij})\partial\ddot{Y}/\partial q_{ij} \qquad 5.2$$

Because allele frequency changes in finite populations are subject to sampling variation, only their expected values can be known. For this reason only the expected value of $\Delta\ddot{Y}$ can be known. Recognizing this, 5.2 is appropriately rewritten in expectation form as follows

$$E(\Delta\ddot{Y}) = \Sigma_i\Sigma_j\, E(\Delta q_{ij})\partial\ddot{Y}/\partial q_{ij} \qquad 5.3$$

With 5.3 as the starting point the task of deriving more informative expressions for $E(\ddot{Y})$ breaks down to the separate tasks of analyzing $E(\Delta q_{ij})$ and $\partial\ddot{Y}/\partial q_{ij}$. It will soon become apparent, and worth remembering, that changes in $E(\Delta q)$ are associated with change in the selection unit, the criterion unit, the selection criterion or selection intensity but that $\partial\ddot{Y}/\partial q$ varies only with the response criterion or the target population. The significance of this is that only $E(\Delta q_{ij})$ must be derived for each selection systems.

• 5.2. Primary Assumptions in Development of More Analytic Expressions

Unless otherwise stated all further algebraic statements of expected responses to selection will involve the following assumptions;

1. diploid meiosis,
2. no multiple alleles, and
3. no epistasis.

It is appropriate to note and emphasize that much of the quantitative genetic theory for selection systems makes these same enabling assumptions. Nevertheless, that theory has been a very valuable aid to understanding and predicting the effects of selection. While it remains to be demonstrated that the last two of the assumptions listed above have been or may be the source of major errors in judgements concerning breeding programs, multiple allelism and epistasis must be accepted as pervasive genetic phenomena. Hence, to assume they are absent is to adopt a nonrealistic genetic model and to risk derivation of expectations that are inadequate for application in the real world. Clearly, the theory to be presented cannot be considered appropriate to auto-polyploids.

• 5.3. The Composition of Value of Genotype

In the absence of epistasis, the effects of genotype at any single locus are unaffected by genotype at other loci. Therefore, assuming no epistasis, the value of total genotype (with respect to any single trait) is appropriately symbolized as

$$Y = \alpha + y_1 + y_2 + ...y_i + ...y_n = \alpha + \Sigma_i y_i \qquad (2.36)\ 5.4$$

where α is a constant, y_i is a quantity that varies with genotype at the i-th locus, and n is the total number of loci (genes). The more specific statements that are required concerning these symbols must take account of the following.

1. Value of genotype (Y) varies depending on the population of environments with respect to which it is defined (section 2.12).
2. The y-values may also vary with the environment population. This must be true for y's if it is for Y's.
3. Each genotype has a separate value (Y) for every trait (phenotypic measure).
4. The y's are trait-specific as well as gene-specific. In the case of a pleiotropic gene, there is a different, nonzero y for each trait affected by the gene.

The two alleles allowed by our assumption of no multiple alleles will be designated by the letters, B and b, with a subscript to specify the locus when required for clarity. The letter B will refer to the allele that, in homozygous state, is the more favorable with respect to the particular trait being considered. Note that in the case of pleiotropic genes, the B-allele need not be more favorable with respect to each of the traits affected.

Total value will be symbolized by the letter, W, and when required, the measure (trait) for which genotypic value is being considered will be indicated by subscript. Now consider value of genotype with respect to total value. We have then

$$Y_W = \alpha_W + \Sigma_i y_{Wi}$$

where Y_W is the genotypic value for W relative to a particular population of environments. (In the context of any breeding program this will be understood to be the TPE unless otherwise indicated),

α_W is the value for W of the genotype that is homozygous for the b-allele at all loci, and

y_{Wi} is the difference in effect on Y_W between the genotype present at the i-th locus and the $b_i b_i$ genotype.

Note that by this definition, y_w for the bb genotype must be zero for all genes and that this is consistent with the definition of α_w which makes $Y_w = \alpha_w$ when the genotype is bb at all loci. Let A be any trait other than W. Then

$$Y_a = \alpha_a + \Sigma_i y_{ai} \qquad 5.5$$

where Y_a, α_a and y_{ai} have meanings relative to A that parallel those of Y_w, α_w, and y_{wi} relative to W.

• 5.4. The Derivative of Ÿ with Respect to q_i

Let the response criterion, whatever it may be, be designated by the letter C. Following 5.5, the value of a genotype with respect to the response criterion is then

$$Y_c = \alpha_c + y_{c1} + y_{c2} + \ldots y_{cn} = \alpha_c + \Sigma_i y_{ci} \qquad 5.6$$

and the population average of genotype values for measure C is

$$\ddot{Y}_c = \alpha_c + \ddot{y}_{c1} + \ddot{y}_{c2} + \cdots\cdots \ddot{y}_{cn} = \alpha_c + \Sigma_i \ddot{y}_{ci} \qquad 5.7$$

Given only two alleles per locus, one symbol suffices to specify the frequencies of both alleles. If q_i is the frequency of B_i the frequency of b_i must be $1-q_i$. Further, the expected frequencies of the locus genotypes can all be expressed in terms of the one allele frequency and only one derivative, $\partial\ddot{Y}/\partial q_i$, per locus is needed in equation 5.3. Accordingly, 5.3 may be written

$$E(\Delta\ddot{Y}_c) = \Sigma_i E(\Delta q_i)\, \partial\ddot{Y}_c/\partial q_i \qquad 5.8$$

for present purposes.

In the absence of epistasis each of the ÿ's in 5.7 is independent of genotype and allele frequencies at all other loci. Consequently, a change in q_i changes $\ddot{y}_{ci}$ (unless the i-th gene has no bearing on the measure C) but has no effect on the magnitude of any other $\ddot{y}_c$ and therefore

$$\partial\ddot{Y}_c/\partial q_i = \partial\ddot{y}_{ci}/\partial q_i \qquad 5.9$$

We now require the composition of $\ddot{y}_{ci}$ as a function of q_i. This depends , of course, on the nature of the T-population. The most common variations, which will be considered here, are:

1. the S-population is a random mating population and is itself the T-population or
2. the T-population is obtained by mating random individuals from the S-population with random individuals from some other population.

Genotypes at the i-th locus and their expected frequencies in the first of these cases along with symbolic values of the y_{ci}'s are listed below:

Genotype	Frequency	y_{ci}
B_iB_i	q_i^2	$2u_{ci}$
B_ib_i	$2q_i(1-q_i)$	$u_{ci} + a_{ci}u_{ci}$
b_ib_i	$(1-q_i)^2$	0

Note that the genotype frequencies listed reflect our assumption of a random mating population; they are the HW frequencies. Note also that the symbolic y-values allow any magnitudes of differential genotype effects and any level of dominance.[2] If a = 1, the effects of BB and Bb are equal; i.e., B is completely dominant; if $a = -1$, b is completely dominant; if $0 < a < 1$, B is partially dominant; if $a > 1$, there is overdominance, etc. Working from the table, we find that

$$\ddot{y} = q^2 2u + 2q(1-q)(u + au) = 2qu + 2q(1-q)au \qquad 5.10$$

Taking the derivative, and attaching the subscripts which, for convenience, were briefly ignored, we obtain (remembering 5.9)

$$\partial\ddot{Y}_c/\partial q_i = \partial\ddot{y}_{ci}/\partial q_i = 2u_{ci} + 2(1-2q_i)a_{ci}u_{ci} = 2[1 + (1-2q_i)a_{ci}]u_{ci} \qquad 5.11$$

Since i can take any value from 1 to n, it is apparent that all of the derivatives required for use in 5.8 will be of the same form as 5.11.

Genotype frequencies are listed below for T-populations obtained by mating random individuals from the S-population with random individuals from some other population. (If this other population is designated as A, the situation being considered can be specified by writing, T= (S x A.)

Genotype	Frequency	y_{ci}
B_iB_i	q_ip_i	$2u_{ci}$
B_ib_i	$q_i(1-p_i) + p_i(1-q_i)$	$u_{ci} + a_{ci}u_{ci}$
b_ib_i	$(1-q_i)(1-p_i)$	0

Here p_i symbolizes the frequency of B_i in the A-population. In this case we find

$$\ddot{y} = pq2u + [q(1-p) + p(1-q)](u + au)$$

$$= (p+q)u + (p+q - 2pq)au$$

and

$$\partial\ddot{Y}_c/\partial q_i = \partial\ddot{y}_{ci}/\partial q_i = u_{ci} + (1-2p_i)a_{ci}u_{ci}$$

$$= [1 + (1-2p_i)a_{ci}]u_{ci} \qquad 5.12$$

This derivative differs from 5.11 by being a function of p_i instead of q_i and by lacking the coefficient 2. The latter reflects the fact that in the first case the expression pertains to the S-population itself while, in the second case, the T-population gets only one-half of its genes from the S-population.

We see, as indicated earlier, that $\partial\ddot{Y}_c/\partial q_i$ varies with the target population but not with any detail of the selection procedure. The effect of change in allele frequencies on $\ddot{Y}$, the average value of genotypes in the T-population, is independent of the way in which allele frequency changes are brought about.

• 5.5. The Expected Change in Allele Frequency

If we assume linear regression of allele frequency in the selection units on the selection criterion, the following regression equation is justified and provides a basis for prediction of Δq_i

$$\hat{q}_{ik} = q_i + (\hat{X}_k - \ddot{X})\beta + d_k \qquad 5.13$$

Here $\hat{q}_{ik}$ is the frequency of B_i in the k-th selection unit,

$\hat{X}_k$ is the average value of the selection criterion for the k-th criterion unit,

$\ddot{X}$ is the average of $\hat{X}$ over the selection units of the population,

β is the coefficient of linear regression of $\hat{q}_i$ on $\hat{X}$ and

d_k is the deviation of $\hat{q}_{ik}$ from the regression line.

The terms <u>genetic population</u> and <u>genetic sample</u>, defined in section 3.1, are relevant here. The regression specified by 5.13 is the regression <u>in the genetic population</u> of the value of $\hat{q}_i$ in any selection unit on that of $\hat{X}$ pertaining to the same selection unit. As defined, the <u>genetic sample</u> is composed of random selection units from the genetic population. Let

$\bar{q}_i$ symbolize the average of $\hat{q}_i$ for the genetic sample and

$\bar{X}$ the average of $\hat{X}$ for the genetic sample.

Then if 5.13 is averaged over all units in the genetic sample, the result is

$$\bar{q}_i = q_i + (\bar{X} - \ddot{X})\beta + \bar{d} \qquad 5.14$$

where $\bar{d}$ is the average of d_k for the genetic sample.

Now consider the selections, the units chosen to be or provide the parents of the following generation. Let $\bar{q}_{si}$ and $\bar{X}_s$ symbolize the averages for the selections of $\hat{q}_i$ and $\hat{X}$, respectively. Then, assuming single-stage selection, an expression for $\bar{q}_{si}$ is obtained by averaging 5.13 over the selections. The result is

$$\bar{q}_{si} = q_i + (\bar{X}_s - \ddot{X})\beta + \bar{d}_s \qquad 5.15$$

where $\bar{d}_s$ is the average of d_k over the selections. Subtracting 5.14 from 5.15, we obtain

$$\bar{q}_{si} - \bar{q}_i = (\bar{X}_s - \bar{X})\beta + \bar{d}_s - \bar{d}$$

The expectation of this expression is

$$E(\bar{q}_{si} - \bar{q}_i) = (\bar{X}_s - \bar{X})\beta \qquad 5.16$$

if $\bar{X}_s$ and $\bar{X}$ (in addition to β) are treated as constants. That is appropriate because, once selection has occurred, their values for the case in point <u>are</u> constants. The expectations of $\bar{d}_s$ and $\bar{d}$ are zero because, given linear regression, the population mean of deviations (from the regression line) associated with any specific value of the independent variable is zero; i.e., the expectation of any random d is zero and so also is the expectation of the d associated with any selected value of $\hat{X}$.

We will now assume that $E[q_{i(t+1)}]$ is equal to $\bar{q}_{sit}$ where t designates time in cycles. This entails assuming (1) that reproduction by the selections is independent of (not correlated with) their genotypes and (2) no selection except that based on the selection criterion. We can then substitute $q_{i(t+1)}$ for $\bar{q}_{si}$ in 5.16 and, recognizing that $E(\bar{q}_i) = q_i$ so that $E(\bar{q}_{it}) = q_{it}$, write

$$E(\Delta q_i) = E[q_{i(t+1)} - q_{it}] = (\bar{X}_s - \bar{X})\beta \qquad 5.17$$

Since the coefficient of linear regression is the covariance between the independent and dependent variables divided by the variance of the independent variable, we obtain from 5.17 that

$$E(\Delta q_i) = (\bar{X}_s - \bar{X})\,\sigma(\hat{q}_i,\hat{X})/\sigma^2(\hat{X}) = \sigma(\hat{q}_i,\hat{X})k/\sigma(\hat{X}) \qquad 5.18$$

where $k = (\bar{X}_s - \bar{X})/\sigma(\hat{X})$ so that $k/\sigma(\hat{X}) = (\bar{X}_s - \bar{X})/\sigma^2(\hat{X})$

Because X is a measured quantity or a function of measured quantities, data accumulates from which $\sigma^2(\hat{X})$ can be estimated; and when selection has been practiced $\bar{X}_s - \bar{X}$, which is called the selection differential, can be known numerically. Equally important, given truncation selection k is inversely related to proportion selected and its approximate expectation, given any specified selection intensity, can be known in advance (see section 6.3.2).[3] The quantity that remains to be examined further is the covariance, $\sigma(\hat{q}_i,\hat{X})$.

Since X, like any other phenotypic quantity, may be viewed as the sum of a genotypic value (or mean genotypic value) and a deviation resulting from environment and interaction of genotype with environment, we write

$$\hat{X} = \hat{Y}_X + \hat{E}_X \qquad 5.19$$

Then expanding $\hat{Y}_X$ in accord with 5.4 and 5.6 we have

$$\hat{X} = \alpha_X + \hat{y}_{X1} + \hat{y}_{X2} + \cdots\cdots \hat{y}_{Xi} + \cdots\cdots \hat{y}_{Xn} + \hat{E}_X$$

Recalling the structure of covariances between a variable and a linear function of variables (Appendix 1), we recognize that

$$\sigma(\hat{q}_i,\hat{X}) = \sigma(\hat{q}_i,\hat{y}_{X1}) + \sigma(\hat{q}_i,\hat{y}_{X2}) + \cdots\cdots\sigma(\hat{q}_i,\hat{y}_{Xi}) + \cdots\cdots \sigma(\hat{q}_i,\hat{y}_{Xn}) + \sigma(\hat{q}_i,\hat{E}_X) \qquad 5.20$$

Assuming the criterion units to be distributed randomly relative to environments of the target population of environments, $\sigma(\hat{q}_i,\hat{E}_X) = 0.0$. It is also true that, given linkage equilibrium, all covariances of $\hat{q}_i$ with the $\hat{y}_X$'s, except $\sigma(\hat{q}_i,\hat{y}_{Xi})$, are zero. This latter fact will appear obvious to some readers but not to others; it is demonstrated for several selection systems in Chapters VI and VIII. Assuming random distribution of criterion units relative to environments, 5.20 reduces to

$$\sigma(\hat{q}_i,\hat{X}) = \sigma(\hat{q}_i,\hat{y}_{Xi}) + \Sigma_{j\neq i}\,\sigma(\hat{q}_i,\hat{y}_{Xj}) \qquad 5.21$$

or to

$$\sigma(\hat{q}_i,\hat{X}) = \sigma(\hat{q}_i,\hat{y}_{Xi}) \qquad 5.22$$

if linkage equilibrium is also assumed. Substituting in 5.18, we have

$$E(\Delta q_i) = [\,\sigma(\hat{q}_i,\hat{y}_{Xi}) + \Sigma_{j\neq i}\,\sigma(\hat{q}_i,\hat{y}_{Xj})]k/\sigma(\hat{X}) \qquad 5.23$$

or, assuming linkage equilibrium

$$E(\Delta q_i) = \sigma(\hat{q}_i,\hat{y}_{Xi})k/\sigma(\hat{X}) \qquad 5.24$$

Remember that,

$$k/(\sigma\hat{X}) = (\bar{X}_s - \bar{X})/\,\sigma^2(\hat{X})$$

and, hence, that the latter can be substituted for the former whenever it is useful to do so.

Before going on we must note that, in real breeding programs, criterion units can ordinarily be assumed randomly distributed relative to the micro-environments of a macro-environment but not relative to all micro-environments of the TPE (target population of environments).[4] This is because the TPE normally has dimensions in both time and space whereas comparisons of criterion units are invariably made at particular points in time and space. Consider a plant breeding program. In each cycle of selection the criterion units are grown for comparison in one or a small number of locations in one or a small number of years whereas the TPE encompasses all the locations of a region that may be very large and an indefinite span of future time. (For related discussion see section 10.2.) As a consequence, to assume that the effective value of $\sigma(\hat{q}_i,\hat{E}_X)$ is its population value in each cycle is not really justified. However, if the one or more macro-environments of the criterion unit comparison can be viewed as random relative to all those of the TPE, the expectation of $\sigma(\hat{q}_i,\hat{E}_X)$ can properly be equated to the population value of that covariance. Thus, on average, 5.21 and 5.22 are correct. It follows that 5.23 and 5.24 are correct as <u>a priori</u> statements concerning the expectation of Δq. However, in each <u>specific</u> case, $E(\Delta q_i)$ is conditioned by the actual macro-environments involved. Symbolically,

$$E(\Delta q_i)_k \neq E(\Delta q_i)$$

but

$$E[E(\Delta q_i)_k] = E(\Delta q_i)$$

Here $E(\Delta q_i)_k$ symbolizes the expectation of Δq_i when the criterion unit comparison is made in the k-th actual set of macro-environments,

$E[E(\Delta q_i)_k]$ is the average of $E(\Delta q_i)_k$ over all sets of macro-environments that are random relative those of the TPE, and

$E(\Delta q_i)$ is the quantity specified by 5.23 and 5.24.

For a more detailed treatment, see Appendix 5.1.

For the most part, our concern will be with the response to selection that is to be expected on the average. However, in particular instances the effect of comparing criterion units in unique macro-environments or the variation in response from cycle to cycle that arises from variation in the macro-environments involved will engage our attention (see, for example, section 11.2.2).

• 5.6. The Expected Change in Average Value of Genotype

Using 5.8 together with 5.11 or 5.12 and 5.23 or 5.24, expressions for $E(\Delta\ddot{Y}_C)$ that are appropriate for specific situations can be put together. For example, assuming that the S-population is also the T-population, the approximation is

$$E(\Delta\ddot{Y}_C) = [k/\sigma(\mathfrak{X})]\Sigma_i\,\sigma(\hat{q}_i,\mathfrak{X})2[1+(1-2q_i)a_{ci}]u_{ci} \qquad 5.25$$

Assuming that the T-population is obtained by mating S-population individuals to individuals from another population, the proper expression is

$$E(\Delta\ddot{Y}_C) = [k/\Sigma(\mathfrak{X})]\Sigma_i[\sigma(\hat{q}_i,\mathfrak{X})][1+(1-2p_i)a_{ci}]u_{ci} \qquad 5.26$$

Remember that $\sigma(\hat{q}_i,\mathfrak{X})$ is given by 5.21 but simplifies to 5.22 in the event of linkage equilibrium.

Note that 5.23, 5.25. and 5.26 were obtained without restrictions on the nature of selection or criterion units. Consequently, all three are applicable with respect to any system of recurrent selection.

In the context of natural selection among individuals, Fisher (1930) concluded that improvement in the "fitness" of a population is proportional to a quantity, W which in the symbolism of this text he described as

$$W = \Sigma_i 2q_i(1-q_i)A'_iA_i.$$

He called this quantity genetic variance. His definitions were such that in the case of selection among individuals of a random mating population, no epistasis, and two alleles per gene,

$$A'_i = \sigma(\hat{q}_i, \hat{X})$$

and

$$A_i = 2[1+(1-2q_i)a_i]u_i.$$

Comstock et al (1949) derived essential equivalents of 5.25 and 5.26 as the bases for theoretical comparison of three procedures for improvement of the cross between two populations. Empig et al (1971) employed the constructs of 5.25 and 5.26 to obtain expectations of genetic improvement specific to more than a dozen of the selection procedures employed in plant breeding.

• 5.7. Genotypic Superiority of Selections

Because the consequences of recurrent selection are the center of interest in this text, primary focus is on cycle to cycle change in allele frequencies and average values of genotypes in the target population. However, there are occasions in which information regarding the difference to be expected between average values of (a) the genotypes of selections and (b) the genotypes of the population from which they came may affect decisions. Attention is restricted here to those instances in which the criterion unit is itself the selection unit because the issue is rarely, if ever, relevant in other situations.

Assuming linear regression of Y_{ci} (= the average, for the i-th selection unit, of genotype values with respect to the response criterion) on $\hat{X}_i$ (= the average, for the i-th selection unit, of the selection criterion)

$$E(\bar{Y}_{cs} - \bar{Y}_c) = (\bar{X}_s - \bar{X})\beta(Y_c, \hat{X}) \qquad 5.27$$

where $\bar{Y}_{cs}$ is, for selections, the average value of genotypes with respect to the the response criterion,

$\beta(Y_c, \hat{X})$ is the coefficient of linear regression of Y_c on $\hat{X}$, and other symbols have significance assigned earlier. From 5.19

$$\hat{X} = Y_X + \hat{E}_X.$$

It follows that

$$\sigma(Y_c, \hat{X}) = \sigma(Y_c, Y_X) + \sigma(Y_c, \hat{E}_X)$$

which reduces to

$$\sigma(Y_c, \hat{X}) = \sigma(Y_c, Y_X) \qquad 5.28$$

if we assume $\sigma(Y_c, \hat{E}_X) = 0.0$ because, in appropriately designed comparisons, the units are distributed at random relative to environments. Using 5.28,

$$\beta(Y_c, \hat{X}) = \sigma(Y_c, Y_X)/\sigma^2(\hat{X})$$

Of course when the response criterion is the selection criterion (C= X), $\sigma(Y_C,Y_X) = \sigma^2(Y_X)$, $\beta(Y_C,\hat{X}) = \sigma^2(Y_X)/\sigma^2(\hat{X})$ and 5.27 becomes

$$E(\bar{Y}_{XS} - \bar{Y}_X) = (\bar{X}_S - \bar{X})\sigma^2(Y_X)/\sigma^2(\hat{X}) = [k/\sigma(\hat{X})]\sigma^2(Y_X) \qquad 5.29$$

• 5.8. Summary of Assumptions

Those employed in justifying 5.8 were

1. constancy of genotype values which implies that the Y's are not frequency dependent and that the target population of environments is constant through time,

2. expected allele frequency changes per cycle are small enough to justify approximation of their effects on $\ddot{Y}$ in the T-population in terms of the derivatives of $\ddot{Y}$ with respect to the q's,

3. no epistasis, and

4. no multiple alleles.

The additional assumptions employed in obtaining 5.11 and 5.12 were

5. diploid meiosis and

6. T-populations obtained via random matings. (This is not to say that only T-populations arising from random matings can be handled by the described procedure. It is just that 5.11 and 5.12 apply only for such populations.)

Further assumptions used in obtaining 5.23 were

7. single stage selection and linear regression of allele frequency in selection units on the selection criterion,

8. reproduction by selections independent of their genotypes,

9. no selection other than that based on the selection criterion, and

10. Criterion units distributed at random relative to environments within macro-environments that are random relative to the macro-environments of the TPE. (This ignores the competition between criterion units that can in some instances be a significant, non-random aspect of environment.)

The reduction of 5.23 to 5.24 involved the assumption of

11. linkage equilibrium in the S-population.

Relative to dioecious species, 5.11, 5.12, 5.23, and 5.24 rest also on assumptions, not explicitly stated earlier, that

12. consequences of sex-linkage can be ignored, i.e., that the proportion of genes located in the X-chromosome is small (there is reason to doubt that this is true in the case of some species, e.g., chicken and drosophila) and

13. the gene effect parameters (u's and a's) and non-genetic variances are equivalent in males and females.

Finally, there has been no mention of variation in either maternal or cytoplasmic effects on the selection criterion. This amounts to the tacit assumption that

14. such variation is relatively insignificant, if not absent, or that it is random relative to the genetic variation among criterion units.

Of these fourteen assumptions 5 and 6 require no justification, though the prediction equations that depend on them should obviously not be employed in the situations where they are not valid (e.g., autopolyploid species or T-populations produced by non-random matings).

6 Mass Selection

In the system commonly called mass selection, individuals are the units of selection and also the criterion units. Stated differently, the selection of this system is among individuals on the basis of some measure of their own phenotypes. In this chapter expressions for the response expectations, $E(\Delta q_i)$ and $E(\Delta \ddot{Y}_c)$, of mass selection will be developed and discussed. Assumptions employed will be those listed in Section 5.8. Linkage equilibrium will be assumed in Section 6.1 but not in Section 6.2.

6.1. Response Expectations, Linkage Equilibrium Assumed

The element of equation 5.18 that is variable among selection systems is $\sigma(\hat{q}_i,\mathcal{X})$. Hence the special requirement in finding the response expectations of any particular selection system is the composition of $\sigma(\hat{q}_i,\mathcal{X})$ specific to that system. Because $\sigma(\hat{q}_i,\mathcal{X}) = \sigma(\hat{q}_i,\hat{y}_{Xi})$ in linkage equilibrium populations (equation 5.22), the task here reduces to finding the composition of $\sigma(\hat{q}_i,\hat{y}_{Xi})$. The bivariate distribution of $\hat{q}_i$ and $\hat{y}_{Xi}$ when the individual is both the selection unit and the criterion unit is shown below.

i-th locus genotype	Probability	$\hat{q}_i$	$\hat{y}_{Xi}$
B_iB_i	q_i^2	1	$2u_{Xi}$
B_ib_i	$2q_i(1-q_i)$	1/2	$u_{Xi} + a_{Xi}u_{Xi}$
b_ib_i	$(1-q_i)^2$	0	0

Note that the list of genotypes possible at the i-th locus provides the base for recognizing the possible pairs of values of $\hat{q}_i$ and $\hat{y}_{Xi}$ and the frequencies in which those pairs occur in the individuals of the S population. The genotypes listed reflect our assumptions of diploid meiosis and no multiple alleles. The HW frequencies shown in the table reflect our further assumptions of random mating, no meiotic drive and no selection other than that based on the selection criterion. Symbolism for y values is that introduced in Section 5.4. The derivation of covariance is straightforward but the means (= expectations) of the two variables will be required and are obtained first. Subscripts are ignored until the final form of the covariance is given.

$$E(\hat{q}) = q^2(1) + 2q(1-q)(1/2) = q \quad 6.1$$

$$E(\hat{y}) = q^2(2u) + 2q(1-q)(u+au) = 2qu + 2q(1-q)au \quad 6.2$$

$$\sigma(\hat{q}_i,\hat{X}) = \sigma(\hat{q}_i,\hat{y}_{xi}) = q^2(1)(2u) + 2q(1-q)(1/2)(u+au) - qE(\hat{y})$$
$$= q_i(1-q_i)[1 + (1-2q_i)a_{xi}]u_{xi} \quad 6.3$$

Now substituting for $\sigma(\hat{q}_i,\hat{X})$ in equation 5.18, we obtain

$$E(\Delta q_i) = [(\bar{X}_s-\bar{X})/\sigma^2(\hat{X})]q_i(1-q_i)[1 + (1-2q_i)a_{xi}]u_{xi} \quad 6.4$$

Using this and equation 5.11 or 5.12 for substitution in equation 5.8, we obtain

$$E(\Delta\ddot{Y}_c) = [(\bar{X}_s-\bar{X})/\sigma^2(\hat{X})]\Sigma_i 2q_i(1-q_i)[1+(1-2q_i)a_{xi}][1+(1-2q_i)a_{ci}]u_{xi}u_{ci} \quad 6.5$$

in the case where S = T (i.e., the S-population is also the T-population) and

$$E(\Delta\ddot{Y}_c) = [(\bar{X}_s-\bar{X})/\sigma^2(\hat{X})]\Sigma_i q_i(1-q_i)[1+(1-2q_i)a_{xi}][1+(1-2p_i)a_{ci}]u_{xi}u_{ci} \quad 6.6$$

in the case where T = S x A (i.e., the T-population is the cross of the S-population with some other). There are two differences between 6.5 and 6.6. First, it is intuitively obvious that the allele frequencies in the populaton designated as A should appear in 6.6 and we see that $[1+(1-2p_i)a_{ci}]$ in 6.6 takes the place of $[1+(1-2q_i)a_{ci}]$ in 6.5. Second, the coefficient 2 appears in 6.5 but not in 6.6. This reflects the fact that each generation in the S-population receives all its genes from S-population parents but the S x A population receives only half its genes from S-population parents, the other half from A-population parents.

When interest is focused on the response to selection reflected by the exact phenotypic measure used as the basis for selection, the selection and response criteria are the same and C = X. Substituting x for c wherever the latter appears as a subscript, 6.5 becomes

$$E(\Delta\ddot{Y}_x) = [(\bar{X}_s-\bar{X})/\sigma^2(\hat{X})]\Sigma_i 2q_i(1-q_i)[1+(1-2q_i)a_{xi}]^2u^2_{xi} \quad 6.7$$

The <u>additive genetic variance</u> of X, in the genetic circumstance assumed in this section, will now be identified as

$$\sigma^2_{gx} = \Sigma_i 2q_i(1-q_i)[1+(1-2q_i)a_{xi}]^2u^2_{xi} \quad 6.8$$

This allows the following modification of 6.7

$$E(\Delta\ddot{Y}_x) = (\bar{X}_s-\bar{X})[\sigma^2_{gx}/\sigma^2(\hat{X})] \quad 6.9$$

Note (1) that $\bar{X}_s-\bar{X}$, the amount by which the average of $\hat{X}$ for <u>selections</u> exceeds the average of $\hat{X}$ for the whole <u>genetic sample</u>, is the quantity that is generally referred to as the <u>selection differential</u>;

(2) that, since $\hat{X}$ in mass selection is an individual value of X, $\sigma^2(\hat{X}) = \sigma^2(X)$, the phenotypic variance of X among individuals; and

(3) that the heritability (in the narrow sense) of X is defined as $\sigma^2_{gx}/\sigma^2(X)$

(see Lush1943, or Falconer, 1960).

With these facts in mind, 6.9 can be verbalized as follows: the expected direct response in mass selection is the product of the selection differential and the heritability of the selection criterion. As noted in Chapter IV, this may be the most familiar theoretical statement of quantitative genetics related to selection programs. The symbolic counterpart of the above verbal statement is

$$E(\Delta\ddot{Y}_x) = (\bar{X}_s - \bar{X})h^2_x \qquad 6.10$$

in which h^2_x = the narrow sense heritability of X.

In this book, the term "heritability" will be employed consistently to mean narrow sense heritability.

Readers will have noted, and perhaps been dismayed by, the fact that equation 6.8 is presented without demonstration that it conforms to original definitions of additive genetic variance. It is tempting to digress long enough to do that but the continuity of my main theme, response to selection, would be disturbed by so doing.

The counterpart of equation 6.8 is identification of the additive genetic covariance between X and C as

$$\sigma_{gcx} = \Sigma_i\, 2q_i(1-q_i)[1+(1-2q_i)a_{xi}][1+(1-2q_i)a_{ci}]u_{xi}u_{ci} \qquad 6.11$$

With this in mind, 6.5 can be rewritten as follows

$$E(\Delta\ddot{Y}_c) = [(\bar{X}_s-\bar{X})/\sigma^2(\hat{X})]\,\sigma_{gcx} \qquad 6.12$$

Here (and in 6.5) $E(\Delta\ddot{Y}_c)$ represents the correlated response in C to mass selection based on X. Correlated response should be considered more important in breeding than direct response since total value (Section 1.4) is rarely the selection criterion employed.

Note from 6.11 that no gene contributes additive genetic covariance unless u_x and u_c are both non-zero; i.e., that in the absence of linkage disequilibrium such covariance arises only from pleiotropy. Note also that, unlike additive variance, additive covariance and, hence, correlated responses can be either positive or negative.

Quantitative genetics has no singular label for the quantity in 6.6 that begins with the summation symbol (Σ). Quantities of this form, which are properties of a pair of populations rather than of one population, will be referred to herein as analogs of additive genetic variance or covariance.

• 6.2. Response Expectations Given Linkage Disequilibrium

In this case it is apparent (equations 5.21, 5.23, and 5.25) that both $\sigma(\hat{q}_i,\hat{y}_{xi})$ and $\sigma(\hat{q}_i,\hat{y}_{xj})$ are required as elements of the response expectations. To obtain the latter the bivariate distribution of $\hat{q}_i$ and $\hat{y}_{xj}$ is required. This means that two locus genotypes must be considered. As the basis for writing the probabilities of two locus genotypes, the probabilities of two locus gamete genotypes will be symbolized as follows[1]

Gamete genotype	Probability
B_iB_j	p
B_ib_j	r
b_iB_j	s
b_ib_j	t

where $p+r+s+t = 1.0$. Given linkage equilibrium we would have $p = q_iq_j$, $r = q_i(1-q_j)$, etc. (Section 2.4.3), but with linkage disequilibrium $p \neq q_iq_j$, $r \neq q_i(1-q_j)$, etc. Note that summing probabilities of gametes that contain B_i yields $q_i = (p+r)$ so that $(1-q_i) = (s+t)$ and, similarly, summing probabilities of gametes that contain B_j yields $q_j = (p+s)$ so that $(1-q_j) = (r+t)$.

The probabilities of offspring genotypes when mating is random can be written in terms of gamete genotype probabilities and the bivariate distribution of $\hat{q}_i$ and $\hat{y}_{xj}$ is as follows

Genotypes	Probabilities	$\hat{q}_i$	$\hat{y}_{xj} - u_{xj}$
B_iB_j/B_iB_j	p^2	1	u_{xj}
B_iB_j/B_ib_j	2pr	1	$a_{xj}u_{xj}$
B_ib_j/B_ib_j	r^2	1	$-u_{xj}$
B_iB_j/b_iB_j	2ps	1/2	u_{xj}
B_iB_j/b_ib_j	2pt	1/2	$a_{xj}u_{xj}$
B_ib_j/b_iB_j	2rs	1/2	$a_{xj}u_{xj}$
B_ib_j/b_ib_j	2rt	1/2	$-u_{xj}$
b_iB_j/b_iB_j	s^2	0	u_{xj}
b_iB_j/b_ib_j	2st	0	$a_{xj}u_{xj}$
b_ib_j/b_ib_j	t^2	0	$-u_{xj}$

Values of $(\hat{y}_{xj}-u_{xj})$, instead of $\hat{y}_{xj}$, are tabulated because I have learned that the algebra to be done in reaching the most useful expression for $\sigma(\hat{q}_i,\hat{y}_{xj})$ is facilitated by working with $(\hat{y}_{xj}-u_{xj})$ instead of $\hat{y}_{xj}$. Recall that coding by addition or subtraction of a constant (Appendix 1) does not affect variance or covariance. Before going on, the following are worth noting:

(1) The ten genotype probabilities are the ten separate terms of $(p+r+s+t)^2$ which sum to 1.0 since $p+r+s+t = 1.0$.

(2) The sum of probabilities of genotypes homozygous for B_i is $p^2 + 2pr + r^2 = (p+r)^2 = q_i^2$. In the same way we find the sum of probabilities of genotypes heterozygous (B_ib_i) at the i-th locus is $2ps+2pr+2rs+2rt = 2(p+r)(s+t) = 2q_i(1-q_i)$ and the sum for genotypes homozygous for b_i is $s^2+2st+t^2 = (1-q_i)^2$. Thus when only the i-th locus is considered the above distribution condenses to that of Section 6.1 from which it is apparent (a) that linkage disequilibrium has no impact relative to $\sigma(q_{u1},\hat{y}_{x1})$ and (b) that, as before

$$E(\hat{q}) = q_i \qquad 6.1$$

Working from the above distribution we find the mean of $(\hat{y}_{xj} - u_{xj})$ to be

$$E(\hat{y}_{xj} - u_{xj}) = u_{xj}(p^2 - r^2 + 2ps - 2rt + s^2 - t^2) + a_{xj}u_{xj}(2pr + 2pt + 2rs + 2st)$$

$$= u_{xj}[(p+s)^2 - (r+t)^2] + 2a_{xj}u_{xj}(p+s)(r+t)$$

$$= (p+s-r-t)u_{xj} + 2(p+s)(r+t)a_{xj}u_{xj} = (2q_j - 1)u_{xj} + 2q_j(1-q_j)a_{xj}u_{xj} \qquad 6.13$$

since $(p+s) = q_j$ and $(r+t) = (1-q_j)$. This result was predictable from 6.2 and the fact noted above that probabilities of single locus genotypes are not affected by linkage disequilibrium.

From the distribution of this section and equations 6.1 and 6.13, and remembering that $q_i = p+r$, we obtain[2]

$$\sigma(\hat{q}_i, \hat{y}_{xj}) = u_{xj}(p^2 - r^2 + ps - rt) + a_{xj}u_{xj}(2pr + pt + rs)$$

$$-q_i[(p+s-r-t)u_{xj} + 2(p+s)(r+t)a_{xj}u_{xj}]$$

$$= u_{xj}[p^2 - r^2 + ps - rt - (p+r)(p+s-r-t)]$$

$$+ a_{xj}u_{xj}[2pr + pt + rs - 2(p+r)(p+s)(r+t)]$$

$$= (pt - rs)[1 + (1-2q_j)a_{xj}]u_{xj}. \qquad 6.14$$

Substitution in 5.21, in terms of 6.5 and 6.14, gives us

$$\sigma(\hat{q}_i, \hat{X}) = q_i(1-q_i)[1 + (1-2q_i)a_{xi}]u_{xi} + \Sigma_{j \neq i}(pt - rs)_{ij}[1 + (1-2q_j)a_{xj}]u_{xj} \qquad 6.15$$

and from 6.15 and 5.18 we obtain

$$E(\Delta q_i) = [(\bar{X}_s - \bar{X})/\sigma^2(\hat{X})]\{q_i(1-q_i)[1 + (1-q_i)a_{xi}]u_{xi}$$

$$+ \Sigma_{j \neq i}(pt - rs)_{ij}[1 + (1-2q_j)a_{xj}]u_{xj}\} \qquad 6.16$$

Finally, substitution of 6.16 and 5.11 in 5.8 yields

$$E(\Delta \ddot{Y}_c) = [(\bar{X}_s - \bar{X})/\sigma^2(\hat{X})]\{\Sigma_i 2q_i(1-q_i)[1 + (1-2q_i)a_{xi}][1 + (1-2q_i)a_{ci}]u_{xi}u_{ci}$$

$$+ \Sigma_i \Sigma_{j \neq i} 2(pt - rs)_{ij}[1 + (1-2q_j)a_{xj}][1 + (1-2q_i)a_{ci}]u_{xj}u_{ci}\} \qquad 6.17$$

for the case where the T-population is the S-population itself. Equation 6.17, like 6.5, pertains to the <u>correlated response</u> in C to mass selection based on X. Equations 6.5 and 6.17 differ by the presence in 6.17 of the terms required when it is assumed that there may be linkage disequilibrium and hence that for any pair of loci, identified as the i-th and j-th, $\sigma(\hat{q}_i, \hat{y}_{xj})$ may be nonzero. Recall now that linkage equilibrium is identified (section 2.4.3) as the situation in which two locus gamete genotype frequencies are $p = q_i q_j$, $r = q_i(1-q_j)$, $s = (1-q_i)q_j$ and $t = (1-q_i)(1-q_j)$ so that $(pt - rs) = 0.0$. Given linkage equilibrium, $pt - rs = 0.0$ for all pairs of loci and 6.17 reduces to 6.5.

The expression for expected <u>direct response</u>, the response in X to selection based on X, is obtained from 6.17 by changing c to x wherever the former appears as a subscript. The result obtained is

$$E(\Delta\ddot{Y}_X) = [(\bar{X}_S-\bar{X})/\sigma^2(\hat{X})]\{\Sigma_i 2q_i(1-q_i)[1+(1-2q_i)a_{xi}]^2u_{xi}^2$$

$$+\Sigma_i\Sigma_{j\neq i}2(pt-rs)_{ij}[1+(1-2q_j)a_{xj}][1+(1-2q_i)a_{xi}]u_{xj}u_{xi}\} \quad 6.18$$

In the special case of linkage equilibrium where every value of (pt-rs) is zero, 6.18 reduces, as it should, to 6.7.

To justify equations 6.9, 6.10 and 6.12 as appropriate in both the linkage equilibrium and linkage disequilibrium situations, we must now identify additive genetic variance of X as[3]

$$\sigma_{gx}^2 = \Sigma_i 2q_i(1-q_i)[1+(1-2q_i)a_{xi}]^2u_{xi}^2$$

$$+ \Sigma_i\Sigma_{j\neq i}\, 2(pt-rs)_{ij}[1+(1-2q_j)a_{xj}][1+(1-2q_i)a_{xi}]u_{xi}u_{xj} \quad 6.19$$

and the additive genetic covariance between X and C as

$$\sigma_{gcx} = \Sigma_i\, 2q_i(1-q_i)[1+(1-2q_i)a_{xi}][1+(1-2q_i)a_{ci})]u_{xi}u_{ci}$$

$$+\Sigma_i\Sigma_{j\neq i}\, 2(pt-rs)_{ij}[1+(1-2q_j)a_{xj}][1+(1-2q_i)a_{ci}]u_{xj}u_{ci}$$

Attention was directed in Section 6.1 to the fact that when linkage equilibrium prevails, additive genetic covariance results solely from pleiotropy. In contrast we see from 6.19 that when linkage disequilibrium is present genetic covariance between traits can also be contributed by pairs of genes that are nonpleiotropic. The requirements are that one gene of the pair must affect X, the other must affect C and the value of (pt-rs) for the pair of genes must be nonzero.

Expressions for $E(\Delta\ddot{Y}_c)$ and $E(\Delta\ddot{Y}_x)$ for the case when the T-population is one obtained by crossing the S-population with some other, when T=S x A, will not be presented but would be obtained as in Section 6.1 using 5.12, instead of 5.11, along with 6.16 and 5.8.

• 6.3. Information and Insights Provided by Response Expectations

Primary questions concerning consequences of recurrent selection have to do with (1) the initial rates of response, (2) the rapidity with which response rate changes are likely to occur, (3) the total response (of the selection criterion) available through recurrent selection and (4) the nature of correlated responses that are possible or probable. The remainder of this chapter will be devoted to preliminary discussion of these and related questions in the context of mass selection when the S-population is the target population.

• 6.3.1. Effects of Linkage Disequilibrium

The differences between expressions obtained when linkage disequilibrium was or was not assumed; e.g., the difference between 6.4 and 6.16, raise such questions as these. What are the sources of linkage disequilibrium? Is the presence of nontrivial amounts of linkage disequilibrium predictable? What are the important effects when nontrivial amounts are present? These and related issues will be discussed after more background has been provided. It will suffice to note here that there are circumstances in which the amount of linkage disequilibrium in random mating populations can reasonably be assumed trivial. In particular, in the absence of selection initial disequilibrium decreases from generation to generation in random mating populations that are infinite in size (Appendix 6.2). This fact provides justification, if any is needed, for assuming linkage equilibrium in the preliminary discussions of consequences of recurrent selection that are presented in the remainder of this chapter.

6.3.2. Rates of Response

Remembering that by definition $k = (\bar{X}_s - \bar{X})/\sigma(\hat{X})$ so that $(\bar{X}_s - \bar{X})/\sigma^2(\hat{X}) = k/\sigma(\hat{X})$, the expected response of an allele frequency in the case of mass selection and linkage equilibrium is from 6.4

$$E(\Delta q_i) = [k/\sigma(\hat{X})]q_i(1-q_i)[1 + (1-2q_i)a_{Xi}]u_{Xi} \qquad 6.20$$

It follows that the expected change in the single locus mean of the response criterion ($\ddot{y}_X$ in the case of direct response) is from 5.11

$$E(\Delta\ddot{y}_{Xi}) = E(\Delta q_i)2[1 + (1-2q_i)a_{Xi}]u_{Xi}$$
$$= [k/\sigma(\hat{X})]2q_i(1-q_i)[1 + (1-2q_i)a_{Xi}]^2u^2_{Xi}$$

and that

$$E(\Delta\ddot{Y}_X) = \Sigma_i E(\Delta\ddot{y}_{Xi}) = [k/\sigma(\hat{X})]2\Sigma_i q_i(1-q_i)[1 + (1-2q_i)a_{Xi}]^2u^2_{Xi} \qquad 6.21$$

Then by 6.8 and 6.10

$$E(\Delta\ddot{Y}_X) = [k/\sigma(\hat{X})]\sigma^2_{gX} = [(\bar{X}_s - \bar{X})/\sigma^2(\hat{X})]\sigma^2_{gX}$$

$$= (\bar{X}_s - \bar{X})[\sigma^2_{gX}/\sigma^2(\hat{X})] = (\bar{X}_s - \bar{X})h^2_X \qquad 6.22$$

6.3.2.1. ALLELE FREQUENCY CHANGES

Obviously, the prime effects of selection are on allele frequencies; changes in the ($\ddot{y}_i$)'s and $\ddot{Y}_X$ are functions of changes in the frequencies of alleles for which $u_i > 0.0$. Recognizing that 6.20 is general in the sense that it is valid for (a) any allele and (b) any selection criterion so long as u, a, and σ are understood to have reference to that criterion, it can be rewritten as follows.

$$E(\Delta q) = kq(1-q)[1+(1-2q)a]u/\sigma \qquad (6.20)\ 6.23$$

Here attention is drawn to the fact that u and σ need not be quantified separately; only the ratio of their magnitudes is relevant. In order to obtain possible values of $E(\Delta q)$, appropriate values of k, a, q, and u/σ must be used in 6.23.

It is intuitively obvious that k will be larger when the proportion selected (p) is smaller. The following relation between k and proportion selected is widely used.

$$E(k) \approx z/p \qquad 6.24$$

In this expression z is the height of the ordinate that divides the standardized normal curve into portions of size p and 1-p, respectively.[4] Assumptions are (1) that X follows the normal distribution, (2) that the genetic sample from which selections are taken is large, and (3) that the units selected are those for which the values of X are the largest (or smallest) in the genetic sample (such selection is commonly identified as truncation selection).

Values of z are available from normal distribution tables, provided in many places of which one of the most readily available is The Handbook of Chemistry and Physics. When the genetic sample is not large, 6.24 is biased upward and Table XX by Fisher and Yates (1963) enables closer apporoximation.to expected values of k, again assuming normal distribution of X in the population from which the genetic sample is drawn. Their table gives expected deviations from the population mean (in standard units) of the largest, next largest, etc., individuals in samples (from normal distributions) where sample size is in the range from 2 to 50. If, for example, we want E(k) when sample size is 30 and p is 0.1 so that the number selected is 3, we average the expected values (2.04, 1.62, and 1.36) of the first, second and third largest from samples of size 30 to obtain E(k) =1.67. To avoid the computation required when using the Fisher and Yates table, see Becker (1984). He tabulated E(k), obtained as described above, for a broader range of sample sizes than covered by Fisher and Yates.

Table 6.1 lists E(k) for various values of p and sample size. More values than required

Table 6.1. E(k) where k is the selection differential divided by the standard deviation of $\hat{X}$, the selection criterion

	Size of the genetic sample					
p	5	10	20	40	50	large[a]
.01	...	...	...	...	...	2.66
.02	...	...	...	...	2.25	2.42
.04	...	...	...	...	2.05	2.16
.05	...	...	1.87	1.96	...	2.06
.10	...	1.54	1.64	1.69	1.70	1.76
.15	...	...	1.47	1.51	...	1.56
.20	1.16	1.27	1.33	1.36	1.37	1.40
.25	...	...	1.22	1.24	...	1.27
.30	...	1.07	1.11	1.13	1.14	1.16
.40	.83	.90	.93	.95	.95	.97
.50	...	.74	.77	.78	.79	.80

[a]Values in this column obtained as z/p using a normal distribution table. Values in other columns obtained using Table XX, Fisher and Yates (1963).

for illustration are presented because the table will be referred to for values required later in dealing with some other problems. The entire range of q from near zero to near 1.0 deserves attention. Values for a that have special interest are - 1.0 (b completely dominant), 0.0 (no dominance, the case of the additive gene) and 1.0 (B completely dominant). To represent the important (theoretically, if not in reality) overdominance class of genes, a = 2 will also be investigated.

Identifying values of u/σ that deserve consideration is more difficult. In the case of significant quantitative traits in economic species adequate estimates of σ are readily available from numerical data but there is virtually no direct information concerning magnitudes of u's. However, perspective concerning possible values of u/σ and $E(\Delta q)$ can be obtained by analysis of hypothetical situations. Consider the F_2 generation from a cross of homozygous lines and any trait (X) of which means in the parental lines are near equal. Though linkage disequilibrium is to be expected in that F_2, it can be assumed that its impact on additive genetic variance, σ^2_{gX}, is trivial (as discussed in section 15.4.5.2). Thus from 6.8 and the fact that expected frequencies of segregating alleles are 0.5, additive genetic variance would be

$$\sigma^2_g \approx nu^2/2 \qquad 6.25$$

if the u-values of the segregating genes were all the same. Dividing σ^2_{gX} by $\sigma^2(X)$ we obtain

$$h^2_X = nu^2/2\sigma^2(X)$$

from which $$u^2/2\sigma^2(X) = 2h^2/n \text{ and } u/\sigma = (2h^2/n)^{1/2}. \qquad 6.26$$

Values of u/σ computed from this expression are presented in Table 6.2. As is apparent from 6.26 they are larger when heritability is larger and smaller when the number of genes is greater. The heritabilities of numerous traits are known reasonably well but very little is known about numbers of genes with effects on the value traits of the important agricultural species.

Table 6.2. Values of u/σ given number (n) of genes and h^2 (when all allele frequencies are 0.5)

	n				
h^2	10	40	100	200	500
.05	.100	.050	.032	.022	.014
.15	.173	.087	.055	.039	.024
.25	.224	.112	.071	.050	.032
.40	.283	.141	.089	.063	.040
.60	.346	.173	.110	.077	.049

To have recognized variation in the magnitude of effect of different genes instead of assuming u to be the same at all loci, procedure in this subsection could have been to assume q = 0.5 at all loci (as we did) but let $1/m_i$ be the fraction of the total additive genetic variance supplied by the i-th locus. We could then have written

$$\sigma^2_{gi} = u_i^2/2 = \sigma^2_g/m_i$$

Then dividing by σ^2

$$u_i^2/2\sigma^2 = \sigma^2_g/m_i\sigma^2 = h^2/m_i$$

and solving for u_i/σ $\quad u_i/\sigma = (2h^2/m_i)^{1/2}$

Compare this with 6.26. Clearly, if m were the same at all n loci, it would be because u was the same at all loci and in that case we'd have m = n. But if u and m are variable among loci, allele frequency can be expected to change more rapidly at loci where u is largest (see 6.22). Thus in the case of traits and noninbred populations where there has been a long history of selection in the same direction (e.g., grain yield in open-pollinated varieties of maize, milk production in dairy cattle or egg production in white leghorn chickens), it is reasonable to suppose a greater probability that genes with large effects (large u-values) have become fixed and that remaining segregation involves a greater proportion of genes with smaller effects. This fact will be found significant when (in a later chapter) we consider evidence that has been presented and arguments that have been advanced relative to numbers of genes that are important as sources of the genetic variance of quantitative traits.

Values of u/σ inferred as outlined above will be treated in the remainder of this chapter as essentially constant over the whole range of allele frequencies, at either the locus in question or at any others. In the absence of epistasis the u's would by definition be constant This leaves σ to be considered. Obviously, the phenotypic variance, σ^2, is potentially variable; change in either its genetic or nongenetic component is possible. If we assume that nongenetic phenotypic variance is constant, it is apparent that σ^2 will be smaller if all allele frequencies are near 0.0 or 1.0 than if they are all near 0.5. However, our greatest concerns will be with low heritability traits for which total genetic variance is no more than one-third of the phenotypic variance. For such traits, the reducton in σ^2 due to reduction of genetic variance could not be greater than one-third and the corresponding decrease in σ would be 19% or less. The non-genetic portion of the phenotypic variance may also vary but not enough to be important when compared with effects on u/σ of gene number(n) of which our information is much less certain.

Table 6.3 shows how $E(\Delta q)$ varies as a function of q, a, and u/σ. Because $E(\Delta q)$ is a linear function of both k and u/σ, its magnitudes associated with other values of k and/or u/σ are easily determined.

The important thing to be learned from tables 6.2 and 6.3 viewed together is that changes in allele frequencies resulting from one cycle of selection may be relatively small. The numbers shown indicate, for example, that it could take quite a few cycles for the frequency of a useful

Table 6.3. Values of E(Δq) for k =1, u/σ = 0.1 and selected values of a and q

	q						
a	.10	.25	.40	.50	.60	.75	.90
-1	.0018	.0094	.0192	.025	.0288	.0281	.0162
0	.0090	.0188	.0240	.025	.0240	.0188	.0090
1	.0162	.0281	.0288	.025	.0192	.0094	.0018
2	.0234	.0375	.0336	.025	.0144	.0000	-.0054

allele to be increased from 0.4 to 0.9. We are led then to inquire how the E(Δq) values would be changed if different values of k,u/σ, or a were assumed. This is easily explored, working from Table 6.3. Note that E(Δq), as given by 6.21, is directly proportional to the product, ku/σ, and remember that the values in Table 6.3 reflect ku/σ= 0.1. If we were concerned with E(Δq) when k = 1.8 and u/σ = 0.07, the product would be 1.8(0.07) = 0.126 and values in the table would have to be multiplied by 0.126/0.1 =1.26. Both equation 6.21 and Table 6.3 indicate that, when q and ku/σ are constant, E(q) is a linear function of a. Thus for values of a not shown in Table 6.3, E(Δq) is easily obtained by interpolation . For example, for q = 0.6, ku/σ = 0.1 and a = 0.75, the appropriate value of E(Δq) is 0.024 + 0.75(.0192 - 0.024) = 0.0204. Other aspects of Table 6.3 that are worthy of notice include the following.

(1) When a = 2.0, then in accord with the [1+(1-2q)a] component of 6.22, E(Δq) = 0.0 when q = 0.75, positive when q = 0.6 and negative when q = 0.9. In general, there is an equilibrium value of q toward which allele frequency is gravitated from either direction.

(2) The inverse relation between values of E(Δq) for a = 1.0 (B dominant to b) and a = -1.0 (b dominant to B).

• 6.3.2.2. GENOTYPIC MEAN OF THE SELECTION CRITERION

We see in 6.21 the often stated proportionality between direct response to mass selection and heritability. The expression says, for example, that response per unit of selection differential will be twice as large if heritability is twice as large. Heritability is of interest in itself but it is made interesting to many more people by this indicated linear dependence on it of direct response to selection. The result has been a large body of literature devoted to methods for estimating heritability, to reporting such estimates for a wide variety of phenotypic measures in various species, and to reporting evidence of how well 6.21 works in practice. These experimental tests indicate (1) that in many instances the predictions conform reasonably with observed responses but (2) that in other instances agreement between observation and prediction is not so good. Failures of deduced theory, an obvious matter for concern may reflect inadequacies of (a) the experimental evidence or (b) the theory itself. Some of the possible explanations are considered in Chapter VII.

In the context of recurrent selection major questions are concerned with (a) the rapidity and direction of change in the rate of response and (b) the total change in $\ddot{Y}_x$ that can be realized.

$E(\Delta\ddot{Y}_x)$ per cycle is proportional to additive genetic variance; hence its change is determined by the change in σ^2_{gx} which in turn is determined by changes in allele frequencies. From

$$\sigma^2_{gx} = \Sigma_i 2q_i(1-q_i)[1+(1-2q_i)a_{xi}]^2 u^2_{xi} \qquad 6.8$$

it is apparent that, in the absence of linkage disequilibrium, additive genetic variance is the sum of contributions from each segregating gene, i.e., that $\sigma^2_{gx} = \Sigma_i \sigma^2_{gxi}$ in which

$$\sigma^2_{gxi} = 2q_i(1-q_i)[1+(1-2q_i)a_{xi}]^2 u^2_{xi}. \qquad 6.27$$

It follows that $$E(\Delta\sigma^2_{gxi}) \approx E(\Delta q_i)2u^2_{xi}(\partial/\partial q)\{q(1-q)[1+(1-2q)a_{xi}]^2\} \qquad 6.28$$

in which $\partial/\partial q$ symbolizes "the derivative with respect to q". From 6.8 the expected change in the total additive genetic variance is

$$E(\Delta\sigma^2_{gx}) = \Sigma_i E(\Delta\sigma^2_{gxi}) \qquad 6.29$$

Of course a and u must be variable among genes and their numerical values are unknowable in the case of quantitative traits. However, for our purpose it will suffice to assume that the magitude of a is the same (0 or 1) for all genes. When a = 0, then

$q(1-q)[1+(1-2q)a_{xi}]^2 = q(1-q)$ and its derivative is $(1-2q)$ and when a = 1

$q(1-q)[1+(1-2q)a_{xi}]^2 = 4q(1-q)^3$ and its derivative is $4(1-q)^2(1-4q)$.

With these derivatives, 6.23, and 6.28 we find when a = 0

$$E[\Delta\sigma^2_{gxi}/\sigma^2_{gxi}] = k(u/\sigma)(1-2q)$$

and when a=1 $$E[\Delta\sigma^2_{gxi}/\sigma^2_{gxi}] = k(u/\sigma)2(1-q)(1-4q).$$

Some values of these ratios are shown in Tables 6.4 and 6.5. Positive values of the ratio indicate that additive genetic variance from segregation of the i-th gene increases as frequency of the B allele is increased by selection and conversely decreases are indicated by negative values of the ratio. For example, we note from table 6.4 that when a = 0, k = 1, u/σ = 0.07, and q =

Table 6.4. $E[\Delta\sigma^2_{gxi}/\sigma^2_{gxi}]$ when a = 0 and k = 1

	q				
u/σ	0.1	0.3	0.5	0.7	0.9
.10	.080	.040	0	-.040	-.080
.07	.056	.028	0	-.028	-.056
.04	.032	.016	0	-.016	-.032

Table 6.5. E $[\Delta\sigma^2_{gxi}/\sigma^2_{gxi}]$ when a = 1 and k = 1

	q				
u/σ	0.1	0.3	0.5	0.7	0.9
.10	.108	-.028	-.100	-.108	-.052
.07	.076	-.020	-.070	-.076	-.036
.04	.043	-.011	-.040	-.043	-.021

0.3 the expected increase in additive genetic variance as the result of one cycle of selection is 2.8 percent. From table 6.5 we see that when a=1, k= 1, u/σ = 0.04, and q = 0.7 the expected decrease is 4.3 percent. Because u, a, and q are without doubt variable among the segregating genes of a random mating population magnitudes of change in total additive variance as the result of selection cannot be predicted though it can be estimated if enough data of the right kind are available (see Chapter XV). However, it is obvious from the content of this section that depending on such variables as heritability, number of contributing genes, selection intensity, and allele frequencies change in additive variance can be slow and hence direct response to recurrent selection can continue through numerous and unpredictable numbers of cycles.

Perspective concerning magnitudes of u/σ is gained by considering magnitudes of u relative to $\ddot{X}$, the mean of the trait on which selection is focused. The relation between $\ddot{X}$ and σ(X) = σ is provided by the coefficient of variation, CV = 100σ(X)/$\ddot{X}$. Thus (u/σ)/(CV/100) = u/$\ddot{X}$. Then, for example, if CV = 15 (as it might be for a quantitative trait), to assume u/σ = 0.1 implies u/$\ddot{X}$ = 0.015, i.e., that substitution of the BB for the bb genotye would increase $\ddot{X}$ by 3 percent.

6.3.3. Total Response Possible

A very general discussion of this issue would require background not yet provided and would introduce more complexity than here justified. However, useful insights can be obtained by considering the special case in which initial allele frequencies are 0.5 at all segregating loci and the values of a (level of dominance) are not < -1 or >1 and are the same at all loci.

6.3.3.1. INFINITE EFFECTIVE POPULATION SIZE

In this limiting case E(Δq) > 0.0 can be assumed until the better allele is homozygous at all loci. The original genotypic mean (when all q's equal 0.5) will be symbolized as $\ddot{Y}_{X0}$ and the greatest possible genotypic mean (when all favorable alleles are homozygous -- when all q's are 1.0) will be symbolized as $\ddot{Y}_{Xm}$. In accord with equations 5.4 and 5.10

$$\ddot{Y}_X = \alpha_X + \Sigma_i \ddot{y}_{Xi}$$

and $$\ddot{y}_{Xi} = u_{Xi} + a_{Xi}u_{Xi}/2 \qquad (\text{when } q_i = 0.5)$$

or $$2u_{Xi} \qquad (\text{when } q_i = 1.0)$$

Because it will be understood that all symbols relate to the measure X, the subscript x will now be omitted. If a_i is the same at all loci as we assume here

$$\ddot{Y}_0 = \alpha + (1+a/2)\Sigma_i u_i$$

and

$$\ddot{Y}_m = \alpha + 2\Sigma_i u_i$$

Then the total response possible (R) is

$$R = \ddot{Y}_m - \ddot{Y}_0 = (1-a/2)\Sigma_i u_i \qquad 6.30$$

This leaves us wondering how large the u's are and how many of them are involved. Stated differently, we need to have the total possible response expressed in units about which we have or can have information. It we multiply and divide the right-hand member of 6.30 by σ_{go} the value of σ_g when all q's are 0.5, we obtain

$$R = (1 - a/2)(\Sigma_i u_i)\,\sigma_{go}/\sigma_{go}$$

Then because $\Sigma_i u_i = n\bar{u}$ and $\sigma^2_{go} = \Sigma_i u^2_i/2$

$$R = (1-a/2)n\bar{u}\sigma_{go}/(\Sigma_i u^2_i/2)^{1/2} \qquad 6.31$$

If all u-values were equal this could be written as follows

$$R = (1 - a/2)nu\sigma_{go}/(nu^2/2)^{1/2} = (1 - a/2)\,\sigma_{go}(2n)^{1/2} \qquad 6.32$$

We can, of course, solve this for any values of a and n that interest us. For example, assuming additive genes ($a = 0$) and $n = 50$, the result obtained is $R = 10\sigma_{go}$. But if $a = 0.0$ and $n = 200$, $R = 20\sigma_{go}$. The relevance of number of effective genes is emphasized.

Related to the foregoing, it is well worth noting that if 6.32 is squared and appropriately rearranged the result is

$$n = R^2/2(1 - a/2)^2\sigma^2_{go} \qquad 6.33$$

which provides, and has been used as, a basis for estimating the number of segregating genes. See Park (1977) for a review of estimators that are variations of 6.33. The earliest of these was due to Wright and reported by Castle (1921).

The assumption that all u's are equal is hardly realistic. It is reasonable to inquire, therefore, what the impact of variation among u's on total possible response would be. The average (expected) value of u_i^2 (see Appendix 1) is

$$E(u_i^2) = \bar{u}^2 + \sigma^2(u)$$

It follows that the expectation of the the sum of n values of $\bar{u}^2$ is

$$E(\Sigma_i u_i^2) = n\bar{u}^2 + n\sigma^2(u)$$

Substituting its expectation for Σu_i^2 in 6.29, we obtain

$$R \approx (1 - a/2)n\bar{u}\sigma_{go}/[n\{\bar{u}^2 + \sigma^2(u)\}/2]^{1/2}$$
$$\approx (1 - a/2)\,\sigma_{go}[2n/(1+v^2)]^{1/2}$$

where $v^2 = \sigma^2(u)/\bar{u}^2$ and 100v is the coefficient of variation of u. Since, by definition, all u's are positive (> 0.0), v must be less than 1.0; hence, $(1 + v^2)^{1/2} < 1.41$. Thus, other things being the same, R could not be reduced by variation in u-values to less than 1/1.41 = 0.71 of what it would be if all the u's were equal.

6.3.3.2. FINITE EFFECTIVE POPULATION SIZE

The preceeding subsection is concerned with total response to selection in the event that all favorable alleles are fixed, i.e., brought to homozygosity. However, in practice selection populations are of finite size and hence, as the consequence of random drilft, the less favorable allele will inevitably be fixed at some loci (see section 2.63).

The probability of fixation of a favorable allele can be approximated using equation 2.34 if a numerical value for s (the selection coefficient) can be supplied. In section 2.6.1, s is defined, following Wright (1931), in terms of relative fitness; and the change in allele frequency from selection when there is no random drift is given by 2.12. In his derivation of 2.33 Kimura took account (necessarily) of both selection and drift using equation 2.12 to measure the effect of selection. For the particular case of additive genes, to which 2.34 pertains, we know that 2.12 becomes

$$\Delta_s = sq(1-q)/2[1-(1-q)s] \qquad 2.13$$

It follows that when we use 2.34 we need to employ a value for s that makes 2.13 correct, or at least roughly correct, even though the selection involved is artificial, not natural selection. This can be done by setting the Δ_s of 2.13 equal to $E(\Delta q)$ for artificial selection and then solving for s. Using 2.13 and 5.24, we obtain

$$sq(1-q)/2[1-(1-q)s] = k\sigma(\hat{q},\hat{y}_X)/\sigma(\hat{X})$$

from which for small values of s such that, for practical purposes, [1-(1-q)s] and 1.0 can be equated. Then solution for s yields

$$s = 2k\sigma(\hat{q},\hat{y}_X)/q(1-q)\sigma(\hat{X})$$

For the special case of mass selection, additive genes and linkage equilibrium we substitute for $\sigma(\hat{q},\hat{y}_X)$ in terms of 6.3 (with $a_X = 0.0$) to obtain $s = 2ku_X/\sigma(\hat{X})$. Remembering that in mass selection $\sigma(\hat{X}) = \sigma(X)$ because the criterion units are individuals, we have

$$s = 2ku_X/\sigma(X). \qquad 6.34$$

Taking differences in symbolism into account, this is the expression given by Falconer (1960) and used widely by others.

Equation 6.34 and others obtained in the same way will be employed extensively when the subject of sufficient effective population size in breeding programs is considered in and beyond Chapter XI.

Simple numerical examples provide insights concerning the utility of 6.34 and 2.34. Consider the long-term response to mass selection with k = 1.2 in the case of genes for which

$u_X/\sigma(X)$ is 0.08 or 0.04 and q_0 (initial frequency of the better allele) is 0.5. Using 6.33 we obtain $s = 0.192$ when $u/\sigma = 0.08$ and $s = 0.096$ when $u/\sigma = 0.04$. Next, if these along with $q_0 = 0.5$ and specific values of N_e are substituted in 2.34, probabilities of fixation are obtained. Six are shown below.

	P_f	
N_e	$s = 0.192$	$s = 0.096$
10	.872	.723
20	.979	.872
30	.997	.947

These numbers give us a sense of how probability of fixation responds to increases in effective population size (N_e). They increase as N_e increases but, given any value of N_e, are smaller when s is smaller. What do the differences among these probabilities of fixation mean with respect to change in average genotypic value? Assuming no epistasis,

$$\ddot{Y} = \alpha + \Sigma_i \ddot{y}_i \qquad 5.8$$

and

$$\ddot{Y}_{\infty} - \ddot{Y}_0 = \Sigma_i(\ddot{y}_{i\infty} - \ddot{y}_{io})$$

where $\ddot{y}_{io}$ is the average of $\ddot{y}_i$ prior to selection and $\ddot{y}_{i\infty}$ is its average when response to selection has ceased. In the case of additive genes, HW frequencies prior to selection, and $q_0 = 0.5$

$$\ddot{y}_{io} = q^2{}_0(2u_i) + 2q_0(1-q_0)u_i = 2q_0u_i = u_i$$

$$E(\ddot{y}_{i\infty}) = P_{fi}(2u_i) + (1-P_{fi})(0.0) = 2P_{fi}u_i$$

and the expected long-term response is

$$E(\ddot{y}_{i\infty}) - \ddot{y}_{io} = (2P_{fi} - 1)u_i \qquad 6.35$$

where P_{fi} = probability of fixation of the the better allele of the i-th gene. Note that the maximum of 6.35 is u_i in accord with 6.28. The expected long-term responses for the six combinations of N_e and s in our example are, by 2.34, as follows.

	$E(\ddot{y}_{i\infty}) - \ddot{y}_{io}$	
N_e	$s = 0.192$	$s = 0.096$
10	.744u	.446u
20	.958u	.744u
30	.994u	.894u

The effects of increases in N_es appear larger when $\ddot{y}$ instead of P_f is considered. This is because when the better allele is brought to homozygosity, y is increased from u to 2u but when the less favorable allele is fixed y is decreased from u to 0.0 and the expected response is the average of these positive and negative increments taking their probabilities (expected frequencies) into account. The following are demonstrated and deserve being remembered.

(1) Sufficient effective population size is a requirement if the full potential of recurrent selection is to be realized.

(2) When N_e is relatively small the genetic change to be expected may be small relative to the full potential.

(3) The effective population size required for a satisfactory approach to maximum long-term response is greater when the s-values of relevant alleles are smaller.

6.3.4. Correlated Responses in the Absence of Linkage Disequilibrium

The expectation of correlated response in measure C to selection based on X is given by 6.5 and 6.12. Additive genetic variance and additive genetic covariance are defined by 6.8 and 6.11, respectively. In accord with the familiar product moment correlation between any two variables, the additive genetic correlation between X and C, any two phenotype measures, will be identified as

$$r_{gcx} = \sigma_{gxc}/\sigma_{gx}\sigma_{gc}$$

Now, for reference purposes, it will be useful to rewrite equations 6.5 and 6.12 in alternative forms.

$$E(\Delta\ddot{Y}_c) = [(\bar{X}_s - \bar{X})/\sigma^2(X)]\Sigma_i 2q_i(1-q_i)[1+(1-2q_i)a_{xi}][1+(1-2q_i)a_{ci}]u_{xi}u_{ci}$$
$$= k\sigma_{gxc}/\sigma(X) = kr_{gcx}\sigma_{gx}\sigma_{gc}/\sigma(X) = kr_{gcx}\sigma_{gc}h_x \qquad 6.36$$

General facts to be noted include the following:

(1) Additive genetic covariance between two traits (when linkage disequilibrium is absent) requires pleiotropy. If present, it reflects the fact that one or more genes affect both X and C

(2) Correlated responses, to selection based on any criterion, may occur in any or all of the other traits of the organism.

(3) When any specific trait other than X is considered, the direction of correlated response (as well as the rate) may possibly change during the course of a recurrent selection program. This is because the rates of change in individual allele frequencies are strong functions of the values of u_x but unaffected by u_c whereas, given the allele frequency changes, the change in $\ddot{Y}_c$ depends on the values of u_c which need not have a high positive correlation with the u_x's. Of course, direction changes would require change in the algebraic sign of the genetic correlation (r_{gcx}) which in turn would require some u_c's opposite in sign to the corresponding u_x's.

In the world of breeding, correlated response is likely to be a critical issue. Let C represent total value of the organism (see section 1.4). It will almost certainly depend on several component traits. For various reasons, one or more measures other than C are likely to be more attractive than C for use as the selection criterion. These are usually components of total value but need not be. Reasons for choosing $X \neq C$ are frequently one or more of the following.

(1) Other measures may be less expensive to obtain. Consider growth rate of swine versus feed required per unit of nutritive value of pork produced. The latter is a more important component of total value but the two measure are correlated and growth rate is much easier and less expensive to determine for the purposes of mass selection.

(2) Other measures believed to be highly correlated genetically with C may be more highly heritable.

(3) Other measures are available earlier in the life of the organism. This becomes a compelling reason in work with long-lived organisms (forest trees and other perennials such as blueberries and apples, dairy cattle, etc.). If selection is based on a measure available early in the life of the individual, the generation interval can be shorter and greater response per unit of calendar time is favored.

A quantitative framework for evaluating these arguments is provided by comparing the expectation of correlated response in C to selection based on X with the expected direct response of C to selection based on C itself. From 6.9 and 6.36 the ratio of these expectations is

$$[E(\Delta\ddot{Y}_{c.x})/L_x]/[E(\Delta\ddot{Y}_{c.c})/L_c] = k_x\sigma_{gxc}\sigma(\hat{C})L_c/k_c\sigma^2_{gc}\sigma(\hat{X})L_x$$

$$= (k_x/k_c)[\sigma_{gxc}/\sigma_{gc}\sigma_{gx}][\sigma(\hat{C})/\sigma_{gc}][\sigma_{gx}/\sigma(\hat{X})](L_c/L_x)$$

$$= (k_x/k_c)r_{gcx}[h_x^2/h_c^2]^{1/2}(L_c/L_x) \qquad 6.37$$

where $\Delta\ddot{Y}_{c.c}$ is direct response in C to selection based on C, $\Delta\ddot{Y}_{c.x}$ is correlated response in C to selection based on X, L_c is selection cycle length when selection is based on C, and L_x is selection cycle length when selection is based on X.

Equation 6.37, which compares correlated response per unit of time with direct response per unit of time, puts the several arguments in perspective. If X is easier to measure, a larger genetic sample and more intense selection are possible without greater cost when X is the selection criterion. Then k_x/k_c would be greater than 1.0. The effect of difference between heritabilities of X and C is shown to be the square root of the ratio of those heritabilities. Finally, the effect of age when measuremaents can be obtained is reflected in L_c/L_x, the ratio of cycle length when C is the selection criterion to cycle length when X is the selection criterion. Of course, the genetic correlation [r_{gcx}] between C and X is crucial.

Now, as an example, suppose in a particular instance (1) that a measurement X is believed to be correlated genetically with total value (C), (2) that in view of costs that k_x/k_c could be 1.5, (3) that $h_x^2 = 0.3$ compared to $h_c^2 = 0.2$ and (4) that L_c/L_x would be 2.0. Then substitution in 6.37 would yield

$$E(\Delta\ddot{Y}_{c.x})/E(\Delta\ddot{Y}_{c.c}) = (1.5)(0.3/0.2)^{1/2}(2)r_{gcx} = 3.67r_{gcx}$$

Hence if $r_{gcx} > 1/3.67 = 0.27$, the ratio would be greater than 1.0. A correlation greater than that is easily possible. However, two warnings are in order.

(1) As noted earlier in this section additive genetic covariance [and therefore r_{gcx}] should be expected to change as recurrent selection proceeds and may even go from positive to negative.

(2) The foregoing speaks only to the ratio of expectations in the cycle at hand. It does not speak to the question of total response achievable by recurrent selection. (This isssue is discussed in greater detail in Chapter XII.)

6.4. Reminders Concerning Assumptions

1. In Section 6.2 linkage equilibrium is not assumed but with that one exception the entire chapter makes all assumptions listed in Section 5.6. In addition, position effects are assumed absent, i.e., that the total genetic effects (y_i + y_j) of coupling and repulsion double heterozygotes are equal.

2. In all of the portions of Section 6.3 that deal with $E(\Delta\ddot{Y}_c)$, considerations are limited to the case where T = S; i.e., the S-population itself is the target (T) population.

3. Procedures for obtaining E(k) assume "normal distribution" of $\mathring{X}$, the selection criterion.

7 Mass Selection—Assumptions and Validity of Theory

Quantitative theory is inevitably a function of assumptions employed in its derivation. Consequences of three assumptions listed in sections 5.8 and 6.4 are examined and discussed in this chapter.

7.1. The Assumption That $\Sigma_i(\Delta q_i)(\partial \ddot{Y}_c/\partial q_i)$ Is an Adequate Approximation to $\Delta \ddot{Y}_c$

Though this assumption might be valid in the absence of epistasis but not when one or another type of epistasis is present, only the non-epistatic situation assumed as part of the base for the theory of Chapters V and VI will be considered in this section. As explained in Chapter V, when (a) epistasis is absent, (b) there are no more than two alleles per gene, and (c) mating is random

$$\ddot{Y}_c = \alpha_c + \Sigma_i \ddot{y}_{ci}, \qquad 5.7$$

$$\partial \ddot{Y}_c/\partial q_i = \partial \ddot{y}_{ci}/\partial q_i, \qquad 5.9$$

and

$$\partial \ddot{y}_{ci}/\partial q_i = 2[1+(1-2q_i)a_{ci}]u_{ci} \qquad 5.11$$

when the T-population is the S-population and

$$\partial \ddot{y}_{ci}/\partial q_i = [1+(1-2p_i)a_{ci}]u_{ci} \qquad 5.12$$

when T = S x A, i.e., the T-population is obtained by mating random individuals from the S-population with random individuals from some other population in which the frequency of B at the i-th locus is p_i. It follows from 5.7 and 5.9 that the approximation error of concern here is

$$\Sigma_i(\Delta q_i)(\partial \ddot{Y}_c/\partial q_i) - \Delta \ddot{Y}_c = \Sigma_i[(\Delta q_i)(\partial \ddot{y}_{ci}/\partial q_i) - \Delta \ddot{y}_{ci}] \qquad 7.1$$

in which $\Delta \ddot{Y}_c$ and $\Delta \ddot{y}_{ci}$ symbolize the changes in $\ddot{Y}_c$ and $\ddot{y}_{ci}$, respectively, in response to one cycle of selection. It is apparent from 7.1 that the error in approximation of $\Delta\ddot{Y}_c$ is the sum over loci of the component errors associated with each gene that affects the response criterion (C). Thus it will be appropriate to focus further attention on the approximation at the level of single loci.

We note from 5.12 that when T = S x A, the derivative of $\ddot{y}_c$ with respect to q is not a function of q, i.e., that $\ddot{y}_{ci}$ in the T-population is a linear function of p_i. This means that $\Delta\ddot{y}_{ci} = (\Delta q_i)(\partial\ddot{y}_{ci}/\partial q_i)$ exactly and that in this case there is no approximation error (unless p_i changes because selection is practiced in the A, as well as the S, population).

In contrast, we note from 5.11 that, when the S-population is the T-population (T = S), $\ddot{y}_{ci}$ is a linear function of q_i when $a_{ci} = 0.0$ but is not when any degree of dominance is present. It is this case (T = S and $a_{xi} \neq 0.0$) that requires attention. Omitting the subscripts i and c, the following will pertain to any single locus and any response criterion, and letting £ indicate E(Δq), the approximation to $\Delta\ddot{y}$

$$£(\partial\ddot{y}/\partial q) = 2£[1+(1-2q)a] \qquad 7.2$$

and the actual value of $\Delta\ddot{y}$ resulting from the change in q is found (using 5.10) to be

$$\Delta\ddot{y} = 2[(q + £)u + (q + £)(1-q-£)au] - 2[qu + q(1-q)au]$$

$$= 2£[1+(1-2q)a]u - 2£^2au \qquad 7.3$$

When 7.2 and 7.3 are compared, the bias (approximation error) of 7.2 is seen to be $2£^2au$. Let this bias, expressed as a fraction of $\Delta\ddot{y}$ be symbolized as D. Then

$$D = £a/\{[1+(1-2q)a] - £a\} \qquad 7.4$$

As noted above and indicated by 7.4, bias is zero when a = 0 and is positive when a > 0 unless the denominator is negative. This suggests that, with the possible exception of overdominance genes, bias would be the greatest in the case of complete dominance of the better allele (a = 1). In that case we see from 7.4 that

$$D = £/[2(1-q) - £] \qquad 7.5$$

Table 6.3 provides information concerning possible magnitudes of £ = E(Δq). However, with greater selection intensity, i.e., k > 1.0, and u/σ greater than the 0.1 assumed for Table 6.3, it is obvious that E(Δq) can be greater than in that table. Using values of .0288 and .0576 (with q = 0.4) for £ = E(Δq) in 7.5 we obtain D = .0295 and .0504. It is apparent that the assumption here investigated can hardly be a source of serious bias.

• 7.2. The Assumption That the Regression of $\hat{q}$ on $\hat{X}$ Is Linear

Consider the realationship between $\hat{q}$ and $\hat{y}$ in the case of mass selection in which individuals are both the selection and criterion units. Magnitudes of the three possible single locus genotypes are shown below.

Genotype	$\hat{q}$	$\hat{y}$
BB	1	2u
Bb	1/2	u + au
bb	0	0

In the case of additive genes (a=0.0), the regression of $\hat{q}$ on $\hat{y}$ is obviously linear. Almost as obviously it is not linear when $a \neq 0.0$. But the regression of $\hat{q}$ on $\hat{y}$ is not the issue. Our concern is whether the regression of $\hat{q}$ on $\hat{X}$ is linear. The linear relation between $\hat{q}$ and $\hat{y}$ when $a = 0.0$ suggests, but does not prove, that the regression of $\hat{q}$ on $\hat{X}$ in the case of additive genes may be linear -- though in fact it is not (see below). The obvious non-linearity in the relation between $\hat{q}$ and $\hat{y}$ when $a \neq 0.0$ suggests non-linearity of the regression of $\hat{q}$ on $\hat{X}$ when one of two alleles is dominant in some degree to the other. For practical purposes the issue to be examined is whether and in what situations, if any, the deviations from linearity are sufficient to undermine the utility of prediction equations that assume linearity.

Equation 6.24 (restated here for reference) tells us that this is most directly approached by examining the regression of $E(\Delta q)$ on k. Our concern must be with (1) how much it deviates

$$E(\Delta q_i) = q_i(1-q_i)[1+(1-q_i)a_{xi}]ku_{xi}/\sigma(\hat{X}) \qquad 6.24$$

from linearity and (2) the extent to which that deviation is affected by the magnitudes of q, a, and the ratio, $u_x/\sigma(\hat{X})$. These issues will be examined at the outset in the framework of truncation selection.

It is intuitively obvious that individuals variable in their i-th locus genotypes may have the same value of X. The regression of $\hat{q}$ on k depends on how the proportions of those genotypes change as X goes through its range from low to high values. Let us imagine an infinite population in which i-th locus genotypes are present in H-W frequencies. Such a population can be viewed as consisting of three sub-populations -- each composed of individuals with similar genotype at the i-th locus -- BB, Bb, or bb. Next consider the values of X associated with the individuals in each of the three subpopulations and the distributions of those X values. If X is the measure of a quantitative trait, the three distributions will be overlapping with differences between their mean values determined by the magnitudes of u_i and $a_i u_i$. The three distributions and relations between them are depicted in Figure 7.1. Distances on the baseline are measured from the mean value of the distribution of individuals having the bb genotype; hence, the mean of that distribution is shown to be 0.0 and the means of the Bb and BB distributions as $u+au$ and $2u$, respectively.

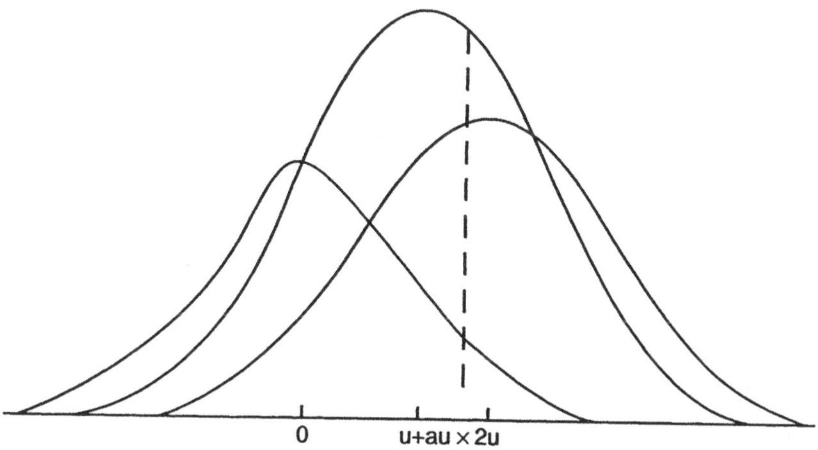

Figure 7.1. Graphical depiction of the overlapping distributions of single-locus genotypes, truncation selection, and the relative frequencies of genotypes among selections.

In the truncation selection process the genetic sample is divided into two groups, i.e., the selections, those to be employed collectively as parents of the next generation and all others. This division is based on the values of X of the individual selection units. Those with the largest (in positive selection) X values constitute the selections. Let x be the value of X that is the point of separation between the groups. Clearly, the selections would be composed of individuals from all three distributions in expected proportions that would vary with the magnitude of x. The expected frequency of B_i in selections would be determined by those proportions. Intuitively, we recognize that when x is larger, the proportion of BB individuals among selections is likely to be larger and hence $\bar{q}_{si}$ also is likely to be larger.

To obtain exact information concerning the regression of $\bar{q}_s$ on k requires exact information concering the nature and parameters of the the three distributions, i.e., information that is rarely, if ever, available. The situation to be examined in what follows is the hypothetical one in which in each of the three sub-populations, X is normally distributed with the same variance in each of the three distributions. Then if all individuals with values of $X > x$ and none for which $X < x$ are selected, the fraction selected is $1 - F(x)$.[1] In Figure 7.1 these fractions are represented by the areas in the three distributions to the right of the ordinate at the point x. In the case of the normal distribution F(x) is tabulated in various texts, e.g., Standard Mathematical Tables published by The Chemical Rubber Company. In such tables x is distance, in standard units, from the mean of the distribution (population). Thus, because the point x shown in figure 7.1 is measured from the mean of the bb sub-population, selection of individuals for which $X > x$ means that expected proportions selected from the three sub-populations are

$P_0 = 1-F(x)$ in the case of the bb sub-population,
$P_1 = 1-F(x-u-au)$ in the case of the Bb sub-population, and
$P_2 = 1-F(x-2u)$ in the case of the BB sub-population.

Then taking acount of the relative sizes of the three sub-populations, the fraction of the whole population that is selected is

$$p = (1-q)^2 P_0 + 2q(1-q)P_1 + q^2 P_2 \qquad 7,6$$

and the expected frequency of the B allele in selections is

$$\bar{q}_s = [2q(1-q)(1/2)P_1 + q^2(1)P_2]/p \qquad 7.7$$

Finally E(k), referred to hereafter as k in this section, is

$$E(k) = z/p \qquad 6.22$$

(see section 6.3.2 for definition of z).

From 6.24 it is apparent that, given (1) linearity of the regression of q on X, (2) a common form of distribution of the three sub-populations, and (3) the same variance of X in each of the three sub-populations, $[E(\Delta q_i) = E(\bar{q}_{si} - q)]/k$ would have the same magnitude whatever the intensity of selection, i.e., whatever the value of p. In this connection, note that tabulated values of F(x) to be employed in obtaining values of p, $\bar{q}_s$, and k using 7.6, 7.7, and

6.22 pertain to populations of infinite size. It follows that quantities to be presented in the tables of this section have no sampling errors, i.e., they are exactly correct except for rounding errors.

Table 7.1 lists values of p, $\bar{q}_s$, k, and $\Delta q/k$ for four equally spaced values of x in one specific case relative to values of q, a, and u. Only $\Delta q/k$ is critical to the issue being examined. The other three are listed to allow readers to check, step by step, the process employed in obtaining values of $\Delta(q)/k$. Judging from the evidence of this table -- the values of $\Delta q/k$ are essentially the same for selection intensities ranging from p = 0.9 to 0.01. It is reasonable to

Table 7.1. Values of Δq/k when q = 0.3, a = 0.5, and u = 0.1

x	p	$\bar{q}_s$	k	$\Delta q/k$
-1.2	.8992	.30487	.196	.0248
0.0	.5322	.31878	.747	.0251
1.2	.1324	.34088	1.618	.0253
2.4	.0104	.36699	2.663	.0252

inquire, though not crucial relative to the linearity issue, whether the values of $\Delta q/k$ are as expected on the basis of theory that has been presented. Rearrangement of 6.24 yields

$$E(\Delta q)/k = q(1-q)[1+(1-2q)a]u/\sigma(\hat{X}) \qquad 7.8$$

This expression calls for a numerical value of $\sigma(\hat{X})$, the variance of X in the entire population, the whole of the three sub-populations identified above. Standard normal distribution tables such as the one employed to obtain results for this presentation pertain to a population in which $\sigma = 1.0$. On the other hand the variance of the entire population in which effects of selection are here being investigated includes variance arising from the differences among single locus genotypes at the locus being considered. That variance is shown in Chapter XY to be

$$\sigma^2(y) = 2q(1-q)[1+2(1-2q)a+(1-2q+2q^2)a^2]u^2 \qquad 7.9$$

Given values of q, a, and u specified in table 7.1, 7.9 yields $\sigma^2(y) = .006489$. Adding this to the intra-population variance implied by the process outlined, yields $\sigma^2(\hat{X}) = 1.006489$ from which $\sigma(\hat{X}) = 1.0032$. Substituting this in 7.8, we obtain $E(\Delta q) = .0251$ instead of the .0252 obtained using $\sigma(\hat{X}) = 1.0$, the intra-population value. Conclusions justified in this particular case are (1) that the regression of q on X is very very close to linear and (2) nothing of practical importance was gained by taking the genetic variance arising from the single locus under consideration.

Tables 7.2 - 7.5 list values of $E(\Delta q)/k$ for different combinations of q, u, and x. The four tables differ with respect to the levels of dominance (a values) assumed.

The equation 7.8 prediction of $\Delta q/k$ is shown <u>in italics</u> for each u,q combination in each table. These enable comparison of actual values to be expected (given normal distribution of X around single locus genotype values) with predictions that assume linear regression of q on X.

Table 7.2. E(Δq)/k as a function of q, u, and x when a = 0.0

u	x	q 0.1	0.3	0.5	0.7	0.9	Range of k
.05	-1.2	.0044	.0105	.0126	.0104	.0048	.22-.19
	0.0	.0045	.0105	.0125	.0104	.0046	.79-.74
	1.2	.0046	.0105	.0126	.0104	.0043	1.68-1.61
	2.4	.0049	.0111	.0129	.0109	.0042	2.73-2.66
Theory		*.0045*	*.0105*	*.0125*	*.0105*	*.0045*	
.10	-1.2	.0086	.0204	.0249	.0215	.0094	.21-.17
	0.0	.0090	.0209	.0250	.0210	.0090	.78-.69
	1.2	.0094	.0214	.0250	.0206	.0087	1.67-1.54
	2.4	.0099	.0213	.0243	.0211	.0074	2.70-2.56
Theory		*.0090*	*.0210*	*.0249*	*.0210*	*.0090*	
.20	-1.2	.0162	.0395	.0496	.0439	.0204	.21-.13
	0.0	.0178	.0415	.0496	.0430	.0185	.77-.58
	1.2	.0196	.0433	.0495	.0403	.0166	1.65-1.40
	2.4	.0214	.0449	.0483	.0372	.0146	2.68-2.40
Theory		*.0179*	*.0417*	*.0495*	*.0417*	*.0179*	

Table 7.3. E(Δq)/k as a function of q, u, and x when a = 0.5

u	x	q 0.1	0.3	0.5	0.7	0.9	Range of k
.05	-1.2	.0061	.0126	.0125	.0088	.0029	.22-.19
	0.0	.0063	.0126	.0125	.0083	.0028	.79-.74
	1.2	.0065	.0127	.0124	.0081	.0028	1.68-1.61
	2.4	.0067	.0126	.0120	.0079	.0024	2.71-2.65
Theory		*.0063*	*.0126*	*.0125*	*.0084*	*.0027*	
.10	-1.2	.0118	.0248	.0255	.0176	.0055	.21-.17
	0.0	.0126	.0251	.0251	.0168	.0054	.78-.68
	1.2	.0133	.0253	.0243	.0161	.0053	1.66-1.53
	2.4	.0140	.0252	.0234	.0154	.0050	2.71-2.55
Theory		*.0126*	*.0251*	*.0249*	*.0168*	*.0054*	
.20	-1.2	.0221	.0485	.0527	.0399	.0125	.20-.12
	0.0	.0252	.0499	.0502	.0341	.0110	.76-.57
	1.2	.0278	.0504	.0471	.0310	.0102	1.63-1.38
	2.4	.0310	.0499	.0436	.0279	.0092	2.65-2.39
Theory		*.0250*	*.0498*	*.0494*	*.0332*	*.0108*	

Table 7.4. E(Δq)/k as a function of q, u, and x when a = 1.0

		q					Range
u	x	0.1	0.3	0.5	0.7	0.9	of k
.05	-1.2	.0078	.0148	.0130	.0065	.00105	.21-.19
	0.0	.0081	.0145	.0124	.0062	.00091	.78-.74
	1.2	.0084	.0147	.0120	.0062	.00083	1.68-1.61
	2.4	.0085	.0141	.0112	.0057	.00079	2.70-2.65
Theory		*.0081*	*.0147*	*.0125*	*.0063*	*.00090*	
.10	-1.2	.0149	.0294	.0265	.0141	.00178	.21-.16
	0.0	.0161	.0294	.0251	.0127	.00183	.77-.68
	1.2	.0173	.0292	.0235	.0117	.00149	1.65-1.53
	2.4	.0186	.0287	.0215	.0102	.00145	2.67-2.55
Theory		*.0162*	*.0293*	*.0249*	*.0126*	*.00180*	
.20	-1.2	.0269	.0574	.0564	.0316	.00398	.20-.12
	0.0	.0317	.0584	.0508	.0262	.00390	.75-.56
	1.2	.0364	.0571	.0449	.0217	.00312	1.61-1.38
	2.4	.0406	.0543	.0389	.0178	.00235	2.64-2.37
Theory		*.0320*	*.0577*	*.0493*	*.0250*	*.00360*	

Table 7.5. E(Δq)/k as a function of q, u, and x when a = 2.0

		q					Range
u	x	0.1	0.3	0.5	0.7	0.9	of k
.05	-1.2	.0110	.0190	.0133	.0028	-.0025	.21-.19
	0.0	.0117	.0189	.0125	.0022	-.0028	.78-.73
	1.2	.0123	.0188	.0117	.0014	-.0028	1.66-1.60
	2.4	.0131	.0185	.0114	.0012	-.0028	2.71-2.64
Theory		*.0117*	*.0189*	*.0125*	*.0021*	*-.0027*	
.10	-1.2	.0205	.0376	.0283	.0067	-.0049	.20-.16
	0.0	.0231	.0377	.0253	.0045	-.0054	.76-.67
	1.2	.0256	.0370	.0224	.0026	-.0057	1.64-1.51
	2.4	.0282	.0354	.0190	.0000	-.0063	2.66-2.54
Theory		*.0232*	*.0374*	*.0248*	*.0042*	*-.0054*	
.20	-1.2	.0354	.0730	.0634	.0207	-.0077	.20-.11
	0.0	.0441	.0744	.0525	.0112	-.0097	.73-.55
	1.2	.0542	.0716	.0410	.0028	-.0123	1.57-1.34
	2.4	.0622	.0643	.0301	-.0027	-.0136	2.56-2.34
Theory		*.0456*	*.0727*	*.0486*	*.0083*	*-.0108*	

In further discussion in this section these will be referred to as the actual and theoretical values, respectively. In Table 7.2, for example, the actual value is .0109 and the theoretical is .0105 when $q = 0.7$, $u = 0.05$, and $x = 2.4$, i.e., the agreement in that instance is very good.

In each of the four tables the range of k is listed for each u,x combination. These indicate selection intensity; k increases as selection intensity increases. Table 6.1 shows that when the genetic sample is "large" and proportions selected are 0.5, 0.1, 0.02, and 0.01, the associated k's are 0.80, 1.76, 2.42, and 2.64, respectively. Thus the 2.68 - 2.40 range shown in the last line of Table 7.2 relates to selection intensities in the area of 0.01 to 0.02.

The significant generalities demonstrated by the content of the four tables are (1) overall good agreement between actual and theoretical values of $\Delta q/k$ when $a = 0.0$, 0.5, or 1.0 and $u = 0.05$ or .10, (2) overall good agreement when k values are in the 0.6 - 1.6 range (which equates to selection intensities in the 55 to 15 percent range) and (3) the largest deviations of actual from theoretical values are associated with the most intense level of selection, with the level of overdominance ($a = 2.0$) investigated, and with $u = 0.10$ or 0..20, or combinations of those conditions.[2]

The phrase "overall good agreement" is ambiguous -- a matter of judgement. To be more explicit, the deviation of actual from theoretical values exceeded ten percent, but was less than twenty percent, in nine of the 120 instances in which u was 0.05 or 0.10 and a was 0.0, 0.5, or 1.0. Among the 120 instances in which x was 0.0 or 1.2 so that k was in the 0.6 to 1.6 range, there were 13 deviations that exceeded 10 percent. Of those 13, eight occurred when a was 2.0 and the four that exceeded 20 percent all occurred when a was 2.0 and q was 0.7.

It must also be emphasized that the situation explored involved normal distribution of the totality of effects (genetic and non-genetic) other than those of genotype at the one locus considered. The impacts of non-normal distributions have not been studied by this author.

• 7.3. The No Epistasis Assumption

As stated in Section 1.3, <u>selection is the only force that can give direction to genetic change</u>. However, in Section 2.9.1, it is noted that given <u>multiple peak</u> epistasis, the progressive change fostered by recurrent selection may or may not promote the genetic change in a population toward synthesis of the "best possible" genotype. These conclusions will be explored here using two of the two gene epistasis models of Figure 2.2. It was shown in Chapter V that

$$E(\Delta q_i) = \sigma(\hat{q}_i,\hat{y}_{xi})k/\sigma(\hat{X}) \qquad 5.24$$

Of the three quantities in this expression k and $\sigma(\hat{X})$, by their definitions, do not change their algebraic sign during the course of recurrent selection. Thus if more than one direction of change is possible in the frequency of any allele that is fostered by selection, it must reflect the possibility of $\sigma(\hat{q}_i,\hat{y}_{xi})$ being positive or negative. It is shown in Chapter VI that

$$\sigma(\hat{q}_i,\hat{y}_{xi}) = q_i(1-q_i)[1+(1-2q_i)a_{xi}]u_{xi} \qquad 6.3$$

when epistasis and linkage disequilibrium are absent. When $0 < a \le 1$, the algebraic sign of $\sigma(\hat{q}_i,\hat{y}_{xi})$ is determined by whether u is a plus or minus quantity.

In the absence of epistasis u_i and a_i, by their definitions, are constants and $u_i > 0.0$. However, it will be shown here that epistasis when present can be such that, depending on allele frequencies, u_i can be positive or negative. In particular this is possible given multiple peak epistasis but not as a consequence of simple epistasis.

7.3.1. Simple Epistasis

As an example consider the two-locus complementary gene model depicted in Figure 2.2, value set (c). The Y values (2 or 0) quantify the joint contribution of the two single locus genotypes to the total genotype value of an individual. Let p and q symbolize the population frequencies of B_1 and B_2 respectively. Then assuming no interaction with genotypes at other loci, HW frequencies of genotypes, and linkage equilibrium, the joint distribution of the frequencies (P and Q) of the better allele of the genes involved and the Y values is shown below.

Genotype	Genotype frequency	P	Q	Y
$B_1B_1B_2B_2$	p^2q^2	1	1	2
$B_1B_1B_2b_2$	$2p^2q(-q)$	1	1/2	2
$B_1B_1b_2b_2$	$p^2(1-q)^2$	1	0	0
$B_1b_1B_2B_2$	$2p(1-p)q^2$	1/2	1	2
$B_1b_1B_2b_2$	$4p(1-p)q(1-q)$	1/2	1/2	2
$B_1b_1b_2b_2$	$2p(1-p)(1-q)^2$	1/2	0	0
$b_1b_1B_2B_2$	$(1-p)^2q^2$	0	1	0
$b_1b_1B_2b_2$	$2(1-p)^2q(1-q)$	0	1/2	0
$b_1b_1b_2b_2$	$(1-p)^2(1-q)^2$	0	0	0

Because of the parallels between the P,Y and Q,Y distributions, it will suffice to obtain $\sigma(Q,Y)$, the covariance between Q and Y, from which $\sigma(P,Y)$ will be apparent. Let $Y_{.2}$, $Y_{.1}$, and $Y_{.0}$ symbolize the averages of Y given the B_2B_2, B_2b_2, and b_2b_2 genotypes, at locus 2. Then

$$Y_{.2} = \{q^2[p^2 + 2p(1-p)](2) + q^2(1-p)^2(0)\}/q^2 = 2p(2-p),$$

$$Y_{.1} = \{2q(1-q)[p^2 + 2p(1-p)](2) + 2q(1-q)(1-p)^2(0)\}/2q(1-q) = 2p(2-p),$$

and

$$Y_{.0} = 0.0.$$

By definition,

$$2u_2 = Y_{.2} - Y_{.0} = 2p(2-p) \text{ and}$$

$$u_2 + a_2u_2 = Y_{.1} - Y_{.0} = 2p(2-p).$$

It follows that

$$u_2 = a_2u_2 = p(2-p) \text{ and } a_2 = 1.0.$$

Substituting in

$$\sigma(\hat{q}_i, \hat{y}_{xi}) = q_i(1-q_i)[1 + (1-2q_i)a_{xi}]u_{xi} \qquad 6.3$$

yields

$$\sigma(Q,Y) = q(1-q)[1+(1-2q)]p(2-p) = 2q(1-q)^2p(2-p).$$

Then by virtue of the parallels between the P,Y and Q,Y bivariate distributions, it follows that

$$\sigma(P,Y) = 2p(1-p)^2q(2-q).$$

Finally, because all possible values of p and q are positive, it is apparent that the algebraic sign of both $\sigma(Q,Y)$ and $\sigma(P,Y)$ is always positive, i.e., that both $E(\Delta p)$ and $E(\Delta q)$ will always be positive when the selection differential is positive and the ideal genotype with respect to these two genes would be progressively approached as the result of positive recurrent selection.

Demonstrated by this example is that by virtue of epistasis, u_i becomes a function of allele frequencies at a locus (or loci) other than the i-th. Intuitively, one supposes that there are forms of epistasis such that a_i, as well as u_i, is a function of frequencies other than those of the i-th locus alleles. The impact is that both $\sigma(\hat{q}_i, \hat{y}_{xi})$ and the single locus contribution to additive genetic variance are functions of allele frequencies at all loci involved in the epistasis system. This means that the basic response expectations, $E(\Delta q)$ and $E(\Delta \ddot{y}_i)$, are approximated by constant functions of values of u_i and a_i that are not constant, but themselves, like q_i, are changed by selection and "drift".

• 7.3.2. Multiple-Peak Epistasis

As an example of multiple-peak epistasis consider the two-locus model depicted in Figure 2.2, value set (g). Again the Y values shown quantify the joint contribution of the two single locus genotypes to the total genotype value of an individual. Making the same assumptions as in our first example, the joint distribution of P, Q, and Y is as shown below.

Genotype	Genotype frequency	P	Q	Y
$B_1B_1B_2B_2$	p^2q^2	1	1	8
$B_1B_1B_2b_2$	$2p^2q(-q)$	1	1/2	4
$B_1B_1b_2b_2$	$p^2(1-q)^2$	1	0	0
$B_1b_1B_2B_2$	$2p(1-p)q^2$	1/2	1	4
$B_1b_1B_2b_2$	$4p(1-p)q(1-q)$	1/2	1/2	3
$B_1b_1b_2b_2$	$2p(1-p)(1-q)^2$	1/2	0	2
$b_1b_1B_2B_2$	$(1-p)^2q^2$	0	1	0
$b_1b_1B_2b_2$	$2(1-p)^2q(1-q)$	0	1/2	2
$b_1b_1b_2b_2$	$(1-p)^2(1-q)^2$	0	0	4

Proceeding as before we find $Y_{\cdot 2} = 8p$, $Y_{\cdot 1} = 2(1+p)$, and $Y_{\cdot 0} = 4(1-p)$. Then

$$u_2 = (Y_{\cdot 2} - Y_{\cdot 0})/2 = 2(3p-1),$$

$$u_2 + a_2u_2 = (Y_{\cdot 1} - Y_{\cdot 0}) = 2(3p-1), \text{ and}$$

$$a_2u_2 = a_2 = 0.$$

Substituting again in 6.3, we obtain in this case

$$\sigma(Q,Y) = q(1-q)[1+(1-2q)a]u = 2q(1-q)(3p-1) \quad 7.10$$

In this example as in the previous one, it is obvious from the parallels between the P,Y and Q,Y that $\sigma(P,Y)$ will be obtained if p and q are replaced by q and p, respectively in $\sigma(Q,Y)$. Thus

$$\sigma(P,Y) = 2p(1-p)(3q-1) \quad 7.11$$

Because the algebraic signs of $E(\Delta q)$ and $E(\Delta p)$ are determined by those of $\sigma(Q,Y)$ and $\sigma(P,Y)$, it is apparent from 7.10 and 7.11 that $E(\Delta q)$ is positive when $p > 1/3$, negative when $p < 1/3$ and that $E(\Delta p)$ is positive when $q > 1/3$ but negative when $q < 1/3$. It follows that if original values of p and q are greater than one-third, fixation of both B_1 and B_2 would be favored by recurrent selection and, conversely, if original values of p and q are less than one-third fixation of both b_1 and b_2 would be favored by recurrent selection. The end result when the original value of p or q is less than one-third and that of the other is greater than one-third would depend on the magnitudes of the deviations of original frequencies from one-third and the "drift" of frequencies[3]. In the end, however, either the 8-unit or the 4-unit "peak" would be approached, i.e., in the event of multiple peak epistasis, the "best possible" population genotype is not guaranteed to be the inevitable consequence of recurrent selection.

8 Progeny Test and Family Systems of Recurrent Selection

Derivations of the covariances of $\hat{q}_i$ with $\hat{u}_{Xi}$ and $\hat{u}_{Yj}$ are presented for various selection systems that have been or might be used by breeders. These covariances and the standard deviation, $\sigma(\hat{X})$, of the selection criterion are the components of the response expectations that vary with the selection unit and/or the criterion unit.[1] Together with equations 5.23 and 5.24, they provide the bases for writing the response expectation, $E(\Delta q)$, specific to various combinations of selection and criterion units. Then the latter, along with equations 5.9 and 5.11 or 5.12, can be used to write $E(\Delta\ddot{Y})$ for specific combinations of selection unit, criterion unit, target population and response criterion; i.e., the expectations of both direct and correlated response for a broad variety of breeding programs. The composition of $\sigma(\hat{X})$ as affected by the nature of the criterion unit is treated in Chapter X.

The assumptions for the derivation of $E(\Delta q_i)$, equation 5.23 of Chapter V, are employed again in this chapter. These are listed in section 5.8. However, linkage equilibrium is only assumed where indicated.

8.1. Progeny Test Selection Systems

In these systems the selection units are individuals and the criterion units are progenies (groups of offspring) of those individuals. Why should individuals be evaluated in terms of the phenotypes of offspring rather than on the basis of their own phenotypes? There are at least three reasons.

(1) When heritability of the selection criterion is low the individual's own phenotype, being principally a function of nongenetic effects, is not a reliable guide to the individual's genotype. In these cases the average of the selection criterion over a group

of offspring can be a better index to the frequencies of useful alleles in the genotype of the parent.

(2) In dioecious species, one or more of the significant components of total value may not be measurable in one of the sexes. Familiar examples are milk production in dairy cattle (as in all mammals), egg production in poultry and litter size in swine. Selection among males for any of these traits must be based on the performance of female relatives (such as daughters, sisters or dams).

A related necessity is encountered in annual plants when the selection criterion is not expressed until after fertilizations have occurred. For example, grain yield of maize is not expressed until after fertilization and consequently, pollen parents cannot be selected on the basis of their own yield of grain.

(3) If there are genes that are overdominant with respect to total value and the target population is a cross between the S-population and some other population, alleles that are superior with respect to phenotype in selection unit individuals may be inferior with respect to total value in the target population. There is then an advantage in selection on the basis of crossbred progeny performance.

Both the logic and the quantitative aspects of these reasons will be clarified by response expectations to be obtained in this section. The essential features of all the progeny test systems to be analyzed are these:

(1) The genetic sample consists of random individuals (or random descendants of such individuals) from a random mating selection population.

(2) Test (criterion unit) progenies are produced by mating each member of the genetic sample with one or more random members of a "test population". This "tester" may be the S-population itself or any other population.

(3) If there are n individuals in the genetic sample, there will be n test progenies. These are grown for comparison in the same macroenvironment (see section 2.12) or set of macroenvironments. Ideally, k individuals from each test progeny would be grown in each of m macroenvironments. Frequently $m = 1$ but if substantial amounts of genotype-environment interaction are anticipated, the value of making $m > 1$ merits investigation. An average value of the selection criterion is obtained for each test progeny and is used as the basis for selection among the individuals of the genetic sample.

(4) The individuals comprising the genetic sample must all be "held" until the selection criterion data from test progenies have been obtained so that those selected will be available for use as parents of a new generation of the S-population. In the case of animals and perennial plants, the selection unit individuals themselves can be saved for the required periods of time though this may involve special costs including some mortality. Holding the individuals themselves is never possible in work with annual plants. The substitute procedure employed in corn breeding is to self-fertilize each selection unit individual and preserve the alleles possessed by the individual in its S_1

progeny (offspring produced by self-fertilization and preservable as seed). Such a progeny is then available for use as a surrogate for each selection in reproducing the S-population. In the case of dairy cattle, progeny testing of bulls has become a highly refined standard practice but technical developments have made it possible and less expensive to preserve large amounts of semen from each bull rather than the animals themselves.

(5) It goes almost without saying that in some species it is practical to use progeny test selection in one sex but not the other. Once more dairy cattle serve as an example. Genotypic value for milk production in the male can only be measured in terms of female relatives; hence progeny testing of bulls is standard. In contrast the number of female offspring per cow would not be sufficient for progeny testing even if heritability were low enough to make it desirable. This statement assumes that getting the required number of daughters by embryo transplant procedures now available would not enable increases in $\Delta\ddot{Y}$ per unit of time.

8.1.1. Justification for Using E ($\hat{y}$) in Derivations of $\sigma(\hat{q}_i,\hat{y})$

In sections 6.1 and 6.2 the joint distributions of $\hat{q}_i$ and $\hat{y}_{Xi}$ and of $\hat{q}_i$ and $\hat{y}_{Xj}$ that were used in derivations of $\sigma(\hat{q}_i,\hat{y}_{Xi})$ and $\sigma(\hat{q}_i,\hat{y}_{Xj})$were specified in terms of the probabilities of genotypes. Thus the covariances obtained were covariances in the infinite population that is conceptually, but not actually, possible <u>via</u> random mating of a finite set of parents. This was appropriate because, as reflected in the logic outlined in section 5.5, actual genetic samples are appropriately viewed as random samples from such an infinite population. Despite the fact that the frequencies of genotypes in a genetic sample will almost always deviate from their probabilities, the expectation of change due to selection in the frequency of an allele is a function of the regression of $\hat{q}$ on $\hat{X}$ <u>in the infinite population.</u>

In the mass selection case, $\hat{y}_i$ or $\hat{y}_j$ is completely determined by the selection unit individual's genotype at the i-th or j-th locus. In contrast when the criterion unit is a relative or a finite group of relatives of the selection unit individual, $\hat{y}$ is <u>not</u> completely determined by the genotype of that individual but the expectation of $\hat{y}$ is. With this in mind, note that $\hat{y} = E(\hat{y}) + \hat{d}$ where $\hat{d}$ is the deviation of $\hat{y}$ from its expectation and that $\Sigma\,\hat{d} = 0.0$ in the genetic sub-population identified with any specific genotype of those possible in selection unit individuals. It follows that the covariance of $\hat{q}$ with $\hat{d}$ in the entire genetic population is zero and hence that

$$\sigma(\hat{q},\hat{y}) = \text{the covariance of } \hat{q} \text{ with } E(\hat{y})$$

and can therefore be obtained from the joint distribution of $\hat{q}$ and $E(\hat{y})$.

8.1.2. Response Expectations When Selection Unit Individuals Are Not Inbred (No Assumption Concerning Linkage Equilibrium)

The joint distribution of $\hat{q}_i$, $E(\hat{y}_{Xi})$, and $E(\hat{y}_{Xj})$ is required as the base for derivation of $\sigma(\hat{q}_i,\hat{y}_{Xi})$ and $\sigma(\hat{q}_i,\hat{y}_{Xj})$. Let v_i be the frequency of B_i in the tester. Then, the probabilities of i-th locus genotypes in test progenies will be functions of v_i and the i-th locus genotypes of selection unit individuals. For example, the BB genotype is formed by the union of two B gametes (one from the selection unit individual, the other from the tester), the probability of B in a gamete from a Bb selection unit is 1/2 and of B in a gamete from the tester is v, and thus the probabiliity of the BB genotype in the test progeny of a Bb selection unit is v(1/2). As another example, when the selection unit is Bb, the Bb genotype in a member of the test progeny results from union of a B gamete from the selection unit with a b gamete from the tester or from union of a b gamete from the selection unit with a B gamete from the tester. The sum of the probabilities of these two events, $(1/2)v + (1/2)(1-v) = 1/2$, is the probability of the Bb genotype in a member of the test progeny of a Bb selection unit individual. The complete set of probabilities (given just two alleles of the gene) is shown below along with values of $E(\hat{y}) = E(\hat{y}_x)$ for each selection unit genotype.

Genotype probabilities in test progeny

Selection unit genotype	BB	Bb	bb	Sum	$E(\hat{y})$
BB	v	1-v	0.0	1.0	$v(2u) + (1-v)(u+au)$
Bb	v/2	1/2	(1-v)/2	1.0	$(v/2)2u + (1/2)(u+au)$
bb	0.0	v	(1-v)	1.0	$v(u+au)$

Using 2u, u+au, and 0.0 as y values for the BB, Bb, and bb genotypes (section 5.4), $E(\hat{y})$ was obtained in the obvious way as the sum of products of the probability and the y value of the three genotypes. Note that the genotype probabilities sum to 1.0 (as they should) in each row of the table above. Now if the $E(\hat{y})$ values are coded by subtracting $vu + (u+au)/2$ from each of them, we obtain the following results.[2]

Selection unit genotype	$E^*(\hat{y})$
BB	$[1+(1-2v)a]u/2 = Z$
Bb	0.0
bb	$-[1+(1-2v)a]u/2 = -Z$

This set of coded values of $E(\hat{y})$ is very convenient because the zero value for the Bb selection unit and the fact that values for the BB and bb selection units are equal though opposite in sign simplify the algebra involved in obtaining the covariance of $\hat{q}_i$ with $\hat{y}_{xi}$ (or with $\hat{y}_{xj}$).

Note that the foregoing derivation of $E^*(\hat{y})$ is a general one in the sense that the result would be similar for any locus (gene) and any trait. Thus, we can specify the required values of $E^*(\hat{y}_{xi})$ in terms of

$$Z_{xi} = [1+(1-2v_i)a_{xi}]u_{xi}/2 \tag{8.1}$$

and those of $E^*(\hat{y}_{xj})$ in terms of

$$Z_{xj} = [1+(1-2v_j)a_{xj}]u_{xj}/2 \tag{8.2}$$

The joint distribution required for obtaining the covariances of $\hat{q}_i$ with $\hat{y}_{xi}$ and $\hat{y}_{xj}$ is shown in Table 8.1. There (as in section 6.2) p, r, s, and t symbolize the probabilities of the

Table 8.1. The joint distribution of $\hat{q}_i$, $*(\hat{y}_{xi})$, and $*(\hat{y}_{xj})$ for progeny test selection.

Selection unit genotypes	Frequencies	$\hat{q}_i$	$*(\hat{y}_{xi})$	$*(\hat{y}_{xj})$
B_iB_j/B_iB_j	p^2	1	Z_{xi}	Z_{xj}
B_iB_j/B_ib_j	$2pr$	1	Z_{xi}	0
B_ib_j/B_ib_j	r^2	1	Z_{xi}	$-Z_{xj}$
B_iB_j/b_iB_j	$2ps$	1/2	0	Z_{xj}
B_iB_j/b_ib_j	$2pt$	1/2	0	0
B_ib_j/b_iB_j	$2rs$	1/2	0	0
B_ib_j/b_ib_j	$2rt$	1/2	0	$-Z_{xj}$
b_iB_j/b_iB_j	s^2	0	$-Z_{xi}$	Z_{xj}
b_iB_j/b_ib_j	$2st$	0	$-Z_{xi}$	0
b_ib_j/b_ib_j	t^2	0	$-Z_{xi}$	$-Z_{xj}$

B_iB_j, B_ib_j, b_iB_j, and b_ib_j genotypes, respectively, in random gametes from the population of gametes produced by a generation of parents in the S-population. In this instance the probabilities symbolized pertain to the gametes involved in the fertilizations giving rise to the genetic population from which selection unit individuals are random members. Recall from section 6.2 that $p+r+s+t = 1.0$, $p+r = q_i$, $s+t = 1-q_i$, $p+s = q_j$ and $r+t = 1-q_j$. Then working from the distribution shown in Table 8.1, the following averages are obtained.

$$E(\hat{q}_i) = (p^2+2pr+r^2)(1) + (2ps+2pt+2rs+2rt)(1/2)$$

$$=(p+r)^2 + (p+r)(s+t) = p+r = q_i \qquad 8.3$$

$$E*(\hat{y}_{xi}) = Z_{xi}[p^2+2pr+r^2-s^2-2st-t^2] = Z_{xi}[(p+r)^2 - (s+t)^2]$$

$$=(p+r-s-t)Z_{xi} =(2q_i-1)Z_{xi} \qquad 8.4$$

and

$$E*(\hat{y}_{xj}) = Z_{xj}[p^2+2ps+s^2-r^2-2rt-t^2] = (p+s-r-t)Z_{xj} = (2q_j-1)Z_{xj} \qquad 8.5$$

Going next to covariances and making substitutions in terms of 8.1, 8.2, 8.3, 8.4, or 8.5 where appropriate, we obtain

$$\sigma(\hat{q}_i,\hat{y}_{xi}) = Z_{xi}(p^2+2pr+r^2) - E(\hat{q}_i)[E*(\hat{y}_{xi})]$$

$$= q_i^2 Z_{xi} - q_i(2q_i-1)Z_{xi} = q_i(1-q_i)Z_{xi}$$

$$= q_i(1-q_i)[1+(1-2v_i)a_{xi}]u_{xi}/2 \qquad 8.6$$

and

$$\sigma(\hat{q}_i,\hat{y}_{xj}) = Z_{xj}[p^2-r^2+ps-rt]$$

$$- E(\hat{q}_i)[E*(\hat{y}_{xj})][p^2-r^2+ps-rt-(p+r)(p+s-r-t)]Z_{xj}$$

$$= (pt - rs)Z_{xj} = (pt -rs)[1+(1-2v_j)a_{xj}]u_{xj}/2 \qquad 8.7$$

For details of algebra see Appendix 6.2.

To illustrate the utility of such explicit expressions for $\sigma(\hat{q}_i,\hat{y}_{Xi})$ and $\sigma(\hat{q}_i,\hat{y}_{Xj})$, let us now compare mass selection and progeny test selection (in which the S-population is used as the tester) with respect to expected amounts of change in allele frequencies. Substituting in 5.23 in terms of 8.6 and 8.7, remembering at the same time that v = q when the S-population is the tester, we obtain

$$E(\Delta q_i) = [k/2\sigma(\hat{X})]\{q_i(1-q_i)[1+(1-2q_i)a_{Xi}]u_{Xi} + \Sigma_{j\neq i}(pt-rs)_{ij}[1+(1-2q_j)a_{Xj}]u_{Xj}\} \qquad 8.8$$

in the case of progeny test selection. Then substituting k for $(\bar{X}_s - \bar{X})/\sigma(\hat{X})$ in 6.16, we obtain

$$E(\Delta q_i) = [k/\sigma(\hat{X})]\{q(1-q_i)[1+(1-2q_i)a_{Xi}]u_{Xi} + \Sigma_{j\neq i}(pt-rs)_{ij}[1+(1-2q_j)a_{Xj}]u_{Xj}\} \qquad 8.9$$

in the case of mass selection. Note that the expression enclosed by curly brackets is the same in 8.8 and 8.9 and would therefore cancel out of the ratio between those two expectations. If we now use m and p as subscripts to indicate mass selection and progeny test selection, respectively, and let L symbolize cycle length in some unit of time (months or years, for example), we find using 8.8 and 8.9 that

$$[E(\Delta q_i)_m/L_m]/[E(\Delta q_i)_p/L_p] = 2(k_m/k_p)[\sigma(\hat{X})_p/\sigma(\hat{X})_m](L_p/L_m) \qquad 8.10$$

This response rate ratio will be considered in greater detail in Chapter XIV but the significant generalities illustrated by 8.10 deserve to be noted here.

(1) Response rate per unit of time is inversely related to cycle length. Hence the ratio of cycle lengths is an important issue in comparative evaluation of selection systems.

(2) Response rate is also inversely related to the standard deviation of $\hat{X}$. In general, $\sigma(\hat{X})$ is smaller when there are more individuals in the criterion unit (see Chapter X for discussion of the composition of $\sigma^2(\hat{X})$ in different selection systems and variations of them). If X is measurable in individuals so that mass selection is possible, greater response rate per unit of time by progeny selection can only result if $\sigma(\hat{X})_p/\sigma(\hat{X})_m$ is sufficiently small.

(3) Response rate is directly proportional to k which is larger when selection is more intense; i.e., when the proportion of the genetic sample that is selected is smaller. It should be noted that when $\sigma(\hat{X})$ is smaller because there are more individuals in the criterion units, the size of the genetic sample is likely to be smaller because of limitations on total number of individuals that can be handled in a program. In this situation k can not be as large in progeny test selection as in mass selection without sacrifice in effective population size and hence in probability of fixation of favorable alleles (section 2.8 and Chapters XI and XIV). Consider a monoecious species such as tobacco and suppose that T individuals can be handled in a program and M must be selected if effective population size is to be as high as desired. Then in mass selection the proportion selected could be M/T. On the other hand, if the number of individuals in test progenies

were n, the proportion selected in progeny test selection could not be smaller than nM/T. This sort of trade off relation between $\sigma(\hat{X})$ and k will usually be encountered in breeding system comparisons.

(4) The 2 in 8.10 is a ratio of the numerical coefficients in 8.8 and 8.9. These are often different in the response expectations associated with different systems. They vary with the nature of the selection unit and the relationship between the selection and criterion units. If the selection units are individuals the coefficient is larger if the individuals are inbred (see next section). If the units are families the coefficient increases with intra-family homogeneity. Finally, the closer the genetic relationship between the selection and criterion units, the larger the coefficient.

8.1.3. Response Expectations When Selection Unit Individuals Are Inbred

It seems probable that if the selection unit individuals were inbred, the covariance of $\hat{q}_i$ with $\hat{X}$ would be affected, that it might be increased and that response per unit of time might in some instances be increased by using inbred selection unit individuals. To investigate this in a preliminary way, we will assume that it is possible to have a genetic sample of equally inbred individuals without generating covariance between $\hat{q}_i$ and $\hat{y}_{xj}$. In that case the covariance between $\hat{q}_i$ and $\hat{X}$ would be just $\sigma(\hat{q}_i,\hat{y}_{xi})$ (see 5.24). Let f be the coefficient of inbreeding of the selection unit individuals. Then the bivariate distribution of $\hat{q}_i$ and $E^*(\hat{y}_{xi})$ required for deriving $\sigma(\hat{q}_i,\hat{y}_{xi})$ is the following.

Selection unit genotype	Frequency	$\hat{q}_i$	$E^*(\hat{y}_{xi})$
B_iB_i	$q_i^2 + fq_i(1-q_i)$	1.0	Z_{xi}
B_ib_i	$2q_i(1-q_i)(1-f)$	1/2	0
b_ib_i	$(1-q_i)^2 + fq_i(1-q_i)$	0.0	$-Z_{xi}$

The single locus genotype frequencies are given by 2.2 and $Z_{xi} = [1+(1-2v_i)a_{xi}]u_{xi}/2$ as explained in section 8.1.2. From the above table we find

$$E(\hat{q}_i) = [q_i^2 + fq_i(1-q_i)](1) + 2q_i(1-q_i)(1-f)(1/2) = q_i,$$

$$E^*(\hat{y}_{xi}) = [q_i^2 + fq_i(1-q_i) - (1-q_i)^2 - fq_i(1-q_i)]Z_{xi} = (2q_i-1)Z_{xi},$$

and

$$\sigma(\hat{q}_i,\hat{y}_{xi}) = [q_i^2+fq_i(1-q_i)](1)Z_{xi} - q_i(2q_i-1)Z_{xi} = q_i(1-q_i)(1+f)Z_{xi}$$

$$= q_i(1-q_i)(1+f)[1+(1-2v_i)a_{xi}]u_{xi}/2 \qquad 8.11$$

The latter reduces, as it should, to 8.6 when f = 0.0. It is apparent from 8.11 that, given $\sigma(\hat{q}_i,\hat{y}_{xj}) = 0.0$ as assumed, both $E(\Delta q_i)$ and $E(\Delta\ddot{Y}_c)$ would be proportional to $(1+f)$.[3] In accord, the appropriate substitutions in 5.24 yield

$$E(\Delta q_i) = q_i(1-q_i)(1+f)[1+(1-2v_i)a_{xi}]u_{xi}k/2\sigma(\hat{X}) \qquad 8.12$$

This result suggests that using inbred selection unit individuals may sometimes be useful procedure and hence that our investigation should be extended to take account of covariance between $\hat{q}_i$ and $\hat{y}_x$ for genes other than the i-th. For two reasons, however, attention is limited to cases where the inbreeding of selection unit individuals is accomplished by self-fertilization.

(1) It was apparent from such papers as those by Weir and Cockerham (1977) and Cockerham and Weir (1973) that a more general treatment would require more space than justified unless the issue were crucial, and

(2) It appears unlikely that increments of inbreeding other than those obtained by self-fertilization have practical significance. When 8.12 is divided by cycle length to obtain the expectation of change in q_i per unit of calendar time, the result is

$$E(\Delta q_i)/L = q_i(1-q_i)(1+f)[1+(1-2v_i)a_{xi}]u_{xi}k/2\sigma(\hat{X})L$$

The largest increment in f that can be achieved by a generation of inbreeding (except by self-fertilization) is 0.25 and there are few, if any, situations in which the increase in L from adding a generation to the cycle would not be 25 percent or more.

Equation 8.13 (derived in Appendix 8.1, where it is numbered A8.1.1) shows the covariance between $\hat{q}_i$ and $\hat{y}_{xj}$ when the selection unit individuals are obtained by n successive generations of self-fertilization without selection beginning with random individuals from a noninbred base population--the population to be improved by selection and designated, therefore, as the S population. In contrast, S_2, for example, will designate individuals or a population of individuals produced by two successive generations of self-fertilization beginning with random individuals from an S population.

$$\sigma(\hat{q}_i,\hat{y}_{xj}) = (pt - rs)_{ij}K_{ij}[1+(1-2v_j)a_{xj}]u_{xj}/2 \qquad \text{(A8.1.1) 8.13}$$

where

$$K_{ij} = [2 - (1-2\Upsilon)^{n+1}/2^n]/(1+2\Upsilon)$$

Here Υ symbolizes the frequency of recombination between the i-th and j-th loci and other symbols are used as in previous sections and chapters. The two locus gamete frequencies symbolized by p, r, s, ant t are those of gametes produced by the parents of the S_0, the noninbred base population. Note that

(1) If individuals of the S_0 population are employed as the selection units of progeny test selection; n = 0.0, K=1.0, and 8.13 reduces , as it should, to 8.7.

(2) If the covariance between $\hat{q}_i$ and $\hat{y}_{xj}$ is zero when S_0 individuals are the selection units, it would also be zero if S_n individuals were employed as the selection units. This is because the (pt-rs) of 8.13 pertains to the S_0 generation of the selection population.

(3) The range of K is from (1+f) to 1.0. Given no recombination between the i-th and j-th loci, Υ = 0.0, its minimum. Then $1+2\Upsilon = 1-2\Upsilon = 1.0$ and $K = 2 - (1/2)^n = 1 + f$ because the inbreeding coefficient of individuals produced by n generations of self-fertilization (see section 2.5.1) is $1 - (1/2)^n$. Given independent assortment, $\Upsilon = 1/2$, its maximum. Then $1 + 2\Upsilon = 2$, $1 - 2\Upsilon = 0.0$ and K = 1.0.

We found (equation 8.12) that, in the absence of covariances between $\hat{q}_i$ and the $\hat{y}_x$ of loci other than the i-th, $E(\Delta q_i)$ would be proportional to (1+f). We can now check whether the same can be said when the possibility of such covariances is taken into account. From 5.23, 8.11 and 8.13 we obtain

$$E(\Delta q_i) = [A + B]k/2\sigma(\hat{X})$$

where

$$A = q_i(1-q_i)[1+(1-2v_i)a_{Xi}]u_{Xi}(1+f)$$

and

$$B = \Sigma_{j\neq i}(pt - rs)_{ij}K_{ij}[1+(1-2v_j)a_{Xj}]u_{Xj}$$

The question to be answered is whether A + B is proportional, or very close to proportional, to (1+f).

First, if the S_0 population were in linkage equilibrium, (pt - rs) would be zero for all $j\neq i$, hence B would be zero and A+B = A would be proportional to (1+f). This is particularly important because it is often (though not always) reasonable to assume a close approach to linkage equilibrium in S_0 populations (see sections 6.3.1 and 11.2.4).

Second, in the absence of linkage equilibrium, some of the relevant values of (pt - rs) would be positive, some negative with the result that B could often be small relative to A. Then whether B were or were not proportional to (1+f), the sum of (A+B) would be almost proportion to (1+f).

Third, if B were substantial but contributed largely by pairs of loci for which linkage is close (Υ small) so that the corresponding K's are near (1+f), then A+B would again be nearly proportional to (1+f).

Only in the unlikely event that B were large relative to A but due, for the most part, to loose linkages, would A+B, and therefore $E(\Delta q_i)$, increase more slowly than 1+f when inbreeding of the selection unit individuals is increased.

In summary, it appears that for practical purposes we can assume that in progeny test selection of both male and female parents

$$E(\Delta q_i) \approx \sigma(\hat{q}_i,\hat{X})_0(1+f)k/2\sigma(\hat{X})$$

where f = the coefficient of inbreeding of the selection unit individuals and $\sigma(\hat{q}_i,\hat{X})_0$ is the value of $\sigma(\hat{q}_i,\hat{X})$ when f = 0.0.

• 8.2. Family Systems

These are systems in which the selection unit is a family of some kind. In those to be considered the same family is also the criterion unit. Selection among families can obviously be combined with selection among individuals by first selecting the families that performed best and then selecting from within those families the individuals that performed best. That two stage combination of family and individual selection will be considered later in this chapter. The single stage system in which the individual is the selection unit but in contrast to mass selection is selected or rejected on the basis of both its own performance and the average performance of the family of which it is a member will be discussed in Chapter XII after optimum weighting of different measures in selection criteria has been considered.

The three family systems to be considered are those in which the selection units are half-sib families, two parent full-sib families, or S_n families. Despite its importance in species where asexual reproduction is normal or feasible, the clone will not be considered here because selection among clones on the basis of their own performance is covered by the mass selection response expectations presented in Chapter VI.

An S_n family is produced by self-fertilization of its single parent; this type of family is obviously restricted to the monoecious species. The full-sib members of S_n families are ordinarily more closely related to each other than the members of two parent full-sib families. When the term, full-sib family, is used without qualification it will refer to two parent full-sib families. The members of S_n families are necessarily inbred with coefficients of inbreeding no less than 0.5. In contrast, the members of two parent full-sib families are not inbred unless the parents of the family are related (see section 2.5) and will be understood to be noninbred unless otherwise stated.

The members of half-sib families have one parent (either the male or the female) in common. The plants grown from the kernels of an open-pollinated ear of corn constitute a half-sib family with the female parent (seed parent) in common. The term polycross is widely used by plant breeders (though not by corn breeders) to designate half-sib families with the female parent in common. In animal breeding the parent common to all members of a half-sib family is ordinarily the male. The structure of half-sib families may, but need not, involve sets of full-sibs. It will be specified herein in terms of the sex of the common parent, the number of parents of the other sex and the number of members in each of the full-sib families contained in the half-sib family. When reasonable to do so, the latter number will be assumed the same for all full-sib sets in the half-sib family.

The possible advantages of family systems arise in some comparisons from the use of families as criterion units and in others from their use as the selection units. When compared to mass selection, they flow from the family as the criterion unit and have received attention in section 8.1 (mass selection and progeny test selection differ in the use of a family as criterion unit in progeny test systems). Using families rather than individuals as the selection units has the disadvantage,except in the case of S_n families, of less genetic variation among families than among individuals. Compared to progeny test selection, the advantage, if any, of family systems has to arise from use of families as the selection units since in both cases the criterion units are families. In this comparison the advantage of using families rather than individuals as selection units lies in length of the selection cycle. In progeny test systems the cycle must include one generation for intermating selections and another one (or more if $f > 0.0$) to produce the test progenies of selection units . These two steps can be combined in half- and full-sib selection though not in the case of S_n families.

The relation between selection and criterion units in family selection can take any of the following forms: (1) identity, (2) partial identity, or (3) division of each family into a criterion unit portion and a selection unit portion. Family selection by itself is rarely used in animal breeding but if it were, data on all members of each family would probably be used as the basis for selection among the families but it is not likely that all members of selected families would be used as parents in continuation of the program. Similar degrees of identity between criterion and selection units are possible in work with perennial plants but not with annuals when the measurements on which selection is based are not available until after the period when flowers can be fertilized. In those cases subdivision of families into a portion used as the criterion unit and a second portion, held as seed, used as the selection unit is the common practice. Once selections have been identified on the basis of the comparative performance of the criterion unit portions of the families, the stored seed enables members of the selected families to be grown and used as parents (or gene source in the case of S_n families) for the genetic sample of the following cycle. Another procedure, sometimes used in work with maize, employs random mating among all families during the performance testing phase of the selection cycle (ordinarily by wind pollination in a field comparison of families isolated by distance from other populations). Then seeds produced by plants of selected families provide the link between successive cycles of the program. This procedure saves the extra generation otherwise required for intermating of different selected families but at the cost of using pollen parents from the total genetic sample instead of from selections only. It amounts to perfomance-based selection of female parents but random selection of male parents.

8.2.1. Half-Sib Families

8.2.1.1. IDENTITY OR PARTIAL IDENTITY BETWEEN SELECTION AND CRITERION UNITS

All members of a half-sib family have one parent that is the same for each of them; this parent is designated here as the "common" parent of the family. The other parent may be the same or different for any two members of the family. Thus the half-sib family may be composed of sets of full-sibs. The parents not the same for all members of the half-sib family are designated herein as "random" parents to reflect the assumption that the mates of each "common" parent are chosen randomly from the selections (or their descendants) of the preceeding cycle.

In practice the structure of families in the genetic sample of any cycle will ordinarily be variable. Let F_p= the number of "random" parents with offspring in the p-th half-sib family and Y_{pm} = the number of offspring of the m-th "random" parent of the p-th family. Then $\Sigma_m Y_{pm} = N_p$ is the number of individuals in the p-th family.

Because of variation in family structure despite intent to make it as uniform as possible and because variable numbers from the best performing families may be used as parents of the next genetic sample we will proceed by deriving the covariance of q_i in an individual family member with the family average of X, the selection criterion.

Consider an offspring of the m-th random parent of the p-th half-sib family. We note first that the average of X for that family (criterion unit) is $\hat{X}_p$ = the average of X-values for the individual in question, for $(Y_{pm}-1)$full-sibs and for $N_p - Y_{pm}$ half-sibs of that individual. It follows that for such individuals

$$\sigma(\hat{q}_i,\hat{X}) = [\,\sigma(q_i,X)_1 + (Y_{pm}-1)\,\sigma(q_i,X)_2 + (N_p-Y_{pm})\,\sigma(q_i,X)_3]/N_p \qquad 8.14$$

where $\sigma(q_i,X)_1$ is the covariance between values of q_i and X for the same individual, $\sigma(q_i,X)_2$ is the covariance between q_i of an individual and the value of X for a full-sib, and $\sigma(q_i,X)_3$ is the covariance between q_i of an individual and the value of X for a half-sib. The first of these covariances, derived in Chapter VI, is

$$\sigma(q_i,X)_1 = q_i(1-q_i)[1+(1-2q_i)a_{xi}]u_{xi}$$
$$+\Sigma_{j\neq i}(pt-rs)_{ij}[1+(1-2q_j)a_{xj}]u_{xj} \qquad (6.15)\ 8.15$$

The joint distribution required for derivation of $(q_i,X)_3$ is, with one exception, that shown in Table 8.1. The exception is that instead of the values of $\hat{q}_i$ for parents of the ten types of test progenies (half-sib families), the quantities needed are the expectations of q_i for members of those families. For each family that expectation is the average of the frequency of B_i in the constant parent and q_i, the expected frequency of B_i in random parents. Clearly, these are $(1+q)/2$, $(1/2+q)/2$, and $q/2$, respectively, for B_iB_i, B_ib_i, and b_ib_i constant parents When coded by subtraction of $q/2$, they become 1/2, 1/4, and 0, in each case one-half of the $\hat{q}_i$ value shown in Table 8.1. For that reason the expressions sought here for $\sigma(\hat{q}_i,\hat{y}_{xi})$ and $\sigma(\hat{q}_i,\hat{y}_{xj})$ are one-half of those given by 8.6 and 8.7 with $v_i = q_i$ in this case because the random parents as well as the constant parents belong to the S population. Consequently,

$$\sigma(q_i,X)_3 = q_i(1-q_i)[1+(1-2q_i)a_{xi}]u_{xi}/4$$
$$+\Sigma_{j\neq i}(pt-rs)_{ij}[1+(1-2q_j)a_{xj}]u_{xj}/4 \qquad 8.16$$

It remains to discover the composition of $\sigma(q_i,X)_2$. To follow the standard procedure of this text would require the joint distributions with respect to two genes, over the 100 possible types of full-sib families, of the frequency of B_i for individuals and the expectations of y_{xi} and y_{xj} for family members. To write these out would be tedious to say the least. Fortunately, there is another way. Let the effect of the two-locus genotype of an individual be expressed as

$$y_{ij} = y_{xi} + y_{xj} = m_{ij} + f_{ij} + w_{ij}$$

in which

m_{ij} = the expected effect of the male parent of the family, f_{ij} = the expected effect of the female parent of the family, and $w_{ij} = (y_{xi} + y_{xj}) - (m_{ij}+f_{ij})$, the genetic segregation effect on the genotype value of an individual. It is apparent (1) that $m_{ij} + f_{ij}$ is the same (constant) for all members of the same full-sib family, (2) that w_{ij} varies randomly among individuals and with respect to both m_{ij} and f_{ij}, (3) that because parents are random members of the S population, there is no covariance between m_{ij} and f_{ij}, and (4) that, ignoring the instances in dioecious organisms in which one or both of the two genes are located in the X-chromosome, the distributions of m_{ij} and f_{ij} must be the same because by design of the selection system each of the parents would be a random member from the S-population.

It follows that

$$\sigma(q_i,m_{ij}) = \sigma(q_i,f_{ij}) = \sigma(q_i,X)_3$$

and that the covariance between the frequency of B_i in any individual and the two-locus genotype effect of any full-sib of that individual is

$$\sigma(q_i,X)_2 = \sigma(q_i,m_{ij}+f_{ij}) = \sigma(q_i,m_{ij}) + \sigma(q_i,f_{ij}) = 2\sigma(q_i,X)_3 \qquad 8.17$$

Substituting in 8.14 in terms of 8.15, 8.16 and 8.17 yields for the covariance between the frequency of B_i in any individual and $\hat{X}$, the average of X for the half-sib family of which it is a member,

$$\sigma(\hat{q}_i,\hat{X}) = (A/4)[4 + 2(V_{pm}-1) + (N_p - V_{pm})]/N_p$$

$$= [1 + (V_{pm}/N_p) + (2/N_p)]A/4 \qquad 8.18$$

where

$$A = q_i(1-q_i)\ [[1+(1-2q_i)a_{xi}]u_{xi} + \Sigma_{j\neq i}(pt-rs)_{ij}[1+(1-2q_j)a_{xj}]u_{xj}$$

Substituting 8.18 in 5.18 yields

$$E(\Delta q_i)_{pm} = [1+ (V_{pm}/N_p) + (2/N_p)](A/4)k/\sigma(\hat{X})_p \qquad 8.19$$

= the expectation of (Δq_i) in the case of any member of a selected family.

Note that, given variation in the structure of the half-sib families (i.e., variation of N_p and the number and sizes of the full-sib families making up the half-sib families), this quantity would be potentially variable except among the members of full-sib families of the same size within a particular half-sib family. This is true because both $\sigma(\hat{X})_p$ and the quantity contained between the square brackets of 8.19 are functions of V_{pm} and N_p. Thus to use 8.19 to obtain the average of $E(\Delta q)$ for individuals employed as parents of the succeeding generation of the S-population would require computation, and then taking the average, of the $E(\Delta q_i)$ for each of those individuals. And in the case of dioecious species such an average would obviously be required for each sex because $E(\Delta q)$ for the cycle is then the unweighted mean of the values of $E(\Delta q)$ for the two sexes. This would be a rather laborious task without an electronic computer and a program written for the purpose. Even then it might not be worthwhile for any practical purpose. Note in this connection that on the assumption that all families would be the same in size and size and structure so that all V_{pm}, F_p, and N_p would be equal to, say, V, F, and N, respectively, 8.18 reduces to

$$\sigma(\hat{q}_i,\hat{X}) = [1 + 1/F + 2/FV](A/4)$$

and 8.19 to

$$E(\Delta q_i) = [1 + 1/F + 2/FV](A/4)k/\sigma(\hat{X}) \qquad 8.20$$

These would obviously be easier to use in exploring the relative efficiency of designs that differ in the numbers and structures of families to be employed or when making a priori comparisons of half-sib family selection with one or another other system. The amount of error involved because of inevitable variation in family structure is not obvious.

Note, finally, that in the limiting case where family size is 1.0 (F = V = 1.0), the selection units are individuals and, with the criterion unit being the same individual , family selection becomes mass selection. It would appear, therefore, that 8.20 should reduce , when F=V=1.0, to 6.16 and this is exactly what happens.

- 8.2.1.2. SELECTION UNITS AND CRITERION UNITS COMPOSED OF DIFFERENT INDIVIDUALS

Attention here will be limited to the case in which the random parents of the criterion unit and selection unit individuals are different so that the common parent is the only source of genetic covariance between individuals of the two sets. Let

q_i be the frequency of B_i in any individual of the selection unit portion of the p-th half-sib family of the genetic population,

$\hat{X}_p$ be the average of X for the criterion unit portion of that half-sib family and

N_p be the number of individuals in that criterion unit portion.

In this case each individual of the selection unit portion of the family is a half-sib of each member of the criterion unit portion; no full-sib relationships would exist. Thus by the logic that led to 8.14 we have in this case that

$$\sigma(q_i,\hat{X}_p) = N_p\sigma(q_i,X)_3/N_p = \sigma(q_i,X)_3$$

which by virtue of 8.16 becomes

$$\sigma(q_i,\hat{X}_p) = q_i(1-q_i)[1+(1-2q_i)a_{xi}]u_{xi}/4$$

$$+ \Sigma_{j\neq i}(pt-rs)_{ij}[1+(1-2q_j)a_{xj}]u_{xj}/4 \qquad 8.21$$

Finally, by 5.18

$$E(\Delta q_i)_p = (\hat{X}_p - \bar{X})\,\sigma(q_i,\hat{X}_p)/\sigma^2(\hat{X}_p) \qquad 8.22$$

in which $\bar{X}$ symbolizes the average of $\hat{X}_p$ over all criterion units of the genetic sample, $E(\Delta q_i)_p$ = $E(\Delta q_i)$ for each individual of the p-th half-sib family, and $\sigma^2(\hat{X}_p)$ = variance of the mean of X for the p-th family. Note that $\sigma^2(\hat{X}_p)$ will be variable among families when size of criterion units(N_p) is variable.

When N_p or the number of individuals to be used as parents varies from family to family a weighted average of 8.22 is required to obtain $E(\Delta q_i)$ for all selections. Weights to be employed would be the number of individuals from the family to be used as parents in regeneration of the S population. When size and structure of the criterion units is not variable all $\sigma^2(\hat{X}_p)$ should be of the same magnitude and so symbolized as $\sigma^2(\hat{X})$. Then if the number of individuals used as parents of the succeeding generation is the same for all families weighting is not required and the expected value of Δq_i for all selections is

$$E(\Delta q_i) = (\bar{X}_s - \bar{X})\,\sigma(q_i,\hat{X})/\sigma^2(\hat{X}) = [k/\sigma(\hat{X})]\,\sigma(q_i,\hat{X})$$

$$=[k/\sigma(\hat{X})]\{q_i(1-q_i)[1+(1-2q_i)a_{xi}]u_{xi}$$

$$+ \Sigma_{j\neq i}(pt-rs)_{ij}[1+(1-2q_j)a_{xj}]u_{xj}\}/4 \qquad 8.23$$

• 8.2.2. Full-Sib Families

Because a full-sib family is the special case of a half-sib family in which F = 1 and V=N the expressions we seek can be obtained as special cases of expressions obtained in section 8.2.1.

• 8.2.2.1. IDENTITY OR PARTIAL IDENTITY BETWEEN SELECTION AND CRITERION UNITS

The covariance between q_i of an individual and the average of X for its family is apparent by reference to the logic by which 8.14 was obtained. Let $N_p = V_p$ be the number of individuals in the p-th family. Then because the individual has $V_p - 1$ full-sibs in its family the covariance in question is

$$\sigma(q_i,\hat{X}_p) = [\sigma(q_i,X)_1 + (V_p - 1)\sigma(q_i,X)_2]/V_p$$

Because $\sigma(q_i,X)_2 = \sigma(q_i,X)_1/2$ (from 8.15, 8.16, and 8.17) this can be written

$$\sigma(q_i,\hat{X}_p) = [\sigma(q_i,X)_1/2][1 + 1/V_p] \quad 8.24$$

and for the selections from the p-th family

$$E(\Delta q_i)_p = [\sigma(q_i,X)_1/2][1 + 1/V_p](\hat{X}_p-\bar{X})/\sigma^2(\hat{X}_p) \quad 8.25$$

When all families are the same size (all $V_p = V$) 8.24 becomes

$$\sigma(q_i,\hat{X}) = [\sigma(q_i,X)_1/2][1+1/V]$$

in which, from 8.15

$$\sigma(q_i,X)_1 = q_i(1-q_i)[[1+(1-2q_i)a_{xi}]u_{xi} + \Sigma_{j\neq i}(pt-rs)_{ij}[1+(1-2q_{ij})a_{xj}]u_{xj}$$

• 8.2.2.2. SELECTION UNITS AND CRITERION UNITS COMPOSED OF DIFFERENT INDIVIDUALS

In this case let V_p be the number of individuals in the criterion unit portion of the family. Then each individual in the selection unit portion has V_p full-sibs in the criterion unit and hence

$$\sigma(q_i,\hat{X}_p) = V_p\sigma(q_i,X)_2/V_p = \sigma(q_i,X)_1//2$$

and for the selections from the p-th familly

$$E(\Delta q_i)_p = [\sigma(q_i,X)_1/2](\hat{X}_p-\bar{X})/\sigma^2(\hat{X}_p) \quad 8.26$$

When V_p is the same for all criterion units and the number of individuals used as parents is the same for each family the expectation of Δq_i for the cycle is

$$E(\Delta q_i) = [\sigma(q_i,X)_1/2](\bar{X}_s - \bar{X})/\sigma^2(\hat{X}) = [\sigma(q_i,X)_1/2]k/\sigma(\hat{X})$$

Otherwise, the best value for $E(\Delta q_i)$ is given by a weighted average of 8.26 (or 8.25 when the selection and criterion units are composed of the same individuals) with weighting as outlined in section 8.2.1.

• 8.2.3. **S_n Families**

When S_n family selection is employed in work with annual plants the selection and criterion units are necessarily composed of different individuals unless selection is to be based on a measure of phenotye that is available before flowering time. On the other hand, in work with perennials (e.g., blueberries or birch trees) it would probably be both practical and desirable to use criterion unit plants as parents of the next generation of the S-population. Nevertheless, only the case where the selection and criterion units are composed of different individuals will be treated here.

Each S_n family is produced by n successive generations of self-fertilization and has a single parent. In the case of S_1 families the parents are noninbred members of the S population; when n>1 the parents have been produced by n-1 generations of self-fertilization and hence are themselves inbred. Whatever the value of n, the inbreeding coefficient of the parents is $1-(1/2)^{n-1}$ as noted in section 2.5.1.

Families are appropriately classified in terms of genotype of parent and the expectation of q_i for any family is the value of q_i in the parent. The probabilities of single locus genotypes in family members and the consequent family expectations of $\hat{y}_x$ (at the i-th or j-th locus) are as follows.

Genotype of parent	Genotype probabilities of offspring BB	Bb	bb	$E(\hat{y}_x)$
B_iB_i	1	0	0	$2u_{xi}$
B_ib_i	1/4	1/2	1/4	$u_{xi} + a_{xi}u_{xi}/2$
b_ib_i	0	0	1	0

Reflecting these facts, the joint distribution required for derivation of $\sigma(\hat{q}_i, \hat{X})$ is shown below.

Parent genotype	Frequency	$E(\hat{q}_i)$	$*(\hat{y}_{xi})$
B_iB_i	$q^2 + hq_i(1-q_i)$	1.0	u_{xi}
B_ib_i	$2q_i(1-q_i)(1-h)$	1/2	$a_{xi}u_{xi}/2$
b_ib_i	$(1-q_i)^2 + hq_i(1-q_i)$	0	$-u_{xi}$

Here $*(\hat{y}_{xi}) = E(\hat{y}_x) - u_{xi}$, $h = 1-(1/2)^{n-1}$, and all other symbols have previously ascribed meanings. The required covariance, obtained in the usual manner, is

$$\sigma(\hat{q}_i, \hat{y}_{xi}) = q_i(1-q_i)[(1+\hat{f}) + (1-\hat{f})(1-2q_i)a_{xi}/2]u_{xi}$$

The frequencies of two locus genotypes in noninbred populations are comparatively simple functions of the frequencies of two locus gamete frequencies as shown in section 6.2. They are less obvious in generations produced by self-fertilization, being functions of the probability of recombination between the two genes in addition to frequencies of two locus genotypes in gametes produced by individuals of the S_0 (non inbred generation). The joint distributions of $\hat{q}_i$ and $*(\hat{y}_{xj})$ in the case of S_n families are shown in Appendix 8.1 and the derivation of $\sigma(\hat{q}_i, \hat{y}_{xj})$ is shown in section A8.1.2 of that Appendix. The procedure of that derivation is standard but the algebra involved requires care and a bit of determination. The result is

$$\sigma(\hat{q}_i, \hat{y}_{xj}) = (pt-rs)_{ij}[(1+\pi) + (1-2q_j)(1-\hat{f})a_{xj}/2] \qquad \text{A8.1.2}$$

Substituting in 5.21 yields

$$\sigma(\hat{q}_i, \hat{X}) = q(1-q)[(1+\hat{f}) + (1-\hat{f})(1-2q)a_{xi}/2]u$$
$$+\Sigma_{j\neq i}(pt-rs)_{ij}[(1+\pi) + (1-2q_j)(1-\hat{f})a_{xj}/2]$$

Here $\pi = [(1-2\gamma)/(1+2\gamma)]1 - \{(1-2\gamma)/2\}^n]$ in which γ symbolizes the probability of recombination between the i-th and j-th genes.

• 8.2.4. Selection among Individuals within Families

Intrafamily selection among individuals on the basis of their own performance is often possible. It can be the only selection practiced as in work by Falconer (1953), it can be employed as a supplement to family selection in a two-stage procedure as proposed by Lonnquist (1964), or it can be combined in a single stage procedure with selection based on family performance as discussed by Lush (1947). The first two of these possibilities, in the case of half- and full-sib families, are treated in this section. The third is discussed in Chapter XII.

8.2.4.1. INTRAFAMILY SELECTION IN THE ABSENCE OF SELECTION AMONG FAMILIES

The key to response expectations in this case is obviously the intrafamily covariance between $\hat{q}_i$ and $\hat{X}$. These are readily obtained from previously derived expressions making use of the fact that the total (among individuals) covariance between q_i and X is the sum of portions between and within families. Thus, using symbols employed in previous sections,

$$\sigma(q_i,X)_h = \sigma(q_i,X)_1 - \sigma(q_i,X)_3 \quad \text{in the case of half-sib families}$$

and

$$\sigma(q_i,X)_f = \sigma(q_i,X)_1 - \sigma(q_i,X)_2 \quad \text{in the case of full-sib families}$$

where $\sigma(q_i,X)_h$ and $\sigma(q_i,X)_f$ symbolize the intrafamily covariance between q_i and X of half- and full-sib families, respectively. It follows from 8.15, 8,16, and 8.17 that

$$\sigma(q_i,X)_h = \sigma(q_i,X)_1(3/4)$$

and

$$\sigma(q_i,X)_f = \sigma(q_i,X)_1(1/2)$$

Remember that $\sigma(q_i,X)_1$ is given by 8.15.

These covariances are averages over families that may be genotypically different because the "constant" parents in the case of half-sib families or the pairs of parents in the case of full-sib families may be genotypically different. In fact, being functions of parental genotype(s), these covariances are variable among families. Nevertheless, when the proportions selected from different families are independent of average family performance as implied by the words "intrafamily selection in the absence of selection among families", response expectations will be proportional to the average, over all families, of the covariance between $\hat{q}_i$ and $\hat{X}$.

8.2.4.2. INTRAFAMILY SELECTION AS A SUPPLEMENT TO SELECTION AMONG FAMILIES

The covariances obtained above assuming no prior selection among families are not exactly correct for selection within selected families [though they have sometimes been treated as sufficient approximations; e.g., by Webel and Lonnquist (1967) and Compton and Comstock (1976)]. The reasons are intuitively obvious. For illustration, consider the linkage equilibrium situation and let families be classified according to parental i-th locus genotype(s). It is obvious, as implied above, that the covariance of q_i and y_{xi} varies among classes of families. For illustration, if B is completely dominant ($a_i = 1.0$) to b, there can be no variation in y_{xi} within the class of half-sib families for which the genotype of the "common" parent is B_iB_i. And, with any level of dominance, there can be no variation of y_{xi} within any full-sib family of which each parent is homozygous (either BB or bb). Lacking variation in y_{xi}, the covariance of q_i and y_{xi} must be zero in such families though nonzero in others. It is also obvious that if there is any response to family selection it must result from difference between relative frequencies of family classes among selections and those in the genetic sample from which the selections were obtained. Taken together, these two facts insure that $\sigma(q_i,y_{xi})$ within selected families. cannot in general be exactly the same as within the unselected genetic sample of families.

On the other hand, situations in which the effects of prior selection among families on the averages of intrafamily covariances are nontrivial are probably rare. It was shown in sections 6.3.3 and 7.2 that, in the case of mass selection, the expected differences in single locus genotype frequencies between selections and the genetic sample are small unless the effect (u) of the allele substitution is large relative to the standard deviation of X. It is apparent that this generalization could be extended to relative frequencies of family classes (as defined above) and hence that, on the average, the difference in $\sigma(q_i, y_{xi})$, or $\sigma(q_i, y_{xj})$, between the genetic sample of families and selected families is likey to be small, as tacitly assumed by authors cited above.

8.3. Summary and Comments

The focus of this chapter has been on the covariance between $\hat{q}_i$ and $\hat{X}$ and how it varies among selection systems. Results obtained are summarized in Table 8.2. the significance of this parameter in relation to responses to selection is quantified by the following expressions from Chapter V.

$$E(\Delta q_i) = \sigma(\hat{q}_i, \hat{X})k/\sigma(\hat{X}) \qquad (5.18)\ 8.27$$

$$E(\Delta\ddot{Y}_c) = [k/\sigma(\hat{X})]\Sigma_i\sigma(\hat{q}_i,\hat{X})2[1+(1-2q_i)a_{ci}]u_{ci} \qquad (5.25)\ 8.28$$

when the target population is the S population.

and

$$E(\Delta\ddot{Y}_c) = [k/\sigma(\hat{X})]\Sigma_i\sigma(\hat{q}_i,\hat{X})[1+(1-2p_i)a_{ci}]u_{ci} \qquad (5.26)\ 8.29$$

when the T population is the cross between the S population and another.

Aspects of the results collected in Table 8.2 that merit particular attention are the following.

(1) Among selection systems (except the $S_n$family system) in which both the selection and criterion units belong to the S-population, linkage disequilibrium has no effect on the relative magnitudes of $\sigma(\hat{q}_i,\hat{X})$. The quantity A (which is affected by linkage. disequilibrium) is the same for each of these systems.

(2) Note that in the case of mass selection $\sigma(\hat{q}_i,\hat{X}) = A$ and hence that $\Sigma_i\sigma(\hat{q}_i,\hat{X})2[1+(1-2q_i)a_{ci}]u_{ci}$ = the quantity defined by 6.19 as the additive genetic variance of X when c = x, and by 6.20 as the additive genetic covariance between traits C and X when the c ≠ x.

Then note further, that because $\sigma(\hat{q}_i,\hat{X})$ equals the product of A and a known constant for all selection systems in which both the selection and criterion units are non-inbred S population individuals, the expectations of direct and correlated response to selection ($\Delta\ddot{Y}_X$ and $\Delta\ddot{Y}_C$ given by 8.28) for all those systems are proportional to additive genetic variance and additive genetic covariance, respectively. As a consequence, the relative effectiveness of any two such systems can be explored without having information about the magnitude of the additive genetic variance (or covariance).

(3) The composition of A' draws attention to the fact that in progeny test selection, values of $\sigma(\hat{q}_i,\hat{X})$ and therefore of $E(q_i)$ are functions of allele frequencies in the "tester". Since the latter are variable in a partially predictable way, attention of the breeder is

drawn to the fact that choice of "tester" may be significant in the design of progeny test programs. Note that in the case of any gene for which $0.0 < a_x \leq 1.0$, $E(\Delta q)$ increases as v decreases. Thus, barring a significant amount of overdominance or dominance of unfavorable alleles, low performance testers (in which the average of v over loci would be smaller) should be more effective testers. This was discussed and investigated by Rawlings and Thompson (1962).[4] Choice of tester when $a_x > 1.0$ at numerous loci is a central feature of breeding systems described and discussed in the next chapter.

Table 8.2. The covariance between $\hat{q}_i$ and $\hat{X}$ (assuming no nongenetic component)

Selection system	$\sigma(\hat{q}_i,\hat{X})$	c
Mass or selection among clonal families	cA	1
Among individuals within half sib families	cA	3/4
Among individuals within full sib families	cA	1/2
Half-sib family (type 1)[a]	cA	1/4
Half-sib family (type 2)	$[1+(1/F)+(2/FY)]cA$	1/4
Full-sib family (type 1)	cA	1/2
Full-sib family (type 2)	$[1+(1/Y)]cA$	1/2
Progeny test (tester = S population)	cA	(1+f)/2[b]
Progeny test (general)	cA'	(1+f)/2
S_n family (type 1)	cA"	1

$A = q_i(1-q_i)[1+(1-2q_i)a_{xi}]u_{xi} + \Sigma_{j\neq i}(pt-rs)_{ij}[1+(1-2q_j)a_{xj}]u_{xj}$

$A' = q_i(1-q_i)[1+(1-2v_i)a_{xi}]u_{xi} + \Sigma_{j\neq i}(pt-rs)_{ij}[1+(1-2v_j)a_{xj}]u_{xj}$

$A'' = q_i(1-q_i)[(1+h)+(1-h)(1-2q_i)a_{xi}/2]u_{xi}$
$+\Sigma_{j\neq i}(pt-rs)_{ij}[(1+\pi)+(1-h)(1-2q_j)a_{xj}/2]u_{xj}$

f = coefficient of inbreeding of selection unit individuals

h = coefficient of inbreeding of the parents of the S_n families

F = number of "random" parents per half-sib family

Y = number of offspring per "random" parent (or full-sib family)

$\pi = [(1-2r)/(1+2r)][1-\{(1-2r)/2\}^n]$

r = the probability of recombination between the i-th and j-th loci

[a] Type 1 specifies that the selection and the criterion units are composed of different random individuals from the family. Type 2 specifies that the same individuals compose both the selection and the criterion unit.

[b] In the expressions for progeny test selection the coefficient, 1+f, is correct when f=0 and when individuals tested are derived by one or more generations of self-fertilization from an S_0 population in linkage equilibrium. It is a close approximation (see section 8.1.3) when such individuals are derived from an S_0 population not in linkage equilibrium. It is not intended for application when tested individuals are inbred by modes other than self-fertilization.

(4) Additive genetic variance and covariance as defined by 6.19 and 6.20 are functions of allele frequencies in a single population. Both A' and 8.29 demonstrate that in systems involving (a) progeny test selection in which the tester is not the S population and/or (b) a target population other than the S population, the genetic parameter to which $E(\Delta\ddot{Y}_X)$ or $E(\Delta\ddot{Y}_C)$ are proportional have the same algebraic forms as additive genetic variance or covariance but are functions of allele frequencies in more than one population. All such parameters are referred to as analogs of additive genetic variance or covariance in this text.

(5) The difference between A and A' indicates that among systems in which criterion units belong to the S-population, the relative effectiveness of the S_n family system should become greater as allele frequencies increase when partial or complete dominance of the better allele is the rule; i.e., when $0.0 < a_X \leq 1.0$ at the majority of loci.

(6) In all systems sufficient negative values of (pt - rs) can cause $\sigma(\hat{q}_i\hat{X})$ to be negative for a portion, though not all, of the genes that affect X (the selection criterion). Thus all questions concerning linkage disequilibrium that were listed in Section 6.3.1 are reemphasized.

(7) It should be noted that the procedures (types 1 and 2) assumed in the case of family selection are extremes with reference to amount of identity between individuals making up the selection and criterion units. Other procedures are practiced that are intermediates between those for which results are given in the table. For example, when half-sib family selection is employed in corn breeding the selection and criterion units, though distinct with respect to individuals, might both be composed of equal numbers of offspring from each of a common set of "random" parents. In that case, if we assume F random parents in all families and N individuals in the criterion unit of each family, each individual of the selection unit would have N/F full-sibs and N - N/F half-sibs in the criterion unit and $\sigma(\hat{q}_i,\hat{X})$ would be (1 + 1/F)A/4.

(8) Procedure for using the content of table 8.2 in comparisons of selection procedures is illustrated in a preliminary way in section 8.1.2 and, more extensively, in Chapter XIV.

9 Selection for Performance in Crossbreds

Hybrid vigor is exhibited by a portion of the traits that are components of total value in most, if not all, of the agricultural organisms. As a consequence, crossbreeding, the mating of individuals from different populations, is extensively used in various species (e.g., maize, sorghum, swine, sheep, and poultry) to supply the plants or animals of commercial production. The required control of matings poses no special difficulty in the case of livestock but whether crossbreeding is employed in a plant species is determined primarily by whether the necessary inter-population fertilizations are economically feasible.

Crossing systems employed are variable among species. Consider maize (corn) and swine for illustration. In corn where reproduction rates are high (many seeds per plant), production costs per individual are low and the performance of crossbred offspring is not much affected by the vigor of seed parents, the commercial material now used in the U.S. and other developed countries is almost exclusively the first (single) cross of a pair of highly inbred, and hence nearly homozygous, lines. In swine on the other hand, reproduction rates are much lower, individuals more costly, and maternal vigor more important. As a result, favored crossing systems employ crossbred females as the mothers of commercial animals, three or more parent populations, and parent populations in which inbreeding is minimal. Such differences have important implications relative to systems for genetic improvement. To illustrate and emphasize, it will suffice to note (a) that in corn the genetic value of a hybrid depends on allele frequencies in only two parent populations as compared to three or more in the case of the products of most crossing systems used in swine production, and (b) that the greater significance of parental vigor and maternal effects means greater complexity of the phenotypic goal in swine breeding.

Detailed attention in this chapter will be limited to selection systems in which the goal is genetic excellence in a single cross (a) between highly inbred lines or (b) between random mating populations in which there is no more inbreeding than the amount dictated by effective population size. Formulas presented are premised on linkage equilibrium (see sections 6.3.1 and 11.2.4 for rationale). They were obtained by assuming (pt-rs) = 0.0 for all pairs of loci in expressions for $\sigma(\hat{q}_i,\hat{X})$ taken, for the most part, from Table 8.2. Thus the more general formulas that take account of linkage disequilibrium are readily obtained in any case by allowing (pt-rs) ≠ 0.0 in the relevant formulas for $\sigma(\hat{q}_i,\hat{X})$.

• 9.1. Relevance of Overdominance

If there were no overdominance, ideal genotypes and hence maximum phenotypic value would be obtainable in a single random mating population. The requirement would be frequency equal to 1.0 for the best allele of every gene having any effect on total value. In contrast, given overdominance in the expression of any portion of the relevant genes, ideal genotypes are heterozygous at each of the overdominance loci and, without asexual reproduction, a population homogeneous for ideal genotypes can only be produced by crossing appropriately complimentary pure lines. This is illustrated by the fact that the mating required to make the probability of B_iB_j in offspring equal to 1.0 is $B_iB_i \times B_jB_j$, the mating required to make $P(B_iB_jC_iC_j) = 1.0$ is $B_iB_iC_iC_i \times B_jB_jC_jC_j$ or $B_iB_iC_jC_j \times B_jB_jC_iC_i$, and so on.

Hull (1945) argued that much of the genetic variation of grain yield in corn is due to segregation of overdominance genes and proposed a system of recurrent selection designed to promote the allele frequency changes in the selection population that would maximize heterozygosity at overdominance loci in the genotypes obtained from crossing the selection population with a specific pure line. Later (Hull, 1952), he called the system <u>recurrent selection for specific combinability</u>. It is now more generally known as <u>recurrent selection for specific combining ability</u> (RSSC). Comstock <u>et al</u> (1949) noted, as Hull himself had, that RSSC would not be a superior procedure with respect to non-overdominant genes of which the best allele is less than completely dominant and described an alternative procedure involving two selection populations that is designed to promote the allele frequency changes at both overdominant and non-overdominant loci that are most useful with respect to the population generated by crossing the two selection populations. It is appropriate to note that this procedure, <u>reciprocal recurrent selection</u> (RRS) is a modification of the one described by Hull.

The substantial increases in productivity of corn that have been achieved via the development and use of hybrids, documented by Russell (1974), Duvick (1977,1984), and Castleberry <u>et al</u> (1984), suggest that if overdominance is an important phenomenon the methods (see section 9.6) that have been used to produce corn hybrids must deal with it reasonably well. However, the improvement to date provides no indication of the improvement still possible by synthesis of ideal genotypes. The issue was considered in section 3.6.1 where it is shown that special procedures for increasing the probability of ideal genotypes would not be required (in species where heterozygous genotypes are reproducible via single crosses of pure lines or asexual reproduction) if the number of overdominance genes were small but would be if that number were in excess of 13 or 14.

In both RSSC and RRS the target population is a single cross, in RSSC that between the S population and the specified pure line, in RRS that between the two S populations. However, it is important to recognize that in practice the performance of parent populations is relevant and that, so far as overdominant genes are concerned, improvement of the single cross, by whatever method, comes at the expense of decreases in genotype values of the parent populations. With overdominance the superior heterozygote is obtained in the single cross when the parents are homozygous and therefore inferior. Given relatively high reproduction rates in homozygous material, as in corn, the associated practical problem would not as great as in species with low reproduction rates in which it could be critical.

To summarize, the prime argument for either RRS or RSSC is that when overdominance is present the ideal genotype, obtainable in the single cross of appropriately complementary homozygous lines, has greater genotypic value than the highest average possible in a random mating population. Unfortunately, this implies that the genotypic values of the homozygous lines required (unless asexual reproduction is possible) as parents of the ideal single cross genotype will be less than what is possible in a random mating population. Conceivably, if the number of parent line individuals required were too great, the average genotype value for the entire system involved when the cultivar is a single cross hybrid could be less than if a highly selected random mating population were the cultivar.

• 9.2. Selection for Specific Combining Ability (RSSC)

RSSC is a progeny test selection system of the class for which the covariance of $\hat{q}_i$ with $\hat{X}$ is given in Table 8.2. The tester employed is the specific pure line (or other population[1]) with which combining ability of the S population is to be improved by the program. If the tester is identified as A, the target population is the S x A cross. As in Chapter VIII, let v_i = the frequency in the tester of the best allele (B_i) of the i-th gene. Then from 5.12

$$\partial\ddot{Y}_c/\partial q_i = [1 + (1-2v_i)a_{ci}]u_{ci} \qquad 9.1$$

and from Table 8.2

$$\sigma(\hat{q}_i,\hat{X}) = (1 + f)q_i(1-q_i)[1 + (1-2v_i)a_{xi}]u_{xi}/2 \qquad 9.2$$

Using 5.24 and 9.2 we obtain

$$E(\Delta q_i) \approx k(1 + f)q_i(1-q_i)[1 + (1-2v_i)a_{xi}]u_{xi}/2\sigma(\hat{X}) \qquad 9.3$$

and from 5.8, 9.1, and 9.3

$$E(\Delta\ddot{Y}_c{:}SA) \approx k(1 + f)\Sigma_i\, q_i(1-q_i)[1+(1-2v_i)a_{xi}][1+(1-2v_i)a_{ci}]u_{xi}u_{ci}/2\sigma(\hat{X}) \qquad 9.4$$

Here, despite the fact that the S x A cross has been recognized as the target population, the fact that 9.4 pertains to response in S x A is specifically indicated. This is because there is reason to be interested in the effect of selection on average genotypic value of both the S x A and the S populations. Using 5.8, 5.11 and 9.3, we obtain expected response in the S population itself.

$$E(\Delta\ddot{Y}_C: S) \approx k(1+f)\Sigma_i q_i(1-q_i)[1+(1-2v_i)a_{xi}][1+(1-2q_i)a_{ci}]u_{xi}u_{ci}/\sigma(\hat{X}) \quad 9.5$$

If the response and selection criteria are the same so that C = X, 9.4 becomes

$$E(\ddot{Y}_X: SA) \approx k(1+f)\Sigma_i q_i(1-q_i)[1+(1-2v_i)a_{xi}]^2 u_{xi}^2/\sigma(\hat{X}) \quad 9.6$$

and 9.5 becomes

$$E(\ddot{Y}_X: S) \approx k(1+f)\Sigma_i q_i(1-q_i)[1+(1-2v_i)a_{xi}][1+(1-2q_i)a_{xi}]\, u_{xi}^2/\sigma(\hat{X}) \quad 9.7$$

Except for differences in symbols employed, 9.3 and 9.6 are in accord with expressions given by Comstock et al (1949). When the tester (A) is a pure line, v must be 0.0 or 1.0 and, accordingly, $[1+(1-2v)a_X]$ must be $(1+a_X)$ or $(1-a_X)$. Consequences with respect to $E(\Delta q)$, from 9.3, are summarized below for loci that are segegating in the S Population.

Tester genotype	v	E(q) when $a_X < 1.0$	$a_X = 1.0$	$a_X > 1.0$
BB	1	> 0.0	0.0	< 0.0
bb	0	>0.0	> 0.0	> 0.0

Aside from exceptions due to random drift, the eventual consequence in the S population should be homozygosity of the B allele when $E(\Delta q) > 0.0$ but homozygosity of the b allele when $E(\Delta q) < 0.0$. Consequences with respect to genotypes in the S x A cross and the S population itself are summarized in Table 9.1 in which the consequencies of three selection systems are compared. These are discussed in section 9.5.

• 9.3. Reciprocal Recurrent Selection (RRS)

RRS, like RSSC, is a progeny test selection system. Selection is practiced in two populations that will be designated here as the S_1 and S_2 populations. In the procedure originally outlined (Comstock et al, 1949) these are used as the testers for each other; S_1 as the tester for selection in S_2, S_2 as the tester for selection in S_1. Considered separately, each of these parts is an RSSC program in which the $S_1 \times S_2$ cross is the target population. Thus expected changes in allele frequencies and genotypic means follow directly from those obtained for RSSC. Let q_{1i} and q_{2i} be the frequencies in S_1 and S_2, respectively, of the better allele (B_i) of the i-th gene. Then to obtain $E(\Delta q_{1i})$, we substitute q_{1i} for q_i and q_{2i} for v_i in 9.3 to obtain

$$E(\Delta q_{1i}) \approx k_1(1+f_1)q_{1i}(1-q_{1i})[1 + (1-2q_{2i})a_{xi}]u_{xi}/2\sigma(\hat{X}_1) \quad 9.8$$

Here k_1, f_1, and $\sigma(\hat{X}_1)$ symbolize the values of k, f, and $\sigma(\hat{X})$ that are relevant to the selection practiced in the S_1 population. Substituting q_{2i} for q_i and q_{1i} for v_i in 9.3 we obtain

$$E(\Delta q_{2i}) \approx k_2(1+f_2)q_{2i}(1-q_{2i})[1 + (1-2q_{1i})a_{xi}]u_{xi}/2\sigma(\hat{X}_2) \quad 9.9$$

Note that 9.8 and 9.9 are parallel expressions; 9.9 becomes 9.8 if 1 is substituted for 2 and 2 for 1 wherever either of these numerals appears as a subscript in 9.9. It is intuitively obvious that this correspondence should exist because (1) the process of selection is the same in the two populations and (2) the roles of the two populations are interchanged in the two portions of the total RRS program.

It is apparent that the frequencies of genotypes in the $S_1 \times S_2$ cross and therefore the average value of its genotypes will be affected by allele frequency changes in both S_1 and S_2. It is appropriate, therefore, to express the change in average genotype value as the sum of two parts.

$$E(\Delta\ddot{Y}_c: S_1S_2) = E(\Delta\ddot{Y}_c: S_1S_2)_1 + E(\Delta\ddot{Y}_c: S_1S_2)_2 \qquad 9.10$$

The portion of this expected genetic improvement that arises from selection in the S_1 population is obtained by substituting q_{1i} for q_i and q_{2i} for v_i in 9.4. The result is

$$E(\Delta\ddot{Y}_c: S_1S_2)_1 \approx$$
$$k_1(1+f_1)\Sigma_i q_{1i}(1-q_{1i})[1+(1-2q_{2i})a_{xi}][1+(1-2q_{2i})a_{ci}]u_{xi}u_{ci}/2\sigma(\hat{X}_1) \qquad 9.11$$

And substituting q_{2i} for q_i and q_{1i} for v_i in 9.4, we obtain

$$E(\Delta\ddot{Y}_c: S_1S_2)_2 \approx$$
$$k_2(1+f_2)\Sigma_i q_{2i}(1-q_{2i})[1+(1-2q_{1i})a_{xi}][1+(1-2q_{1i})a_{ci}]u_{xi}u_{ci}/2\sigma(\hat{X}_2) \qquad 9.12$$

The expresssion for <u>direct response</u>, the expected change in average genotype value with respect to X, is obtained by substituting x for c wherever the latter appears as a subscript in 9.10, 9.11, and 9.12 and then combining 9.11 and 9.12 as indicated by 9.10. The result is

$$E(\Delta\ddot{Y}_x: S_1S_2) \approx k_1(1+f_1)\Sigma_i q_{1i}(1-q_{1i})[1+(1-2q_{2i})a_{xi}]^2u_{xi}^2/2\sigma(\hat{X}_1)$$
$$+ k_2(1+f_2)\Sigma_i q_{2i}(1-q_{2i})[1+(1-q_{1i})a_{xi}]^2u_{xi}^2/2\sigma(\hat{X}_2) \qquad 9.13$$

Except for differences in symbols and for recognition of the possibilities that $k_1 \neq k_2$ and $f_1 \neq f_2$, this is the expression given by Comstock <u>et al</u> (1949). The corresponding expression for the expected change in mean genotype value of the S_1 population, obtained by substituting q_{1i} for q_i and q_{2i} for v_i in 9.7 is

$$E(\Delta\ddot{Y}_x: S_1) \approx$$
$$k_1(1+f_1)\Sigma_i q_{1i}(1-q_{1i})[1+(1-2q_{2i})a_{xi}][1+(1-2q_{1i})a_{xi}]\ u_{xi}^2\ /\sigma(\hat{X}_1)$$

And substituting q_{2i} for q_i and q_{1i} for v_i in 9.7, we obtain

$$E(\Delta\ddot{Y}_x: S_2) \approx$$
$$k_2(1+f_2)\Sigma_i q_{2i}(1-q_{2i})[1+(1-q_{1i})a_{xi}][1+(1-2q_{2i})a_{xi}]\ u_{xi}^2\ /\sigma(\hat{X}_2)$$

9.3.1. Allele Frequency Changes Promoted by RRS

The long run potential of RRS is inherent in the expected directions of allele frequency changes. Given linkage equilibrium and positive selection differentials so that k's are positive, it is apparent from 9.8 and 9.9 that the expectations of Δq_{1i} and Δq_{2i} are positive or negative depending on whether $[1+(1-2q_{2i})a_{xi}]$ and $[1+(1-q_{1i})a_{xi}]$ are positive or negative.

9.3.1.1. OVERDOMINANCE LOCI

When $a>1.0$, $[1+(1-2q)a] = 0.0$ when $q = (1+a)/2a = q_e$, is positive when $q < q_e$, and is negative when $q > q_e$ (see section 6.3.3). It is apparent that what happens at these loci depends on the initial frequencies of B in the two selection populations. Consider first the case where $q_1 < q_e$ and $q_2 > q_e$. Then $[1+(1-2q_1)a] > 0.0$ so that $E(\Delta q_2)$ is positive and $[1+(1-2q_2)a]$ is negative so that $E(\Delta q_1)$ is negative. The long run consequence, apart from exceptions due to random drift, would be trends upward in q_2 and downward in q_1 until fixation of B in S_2 and of b in S_1. This course of events is depicted in Figure 9.1.

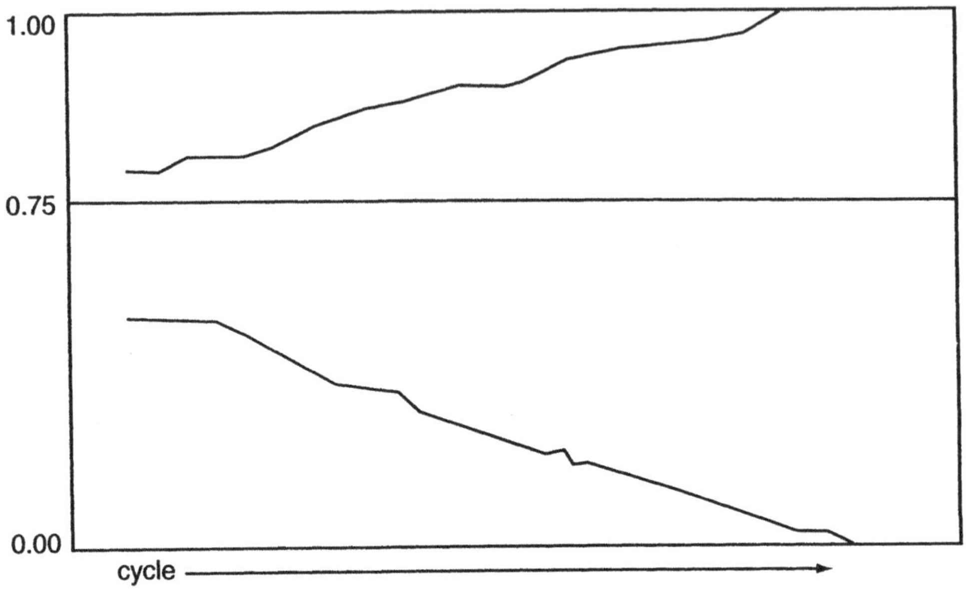

Figure 9.1. Diagrammatic generalization of the frequency pathways of the B allele of an overdominant gene in the two S populations of reciprocal recurrent selection when the initial frequency of the allele is greater than q_e in one S population and less than q_e in the other S population. For the purpose of this diagram, $a = 2.0$ so that $q = (1+a)/2a = 0.75$.

In other instances q_1 and q_2 may both be smaller (or larger) than q_e. When they are both smaller, so that $E(\Delta q_1)$ and $E(\Delta q_2)$ are both > 0.0, the trend over generations will be upward for both q_1 and q_2 until one of them reaches q_e --- it is unlikely that both would reach q_e at the same time. Suppose that both q_1 and q_2 have been increasing during x cycles of selection and that at the end of that time $q_1 \geq q_e$ and q_2 is still $< q_e$. Then in cycle $(x+1)$, $E(\Delta q_1)$ would still be positive but $E(\Delta q_2)$ would be zero or negative, making it likely that, in cycle $(x+2)$, q_1 would be greater and q_2 less than q_e. After that, as outlined in the preceeding paragraph, fixation of B in S_1 and of b in S_2 is the probable end result. This course of events is depicted in Figure 9.2.

When initial values of q_1 and q_2 are both $> q_e$, both q_1 and q_2 should trend downward until one reaches or falls below q_e. After that the force of selection would favor further decrease in that allele frequency and a reversal of the downward trend of the frequency of B in the other selection population.

- 9.3.1.2. COMPLETE DOMINANCE LOCI

When $a = 1.0$, $[1+(1-2q)a] = 0.0$ if $q = 1.0$ but is positive whenever $q < 1.0$. Consequently, $E(\Delta q_{1i}) > 0.0$ unless $q_{2i} = 1.0$ and $E(\Delta q_{2i}) > 0.0$ unless $q_{1i} = 1.0$. The associated expectation for any single complete dominance locus is an increase in the frequency of

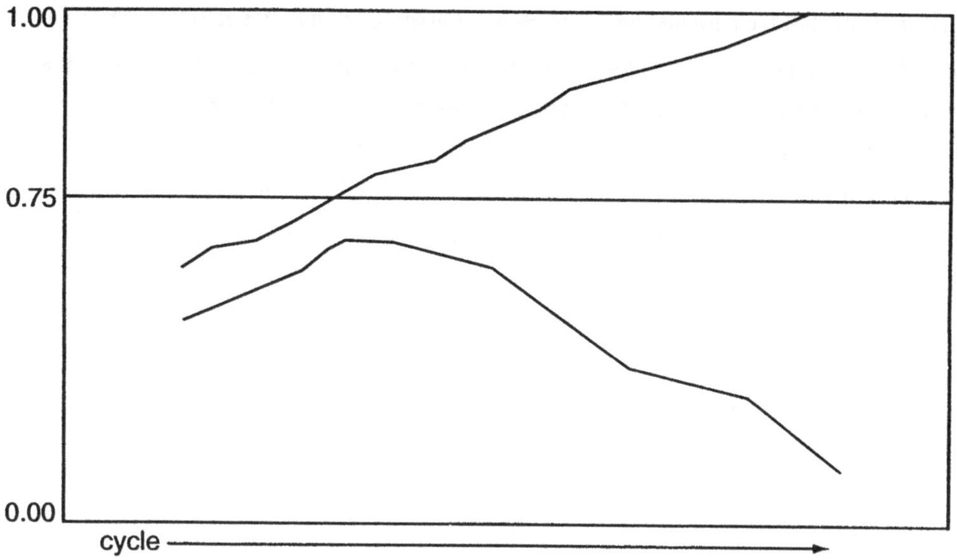

Figure 9.2. Diagrammatic generalization of the frequency pathways of the B allele of an overdominant gene in the two S populations of reciprocal recurrent selection when the initial frequency of the B allele is less than q_r in both S populations. For the purpose of this diagram, a = 2.0 so that $q_r = (1+a)/2a = 0.75$.

B in both selection populations until it reaches 1.0 in one of them. After that the selection practiced would have no further impact on the frequency of B in the other selection population in which, as a consequence of random drift, either B or b might eventually become fixed. After fixation of B in S_1 or S_2, genotypes in the target ($S_1 \times S_2$) population would be either BB or Bb.

9.3.1.3. PARTIAL DOMINANCE LOCI

Since $[1+(1-2q)a] > 0.0$ for all possible values of q when $-1 < a < 1.0$, $E(\Delta q)$ is always positive. The expected final result, therefore, is fixation of the favorable allele (B) in both selection populations unless it was completely absent (q = 0.0) from one, or both, of them at the outset.

The probable consequences of these expected allele frequency changes are summarized in Table 9.1 and discussed in section 9.5.

9.3.2. Full-Sib Reciprocal Recurrent Selection

The RRS procedure outlined by Comstock et al (1949) calls for use of several plants from the tester population in production of each test progeny. More specifically it was suggested that each plant (selection unit individual) should be mated with four or five random plants of the tester and the seed from these four or five matings bulked into one lot of seed. Jones et al (1971) noted that the test progenies produced from such seed lots are half-sib families and identified RRS following the original outline of procedure as half-sib RRS.

Hallauer and Eberhart (1970) proposed a version of RRS for maize breeding in which each test progeny is a full-sib family produced by mating an individual from population S_1 with just one individual from population S_2 and each test progeny provides the selection criterion basis for evaluation of both of its parents[2]. This version is referred to as full-sib RRS. It is obviously not adaptable to species (e.g., cattle or sheep) in which offspring per female are too few for adequate progeny testing. Assuming n selection unit individuals in each population, the number of test progenies to be compared would also be n, as compared to the 2n required by the half-sib RRS procedure. This advantage was not noted by Hallauer and Eberhart who emphasized instead the utility of full-sib progeny performance as a basis for selection of plant pairs from which to derive pure lines likely to perform well in single crosses. Jones et al (1971) considered the potential efficiency of using one progeny as the base for evaluation of a selection unit individual in both of the RRS populations and outlined the following theory base for comparing full-sib and half-sib RRS.

Because improvement of the $S_1 \times S_2$ cross flows from the allele frequency changes generated by the selection practiced, expected improvement is greater when $E(\Delta q)$ is greater. Thus improvement expected from full-sib RRS relative to that expected by half-sib RRS is indicated by $E(\Delta q_{1i})_F/E(\Delta q_{1i})_H$ and/or $E(\Delta q_{2i})_F/E(\Delta q_{2i})_H$ in which F and H as subscripts specify full-sib RRS and half-sib RRS, respectively. Assuming that the inbreeding (f) of selection unit individuals would be the same in either case, we have from 9.8

$$[E(\Delta q_{1i})_F/E(\Delta q_{1i})_H] = [k_{1F}/k_{1H}][\sigma(\hat{X}_1)_H/\sigma(\hat{X}_1)_F] \qquad 9.14$$

Note that the right-hand member of 9.14 contains no quantity specific to a particular gene and hence that the ratio of expected allele frequency changes is the same for all genes. An exactly parallel expression could obviously be written concerning allele frequency changes in the S_2 population and the information provided by it and by 9.14 would be the same.

An appropriate (though not the only) basis for comparing the procedures is to assume that the total of test progenies and the number of selections would be the same in either case. Let 2T be the total number of test progenies and M the number of selections in each of the RRS populations. The corresponding numbers of selection unit individuals in each of the selection populations would be T and 2T in the case of the half-sib and full-sib procedures, respectively, and proportions selected would be M/T and M/2T. Note that the fraction selected in half-sib RRS would be double the fraction selected in full-sib RRS. Some examples of the associated k-values, taken from Table 6.1 and assuming T> 50, are shown below.

Proportions selected				
half-sib	full-sib	k_H	k_F	k_F/k_H
.50	.25	.80	1.27	1.59
.30	.15	1.16	1.55	1.34
.20	.10	1.40	1.76	1.26
.10	.05	1.76	2.06	1.17
.04	.02	2.16	2.42	1.12
.02	.01	2.42	2.64	1.09

The ratio of k_F to k_H favors full-sib RRS in all instances, as anticipated, but the advantage is greatest when the selection intensity is low. Balanced against this advantage is the fact, noted by Jones et al, that $\sigma(\hat{X})_H$ will ordinarily be less than $\sigma(\hat{X})_F$ because the sampling variance of test progeny means is reduced by increasing the number of tester parents from which each test progeny is produced.

9.3.3. RRS Using Pure Line Testers

Darrah et al (1972) and Horner et al (1973) observed that genetic variance among maize test-cross progenies was about twice as large when inbred line testers were used as when a non-inbred population tester was used. Russell and Eberhart (1975) cited those results, pointed out that they were to be expected, and argued that therefore genetic gains in RRS would be greater if inbred lines extracted from the selection populations were used as testers instead of the selection populations themselves. Specifically, they proposed that an inbred line extracted from a previous generation of S_1 be used in place of S_1 as the tester for S_2 individuals and that an inbred line from S_2 be used in place of S_2 as the tester for S_1 individuals.

The question raised is whether, without exception, response to selection is proportional to the genetic variance among criterion units. Recall (from Chapter V) that $E(\Delta q_i)$ varies with the selection system but that $\partial \ddot{Y}_c/\partial q_i$, which depends on the nature of the target population, is not affected by the nature of the selection system. Thus it is apparent that

$$E(\Delta \ddot{Y}_c) = \Sigma_i E(\Delta q_i) \partial \ddot{Y}_c/\partial q_i$$

is proportional to expected changes in allele frequencies whether it is, or is not, proportional to genetic variance among criterion units. Comstock (1979) pointed this out and showed that $E(\Delta q_i)$ is almost the same when inbred line testers are employed as when the populations are used as testers. The slight deviation from equality arises from the fact that, because genetic variance among criterion units is larger, $\sigma^2(\hat{X})$ is larger so that $E(\Delta q_i)$ is smaller when inbred line testers are employed. The important point is that the expectation of $\sigma(\hat{q}_i,\hat{X})$ is the same in both cases.

The issue can be examined in terms of the selection in either S_1 or S_2; we will do so in terms of the selection in S_1. From Table 8.2

$$\sigma(\hat{q}_{1i},\hat{X}) = (1+f_1)q_{1i}(1-q_{1i})[1+(1-2v_i)a_{xi}]u_{xi}/2 \qquad 9.15$$

in the case of progeny test selection. The only quantity in 9.15 that varies with the tester is v_i. When the S_2 population itself is the tester $v_i = q_{2i}$. When the tester is an homozygous line extracted from S_2, v_i will be 1.0 or 0.0. It will be 1.0 when the more favorable allele (B) is homozygous at the locus in question and will be 0.0 when the less favorable allele (b) is homozygous at the locus. Whether the more or less favorable allele is present in the particular line used would be a chance event. Let the probabilities of the BB and bb genotypes in the inbred line tester be P and (1-P), respectively. The alternative possibilities, their probabilities and the associated values of $[1+(1-2v)a_x]$ are shown below.

Genotype of pure line tester	Probability	v	$[1+(1-2v)a_x]$
BB	P	1	$1-a_x$
bb	1-P	0	$1+a_x$

Averaging over all possible inbred lines, yields

$$E[1+(1-2v)a_x] = P(1-a_x) + (1-P)(1+a_x) = [1+(1-2P)a_x] \quad 9.16$$

for every locus. It follows from 9.15 and 9.16 that the expectation of $\sigma(\hat{q}_{1i},\hat{X})$ when an inbred tester is used is

$$E\sigma(\hat{q}_{1i},\hat{X}) = (1+f_1)q_{1i}(1-q_{1i})[1+(1-2P_i)a_{xi}]/2 \quad 9.17$$

Comparing 9.15 and 9.17 and, remembering that if the S_2 population is the tester $v_i = q_{2i}$, we see that $E[\sigma(\hat{q}_{xi},\hat{X})]$ is the same when inbred line testers are used as when the S_2 population itself is used if $P_i = q_{2i}$. This would be the case (see section 2.10.2) if allele frequencies in S_2 were the same at the time of selection in S_1 as they were when the inbreeding of lines, that might (as alternatives to S_2 itself) be used as the tester for S_1, was initiated and if lines were developed without selection. However, during the time period required to produce an inbred line, allele frequencies in S_2 would change (1) as the result of drift and (2) because in an efficient RRS program at least one cycle of selection in S_2 would have been conducted. The impact of the consequent differences between P's and q_2's, though it might be small, would be to reduce the relative effectiveness of inbred testers because, to the extent that the P's deviate from the q_2's, selection in S_1 using pure line testers would deviate from selection for combining ability with S_2.

• 9.4. Selection Based on Performance in S Populations (RSP)

In both RSSC and RRS selection is based on the performance of crossbred test progenies, i.e., the criterion unit individuals are not members of the S population itself. How much is gained by the direct selection for crossbred performance that is the distinctive feature of those systems? The alternative considered in this section is selection in two populations, S_1 and S_2, using any systems in which the criterion unit individuals are non-inbred members of the selection population itself. Such a two population program will be designated RSP. Suppose, for example, that full-sib (type 1) family selection is practiced in each of the S populations. Then, from Table 8.2

$$\sigma(\hat{q}_i,\hat{X}) = q_i(1-q_i)\,[1+(1-2q_i)a_{xi}]u_{xi}/2$$

It follows, from 5.18 that

$$E(\Delta q_{1i}) = k_1q_{1i}(1-q_{1i})[1+(1-2q_{1i})a_{xi}]u_{xi}/2\sigma(\hat{X}_1) \quad 9.18$$

and

$$E(\Delta q_{2i}) = k_2q_{2i}(1-q_{2i})[1+(1-2q_{2i})a_{xi}]u_{xi}/2\sigma(\hat{X}_2) \quad 9.19$$

The derivatives of $\ddot{Y}_c$ with respect to q_i that we require, are from 5.11 and 5.12

$$\partial\ddot{Y}_{c1}/\partial q_{2i} = 2[1+1-2q_{1i})a_{ci}]u_{ci} \quad 9.20$$

$$\partial\ddot{Y}_{c2}/\partial q_{2i} = 2[1+(1-2q_{2i})a_{ci}]u_{ci} \quad 9.21$$

$$\partial\ddot{Y}_{c12}/\partial q_{1i} = [1+(1-2q_{1i})a_{ci}]u_{ci} \quad 9.22$$

and

$$\partial\ddot{Y}_{c12}/\partial q_{2i} = [1+(1-2q_{2i})a_{ci}]u_{ci} \qquad 9.23$$

where $\ddot{Y}_{c1}$, $\ddot{Y}_{c2}$, and $\ddot{Y}_{c12}$ are averages of $\ddot{Y}_c$ in the S_1, S_2, and $S_1 \times S_2$ populations, respectively.

Finally, assuming that the response criterion is the selection criterion and using 9.18 through 9.23 for substitution in 5.8, we obtain for direct responses (C = X)

$$E(\Delta\ddot{Y}_X: S_1) = k_1\Sigma_i q_{1i}(1-q_{1i})[1+(1-2q_{2i})a_{Xi}]u_{Xi}^2/\sigma(\hat{X}_1)$$

$$E(\Delta\ddot{Y}_X: S_2) = k_2\Sigma_i q_{2i}(1-q_{2i})[1+(1-2q_{2i})a_{Xi}]u_{Xi}^2/\sigma(\hat{X}_2)$$

and remembering 9.10,

$$E(\Delta\ddot{Y}_X: S_1S_2) = k_1\Sigma_i q_{1i}(1-q_{1i})[1+(1-2q_{1i})a_{Xi}][1+(1-2q_{2i})a_{Xi}]u_{Xi}^2/2\sigma(\hat{X}_1)$$
$$+k_2\Sigma_i q_{2i}(1-q_{2i})[1+(1-2q_{2i})a_{Xi}][1+(1-2q_{1i})a_{Xi}]u_{Xi}^2/2\sigma(\hat{X}_2) \qquad 9.24$$

It is apparent from 9.18 and 9.19, as it was from 6.24 in the case of mass selection, that (1) $E(\Delta q)$ is always positive at each locus where $-1 \le a_X \le 1.0$, but (2) at overdominance loci $E(\Delta q)$ is positive when $q < (1+a)/2a$ but negative when $q > (1+a)/2a$. Long run consequences are summarized in Table 9.1 for comparson with those of RSSC and RRS.

It is important to note (see Table 8.2) that for all selection systems in which criterion unit individuals are non-inbred members of the S population $\sigma(\hat{q}_i,\hat{X})$, and hence $E(\Delta q_i)$, is proportional to $A = q_i(1-q_i)[1+(1-2q_i)a_{Xi}]u_{Xi}$ if the S population is in linkage equilibrium. Thus, if the same selection criterion is used in all cases, the long run consequences of mass selection, half- or full-sib family selection, and progeny test selection (if the S population itself is used as the tester) would be as shown in Table 9.1.[3] On the other hand, in the case of $S_{n(n-1)}$ family selection, in a population at linkage equilibrium, $E(\Delta q_i)$ is proportional to $q_i(1-q_i)[1+(1-2q_i)a_{Xi}/2]u_{Xi}$ and is therefore not negative unless $q > (2+a)/2a$. The consequences, with respect to the performance of crossbreds, of selection among families produced by inbreeding is considered in more detail in section 9.6.2.

• 9.5. Ultimate Allele Frequencies and Genotypes

Table 9.1 summarizes what the long run consequencies of the selection systems considered in sections 9.2, 9.3, and 9.4 would be (assuming whenever relevant that each S population is initially heterozygous at the B,b locus) if the force of selection as the determinant of ultimate allele frequency were never over-ridden by random drift. For example, given partial dominance and RSSC with the BB genotype in the A (tester) line, $E(\Delta q) > 0.0$ when $0.0 < q < 1.0$ and it was assumed accordingly that q would continue to increase in the S population until it reached 1.0. Then all genotypes in S would be BB and, with genotype in the A line being BB, genotypes in the S x A cross would all be BB. As a second example, given overdominance and RSSC with the BB genotype in the A line, $E(\Delta q) < 0.0$ when $0.0 < q < 1.0$ and consequently q would decrease in the S population until it reached 0.0. Then all S population genotypes would be bb and all genotypes in the S x A cross would be Bb. In the special situations involving complete

Table 9.1. Ultimate allele frequencies and genotypes as functions of selection system and level of dominance.

Level of dominance	Type of selection[a]	Final allele frequency S_1 (or S)	S_2 (or A)	Genotype when response to selection is complete S_1 (or S)	S_2 (or A)	$S_1 \times S_2$ (or S x A)
0< a<1	RSSC	1.0	1.0	BB	BB	BB
	RSSC	1.0	0.0	BB	bb	Bb
	RRS	1.0	1.0	BB	BB	BB
	RSP	1.0	1.0	BB	BB	BB
a = 1	RSSC	*	1.0	***	BB	BB or Bb
	RSSC	1.0	0.0	BB	bb	Bb
	RRS	1.0	**	BB	***	BB or Bb
	RRS	**	1.0	***	BB	BB or bb
	RSP	1.0	1.0	BB	BB	BB
a>1	RSSC	0.0	1.0	bb	BB	Bb
	RSSC	1.0	0.0	BB	bb	Bb
	RRS	1.0	0.0	BB	bb	Bb
	RRS	0.0	1.0	bb	BB	Bb
	RSP	(1+a)/2a	(1+a)/2a	****	****	****

*E(Δq) = 0.0; final q determined by initial q and drift.

** E(Δq) = 0.0 after q = 1.0 in counterpart population; final q determined partly by drift.

*** Genotype determined by final q in the population.

**** Hardy-Weinberg frequencies of BB, Bb, and bb [q_e^2: $2q_e(1-q_e)$: $(1-q_e)^2$].

[a] RSSC = recurrent selection for specific combining ability, RRS = reciprocal recurrent selection, and RSP = recurrent selection based on performance in the S populations.

dominance in which E(Δq) = 0.0 in the S population of RSSC or in one of the S populations of RRS because the genotype of (a) the A line of RSSC or (b) one of the S populations of RRS is BB, drift is postulated to be the source of change in allele frequency (though natural selection could also be partially responsible). Values of E(Δq) are indicated for each selection system in the preceeding sections.

The final column of Table 9.1 indicates that, aside from exceptions due to drift or the initial absence of an allele in an S population, the eventual genotype of the population cross [(S x A) when the selection system is RSSC, $S_1 \times S_2$ otherwise] is the optimum one in all but two situations. Given partial dominance, the best genotype is BB but the final genotype in the case of RSSC with the bb genotype in the A line is Bb. Given overdominance, the optimum genotype is Bb but the final genetic composition of the $S_1 \times S_2$ cross in the case of RSP includes all three genotypes with the probability of Bb not larger than 0.5. Thus, on the basis of eventual genotypic value of the population cross, RSSC is inferior with respect to partially dominant genes, RSP is inferior with respect to overdominance genes and RRS is equal to the better of the other two systems whatever the level of dominance. The implication, emphasized by Comstock et al (1949), is that by the criterion of potential for improvement of the cross between two populations, RRS should be the choice whenever there is reason to believe that all levels of dominance may be important in the genetics of total value.[4]

Would this apparent advantage of RRS disappear if the effects of random drift were evaluated? Obviously not, if effective population size is, in all cases, sufficient to make the consequences of random drift small relative to those of selection. While that would be the case in any well designed recurrent selection program (see Chapter XIII), the required resources(the cost) are not the same for all selection systems. For that reason, the issue is considered further in Chapter XIV.

Collectively, the last three columns of Table 9.1 indicate how eventual genotypes in S populations would compare with those in the S x A or $S_1 \times S_2$ cross if there were no random drift and the same alleles were present at the outset in all S populations. Note the following features.

(1) Given RSP, the final genotype or distribution of genotypes is the same for the S populations as for the $S_1 \times S_2$ cross whatever the level of dominance. A significant implication is that in the long run heterosis in the $S_1 \times S_2$ cross should be diminished by RSP.

(2) Given selection based on crossbred performance (RSSC or RRS), whether the ultimate genotypes of the S population(s) and the S x A or $S_1 \times S_2$ cross are comparable depends on level of dominance. Given partial dominance of the better allele the eventual genotype in the S population(s) is always BB, the same as that of the $S_1 \times S_2$ cross of RRS and somewhat better, all loci considered, than the S x A cross of RSSC. The outcome is distinctly different in the event of overdominance. Then the eventual S population genotypes are homozygous (BB or bb) while those in the population cross are all Bb. So far as overdominant genes are concerned genotypic value in the cross would be greater than in the S population(s). That this should be so is, of course, intuitively obvious and was forecast in section 9.1. It is no surprise that long run consequences in the case of complete dominance are intermediate between those for partial dominance and overdominance. In that case, the eventual genotype in S populations would sometimes be bb but the majority of the loci would be BB or Bb, as good as those of the S x A or $S_1 \times S_2$ cross.

The comparisons summarized above would be modified in some degree by random drift. Both the direction and magnitude of the drift effect would vary depending on type of selection and level of dominance. However, given adequate effective population size by the probability of fixation criterion outlined in Chapter XI, the magnitude of change from what is shown in Table 9.1 would be relatively unimportant. Consequences in the case of selection based on crossbred performance are clearly of greatest concern. We noted above that eventual genotypes produced by RSSC or RRS would be as good in the S population(s) as in the cross in the case of partial dominance but distinctly better in the cross in the case of overdominance. This differential associated with level of dominance would be reduced by random drift. Consider RRS. The possible genotypic outcomes are listed below.

Outcome	S_1	S_2	$S_1 \times S_2$
1	BB	BB	BB
2	BB	bb	Bb
3	bb	BB	Bb
4	bb	bb	bb

Given partial dominance and no drift, the long run consequence is outcome 1. Outcomes 2 - 4 are possible by drift. Remembering that when $a > 0.0$ the average value of the BB and bb genotypes is less than that of the Bb genotype, it is apparent that, while genotypes would be as good in the S populations as in the cross if there were no drift, the average value of genotypes should normally be a little less in the S populations than in the cross as the result of drift. Now consider overdominance. In the absence of drift the long run consequence of RRS would be outcome 2 or 3. It would then be 1 and 4 that would result only by drift. In those cases the S populations are the same as those in the $S_1 \times S_2$ cross. Thus the effect of drift is to reduce the difference between the S populations and the cross in average value of genotypes at overdominance loci. Extension of this sort of analysis to the effect of drift in the casse of RSSC yields essentially the same result; decrease in the average value of genotypes of the S population relative to the average for the S x A cross in the case of partial dominance and the reverse effect in the case of overdominance.

It is obvious from the foregoing that data obtained using these selection systems may be critical relative to levels of dominance in the genetics of quantitative traits. This has long been apparent and numerous selection investigations of these types have been conducted and reported. Examples are Bell et al 1955), McNew and Bell (1976), Enfield (1977) and Moll et al (1978).

• 9.6. The Inbred-Hybrid (IH) System

This system has been employed in various species but is best known for its role in corn (maize) improvement and will be discussed here in the context of maize breeding. Corn hybrids, produced by crossing highly inbred lines,[5] have much to do with the present productivity of world agriculture and most of those that are or have been important in maize production have been developed using the IH system.

The essential features of the system are (1)the formation, typically by self-fertilization, of potentially useful inbred lines and (2) the selection of superior hybrids from those that might be formed using the lines available at any point in time. Selection during line formation, as well as among hybrids, is characteristic. Sprague and Eberhart (1977) state, "Inbreeding normally is accompanied by selection. Plants to be inbred are selected for vigor, freedom from disease, and other desired attributes". They go on to say that selection is practiced "both among and within progenies". While the nature and amount of such selection varies substantially from any one breeder or program to another, it is considered essential because inbreeding depression is severe and hence cost of producing hybrid seed from unselected inbred line parents could be prohibitive. Further, the selection practiced among inbred lines results in better hybrids, i.e., on average phenotypically superior lines produce phenotypically superior hybrids. Both of these facts were emphasized by Hayes (1963) who noted that as the reproductive performance (seed and pollen production) of inbred lines was improved, it became feasible for hybrid seed producers to substitute single crosses for double or three-way crosses and thereby provide farmers with more productive hybrids.

Selection among hybrids is complicated by the number that can be formed using relatively few inbred lines, a fact that has been noted again and again in the breeding literature, e.g., Hayes and Immer (1942) and Allard (1960). Not counting reciprocals, the number of single crosses that can be made using n lines is $n(n-1)/2$ and the number of double crosses is $n(n-1)(n-2)(n-3)/8$. Some numerical examples are shown below.

	Number of possible hybrids	
Number of inbred lines (n)	Single cross	Double cross
15	105	4095
30	435	82215
60	1770	1,462,905

It is quite clear that all possible hybrids can hardly be grown as the basis for selection among them. The problem is less acute now that the performance level of inbred lines is such that the use of single cross hybrids is practical, but it is still true that actual testing of all possible hybrids would take money and effort that would probably be more usefully expended elsewhere in the total breeding program. In practice the response to this problem has been relatively intense selection of the lines to be used as parents of the sample of hybrids from which those that perform best will be recommended for use as cultivars. The line selection process is commonly sequential; the first criterion being their own attributes and the second their "general combining ability".[6]

9.6.1. General Considerations

9.6.1.1. NEED FOR RECURRENT SELECTION

There is abundant evidence indicating that total value in maize (and probably any economic species in which the IH system is or might be used) is affected by many genes and hence that synthesis of ideal genotypes cannot logically be expected without recurrent selection.

That this has long been recognized by breeders is evident from their use of selected inbred lines in developing new non-inbred populations from which to extract further pure lines of which a portion are expected to be superior to their progenitors. However, much that individuals have done in this arena has not been focused in systematic recurrent patterns for continuous improvement of one or more defined populations to be used as the source or sources of inbred lines in successive cycles of their IH programs. The importance of integrating some well chosen form of recurrent selection into IH programs has been emphasized by various authors, e.g., Comstock (1964), Sprague and Eberhart (1977) and Hallauer and Miranda (1981).

9.6.1.2. THE TWO PHASES OF THE IH SYSTEM

They will be called (1) the source population improvement phase and (2) the cultivar search phase. The first of these involves the selection of lines, families, or individuals to be used in regenerating the source population (or populations) and is thereby responsible for the changes, from one cycle to the next, in the frequencies of useful alleles in the source population(s). The second has to do with identifying the best, or near best, of the hybrids possible using the available inbred lines. Cycle by cycle the allele frequency changes in response to phase 1 selection are cumulative, i.e., the change in each allele frequency that is achieved during n cycles is the sum of the changes achieved in the n individual cycles. In contrast, the superiority of cultivar search selections (significant in terms of cultivar performance) contributes nothing that is cumulative beyond that resulting from the first phase selection unless the cultivar search is itself a component of the population improvement phase --which it can be. However, length of cycle is thereby increased so that genetic gain per unit of time may well be decreased though increases in frequencies of favorable alleles per cycle are increased.

9.6.1.3. EFFECTIVE POPULATION SIZE

The expected effect of selection is increase in frequency of useful alleles and allele frequency changes are cumulative in successive cycles of recurrent selection. However, as noted in section 2.8, recurrent selection does not insure eventual fixation of each most favorable allele. Instead, other things being equal, the probability of fixation of such alleles is greater when N_e (effective population size) is greater and approaches 1.0 as N_e becomes very large. The obvious corollary is that there is always a finite probability (which becomes smaller as N_e is increased) that an allele other than the most favorable will become fixed, because in real situations N_e can never be infinite. Clearly then, N_e must be sufficient if a reasonable approach to the full potential of recurrent selection is to be realized. The problem of decisions regarding effective population size is discussed in Chapter XI. The facts to be noted here are

(a) Assuming no variation of genetic relationships among pairs of selection units and that M selections are to be used in regeneration of the source population, the effective size of that population is larger when M is larger (see Appendix 11.1) and

(b) the optimum selection intensity within single cycles is determined in part by its bearing on effective size of the source population.

• 9.6.1.4. GENOTYPE DISTRIBUTIONS IN THE SOURCE POPULATION, PURE LINES, AND SINGLE CROSSES

Consider (1) a single random mating (consequently non-inbred) source population in which there is no linkage disequilibrium, (2) a population of pure lines, each tracing to a different random S_0 plant, developed without any selection[7] and (3) a population of single crosses between the members of random pairs of such lines. The probabilities of single locus genotypes in these population are as follows

	Probability in		
Genotype	Source population	Pure lines	Single crosses
BB	q^2	q	q^2
Bb	$2q(1-q)$		$2q(1-q)$
bb	$(1-q)^2$	1-q	$(1-q)^2$

By virtue of random mating, genotype probabilities in the source population equate to the Hardy-Weinberg frequencies of a random mating population infinite in size. Because the pure lines are formed without selection the probabilities of the BB and bb genotypes are the frequencies of the B and b alleles in the source population. Finally, the probabilities of single cross genotypes are determined by the probabilities of genotype combinations in pairs of lines For example, the probability that the genotypes of two random lines will both be BB is q^2 and hence the probability of the BB genotype in single crosses is q^2.

When more than one locus is considered it can be shown (see Appendix 9.1) that when there is no linkage disequilibrium in the source population there will be none in a population of unselected pure lines derived from that source population. This plus the equivalence of allele frequencies in these two populations has the consequence that in gametes produced by the two populations the arrays of total genotypes are identical. The result is that the genotype frequencies of individuals in the source population and of single crosses in the population of single crosses are identical. This in turn means that in each cycle the average value of genotypes is the same in the two populations. Selections from the single cross population will of course be genotypically superior to that average.

• 9.6.1.5. THE CASE FOR TWO-POPULATION, AS OPPOSED TO SINGLE-POPULATION, PROGRAMS

Because the best hybrids have usually been obtained using pure line parents derived from different source populations (emphasized, for example, by Hayes and Immer, 1942; Hayes, 1963; and Sprague and Eberhart, 1977) IH programs involving recurrent selection for the improvement of two line source populations may be superior.

• 9.6.1.6. LETHAL ALLELES AND NATURAL SELECTION

It is often convenient, for theoretical purposes to assume a population of pure lines formed without selection from random members of a source population; this will be done at some points in what follows. However, while artificial selection can be avoided, natural selection cannot. This is most obvious in the case of recessive lethal alleles. By definition, when homozygous they cause the organism to die. Hence surviving lines never carry such alleles and, whatever their frequencies in the source population, those frequencies are reduced to zero in any population of pure lines. By analogy, it is apparent that the frequency of any deleterious but non-lethal allele (though not constrained to zero) is not likely to be as high in a population of pure lines as in their source population even though no artificial selection was exerted during formation of the lines.

• 9.6.1.7. MULTI-STAGE SELECTION

Artificial selection in the source population improvement phase of the IH system is not restricted to one stage per cycle. Selection among lines is ordinarily practiced in more than one of the successive generations of inbred line formation and, throughout, the intra-line selection of S_n plants self-fertilized to provide S_{n+1} lines inevitably favors those judged to be superior in one or more significant aspects of phenotype. The final component of selection is that among single crosses of inbred lines.

The borderline between the two phases of the IH system depends on the materials employed to regenerate source populations. For example, those materials could be S_1 families. In that case the two selection stages contributing to increases in frequencies of useful alleles would only be that among S_0 plants to be parents (by self-fertilization) of S_1 families followed by selection among S_1 families of those to be intermated in regeneration of the source population. All the rest of the selection of the system would then be part of the cultivar search phase. Obviously, if the materials employed in regeneration of the source population were families derived by two or more successive self-fertilizations, more stages of artificial selection would be included in the source population phase of the program. This inherent multi-stage complexity is at least partially responsible for the fact that no very neat treatment of response expectations has ever been provided for the IH system by quantitative genetics.

• 9.6.2. Theoretical Approach

Because the IH system employs multi-stage selection it differs in a significant way from all the systems that have so far been discussed (in this and preceding chapters) in which selection has been assumed to occur at only one stage in each cycle. The theory for single-stage recurrent selection systems assumes restoration of Hardy-Weinberg frequencies of single locus genotypes by random mating among selections following each round of selection. HW frequencies cannot appropriately be assumed in stages beyond the first in the case of multi-stage selection. Clearly, if the first stage of selection is to have a genetic impact it can only result from changes in genotype frequencies--changes such that the after selection frequencies of genotypes are most probably not HW frequencies. Unfortunately, though the problem posed has been considered by some authors [see, for example, Pederson and Matzinger (1972)], no theory as generally applicable as that for single stage systems of selection is available for multi-stage systems. Note, however, that within the IH system the selection among either lines or hybrids could be of the single stage variety. Theory presented here will (1) deal separately with lethal alleles, (2) employ single stage theory as the basis for inference concerning responses that can be expected more generally, and (3) focus on a cyclic, one source population system for producing superior single cross hybrids to be used as cultivars.[8]

Despite the observation, alluded to earlier, that the best hybrids have been obtained most frequently by crossing lines from different source populations, restricting attention to the case of one source population will suffice because (a) the potentials of selection procedures that can be used in the development of superior lines are the same whether lines are being produced from one or two source populations (as in the RSP system defined in section 9.4) and (b) the comparative merits of one and two population programs are examined in Chapter XIV.

9.6.3. Selection Within and Among Lines—the First Phase of the IH System

Assuming inbreeding by self-fertilization (the mode usually employed in maize breeding) the development of a sample of lines proceeds more or less as follows

1. A series of S_1 families is produced. The individuals chosen to be parents of those families are ordinarily vigorous, erect plants that are free from visually evident disease. Other attributes such as desirable ear height and two or more ear shoots may also be required. The number of individuals grown per family varies but will frequently be in the range of 10 - 20.

2. Self-fertilization in several successive generations (say 5 or 6) is typical. The number of families may be about the same in each generation but may be reduced in later generations as the result of inter-line selection in preceding generations.

3. In the S_1 and subsequent generations the selection practiced may be a combination of selection among individuals and selection among families. Plants that qualify so far as their own phenotypes are concerned may be rejected because their sibs as a group are not good enough.

4. The traits on which selection is based may be much the same, or may vary, from generation to generation.

9.6.3.1. THE CONSEQUENCE OF ELIMINATING LETHAL ALLELES (NATURAL SELECTION)

As shown in section 2.6.2, alleles that are lethal when homozygous, but not when heterozygous, remain in large random mating populations at low frequencies determined by the opposing forces of selection and mutation (from the non-lethal to the lethal allele of the gene in question). In contrast, by the definition of a lethal allele, there can be none of them in a pure line; they could only be present in homozygous state in which case the line is automatically eliminated.

Because better hybrids are obtained when pure lines from different source populations are crossed the effect of elimination of lethals is appropriately determined by comparing the genotypic means of (1) the population obtained by mating random members from two non-inbred source populations and (2) the population of hybrids obtained by crossing random lines extracted from the two source populations without selection except for the elimination (by their own effects) of all lethal alleles. In doing so we build on arguments employed by Crow (1948, 1952) and Fisher (1949).

Consider those lethals that are completely recessive in all their effects, i.e., those genes of which viable genotypes, the lethal heterozygote and the normal homozygote, evoke exactly the same distribution of phenotypes.[9] The expected frequency of such a lethal in a large, random mating population has been shown (equation 2.20) to be very close to $(m)^{1/2}$ where m symbolizes the rate of mutation to the lethal. Note that this is the frequency expected in both of the source populations and hence that the probability of the homozygous lethal genotype in any zygote from crossing random members of the two populations is the square of $(m)^{1/2}$ (= m) and the probability of a non-lethal genotye at the locus in question is 1 - m. It follows that the probability of a total genotype that is viable (lethal allele absent at every locus) is

$$P = (1-m_1)(1-m_2)(1-m_3)\cdots = \pi_i(1-m_i)$$

the product of the $(1-m_i)$ values of all genes. From this we see that the probability of a non-viable genotype is

$$Q = 1 - P = 1 - \pi_i(1-m_i).$$

When expanded,

$$\pi_i(1-m_i) = 1 - \Sigma_i m_i + \Sigma_i\Sigma_{j>i} m_i m_j - \Sigma_i\Sigma_{j>i}\Sigma_{k>j>i} m_i m_j m_k + \cdots\cdots$$

Because spontaneous mutation rates are very small quantities (usually $<10^{-4}$) and because of the alternating algebraic signs of the successive terms of the expansion, the total of terms involving the product of two or more m's is very small relative to $\Sigma_i m_i$. Hence

$$Q = 1 - \pi_i(1-m_i) \approx \Sigma_i m_i = n\ddot{m}$$

where $\ddot{m}$ is the average mutation rate of genes from normal alleles to the recessive lethal allele and n is the number of genes.

If the average values of viable and non-viable genotypes are $\ddot{Y}_V$ and 0.0, respectively, the population averages of genotype value in the cross of the non-inbred source populations and the hybrid population obtained by interpopulation matings between pure lines are $(1-Q)\ddot{Y}_V$ and $\ddot{Y}_V$, respectively.[10] The increase due to elimination of lethals would then be

$$\Delta\ddot{Y} = Q\ddot{Y}_V \approx n\ddot{m}\,\ddot{Y}_V \qquad 9.25$$

and as a fraction of the expectation in the cross of the non-inbred source populations this would be

$$\Delta\ddot{Y}/(1-Q)\ddot{Y}_V = Q/(1-Q) \approx n\ddot{m}/(1-n\ddot{m})$$

This should be viewed as the <u>upper limit</u> of improvement resulting from the elimination of lethal alleles that have no effects when present in heterozygotes. Fisher (1949) pointed out that because the utilization of the nutritive resource of the soil by plants is not likely to be reduced in proportion to the fraction of seed that does not produce viable plants (the surviving, healthy plants capitalize on some if not most of the soil resource unused by the non-viable plants), the fractional increase in yield must ordinarily be less than $n\ddot{m}$.

Suppose that the number of genes (n) having a bearing on total value is 5000 and that the average rate of mutation to a lethal allele ($\ddot{m}$) is 10^{-5}. Then $n\ddot{m}/(1-n\ddot{m}) = 1/19 = 0.053$ and the actual improvement from elimination of lethals would, according to Fisher's argument, be less than that amount. It is obvious of course that $n\ddot{m}/(1-n\ddot{m})$ is larger or smaller if $n\ddot{m}$ is larger or smaller.

The impact on the average value of genotypes is, of course, different from that shown above when lethal alleles are not completely recessive in all respects. Let d be a lethal allele, D be any other allele of the same gene, $\ddot{Y}$ be the average value (in a random mating population) of genotypes in which no locus is heterozygous for a lethal not recessive in all its effects, $K_i\ddot{Y}$ be the average value of genotypes heterozygous for d at the i-th locus, and $(\pi_i K_i)\ddot{Y}$ the average value of genotypes heterozygous for a lethal allele at more than one locus.[11] The instances in which the recessive lethal (d) is not recessive in all respects can be divided into those in which (1) $K \neq 1.0$ but the DD and Dd individuals have the same fitness, i.e., the same probability of reproduction and (2) there is a difference in fitness between the DD and Dd individuals and K may or may not be equal to 1.0.

Now consider any class of lethals for which K and m (the rate of mutation to that lethal allele) are constant from gene to gene. Then with respect to each gene there are three genotypes to be considered---DD, Dd, and dd. Assuming linkage equilibrium as a consequence of a history of random mating and that the frequency of the lethal allele is the same at all loci because K and m are the same at all loci, the probabilities of the various classes of total genotypes are given by the separate terms of the expansion of

$$(p + r + s)^n = [(p + r) + s]^n$$

where $p = (1-q)^2$, q is the equilibrium frequency of the lethal allele, $r = 2q(1-q)$, $s = q^2$, and n = number of genes. For example, p^n is the probability of the total genotype that is DD at all loci and $p^{n-2}r^2\, n(n-1)/2$ is that of the genotype that is DD at n-2 loci and Dd at the other 2 loci. Because all genotypes that are dd at any one or more loci are non-viable, the probabilities of the classes of viable genotypes are given by expansion of $(p+r)^n$. Finally, the genotypic value of the class of genotypes in which there are x loci with the Dd genotye and therefore n - x with the DD genotype would be $K^x(\ddot{Y})$. It follows that E(Y) for the random mating population would be

$$E(Y) = \ddot{Y}[p^n + p^{n-1}rKn + p^{n-2}r^2K^2n(n-1)/2 + p^{n-3}r^3K^3n(n-1)(n-2)/6 + p^{n-4}r^4K^4n(n-1)(n-2)(n-3)/24 + \cdots] = \ddot{Y}(p + rK)^n \qquad 9.26$$

Because recessive lethals will have been eliminated from pure lines all single locus genotypes in hybrids from inter-population crosses of pure lines must be DD. Thus the mean value of the total genotypes of such crosses would be $\ddot{Y}$ and from this and 9.26 the increase in value from the elimination of lethals would be

$$\Delta\ddot{Y} = \ddot{Y}[1-(p+rK)^n] \qquad 9.27$$

In the special case where effects of the DD and Dd genotypes are the same at all loci (i.e., K=1 and the fitnesses of DD and Dd are equal) $q = m^{1/2}$ and

$$(p+rK)^n = (p+r)^n = [(1-m^{1/2})^2 + 2(1-m^{1/2})m^{1/2}]^n = (1-m)^n$$

Expanding, we obtain (because m can be assumed small enough so that terms beyond the second are inconsequential relative to mn)

$$(p+rK)^n = [1-mn + m^2n(n-1)/2 - m^3n(n-1)(n-2)/6 + \cdots] \approx 1-mn \qquad 9.28$$

and from 9.27 and 9.28 we obtain $\Delta\ddot{Y} \approx \ddot{Y}mn$ in accord with 9.25.

Consider now the case of genes for which $K \neq 1.0$ but the fitnesses of the DD and Dd genotypes are equal, i.e., the lethal allele is completely recessive with respect to fitness. It is then obvious from 9.27 that $\Delta\ddot{Y}$ would be greater if $K < 1.0$ than when $K = 1.0$. On the other hand, given $K > 1.0$, $\Delta\ddot{Y}$ would be less than when $K = 1.0$ and could be negative in which case the value of genotypes would be reduced by elimination of the lethal alleles. Note that $K > 1.0$ when Y is greater for the Dd than for the DD genotype, i.e., when the gene is overdominant with respect to total value. Thus if there were enough of such overdominance the elimination of lethals via production and use of pure lines would be counter-productive.

Finally, when there is a difference in fitness between the DD and Dd genotypes the equilibrium frequency of the d-allele in a random mating poulation is lower when Dd is less fit than DD (see 2.21 and 2.22) but greater when Dd is more fit than DD (see 2.16). Thus, for any specific value of K, the impact of eliminating the d-allele is less when DD is more fit than Dd but greater when Dd is more fit than DD.

In summary, theory together with what may be reasonable postulates concerning number (n) of genes bearing on total value and the average of rates of mutation from non-lethal to lethal alleles ($\ddot{m}$) suggests that the elimination of lethal alleles achieved in pure lines is not likely to mean more than (if as much as) a five percent improvement of hybrids or synthetic random breeding populations derived from pure lines [as emphasized by Crow (op. cit.) and Fisher (op. cit.)]. Moreover, whatever the beneficial effect of the elimination of lethals, it can only be a one time achievement. Lethal alleles eliminated can recur and be eliminated again but after the gain from their first elimination, the benefit is only to regain what was previously gained.

- 9.6.3.2. **ARTIFICIAL SELECTION AMONG LINES**

Within this section, only $E(\Delta q)$, the expected response in allele frequencies, is considered. In general, $\Delta\ddot{y}_X$ will be greater when Δq is greater. However, traits other than X, the one on which selection is based may or may not be improved by increase in Δq. (see 5.8, 5.11, and 5.12.)

Consider, as outlined in section 9.6.1.4, a random mating source population in which there is no linkage disequilibrium and then that a population of pure lines, each tracing to a different random S_0 plant, is derived without any selection. Selections from this population of lines will be intermated to provide the genetic base for the version of the source population to be employed in the following cycle of the program.

If lines are selected on the basis of their own performance, we obtain from Table 8.2 that

$$\sigma(\hat{q}_i,\hat{X}) = 2q_i(1-q_i)u_{Xi} + \Sigma_{j\neq i}(pt-rs)_{ij}(1+\pi_{ij})u_{Xj} \qquad 9.29$$

in which all values of (pt-rs) are zero because there was no linkage disequilibrium in the source population.[12] Thus, 9.29 reduces to

$$\sigma(\hat{q}_i,\hat{X}) = 2q_i(1-q_i)u_{Xi}$$

where q_i is the frequency of B_i in both the source population and the population of pure lines. Substituting this covariance in 5.24 yields

$$E(\Delta q_i) = [k/\sigma(\hat{X})]2q_i(1-q_i)u_{Xi} \qquad 9.30$$

Instead of selecting pure lines (from the population specified above) on the basis of their own performance, they can be selected on the basis of estimates of their "general combining ability". In that system the lines are the selection units and progenies, obtained by mating each line with a tester population that is the same for all lines, are the criterion units. It is the system called progeny test selection in Chapter VIII. From Table 8.2 we see that in this case (because f = 1.0 and all values of pt-rs are zero)

$$\sigma(\hat{q}_i,\hat{X}) = q_i(1-q_i)[1+(1-2v_i)a_{xi}]u_{xi}$$

where v_i is the frequency of B_i in the tester population. Substituting in 5.24 yields

$$E(\Delta q_i) = [k/\sigma(\hat{X})]\, q_i(1-q_i)[1+(1-2v_i)a_{xi}]u_{xi} \qquad 9.31$$

Before going on, it is worth noting that either of these modes of selection can be employed at any earlier point of the line formation process. Remember, however, that effective population size of the source population in cycles beyond the first is circumscribed by the number of lines(all having different S_0 ancestors) from which it is regenerated and that this imposes a limit on the total amount of selection among lines. If such selection is practiced early, the amount of it that can be practiced later is correspondingly reduced. Instead of selection among lines delayed until they are nearly homozygous, consider selection of the S_0 plants from which lines will be developed via selection among S_1 families or by progeny test selection among S_0 plants. The corresponding $\hat{q}_i,\hat{X}$ covariances (from Table 8.2, remembering our assumption of no linkage disequilibrium in the source population and that f = 0.0 in S_0 plants) are

$$q_i(1-q_i)[1+(1-2q_i)a_{xi}/2]u_{xi}$$

and

$$q_i(1-q_i)[1+(1-2v_i)a_{xi}]u_{xi}/2 \text{ , respectively.}$$

By substitution of these in 5.24, we obtain

$$E(\Delta q_i) = [k/\sigma(\hat{X})]q_i(1-q_i)[1+(1-2q_i)a_{xi}/2]u_{xi} \qquad 9.32$$

in the case of selection among S_1 families and

$$E(\Delta q_i) = [k/\sigma(\hat{X})]q_i(1-q_i)[1+(1-2v_i)a_{xi}]u_{xi}/2 \qquad 9.33$$

in the case of progeny test selection among S_0 plants.

Finally, consider selection among single crosses that have been produced using random pairs of pure lines as parents. This is not selection among single lines but can be viewed as selection among pairs of lines. As noted in section 9.6.1.4, linkage disequilibrium is not expected in a single cross population derived using random unselected pure lines as parents. Consequently, $\sigma(\hat{q}_i,\hat{X}) = \sigma(\hat{q}_i,\hat{y}_{xi})$. Using probabilities of single locus genotypes shown in section 9.6.1.4, the bivariate distribution of $\hat{q}_i$ and $\hat{y}_{xi}$ is the following

Genotype	frequency	$\hat{q}_i$	$\hat{y}_{xi}$
B_iB_i	q^2	1.0	$2u_{xi}$
B_ib_i	$2q(1-q)$	1/2	$u_{xi}+a_{xi}u_{xi}$
b_ib_i	$(1-q)^2$	0	0

Note that all essential features are identical to those of the bivariate distribution shown in section 6.1. This is, of course, as it should be because in both cases the distribution pertains to single locus genotypes in HW frequencies. It follows that $E(\Delta q)$ for selection among the single crosses here being considered must be the same in form as derived in section 6.1 for mass selection. The expectation of Δq, given by 6.4, is restated (and renumbered) below.

$$E(\Delta q_i) = [(\bar{X}_s - \bar{X})/\sigma^2(\hat{X})]q_i(1-q_i)[1+(1-2q_i)a_{xi}]u_{xi} \quad (6.4)\ 9.34$$

It is imperative, of course, to remember that $\sigma^2(\hat{X})$ in selection among single crosses can be made much smaller than the $\sigma^2(\hat{X})$ for individuals that appears in 6.4. To parallel other expressions for $E(\Delta q_i)$ in this section 9.34 can, of course, be rewritten as follows

$$E(\Delta q_i) = [k/\sigma(\hat{X})]q_i(1-q_i)[1+(1-2q_i)a_{xi}]u_{xi} \quad 9.35$$

9.6.3.3. ARTIFICIAL SELECTION WITHIN LINES

By virtue of the emphasis that has been placed on effective population size and hence on the number of S_0 ancestors to which the finally selected lines trace their ancestry, lines have for our purposes been identified as random (unrelated) if they are descended from different S_0 plants. Potentially then, there is no limit to the number of sub-lines within any single line. For example, a sub-line could be initiated from each of any numer of S_1 offspring of a single S_0 plant and from each of any number of S_2 offspring of each of those S_1 plants. Obviously, within line selection could take any of various forms. Despite this, only selection among individual members of the same family (produced by self-fertilization) will be considered in this subsection.

In families obtained by self-fertilization there can be no genetic variation at loci homozygous in the parent. At loci heterozygous in the parent H-W frequencies of genotypes (1/4,1/2,1/4) are expected whatever the selection in preceeding generations may have been. Thus single-stage selection theory is applicable to the within family selection in successive generations of line formation. The selection practiced is among individuals and almost always on the basis of their own phenotypes, i.e., mass selection. In that case, and remembering that allele frequencies are 0.5 at all segregating loci, we have from 6.15

$$\sigma(\hat{q}_i,\hat{X}) = u_{xi}/4 + \Sigma_{j\neq i}(pt-rs)_{ij}[1+(1-2q_j)a_{xj}]u_{xj} \quad 9.36$$

in the case of families in which the i-th gene is segregating.

Account must now be taken of the fact that the expected portion of S_1 families segregating for any single gene is $2q(1-q)$ and that, in general, the portion of S_n families expected to be segregating for any one gene is $2q(1-q)(1/2)^{n-1}$ where q is the frequency of B in the S_0 (the source population). Thus disregarding the linkage disequilibrium term of 9.36

$$\sigma(\hat{q}_i,\hat{X}) = (1/2)q_i(1-q_i)u_{xi} \quad 9.37$$

for selection within all S_1 families (those not segregating at the i-th locus as well as those segregating) and

$$\sigma(\hat{q}_i,\hat{X}) = (1/2)^n q_i(1-q_i)u_{xi} \quad 9.38$$

for selection within S_n families. Substituting in 5.24 in terms of 9.38 we obtain for selection within families in the S_n generation

$$E(\Delta q)_n = [k_n/\sigma(\hat{X})_n](1/2)^n q_i(1-q_i)u_{xi} \quad 9.39$$

and for selection in all generations from the S_1 to the S_n, the total change to be expected in allele frequency is

$$\Sigma_n E(\Delta q)_n = \Sigma_n[k_n/\sigma(\hat{X})](1/2)^n q(1-q)u_x \quad 9.40$$

and, in the special case where the expectation of $[k/\sigma(\hat{X})]$ is the same in all generations

$$\Sigma_n E(\Delta q)_n = [k/\sigma(\hat{X})]q(1-q)u_X \Sigma_n (1/2)^n \approx [k/\sigma(\hat{X})]q(1-q)u_X \qquad 9.41$$

because $\Sigma_n(1/2)^n$ approaches 1.0 as its limit when n becomes large. For example, given n=5, $\Sigma_n(1/2)^n = 1/2 + 1/4 + 1/8 + 1/16 + 1/32 = 31/32$.

Linkage disequilibrium was ignored in equations 9.37 - 9.41 though it is, in fact, inevitable within families by self-fertilization. However, it has no impact on $E(\Delta q)$ in this instance because for every pair of genes segregating in a source population at linkage equilibrium, the balance between coupling and repulsion is such that disequilibrium effects sum to zero in $\sigma(\hat{q}_i, \hat{X})$ within S_n families. This is demonstrated in Appendix 9.2.

9.6.3.4. COMMENTS

When 9.30 is compared with 9.32 it becomes apparent that expected allele frequency response to selection among lines tracing to different S_0 ancestors whould be about twice as large when delayed until lines are homozygous. Using subscripts p and 1 to identify variables of 9.30 and 9.32, respectively, we obtain

$$E(\Delta q)_p/E(\Delta q)_1 = 2[(k_p/k_1)][\sigma(\hat{X})_1/\sigma(\hat{X})_p]/[1+(1-2q)a_X/2]$$

If the size of the genetic sample and number selected are the same in both cases, $k_p/k_1 = 1.0$. Then, assuming that $\sigma(\hat{X})_p = \sigma(\hat{X})_1$, the ratio simplifies to[13]

$$E(\Delta q)_p/E(\Delta q)_1 = 2/[1+(1-2q)a_X/2] \qquad 9.42$$

The quantity, $(1-2q)a_X$, should in general be negative because (1) the source population frequencies of favorable alleles are expected to be larger than 1/2 more often than not and (2) the fact of inbreeding depression suggests that a majority of favorable alleles are at least partially dominant ($a>0.0$). If $(1-2q)a_X$ is negative the right hand member of 9.42 is greater than 2.0 and this would partly or completely offset $\sigma(\hat{X})_p$ being larger than $\sigma(\hat{X})_1$. Comparing 9.31 and 9.33 the same general result is obtained. Other things being equal, the expected allele frequency response to selection among lines on the basis of general combining ability estimates increases when delayed until lines are more highly inbred. Again the response appears to be about twice as great for selection among pure lines as for selection among S_0 plants. On the other hand, it is apparent from 9.39 that the response to within family selection is greatest in the case of S_1 families and becomes smaller (by 50%) in each succeeding generation. This strongly suggests that selection within S_3 or more advanced generation families is not worth the effort. It appears that within family selection is best concentrated in no more than the first two generations with between line selection being delayed until later.

Another contrast worth noting is that in selection within families the criterion (and selection) units are individual plants while the criterion units in both forms of selection among lines are families. Consequently, while the selection criterion in selection within families is ordinarily a score based on visual evaluation of one or more agronomic traits of the plants, it can be total value (W) in either form of selection among lines.[14] Stated differently, the value of within family selection hinges on correlated response but that of selection among lines can be a function of direct response. Sections 6.37 and 12.4 are both relevant in this connection. The former provides the basis for predicting relative magnitudes of correlated and direct responses.

The latter makes the point that positive selection coefficients for all favorable alleles can only be assured by the use of total value as the selection criterion. However, harvestable yield, proposed in footnote 14 as a practical measure of total value in maize, would not reflect the pollen production of lines not grown in isolation. In practice, lines deficient in pollen production would have to be eliminated on the basis of visual evaluation.

Expressions 9.30, 9.31, and 9.35 enable comparisons of three forms of selection among pure lines (selection based on performance of the lines themselves, selection based on "general combing ability", and selection among pairs of lines based on single cross performance) as procedures for increasing allele frequencies in the source population.

Cycle length does not appear as a variable in this section (9.6.3). However, when comparing variants of the IH system with each other (as in section 9.6.1.2) or with other selection systems response expectations must be stated as functions of time. Examples are provided in Chapter XIV.

Finally, if more than one of the types of selection that have been considered are to be employed within single cycles of the IH system, whether the response expectations shown above are additive comes into question. For example, given selection within families in the S_1 and S_2 generations and between lines in the S_6, would $E(\Delta q)$ for the whole process be reasonably approximated by the sum of the three quantities obtained using 9.30 and 9.39? Intuition suggests that the true expectation would be over-estimated a little by that sum. However, depending on how the response expectations for variants of the IH system compare with those for other recurrent selection systems this issue may or may not be critical.

• 9.6.4. Selection among Line Crosses—The Cultivar Search Phase of the IH System

The objective in this phase is to identify elite crosses, i.e., those for which genotype value excels by as much as possible the average for the entire sample of crosses that could be generated using all of the lines developed in the program. It is well known that the genotype values of single cross hybrids are (a) positively correlated with the average genotype values of their pure line parents but (b) that this correlation is not perfect as it would be if all gene effects were additive. The terms "general combining ability" and "specific combining ability", familiar to plant breeders, recognize and reflect the existence of this positive but incomplete correlation. Because specific combining abilty exists, i.e., the performance of hybrids is not perfectly predictable from the performance of their pure line parenats, the best of any sample of hybrids can only be identified via performance comparisons that are extensive enough to make standard errors of the averages for individual hybrids relatively small. It goes without saying that unless standard errors are extremely small the hybrid for which the observed phenotypic mean is highest may not be the very best of the sample. On the other hand, given reasonably small standard errors, it would be among the best, not greatly inferior to the the very best and substantially better than the average for all those compared. Recognizing these facts and that as long as its pure line parents are maintained any selected hybrid can be reproduced as extensively as desired for commmercial use, interest centers on the "genotypic superiority of selections" for which the expectation is given by 5.29, shown below for the case where X = W.

$$E(\bar{Y}_{WS} - \bar{Y}_W) = (\bar{W}_S - \bar{W})\sigma^2(Y_W)/\sigma^2(\hat{W}) = [k/\sigma(\hat{W})]\sigma^2(Y_W) \qquad (5.29)\ 9.43$$

In this expression $(\bar{Y}_{WS} - \bar{Y}_W)$ is the expected amount by which selections are genotypically superior, with respect to total value, to the average of the unselected line cross population from which they came; $(\bar{W}_S - \bar{W})$ is the selection differential; $\sigma^2(Y_W)$ is the total variance of genotype values; and $\sigma^2(\hat{W})$ is the phenotypic variance of the performance means employed as the basis for selection among the single crosses.

It is natural to question whether the selection of both (a) lines for use in the regeneration of the source population and (b) the hybrids to be recommended for use as cultivars could reasonably be based on the same field comparison of single crosses. For the following two reasons it appears that would not be optimum procedure.

(1) Comparison of 9.30 and 9.35 suggests that $E(\Delta q)$ would be larger if selection among lines were based on their own performance. Strongly favoring that inference is the fact that the 2 of 9.30 is replaced by $[1+(1-2q)a]$ in 9.35. Because $q > 0.5$ and $a > 0.0$ are expected for a majority of loci the expectation of $2/[1+(1-2q)a]$ is without doubt > 2. However, $k/\sigma(\hat{X})$ would be larger, though hardly twice as large, in 9.35 than in 9.30 because (a) almost certainly, $\sigma(\hat{X})$ would be less in single cross comparisons and (b) if the number of units to be compared were equal, the effective population size requirement would allow more intense selection among single crosses than among lines (each single cross equates to two lines).

(2) Cycle length in phase 1, the recurrent selection for increased frequency of best alleles in the source population, would be increased.

9.7. Summary Comments

The unique attribute of inbred-hybrids is that their genotypes, and thus the levels of heterosis captured in superior hybrids, are reproducible through time. That attribute, capitalized on in the cultivar search phase of I-H programs, enables identification of single cross hybrids that are superior (by as much as 15 - 20 percent) to the source population or the F_1 cross of two source populations from which they are derived. Because the reproducibility of hybrid genotypes flows from the genetic constancy over generation time of pure lines, it is obvious that inbreds will remain a crucial lynchpin in maize breeding for the foreseeable future.

The essential elements in cycles of I-H breeding programs are (1) the source population or populations from which near homozygous lines are derived, (2) a population of single crosses between lines, (3) selection focused on increasing the frequencies of useful alleles in the source populations of following cycles, (4) selection aimed at increasing the average genotype value of lines and thereby of line crosses, and (5) selection focused on identifying the very best line crosses for use as cultivars in commercial agriculture. In the absence of continuing increases in the frequencies of superior alleles in source populations, expectations of the genotype values of lines and line crosses would remain the same and genetic gain in selected line crosses would taper off rapidly in successive program cycles.

Obviously, the source population improvement phase of I-H programs is crucial with respect to total genetic gain that can be expected via any series of cycles of such programs. However, it is equally obvious that the investment in the cultivar search phase of the program needs to be adequate. Thus the design decisions in the case of I-H programs should be focused on (1) maximizing the expectation of $E(\Delta q)$ per unit of time consistent with (2) optimum allotment of resources (manpower, land, etc) to the two phases of the program.

Theoretical comparison of single stage selection systems is discussed in detail in section 14.1. The criterion employed is $E(\Delta q)$ per year, subject to a requirement concerning effective size of the S population. Allele frequency changes are, of course, functions of the substitution values (u's) of alleles with respect to the selection criterion ,X, [see any of the $E(\Delta X)$ expressions in sections 9.6.3.2 and 9.6.3.3]. Thus if X is not equal to total value (W), $E(\Delta q)$ may or may not be positive in the case of all useful alleles. It is shown in section 12.4 that positive changes in the frequencies of all alleles favorable with respect to total value can only be expected when selection is based on total value itself (when X=W). The necessary condition is linkage equilibrium in the S population--(pt-rs) = 0.0 for all pairs of genes with effects on total value. Thus if grain yield can be determined and adjusted in ways that make the result acceptable as a measure of total value, the utility of selection based on criteria for which the genetic correlation with total value is positive but not close to perfect can be questioned. It is worth remembering that (a) if there is positive genetic correlation between total value and phenotypic traits visible in single plants, those traits will be improved by selection based on total value and (b) there is a cost (which may or may not be great) of selection among plants that is based on observations of such traits.

It may be that restriction of the source population improvement phase of I-H programs to selection among S_1 or S_2 families, complemented by intra-family selection among individual plants judged desirable in the context of its time requirement, would be optimum. While $E(\Delta q)$ per cycle would not be maximum, $E(\Delta q)$ per year might be maximized as a consequence of the reduction of time required per cycle. Probably time required per cycle in the source population improvement phase would then be quite a bit less than required in the cultivar search phase. The latter could be repeated as often as the source population improvement phase or, cost considered, less often.

Whether to employ one or two source populations is a primary decision required in the design of IH programs (see section 9.6.1.5). The two population alternatives are RRS (reciprocal recurrent selection) and RSP (in which selection in both populations is based on intra-population performance). RRS would be superior if overdominance were important in the genetics of total value. However, because selection in RRS is of the progeny test variety but selection in RPS is not so restricted, comparison of 9.30 with 9.31 and of 9.32 with 9.33 indicates that RSP can be much more effective when there is little if any overdominance. The remaining question is whether RSP can be superior to the single population alternative. That question is examined in section 14.3.1 where it is shown that the answer depends on the amount of heterosis exhibited by the population obtained when random members of the two source populations are mated (see sections 14.3.1.1 and 14.3.1.2).

10 The Variance, $\sigma^2(\hat{X})$, of the Selection Criterion

The quantity on which selection is based in a single stage system of artificial selection is identified as the selection criterion and symbolized as X in this manuscript. Its value or average value for any criterion unit is symbolized as $\hat{X}$. The basic propositions,

$$E(\Delta q_i) \approx (\bar{X}_s - \bar{X})\beta(\hat{q}_i, \hat{X}) = [k\sigma(\hat{X})][\sigma(\hat{q}_1, \hat{X})/\sigma^2(\hat{X})] \qquad 5.18$$

and

$$E(\Delta \ddot{Y}_c) \approx \Sigma_i E(\Delta q_i)\,(\partial \ddot{Y}_c/\partial q_i) \qquad 5.8$$

establish the variance of $\hat{X}$ as a primary component in our theory for artificial selection. $\sigma^2(\hat{X})$ enters $E(\Delta q_i)$ <u>via</u> both the selection differential and the coeffcient of regression of $\hat{q}_i$ on $\hat{X}$. The net effect is that our approximations to $E(\Delta q_i)$ and $E(\Delta \ddot{Y}_c)$ are, in all instances, inversely proportional to $\sigma(\hat{X})$. It follows, because $\sigma(\hat{X}) = [\sigma^2(\hat{X})]^{1/2}$, that the composition of $\sigma^2(\hat{X})$ is an element of the information required for optimizing selection system procedures.

10.1. The Formulation of Statements Concerning the Composition of $\sigma^2(\hat{X})$

Three steps are involved. First a mathematical model is specified. This is a symbolic statement regarding the composition of X. It is supplemented, ordinarily, by one or more assumptions concerning the distributions of the variable components of X. The model is usually linear in form. Linear models appear satisfactory in most cases and are easier to work from. Only linear models are discussed herein. Second, if the criterion units are families, rather than individuals, an expression for $\hat{X}$ is derived from the one chosen for X. The final step is to write the expression for $\sigma^2(\hat{X})$ that is dictated by the composition of $\hat{X}$ (see Appendix 1, section A1.5.2) and the assumptions stated as adjuncts to the model.

Obviously, the mathematical model is crucial. If the sources of variation in X are not reflected appropriately in the model, the derived expression for $\sigma^2(\hat{X})$ cannot be completely valid and may be seriously misleading. In general, appropriate models vary with the nature of the organism and the way in which criterion units are compared. Examples will be discussed to illustrate important principles. Terms and symbols defined in section 2.13 will be employed extensively in this chapter.

• 10.2. Models for Plants

Consider individuals produced from random matings and assume (1) that each individual results from a separate random mating and (2) that the micro-environment of each individual is a random member of the total population of environments with respect to which the values of genotypes are defined. Then the following, from section 2.13 is a reasonable model.

$$X_{jkp} = \ddot{X} + z_k + f_j + (fz)_{jk} + e_{jkp} \quad (2.49) \quad 10.1$$

where

X symbolizes a measurement on an individual and its subscripts identify the specific individual (the k-th) and the micro-environment (the p-th within the j-th macro-environment) in which the individual developed,

$\ddot{X}$ is the average of X for the entire genotype-environment population,

z_k is the effect of the k-th genotype, i.e., the genotype of the k-th individual,

f_j is the effect of the j-th macro-environment,

$(fz)_{jk}$ is the effect of interaction of the k-th genotype with the j-th macro-environment

and

e_{jkp} is a residual that includes all components of X not specifically identified above. For more explicit definitions of these symbols see section 2.13. The several effects (z, f, fz, and e) are, of course, effects relative to the quantity X.

Comstock and Moll (1963) demonstrated that, by virtue of the definitions of effects, covariances among z, f, fz, and e are all zero when considered with respect to the entire genotype-environment population. It follows that in the population of X-values that is sampled when data are collected on a finite number of individuals grown in random members of the entire environmental population, the covariances among z, f, (fz), and e are all zero. Then with 10.1 as our mathematical model

$$\sigma^2(\hat{X}) = \sigma^2(z) + \sigma^2(f) + \sigma^2(fz) + \sigma^2(e) \quad 10.2$$

when the criterion units are individuals as in mass selection. Here

$\sigma^2(z)$ is the variance of values of the genotypes of the genetic population,

$\sigma^2(f)$ is the variance of effects of the macro-environments of the environment population,

$\sigma^2(fz)$ is the variance of the entire population of effects of interaction of genotype with macro-environment (the number of these interaction effects is the product of the number of genotypes in the genetic population and number of macro-environments in the environment population, the TPE, and

$\sigma^2(e)$ is the variance of the entire population of e's (the population that includes any e that could be included in the X-value observed when an individual possessing a random genotype is grown in a random micro-environment).

Actually, in practice, individuals would never be distributed as assumed above. There are two major reasons. First, because the target population of environments visualized by the breeder always encompasses future years, to do so would be impossible as well as impractical. Second, growing the members of the genetic sample in different (random) macro-environments would increase the non-genetic portion of $\sigma^2(\hat{X})$ by inclusion of $\sigma^2(f)$ and, because response expectations are inversely proportional to $\sigma(\hat{X})$, would decrease the response to selection. The normal procedure when the criterion units are individuals is to grow and observe them all in the same macro-environment. Symbolically, no change is required in the model; 10.1 is still appropriate, but when all individuals are grown in the same macro-environment f_j, being the same for all individuals, is a constant instead of a variable as before and hence not a source of variance among the observed X-values. With this in mind, we obtain from 10.1 that

$$\sigma^2(\hat{X}) = \sigma^2(z) + \sigma^2(fz) + \sigma^2(e)$$

if it is assumed that the covariances among z, (fz), and e are all zero. Here $\sigma^2(fz)$ and $\sigma^2(e)$ symbolize the variances of the (fz)'s and e's encountered in the particular macro-environment and may be somewhat different from the quantities represented by the same symbols in 10.2. The assumption that covariances among the three effects are all zero is ordinarily made (often without being specified) and will be in this book unless the contrary is indicated. However, Comstock and Moll (op. cit.) emphasized that, while the covariance of z with (fz) is constrained to zero when averaged for all macro-environments, it may be non-zero in any single macro-environment.[1]

When the criterion units are families of one sort or another, it is possible to compare them in more than one macro-environment, growing a different sample from each family in each macro-environment. An example familiar to plant breeders is the comparison of a series of families at each of two or more locations. The progeny testing of dairy bulls provides a parallel example from animal breeding except that the progenies (the criterion units) are ordinarily not distributed so uniformly to the same macro-environments. Variations of 2.51 from section 2.12 (restated below in a slightly modified form) provide appropriate models.

$$X_{ijm} = \ddot{X} + \bar{z}_i + f_j + (f\bar{z})_{ij} + e_{ijm} \qquad (2.51) \quad 10.3$$

where $\ddot{X}$ and f_j have the same meanings as in 10.1,

$\bar{z}_i$ is the average effect of genotypes possible in the i-th family,

$(f\bar{z})_{ij}$ is the effect of interaction between the i-th family and the j-th macro-environment, and

e_{ijm} is the residual effect associated with the m-th individual member of the i-th family that is grown in the j-th macro-environment.

More explicit definitions of these symbols are provided in section 2.13. The variation of 10.3 that is appropriate in a specific case depends on (1) whether the relationship between family members is the same for all pairs of members,[2] (2) the structure of the field comparison and (3) whether a value of X is obtained for each individual or the separate values of X pertain to collections of individuals, e.g., those in the same plots of a field trial in the case of plants.

Consider selection among S_1 families in a tree species and assume (1) that the families are the criterion, as well as the selection, units, (2) that p members of each family are grown in each of n macro-environments (the same ones for all families), (3) that all individuals grown in the same macro-environment are distributed randomly relative to micro-environment variation, and (4) that random macro-environments can be employed in the comparison. Then 10.3 is an appropriate model and the corresponding expression for $\hat{X}_i$ is

$$\hat{X}_i = [pn\ddot{X} + pn\bar{z}_i + p\Sigma_j f_j + p\Sigma_j(f\bar{z})_{ij} + \Sigma_j\Sigma_m e_{ijm}]/pn$$

$$= \ddot{X} + \bar{z}_i + \bar{f} + \Sigma_j(f\bar{z})_{ij}/n + \Sigma_j\Sigma_m e_{ijm}/pn \qquad 10.4$$

where $\bar{f}$ is the average of the effects on X of the n macro-environments of the comparison. From 10.4 the variance of $\hat{X}$ is

$$\sigma^2(\hat{X}) = \sigma^2(\bar{z}) + \sigma^2(f\bar{z})/n + \sigma^2(e)/pn \qquad 10.5$$

if the variances of the $f\bar{z}$'s and e's are the same in all macro-environments or if $\sigma^2(f\bar{z})$ and $\sigma^2(e)$ are defined as the averages of the variances of the $f\bar{z}$'s and e's in the macro-environments of the comparison.

Now consider the following variation in the structure of the field comparison visualized immediately above. Suppose that, instead of a completely randomized arrangement within macro-environments, the standard randomized block arrangement is employed with one tree from each family being distributed randomly within each of p blocks. An appropriate model would then be the following modification of 10.3.

$$X_{ijm} = \ddot{X} + \bar{z}_i + f_j + (f\bar{z})_{ij} + B_{jm} + e_{ijm} \qquad 10.6$$

where B_{jm} represents the effect of the m-th block in the j-th macro-environment,

e_{ijm} represents the residual effect associated with the member of the i-th family grown in the m-th block in the j-th macro-environment, and other symbols have the same meaning as in 10.3. Then

$$\hat{X}_i = [pn\ddot{X} + pn\bar{z}_i + p\Sigma_j f_j + p\Sigma_j(f\bar{z})_{ij} + \Sigma_j\Sigma_m B_{jm} + \Sigma_j\Sigma_m e_{ijm}]/pn$$

$$= \ddot{X} + \bar{z}_i + \bar{f} + \Sigma_j(f\bar{z})_{ij}/n + \bar{B} + \Sigma_j\Sigma_m e_{ijm}/pn \qquad 10.7$$

where $\bar{B}$ is the average effect of all the pn blocks. Both $\bar{f}$ and $\bar{B}$ are constants relative to families and hence do not contribute to the variance of $\hat{X}$ which from 10.7 is

$$\sigma^2(\hat{X}) = \sigma^2(\bar{z}) + \sigma^2(f\bar{z})/n + \sigma^2(e)/pn \qquad 10.8$$

Though 10.8 appears identical to 10.5, there is a difference. In 10.8, $\sigma^2(e)$ symbolizes the within block variance of e instead of the total within macro-environment variance of e symbolized in 10.5. Because the first of these variances is expected to be smaller than the second, the randomized block structure would usually be employed in the sort of comparison being discussed.

More often than not the unit in plant family comparisons is a "plot" within which several (or numerous) family members are observed as a group with one value of X being obtained and recorded per plot. Then if all else is as assumed in the immediately preceding case, 10.6, 10.7, and 10.8 are all applicable if e is redefined as the "plot effect" on the value of X specified by its subscripts in 10.6 and $\sigma^2(e)$ as the variance of such plot effects.

In the preceding examples involving comparisons of family units in more than one macro-environment it has been assumed that random macro-environments could and would be employed. However, if temporal variation in environment is significant and relatively consistent over the geographical expanse of the environment population, the macro-environments associated with any specific time interval comprise a sub-set within which the variation is less than among random macro-environments from the entire environment population. Then a truly random sample of macro-environments would not be available within any finite period of time and the assumption of random macro-environments for family unit comparisons would not be realistic for practical programs. Recognizing this many (probably the majority) who work with annual plants have adopted a model that cross-classifies macro-environments by year and location. The environments encountered in a specific year and location are identified as the micro-environments of a specific macro-environment and the effect of that macro-environment is described as the sum of an effect of the location, an effect of the year, and an effect of interaction between the year and the location.[3] Year effects, by virtue of their definition, are constants, the same at every location, and location effects, again by virtue of definition, are the same in all years.[4] Symbolically, the model states that

$$X_{ijkm} = \mu + \bar{z}_i + a_j + b_k + (ab)_{jk} + (\bar{z}a)_{ij} + (\bar{z}b)_{ik} + (zab)_{ijk} + B_{jkm} + e_{ijkm} \qquad 10.9$$

where

a_j is the effect of the j-th location,

b_k is the effect of the k-th year,

$(ab)_{jk}$ is the effect of interaction between the j-th location and the k-th year,

$(\bar{z}a)_{ij}$ is the effect of interaction between the i-th family and the j-th location,

$(\bar{z}b)_{ik}$ is the effect of interaction between the i-th family and the k-th year,

$(\bar{z}ab)_{ijk}$ is the second order interaction effect of the i-th family, j-th location, and k-th year,

e_{ijkm} is the residual effect associated with the observation unit (plot or individual) identified by the subscripts, and other symbols have meanings indicated earlier in this section. As in previous examples, uniformity of relationship between family members is assumed. The difference between this model and 10.6 is only that here the effects of macro-environment and of interaction between family and macro-environment are subdivided; 10.9 is 10.6 with $a_j + b_k + (ab)_{jk}$ substituted for f_j and $(\bar{z}a)_{ij} + (\bar{z}b)_{ik} + (\bar{z}ab)_{ijk}$ substituted for $(f\bar{z})_{ij}$. Assuming that the families are compared in r replications (blocks) at each of v locations in each of t years, we obtain from 10.9 that

$$\hat{X}_i = \ddot{X} + \bar{z}_i + \bar{a} + \bar{b} + (ab) + \Sigma_j(\bar{z}a)_{ij}/v + \Sigma_k(\bar{z}b)_{ik}/t + \Sigma_j\Sigma_k(\bar{z}ab)_{ijk}/vt + \bar{B} + \Sigma_j\Sigma_k\Sigma_m e_{ijkm}/rvt \qquad 10.10$$

where $\bar{a}$, $\bar{b}$, and (ab) are the respective averages of the effects of locations, years, and interactions between years and locations involved in the comparison. Each of these, as well as

$\bar{B}$ and $\ddot{X}$, are constant relative to families. With that in mind, we obtain from 10.10 that

$$\sigma^2(\hat{X}) = \sigma^2(\bar{z}) + \sigma^2(\bar{z}a)/v + \sigma^2(\bar{z}b)/t + \sigma^2(\bar{z}ab)/vt + \sigma^2(e)/rvt \qquad 10.11$$

A slightly different result is obtained for the case in which the locations in which comparisons are made are different from one year to the next when $t > 1$. Then two subscripts are required to identify the location effect and a_{jk} would be the effect of the j-th location in the k-th year, $(\bar{z}a)_{ijk}$ the effect of interaction between the i-th family and the j-th location in the k-th year and $(\bar{z}ab)_{ijk}$ the effect of second order interaction between the i-th family, the j-th location in the k-th year and the k-th year. The expression for $\hat{X}_i$, would be the same as 10.10 except that $\Sigma_j(\bar{z}a)_{ij}/v$ would be replaced by $\Sigma_j\Sigma_k(\bar{z}a)_{ijk}$. As a result of that difference the variance of $\hat{X}$ would be

$$\sigma^2(\hat{X}) = \sigma^2(\bar{z}) + \sigma^2(\bar{z}a)/v + \sigma^2(\bar{z}b)/t + \sigma^2(\bar{z}ab)/vt + \sigma^2(e)/rvt \qquad 10.12$$

instead of the quantity indicated by 10.11.

When biennnial or perennial plants are considered, 10.9 is still an appropriate model if the time inteval is understood to consist of two or more consecutive years instead of a single year. Then using "time" to indicate a set of consecutive years, b_k would be the effect of the k-th "time", $(\hat{z}b)_{ik}$ the effect of interaction between the i-th family and the k-th "time", and $(\hat{z}ab)_{ijk}$ the effect of second order interaction involving the i-th family, the j-th location and the k-th "time". Conceivably, a series of families might be compared in two "times" having no year in common if the interval required to obtain values of X were no more than two or three years but it is unlikely that this would be done if the required time interval were over three years. If comparisons were made in two or more "times" having years in common, the usual assumption that interaction effects are independently distributed would hardly be justified; hence neither 10.10 nor 10.11 would be exactly correct as they otherwise could be.

When the criterion units are half-sib families as they are in selection among half-sib families on the basis of their own performance or may be in progeny test selection, another modification of the basic model is appropriate and useful. Using the terms "common " parent and "random" parent as they were used in section 8.2.1,[5] let c_i be the effect of genotype of the common parent on the average genotype value of the i-th family and r_{ij} be the effect of genotype of the j-th random parent on the value of genotype of each of its offspring members of the i-th family and consider a population of half-sib families in which each family is composed of nF individuals (n from each of F random parents). Symbolically, the mean of the i-th family would be

$$\bar{z}_i = [Fnc_i + nr_{i1} + nr_{i2} + \cdots\cdots nr_{iF}]/Fn.$$

It follows, assuming that the random parents are indeed random relative to both the common parent and each other, that the variance among family means would be

$$\sigma^2(\bar{z}) = [F^2n^2\sigma^2(c) + Fn^2\sigma^2(r)]/n^2F^2 = \sigma^2(c) + \sigma^2(r)/F \qquad 10.13$$

Intuition suggests that the interaction variances of 10.11 should be affected by the structure of half-sib families in more or less the same way as 10.13 shows $\sigma^2(\bar{z})$ to be affected. It can, in fact, be shown following the principle by which $(f\bar{z})$ was defined, see 2.50, that, in the case of half-sib families in which each random parent has the same number of offspring and the number of random parents is the same,

$$\sigma^2(\bar{z}a) = \sigma^2(ca) + \sigma^2(ra)/F \qquad 10.14$$

where $\sigma^2(ca)$ is the variance of the effects of interaction between common parent and location and

$\sigma^2(ra)$ is the variance of effects of interaction between random parent and location.

The analogous expressions for $\sigma^2(b)$ and $\sigma^2(\hat{z}ab)$ should be apparent.

For work with plants, use of 10.13, 10.14, and their analogs relative to $\sigma^2(\hat{z}b)$ and $\sigma^2(\hat{z}ab)$ is justified in the usual case where seed lots from which half-sib families are grown are bulks of roughly the same numbers of seeds from each random parent. Substituting $\sigma^2(c) + \sigma^2(r)/F$ for $\sigma^2(\hat{z})$, $\sigma^2(ca) + \sigma^2(ra)/F$ for $\sigma^2(\hat{z}a)$, etc. in 10.11 we obtain

$$\sigma^2(\hat{X}) = \sigma^2(c) + \sigma^2(r)/F + [\sigma^2(ca) + \sigma^2(ra)/F]/v + [\sigma^2(cb) + \sigma^2(rb)/F]/t + [\sigma^2(cab) + \sigma^2(rab)/F]/vt + \sigma^2(e)/rvt \qquad 10.15$$

where $\sigma^2(cb)$, $\sigma^2(cab)$, $\sigma^2(rb)$, and $\sigma^2(rab)$ have meanings that are apparent by analogy from the definitions of $\sigma^2(ca)$ and $\sigma^2(ra)$ stated above. From 10.15, supplemented by 10.11,10.13, and 10.14, we note the following.

(1) As F, the number of random parents is increased, the genetic variance among families decreases. This, of course, comes as no surprise; it is obvious from statistical principles that it should be true.

(2) As F is increased there is a decrease in each of the genotype-environment interaction variances that is proportional to the decrease in $\sigma^2(\bar{z})$. That this should be so was perhaps less obvious in advance but in retrospect appears entirely reasonable.

10.3. Models for Animals

A sufficiently general model would be appropriate for either plants or animals. However, (1) maternal effects are important relative to many traits of animals but are ordinarily considered trivial as a source of variation in the quantitative traits of plants, (2) effects of interaction between genotype and environment, which contribute substantially to the variance of productivity traits of plants, are more often than not assumed quite unimportant with respect to animal traits, (3) the systematically replicated family unit comparisons that are so very common in plant breeding are quite rare in animal breeding and (4) the individual is the unit observation more frequently in animal breeding than in plant breeding. For all of these reasons, models actually used are ordinarily different in the two areas. At the same time those to be shown for animals will be seen to be straight-forward modifications of one or another models shown in section 10.2. A small number of examples will suffice to illustrate the general nature of models for animals and the reasons for variation in the variance of $\hat{X}$.

The following expression contains more detail than is usual in models for measurements on individual animals but, for that reason, will serve to link all the example situations to be discussed.

$$X_{knq} = \ddot{X} + z_k + m_k + f_n + (fz)_{kn} + p_{nq} + e_{knq} \qquad 10.16$$

Here X symbolizes a measurement on an individual and its subscripts identify the individual (the k-th) and enclosure (the q-th within the n-th macro-environment) in which the animal was contained,

z_k is the effect of the genotype of the k-th individual,

m_k is the maternal effect of the dam of the k-th individual,

f_n is the effect of the n-th macro-environment,

$(fz)_{nk}$ is the effect of interaction between the k-th genotype and the n-th macro-environment,

p_{nq} is the effect of the q-th enclosure (pen) in the n-th macro-environment, and

e_{knq} is the effect of the within pen micro-environment of the k-th individual.

This expression is a modification of 2.49. The maternal effect, normally considered of no consequence in plants, and the enclosure effect have been added. The enclosure, necessitated by the mobility of animals, parallels the "block" recognized in 10.6. The reality of enclosure (pen) effects is sometimes questioned on the grounds that pens can be so similar in all respects that no differential effects should exist. However, they are inevitable though they need not always be large, if for no other reasons than (1) infections are more likely to be transmitted from one animal to another in the same pen than from pen to pen and (2) competitive interactions among animals will vary from pen to pen.

When the criterion units are individuals, i.e., in the case of mass selection, the variance of $\hat{X}$ can be deduced directly from 10.16 but will depend in its details on the distribution of animals relative to pens and macro-environments and the assumptions made concerning covariances among effects. Consider first the situation in which the selection practiced is always between animals raised in the same enclosure, pasture lot, or pen and managed in the same way (so that they are also in the same macro-environment). This does not mean that all selections will necessarily come from the same macro-environment and pen but rather that the basis for their selection is their performance relative to others grown with them in the same enclosure, not their performance relative to others in the genetic sample that have been maintained in other pens or macro-environments. Then because f and p are constants in the values of $\hat{X}$ being compared as the basis for selection

$$\sigma^2(\hat{X}) = \sigma^2(z) + \sigma^2(m) + \sigma^2(fz) + \sigma^2(e)$$

if the covariances among z, m, (fz), and e are all assumed to be zero. Next, suppose that selection is among animals raised in the same macro-environment but in several pens instead of just one. Then f would be the same for all animals being compared but p, though constant for animals in the same pen would not be for all being compared. The variance would be

$$\sigma^2(\hat{X}) = \sigma^2(z) + \sigma^2(m) + \sigma^2(fz) + \sigma^2(p) + \sigma^2(e)$$

if, once more, covariances were assumed equal to zero. Finally, suppose selection were among all animals without regard to the macro-environment in which they were raised so that f, as well as p, would be variable among $\hat{X}$ values compared. Then, assuming covariances to be zero,

$$\sigma^2(\hat{X}) = \sigma^2(z) + \sigma^2(m) + \sigma^2(f) + \sigma^2(fz) + \sigma^2(p) + \sigma^2(e). \qquad 10.17$$

It is quite clear that $\sigma^2(f)$ could be very large relative to other components of 10.17 and may appear, therefore, that the last of the three variations of $\sigma^2(\hat{X})$ was not worth listing. However, until the last few decades, most of the selection among males in the pure breeds (the S populations of most animal breeding) was among animals from different macro-environments. Farmer-breeders, to avoid inbreeding within their herds, bought their sire replacements from each other and in the selection process compared animals raised in various locations and under different husbandry systems. Recognition of the potentially large non-genetic variance because of the inclusion of $\sigma^2(f)$ has finally led to the development of centralized test-stations and self-contained breeding operations.

To set the stage for treatment of families as criterion units, 10.16 will be modified further by subdivision of the effect (z) of genotype, as when half-sib families were considered in section 10.2. Here because paternal half-sibs are common and maternal half-sibs rare in animal breeding comparisons, the "common parent" will be equated throughout with the male parent and "random parent" with the female parent and symbols will be employed that are more frequently seen in animal breeding literature. Let

$$z_{ijk} = s_i + d_{ij} + w_{ijk} \qquad 10.18$$

where s_i = the effect of genotype of the i-th sire,

d_{ij} = the effect of genotype of the j-th dam when mated to the i-th sire, and

w_{ijk} = the deviation of z_{ijk} from $(s_i + d_{ij})$, i.e., the deviation of the value of the genotype of the k-th offspring of the j-th dam mated to the i-th sire from the expectation of genotypic values for members of the identified full-sib family.

Substituting in 10.16, we obtain

$$X_{ijknq} = \ddot{X} + s_i + d_{ij} + m_{ij} + f_n + (fz)_{nijk} + p_{nq} + w_{ijk} + e_{ijknq} \qquad 10.19$$

Proper subscripts to the pen effect vary with the exact ways in which pens are employed. Those that appear in 10.19 and 10.21 would be appropriate if the assignment of animals to pens was to be at random and without regard to family membership.

Applying the principles employed in subdividing the $(fz)_{jk}$ of 2.49 into the $(f\bar{z})_{ij}$ and $[(fz)_{ijk} - (f\bar{z})_{ij}]$ of 2.51 and 2.52, see section 2.13, $(fz)_{nijk}$ of 10.23 can be subdivided as follows

$$(fz)_{nijk} = (fs)_{ni} + (fd)_{nij} + (fw)_{nijk} \qquad 10.20$$

Then, substituting in 10.19 in terms of 10.20, we obtain

$$X_{ijknq} = \ddot{X} + f_n + s_i + (fs)_{ni} + d_{ij} + m_{ij} + (fd)_{nij} + p_{nq} + w_{ijk} + (fw)_{nijk} + e_{ijknq}. \qquad 10.21$$

Assuming half-sib families as criterion units, consider first the situation in which all animals are grown in the same macro-environment and enclosure, e.g., beef cattle or sheep managed as one herd or flock on the range. Then, from 10.21 the average of the selection criterion for the i-th half-sib family would be

$$\hat{X}_i = [\Sigma_j \Sigma_k X_{ijk}] = \ddot{X} + f + p + s_i + (fs) + \Sigma_j n_{ij}[d_{ij} + m_{ij} + (fd)_{ij}]/N_i + \Sigma_j \Sigma_k[w_{ijk} + (fw)_{ijk} + e_{ijk}]/N_i \qquad 10.22$$

where n_{ij} is the number of offspring of the j-th dam mated to the i-th sire and

$N_i = \Sigma_j n_{ij}$ = number of members of the i-th family.

Let

$$\ddot{X} = \ddot{X} + f + p,$$

$$S_i = s_i + (fs)_i, \quad 10.23$$

$$D_{ij} = d_{ij} + m_{ij} + (fd)_{ij}, \quad 10.24$$

and $W_{ijk} = w_{ijk} + (fw)_{ijk} + e_{ijk}$. 10.25

Then 10.22 can be written in the following condensed form

$$\hat{X}_i = \ddot{X} + S_i + \Sigma_j n_{ij} D_{ij}/N_i + \Sigma_j \Sigma_k W_{ijk}/N_i \quad 10.26$$

and for families similar in structure, i.e., families exactly alike with respect to to their distributions of n_{ij}'s, the variance of $\hat{X}$ would be

$$\sigma^2(\hat{X}) = \sigma^2(S) + \sigma^2(D)\Sigma_j(n_{ij})^2/N^2 + \sigma^2(W)/N$$

where $N = \Sigma_j n_{ij}$, the number of individuals in each half-sib family.

As a variation of the above, consider the situation in which the offspring of each dam are raised in a separate random pen, as when litters of pigs are raised separately, each in a different pen. Then, instead of a pen effect that is constant for all individuals, there would be as many pen effects as dams and the average of X for a half-sib family would be

$$\hat{X}_i = \ddot{X} + S_i + \Sigma_j n_{ij}(D_{ij} + p_{ij})/N_i + \Sigma_j \Sigma_k W_{ijk}/N_i \quad 10.27$$

where p_{ij} is the effect of the pen in which offspring of the j-th dam mated to the i-th sire are raised.

In this case, differences among pen effects contribute to the variance of $\hat{X}$ which for families with the same structure is

$$\sigma^2(\hat{X}) = \sigma^2(S) + [\sigma^2(D) + \sigma^2(p)]\,\Sigma_j(n_{ij})^2/N^2 + \sigma^2(W)/N \quad 10.28$$

In another variation, each individual would be maintained in a separate pen. This might occur when pullets are kept in separate cages during observation of egg production. Then the mean for the i-th half-sib family would be

$$\hat{X}_i = \ddot{X} + S_i + \Sigma_j n_{ij} D_{ij}/N_i + \Sigma_j \Sigma_k (W_{ijk} + p_{ijk})/N_i \quad 10.29$$

and the variance would be

$$\sigma^2(\hat{X}) = \sigma^2(S) + \sigma^2(D)\Sigma_j(n_{ij})^2/N^2 + \sigma^2(W)/N + \sigma^2(p)/N$$

There are, of course, other ways in which families can be distributed to environments. For example, each half-sib family could be assigned to a different pen or each could be divided between two or more pens in which they would not be co-mingled with members of other half-sib families.[6] In each case the family mean value derived from 10.19 and the consequent expression for the variance of $\hat{X}$ would vary with and reflect the procedural detail.

It should be noted that all expressions presented for half-sib families apply as well for full-sib families. The full-sib family is the special case in which the number of female parents of family members is 1.0. In that case 10.27, for example, collapses to

$$\hat{X}_i = \ddot{X} + S_i + D_{ij} + p_{ij} + \Sigma_k W_{ijk}/n$$

and 10.28 to

$$\sigma^2(\hat{X}) = \sigma^2(S) + \sigma^2(D) + \sigma^2(p) + \sigma^2(W)/n$$

where n is number of offspring per family.

Note also that in the special case where the number of offspring is the same for each female parent and the number of female parents is the same for each half-sib family, the $\Sigma_j(n_{ij})^2/N^2$ of 10.26, 10.27, and 10.29 becomes

$$\Sigma_j(n_{ij})^2/N^2 = FY^2/F^2Y^2 = 1/F$$

where F = number of female parents per half-sib family and

Y = number of offspring per female parent as in section 8.2.1.

10.3.1. Adjustments for Tangible Sources of Non-Genetic Variation

Variation in the performance of animals being compared is more likely to be due in part to such measurable causative variables as litter size or age and weight at the beginning of test periods than is plant variation. The variation from such sources is often considered sufficient to make adjustments of the raw data worthwhile. A familiar example is the adjustment of annual milk production in dairy cattle to the so-called "mature-equivalent" basis. The obvious purpose of such adjustments is to reduce the variance of X by reduction of the non-genetic portion of its variance. Expressions such as 10.16 serve as models for either original or adjusted values of X.

10.3.2. Covariances between Effects

The several expressions for the variance of $\hat{X}$ when the criterion units are individuals were all obtained assuming absence of covariance among the variable effects of10.16. It is proper to note that at least three of these can be non-zero. As noted in section 10.2, Comstock and Moll (1963) emphasized that the covariance between effects of genotype and the interactions of genotypes with a specific macro-environment need not be zero. In this instance then, $\sigma(z,fz)$ may be non-zero. Next consider z and m. Both of these effects can be assumed partly determined by genotype of the dam. Obviously, there may therefore be covariance between them. Finally, when animals occupy the same enclosure, each becomes a part of the environment of the others. If the more vigorous "bully" or take advantage of the less vigorous and vigor is determined partly by genotype, there is reason to expect covariance between z and e when animals are not all kept in separate pens.

The expressions that have been presented for variance of $\hat{X}$ when the criterion units are half-sib families assume absence of covariances among S, D, W, and p. This appears reasonable if female parents are random relative to male parents and pen assignments (when more than a single enclosure is employed) are random. On the other hand, the variances of S, D, and W may be affected by covariances of component effects that deserve to be recognized. From 10.23, 10.24, and 10,25 it is apparent that

$$\sigma^2(S) = \sigma^2(s) + \sigma^2(fs) + 2\sigma(s,fs)$$

$$\sigma^2(D) = \sigma^2(d) + \sigma^2(m) + \sigma^2(fd) + 2\sigma(d,m) + 2\sigma(d,fd) + 2\sigma(m,fd)$$

and $$\sigma^2(W) = \sigma^2(w) + \sigma^2(fw) + \sigma^2(e) + 2\sigma(w,fw) + 2\sigma(w,e) + 2\sigma(fw,e)$$

Which of the seven covariances in these three expressions may be non-zero? If $\sigma(z,fz)$ can be non-zero (see earlier reference to Comstock and Moll), it is obvious from 10.18 and 10.20 that the same is true for $\sigma(s,fs)$, $\sigma(d,fd)$, and $\sigma(w,fw)$. For the same reason that covariance can be expected between z and m, covariance between d and m is equally to be expected. Whether it is large or small relative to $\sigma^2(d)$ and $\sigma^2(m)$ in particular cases probably varies with the trait or function of traits employed as the selection criterion and designated as X. Finally, the argument for covariance between the z and e of 10.16 applies with equal force to the w and e components of W. Hence covariance between w and e cannot be ruled out.

The covariances discussed in the preceding paragraph are between effects that are both contained by S, D, or W. Consequently, when only the total magnitudes of $\sigma^2(S)$, $\sigma^2(D)$, and $\sigma^2(W)$ are important, whether they are affected by one or more of these covariances may be irrelevant. However, for logical completeness we should recognize that if covariances between the z and e of 10.16 and the w and e of 10.21 and 10.25 are possible, then covariance between d and e is equally possible when animals from different dams are kept in the same pens and covariance between s and e is possible when animals by different sires are kept in the same pens.

Practical ends will not always be served by taking account of possible covariances. On the other hand, to ignore the fact that covariances can be non-zero and that particular ones are likely to be non-zero may, in some instances, be costly with respect to judgements or interpretations.

• 10.4. The Composition of $\sigma^2(\hat{X})$ and Comparisons of Procedures

The expected genetic change per unit of time should obviously be considered when a choice is to be made between selection systems or variations of the same system. Comparisons can be made in terms of either $E(\Delta\ddot{Y})$, the expected change in the average value of genotypes or $E(\Delta q)$, the expected change in the frequency of an allele. The expected change in $E(\Delta q)$ per cycle is given by 5.18. Dividing by cycle length yields expected change per unit of time

$$E(\Delta q_i)/L = [k/\sigma(\hat{X})]\,\sigma(\hat{q}_i,\hat{X})/L$$

where L symbolizes cycle length. It follows that

$$[E(\Delta q_i)_1/L_1]/[E(\Delta q_i)_2/L_2]$$
$$=[k_1/k_2][\sigma(\hat{X}_2)/\sigma(\hat{X}_1)][\sigma(\hat{q}_i,\hat{X}_1)/\sigma(\hat{q}_i,\hat{X}_2)][L_2/L_1]$$

where 1 and 2 as subscripts identify quantities pertaining to procedures 1 and 2. Any of the four ratios of which the right hand member is composed can be larger or smaller than 1.0. $\sigma(\hat{X}_2)/\sigma(\hat{X}_1)$, the ratio of the standard deviations of $\hat{X}$, will always have a bearing on the relative rates of allele frequency change promoted by two procedures but it may or may not be dominant with respect to the outcome of the comparison. Consider the following examples.

• 10.4.1. Full-Sib vs. Half-Sib Reciprocal Recurrent Selection

These variations of RRS were distinguished in section 9.3.2. Assuming that the inbreeding of selection unit individuals and (by implication) that cycle length would be the same for the two procedures it was shown, as Jones et al (1971) had shown, that with respect to allele frequency change in either of the S populations

$$E(\Delta q_i)_F/E(\Delta q_i)_H = [k_F/k_H][\sigma(\hat{X})_H/\sigma(\hat{X})_F] \qquad (9.14)\ 10.30$$

Here F and H as subscripts indicate full-sib and half-sib RRS, respectively. It was noted that in the full-sib version, each test progeny serves as the criterion unit for a selection unit in each S population and hence that the genetic samples can be twice as large in full-sib RRS though the number of test progenies to be field tested is no greater. It follows that if number of selections is the same in the two cases $k_F/k_H > 1.0$. Values of this ratio, tabulated in section 9.3.2, range from 1.09 to 1.59 as proportion selected goes from .02 to .50 in half-sib RRS (and, accordingly, from .01 to .25 in full-sib RRS). The question posed by 10.30 is whether the advantage resulting from the greater selection intensity possible with full-sib RRS can be more than offset by $\sigma(\hat{X})_H < \sigma(\hat{X})_F$. Rearranging 10.15, we find[7]

$$\sigma^2(\hat{X}) = \sigma^2(c) + \sigma^2(ca)/v + \sigma^2(cb)/t + \sigma^2(cab)/vt + [\sigma^2(r) + \sigma^2(ra)/v + \sigma^2(rb)/t + \sigma^2(rab)/vt]/F + \sigma^2(e)/pvt \qquad 10.31$$

Remembering that F = 1.0 in full-sib RRS and F> 1.0 in half-sib RRS, it is apparent from 10.31 that $\sigma^2(\hat{X})_H/\sigma^2(\hat{X})_F$ is less than 1.0. The actual magnitude of the ratio will depend on F, p, v, t and the relative magnitudes of the variances that contribute to $\sigma^2(\hat{X})$. On the basis of data from Nebraska field trials at one location (v = 1) in one year (t = 1), Jones and co-workers listed estimates as follows for variances of grain yield of corn in kgs/plant

$$\sigma^2(m)\ [=\sigma^2(c) + \sigma^2(ca) + \sigma^2(cb) + \sigma^2(cab)] = 0.00014$$

$$\sigma^2(f)\ [=\sigma^2(r) + \sigma^2(ra) + \sigma^2(rb) + \sigma^2(rab)] = 0.00027$$

and $\sigma^2\ [=\sigma^2(e)] = 0.0010.$

Substituting these in 10.31 yields 1.22 as an estimate of $\sigma(\hat{X})_F/\sigma(\hat{X})_H$, i.e., 0.82 as an estimate of $\sigma(\hat{X})_H/\sigma(\hat{X})_F$, when p (number of replications) = 4 and v = t = 1 in field trial comparisons of test progenies, F = 5 in half-sib RRS and the selection unit individuals are not inbred (f = 0.0). They conclude accordingly that "the selection differential would need to be about 1.2 times greater for full than for half-sib RRS to give similar response".[8] Clearly, the conclusions would be at least slightly different if specifics were different. The generalization that is apparent from 10.31 is that, whatever the design of the field comparison (i.e., the values of v, t, and p), the reduction of $\sigma^2(\hat{X})_H$, and therefore of $\sigma^2(\hat{X})_H/\sigma^2(\hat{X})_F$, by increasing F is greater when the ratio of

$$[\sigma^2(r) + \sigma^2(ra)/v + \sigma^2(rb)/t + \sigma^2(rab)/vt] \text{ to}$$

$$[\sigma^2(c) + \sigma^2(ca)/v + \sigma^2(cb)/t + \sigma^2(cab)/vt + \sigma^2(e)/pvt]$$

is greater.

• 10.4.2. The Number of Random Parents in Progeny Test Selection

The preceeding discussion of full and half-sib RRS provides a background for examination of this issue in a broader context. Number of random parents is an issue in some programs but not in others depending on (1) the nature of the "tester" and (2) the prolificacy of random parents. Consider recurrent selection for specific combining ability (RRSC) with a "tester" that is a homozygous line. Then random parents are uniform genetically and the number employed in the production of test progenies has no bearing on the variance of $\hat{X}$. The role of prolificacy is clear when progeny testing of bulls is considered. Cows are then the random parents. Because they rarely produce more than one calf per pregnancy, the number of random parents is usually equal to family size and the design issue becomes that of optimizing number of offspring on which the progeny test is based (to employ more than one offspring per female would mean extending the cycle length of the program).

When random parents are both prolific and genetically variable 10.31 constitutes a required portion of the theoretical basis for decision. Consider again the test design (one location, one year, and four replications) and the variance component estimates used by Jones et al (op. cit.) in their comparison of full and half-sib RRS. Suppose the half-sib version is to be used so that number of random parents per test progeny (F) must be decided. With F = 5, and assuming that variances were correctly estimated,

$$\sigma^2(\hat{X}) = .00014 + .00027/5 + .0010/4 = .000444$$

Values of $\sigma^2(\hat{X})$ and $1/\sigma(\hat{X})$ are listed below assuming different values of F.

F	$\sigma^2(\hat{X})$	$1/\sigma(\hat{X})$
2	.000525	43.67
3	.000480	45.66
4	.000458	46.73
5	.000444	47.39
10	.000417	49.02

Response expectations are proportional to $1/\sigma(\hat{X})$. Thus using five rather than three random parents would result in a 3.8 percent increase. In the case of maize the costs would include a two-thirds increase in plants to be grown, pollinated and harvested to obtain test-progeny seed. There would be various alternatives for use of the extra labor and land required. Quite possibly one or more would have a larger payoff.

• 10.4.3. Optimum Design of Field Trial Comparisons

This issue is treated more thoroughly, i.e., in the context of interrelations with other significant variables, in Chapter XIV. The purpose here is only to draw attention to the two most obvious implications of the composition of $\sigma^2(\hat{X})$. It will suffice to consider 10.11 which pertains, most specifically, to annual plants and criterion units of which all members are equally related.

Other things being equal, it is obviously desirable to make $\sigma^2(\hat{X}) - \sigma^2(\bar{z})$, the nongenetic portion of the variance of $\hat{X}$, as small as possible, i.e., to make $\hat{X}$ as precise as possible as a measure of $\bar{z}$. With this in mind, the first thing to meet the eye is that an increase in either the number of locations (v) or the number of years (t) decreases three of the four error variance terms whereas only $\sigma^2(e)/rvt$ is decreased by increasing the number of replications. Thus, if differential costs were not involved, decreases in error variance would always be achieved more efficiently by increasing v, t, or both than by increasing r (as noted and emphasized by Sprague and Federer, 1951).[9]

However, to increase t is to increase cycle length. Thus, judged in terms of effect on response expectations per unit of time, increasing t would rarely, if ever, be as beneficial as increasing v. It would require $\sigma^2(\hat{z}b)$ much larger than $\sigma^2(\hat{z}a)$ and a long cycle when t = 1 so that increasing t to two would not result in a large fractional increase in cycle length.

• 10.5. The Actual Magnitudes of Variances

The variance of $\hat{X}$ can alway be estimated directly if a set of mean values is available for the criterion unit and selection criterion in question. However, it is apparent from 10.12, 10.15, and 10.31 that when the criterion units are families, $\sigma^2(\hat{X})$ is a function of the design parameters (number of replications, v, t, and F). Hence to explore the consequences of changing any of those quantities, estimates of the separate variances contributing to $\sigma^2(\hat{X})$ are required. Estimates of these variance components are obtained from analyses of variance (See Chapter XV). Early examples in the literature include Stonaker and Lush (1942), Baker et al (1943), Robinson et al (1949), Sprague and Federer (1951) and Rojas and Sprague(1952). Some problems in the estimation of variance components were discussed by Comstock and Robinson (1952) and Comstock and Moll (op. cit.).

11 Effective Population Size in Recurrent Selection Programs

The goal of recurrent selection is to make successive increases in the average value of target population genotypes via successive changes of allele frequencies in the S population(s). In the case of non-overdominant genes, the long-term goal is fixation ($q = 1.0$) of the superior allele so that only the BB genotype is present. With respect to these genes, it is obviously desirable that the probabilities of fixation (see section 2.8) be high for all superior alleles. When a significant portion of the genes with effects on total value are overdominant, populations in which the best genotype is present at all loci can only be obtained by crossing homozygous populataions (lines) that are complimentary with respect to alleles of the overdominant genes. The production of such pairs of populations is the goal of selection for combining ability (as in RRSC and RRS). Again it is desirable that probabilities of fixation be high for superior alleles, though so far as any overdominant gene is concerned, which allele is superior in either population depends on which is at highest frequency in the other. In summary, probability of fixation, which is critical with respect to non-overdominant genes in all systems of recurrent selection, is equally critical in the case of overdominant genes in selection for combining ability.

The effective size (see section 2.7) of populations bears on both probabilities of fixation and the net response to selection per cycle. Both are reduced when effective size (N_e) is reduced and both are functions of selection intensity as well. Unfortunately, in programs finite in scope (as all must be) there is a necessary trade-off between selection intensity and N_e. Unless program size is increased more intense selection means an increase in selection differential but a decrease in N_e.

Our discussion of optimum effective population size will be facilitated by distinguishing between long-term and short-term recurrent selection programs as follows:

(1) let long-term mean that the aim is to produce an ideal, or nearly ideal, genotype by

continuing recurrent selection until further response is not obtained and

(2) let short-term denote any program, not to be continued indefinitely, of which the aim is to increase $\ddot{Y}$, the mean value of genotype, as much as possible within a limited time period, e.g., five, ten, or twenty years.

In the first instance, it is clear that N_e should be large enough to assure probabilities of fixation near unity for all useful alleles and that these probabilities are more critical than the initial rates of improvement achieved. Conversely, in short-term programs the expected total change in $\ddot{Y}$ is paramount and optimum N_e is more appropriately judged in terms of impact on response to selection during an arbitrary period of time than in terms of probabilities of fixation. Both will receive attention but because the primary focus of this text is on synthesis of ideal genotyples, probability of fixation will be considered first.

11.1. Probability of Fixation, P_f

To be certain that an ideal genotype would be produced by recurrent selection would require $P_f = 1.0$ for the best allele of each gene having an effect on total value. That is not possible because it would require infinite effective population size. It can be approached, however, by making N_e large enough to make P_f equal to or greater than some arbitrary value that is not much smaller than 1.0. The remainder of this section is devoted to the following question---how large must N_e be to make $P_f \geq P_a$ where P_a is an arbitrary value, such as 0.96 or 0.98, that is close to 1.0?

Unfortunately, for reasons to be reviewed later in this section, precise answers to this question are never possible. At the same time, decisions concerning effective population size cannot be avoided. Clearly, the answers available, rough though they must be, should guide those decisions if major errors are to be avoideed.

Our question will be examined using 2.34 which means that results obtained will be most nearly correct relative to the very special case of two allele additive genes (no dominance, no epistasis) for which s, the selective value, is constant over time. How P_f values may be affected by deviations from the special situation to which 2.34 is applicable is discussed in section 11.2 For ease of reference 2.34 is repeated here and assigned an alternate number.

$$P_f = (1 - e^{-2N_e s q_0})/(1 - e^{-2N_e s}) \qquad (2.34) \quad 11.1$$

where s = the selective value of the allele, defined in section 2.6.1 and discussed in the context of artificial selection in section 6.3.3.2, and

q_0 = the population frequency of the allele when recurrent selection is initiated.

A general appreciation for the relationship between N_e and probability of fixation is obtained by using 11.1 to compute values of P_f associated with various assumptions concerning q_0, s, and N_e. Some values so obtained are shown in Table 11.1. As anticipated, P_f increases when N_e is larger, when s is larger and when q_0 is larger. Note that the smallest value of the product $N_e s$, associated with $P_f > 0.9$ is 6.4 and also that the denominator of 11.1 approaches 1.0 as $N_e s$ increases, e.g., when $N_e s = 2.5$, $1 - e^{-2N_e s} = 0.993$. Clearly, when N_e is large enough to make P_f large (say, ≥ 0.92), the denominator of 11.1 can be ignored. Then taking logarithms and

Table 11.1. Probability of fixation as a function of N_e, s, and q_o

		Effective Population Size (N_e)			
s	q_0	2	8	32	128
.05	.1	.109	.140	.285	.722
	.3	.322	.387	.643	.978
.1	.1	.119	.185	.474	.923
	.3	.343	.478	.855	1.000
.2	.1	.140	.285	.722	.994
	.3	.387	.643	.978	1.000

substituting any arbitrary probability for P_f, we obtain

$$-2N_e s q_0 = \ln(1 - P_a)$$

$$N_r = -\ln(1 - P_a)/2sq_0 \qquad 11.2$$

where N_r = the effective population size that is required to make $P_f = P_a$.

Table 11.2 lists such requirements for a few values of sq_0 and P_a.

Table 11.2. Required effective population size, N_r

sq_0	$P_a = 0.92$	$P_a = 0.96$	$P_a = 0.98$
.1	12.6	16.1	19.6
.06	21.0	26.8	32.6
.01	126.2	160.9	195.6

It is apparent from equation 11.2 and Table 11.2 that, with respect to any single value of P_a, N_r is proportional to $1/sq_0$ and becomes large as sq_0 becomes small. The values shown raise immediate questions concerning the magnitudes of s, q_0, and N_e that are characteristic of the populations and procedures most frequently employed in actual breeding programs. Of very particular importance is the potential genetic improvement inherent in superior alleles for which sq_0 is small. It is these that are critical relative to satisfactory decisions concerning effective population size since, as emphasized above, N_r becomes large as sq_0 decreases.

• 11.1.1. Effective Population Size

The definition (section 2.7) of effective population size indicates the procedure for deriving expressions for N_e that are appropriate in various situations. The process is illustrated by derivations presented in Appendix 11.1.

An extension of what is contained in section 2.7 and Appendix 11.1 is required because in many recurrent selection systems the selection cycle involves both (1) more than one generation of the organism and (2) variation in the population mating structure between generations within cycles. Kimura's expressions (equations 2.33 and 2.34) for probability of fixation reflect the joint effects of $E(\Delta q)$, determined by s and h, and the variance of q around its expectation. Because his derivation assumed $\sigma^2(\Delta q) = q(1-q)/2N_e$, it is apparent that, if his results are to be employed in cases where s is obtained as a function of $E(\Delta q)$ over the period of a cycle (see section 11.1.2), the appropriate value of N_e must be the one that makes the variance of Δq from cycle to cycle equal to $q(1-q)/2N_e$. This variance will be affected by drift in each generation of the cycle. It follows that the appropriate value of N_e, its cycle value, must be a function of mating structure in each of the generations of the cycle. Let the change in allele frequency during the whole selection cycle be

$$\Delta q_i = \Sigma_k (\Delta q_i)_k \qquad 11.3$$

where $(\Delta q_i)_k$ is the change in the frequency of B_i during the k-th generation of the cycle. Because the variance of concern arises from random sampling of the gametes produced by parents, there can be no covariance between the Δq's of any pair of generations. Thus

$$\sigma^2(\Delta q_i) = \Sigma_k \sigma^2(\Delta q_i)_k \qquad 11.4$$

where

$$\sigma^2(\Delta q_i)_k = q_{ik}(1 - q_{ik})/2N_{ek} \qquad 11.5$$

and N_{ek} is effective population size in the k-th generation of the cycle. Though ordinarily q_{ik} will not be precisely the same from generation to generation of the cycle, its variation can, for our purpose, be ignored because it is not likely to be great and the resulting variation in $q(1-q)$ is necessarily still less. Ignoring the within cycle variation in $q_{ik}(1 - q_{ik})$, we obtain from 11.4 and 11.5 that

$$\sigma^2(\Delta q) = q(1 - q)/\Sigma_k(1/2N_{ek}) \qquad 11.6$$

Setting 11.6 equal to $q(1 - q)/2N_e$ and solving for N_e, we find

$$N_e = 1/\Sigma_k(1/N_{ek})$$

as given by Venkovsky and Godoi (1978).

The key steps in obtaining this cycle value of N_e are those of deriving the two or more values of $\sigma^2(\Delta q)_k$. Once those are obtained, N_e for the cycle can be obtained directly by setting 11.4 equal to $q(1 - q)/2N_e$ and solving to obtain

$$N_e = q(1 - q)/2\Sigma_k \sigma^2(\Delta q)_k$$

The several N_e's need not be computed.

• 11.1.2. Effective Selective Values in Artificial Selection

Whatever the selection system, the effective selective value of the superior alleles can be approximated as outlined in section 6.3.6. The steps are

(1) set $E(\Delta q)$ for the system in question equal to $E(\Delta q)$ in the case of natural selection and an additive gene and

(2) solve the resulting equation for s.

The general expression, 5.18, for approximation of $E(\Delta q)$ in any single-stage system of artificial selection is

$$E(\Delta q) \approx k\sigma(\hat{q},\hat{X})/\sigma(\hat{X}) \qquad 11.7$$

and $\sigma(\hat{q},\hat{X})$ is shown in Table 8.2 for various selection systems. The expectation of allele frequency change given natural selection and an additive gene is

$$E(\Delta q) = (1/2)sq(1-q)/[1 - (1-q)s] \qquad (2.13) \quad 11.8$$

For our purpose the quantity $[1 - (1-q)s]$ can be ignored because (1) it approaches 1.0 when s is small and (2) it is the P_f for the useful alleles with the smallest s-values worthy of considering that is critical relative to effective population size in long-term recurrent selection programs. Equating 11.7 with the numerator of 11.8 and solving for s, we obtain

$$s' \approx 2k\sigma(\hat{q},\hat{X})/q(1-q)\sigma(\hat{X}) \qquad 11.9$$

This quantity is symbolized as s' to distinguish it from the s of 11.8, 2.14, and 2.15. The latter is the difference in relative reproductive rate between the BB and bb genotypes while the former is defined to make $E(\Delta q) = s'q(1-q)/2$. It is particularly important to remember (1) that probability of fixation is a function of the product, $N_e s'$, (2) that $\sigma(\hat{q},\hat{X})$ varies among selection systems, and (3) that $\sigma(\hat{X})$ is a function of procedure followed in obtaining the values of $\hat{X}$ on which selection is based. As a consequence, in comparing selection systems or procedures, the value of $N_e s'$ required to insure a specific fixation probability is more to the point than N_r, the required value of N_e. From 11.2 it is apparent that

$$(N_e s')_r \approx -\ln(1-P_a)/2q_0 \qquad 11.10$$

The s'-approximations shown in Table 11.3 were obtained from 11.9 using expressions for $\sigma(\hat{q},\hat{X})$ shown in Table 8.2. The linkage equilibrium form of $\sigma(\hat{q},\hat{X})$ was used in all

Table 11.3. s' Approximations for various selection systems

Selection system	s'
Mass	$2k[1+(1-2q)a_X]u_X/\sigma(\hat{X})$
Half-sib family (type 1)	$k[1+(1-2q)a_X]u_X/2\sigma(\hat{X})$
Half-sib family (type 2)	$k[1+(1/F)+(2/FV)][1+(1-2q)a_X]u_X/2\sigma(\hat{X})$
Full-sib family (type 1)	$k[1+(1-2q)a_X]u_X/\sigma(\hat{X})$
Full-sib Family (type 2)	$k[1+(1/V)][1+(1-2q)a_X]u_X/\sigma(\hat{X})$
Progeny test (tester = S population)	$k(1+f)[1+(1-2q)a_X]u_X/\sigma(\hat{X})$
Progeny test (general)	$k(1+f)[1+(1-2v)a_X]u_X/\sigma(\hat{X})$
S_n family (type 1)	$2k[(1+h)+(1-h)(1-2q)a_X/2]u_X/\sigma(\hat{X})$

Note: f = the inbreeding coefficient of selection unit individuals, h = the inbreeding coefficient of parents of the families, v = frequency of the B-allele in the tester, and F and V have meanings specified in section 8.2.1.

cases. For illustration consider progeny test selection in which the "tester" is the S-population itself. From Table 8.2 and assuming linkage equilibrium

$$\sigma(\hat{q},\hat{X}) = (1+f)q(1-q)[1+(1-2q)a_X]u_X/2$$

Substituting in 11.9, we obtain

$$s' \approx k(1+f)[1+(1-2q)a_X]u_X/\sigma(\hat{X})$$

In the case of mass selection and additive genes ($a_X = 0.0$), the s'-approximation is $2ku_X/\sigma(\hat{X})$

which, taking symbol differences into account, is the widely employed value given by Falconer (1960). Note that, for additive genes ($a_X = 0.0$), s' can be nearly constant from cycle to cycle because it is not a function of q.

Each s'-value shown is proportional to u_X and it is obvious that this should be. If the allele present has no impact on the selection criterion, there is in the event of linkage equilibrium no reason other than drift why allele frequency should be different in selections than in units not selected. In contrast, if the allele substitution effect is large, the frequency of the more favorable allele should on average be proportionately larger in selected than in non-selected units. Because s' is a function of u_X assumptions concerning the magnitude of u_X must always be made when 11.10 is employed to provide information concerning adequate effective population size. Unfortunately, with the exception of genes having observable effects such that individuals can be classified according to genotype at the locus, direct estimates of u-values cannot be made. As a result, knowledge concerning ranges of u-values is very limited. For comments concerning magnitudes of $u_X/\sigma(X)$ where $\sigma^2(X)$ is phenotypic variance among individuals, see section 6.3.2.1, and for sources of information about u-values see section 15.5.

11.1.3. Initial Allele Frequency (q_o)

As indicated by 11.2, N_r, the effective population size required to make the probability of fixation as high as desired, is inversely proportional to q_0. Thus, given variation in the initial frequencies of favorable alleles, it is the low end of those frequencies that is critical with respect to N_r. For example, if $P_f = P_a$ in the case of an allele for which $q_0 = 0.2$ and $s = 0.1$, it will be $> P_a$ in the case of all alleles for which $sq_0 > 0.02$.

The frequencies of alleles are affected by selection, drift, mutation and migration (section 2.6). Thus population allele frequencies at a specific point in time are products of past events and what can or should be inferred concerning them at the beginning of a recurrent selection program hinges on the known history of the population. Consider the following examples.

11.1.3.1. POPULATIONS FROM THE CROSS OF TWO PURE LINES

By definition a pure line is one in which, locus by locus, all individuals are homozygous for the same allele. Any two lines will possess the same allele of some genes but different alleles of others. Thus, in the F_1 generation of a two line cross, allele frequencies will be 0.5 unless equal to 1.0 or 0.0. It follows that in an F_2, allele frequencies at segregating loci would all be 0.5 except for deviations due to drift, selection and mutation. The variance of drift deviations would be (0.5)(0.5)/2N and therefore small whenever the number (N) of F_2 individuals is reasonably large. Effects of selection and mutation will ordinarily be trivial.

11.1.3.2. POPULATIONS FORMED BY INTERCROSSING n LINES

In this case the "best" allele of any gene that is present in one or more of the n lines may be present in only one of the lines or in any larger number up to all n. Assuming procedure such that the expected gene contribution to the population is the same for all lines and that the "best" allele was present in x of the n lines, the frequency of the allele in the derived population would be in the neighborhood of x/n, i.e., $1/n$ if the allele had been present in only one of the n lines. It follows that effective population size should, in such cases, be large enough to make P_f large for useful alleles with $q \approx 1/n$.

To argue that there would be few favorable alleles with q_0 as small as $1/n$ and hence that N_e would not need to be geared to initial frequencies that low would be inconsistent with the fact that n lines had been employed. The usual logic for using n, instead of fewer, lines is that each may possess useful alleles not present in any of the others. If the final product of the recurrent selection program is to possess and benefit from useful alleles present in only one of the parent lines, N_e must be sufficient to make P_f high for such alleles.

11.1.3.3. BROAD-BASED FOUNDATION POPULATIONS

The distribution of useful alleles relative to the sub-populations of a species is an aggravating unknown of plant and animal breeding. However, the investments that have gone into the development and maintainence of germplasm collections (see, for example, Harlan, 1975) indicate that many geneticists are convinced that each sub-population may possess at least one useful allele not to be found elsewhere in the species. If their view is correct, the potential for ultimate improvement through recurrent selection should be greater in the case of foundation populations formed using inputs from numerous sources. The implications relative to allele frequencies are much the same as when base populations are formed using pure lines. Let P_k be the proportion of the foundation populataion genes contributed by the k-th source. Then the initial frequency of any allele coming only from that source will be in the neighborhood of $P_k q_k$ where q_k is the frequency of the allele in the k-th source population. Thus, if a population is formed by compositing material from a number of sources, initial allele frequencies at least as low as the lowest of the associated P_k's must be taken into account.

11.1.3.4. OLD POPULATIONS

Nothing can be said with complete certainty concerning allele frequencies in such populations as the breeds of livestock and open-pollinated varieties of maize. Because of selection in the past it is reasonable to suppose that numerous useful alleles have been fixed and that of those segregating, many are at high frequencies. For example, the large increases in milk production of dairy cattle and egg production in some varieties of poultry that have resulted from past recurrent selection must reflect high frequencies of many alleles favorable to those traits. On the other hand, there are at least three reasons for possible low frequencies of a portion of the superior alleles in old populations.

(1) Environments change because of changes in pest organisms, soil erosion or modifications of management practices (nutrition, housing, pest control practices, etc.) and the rank order values of alleles of the same gene may be altered as a consequence. Thus alleles now favorable may only recently have become so. They could be at low frequencies because their selective coefficients had been negative until quite recently.

(2) In the case of pleiotropic genes that affect two or more value traits, the selective

coefficient of an allele that is superior with respect to total value may have been negative in the past because selection was based on a criterion other than total value, perhaps just one value trait (see section 12.4).

(3) An allele superior to all other alleles of the same gene may be at low frequency because it arose by mutation in a relatively recent generation and could not yet have been brought to high frequency by selection. The prevailing opinion has been that there are not enough useful alleles arising by mutation to provide the basis for more than a trivial amount of genetic improvement; said differently, that consequential response to selection within, say, one or two human generations is limited to what is possible by changing the frequencies of alleles already present. However, Hill (1982) after a theoretical examination of the issue concluded that "there is . . . sufficient evidence to suggest that we should take mutation more seriously in designing our breeding programmes and consider not just how to utilize existing variation but how to obtain and utilize new variation." It is shown in Appendix 11.2 that, in accord with Hill, the potential for genetic improvement inherent in mutations that occur during each of the successive cycles of recurrent selection may be substantial.

• 11.2. Comments Concerning Non-validity of Assumptions

With respect to total value, let Y_{wi} be the value of an ideal genotype, $\ddot{Y}_{wo}$ be the average of genotype value in the foundation population for a recurrent selection program and $E(\ddot{Y}_{w\infty})$ be the expected average genotype value when there is no further response to the selection of that program.

In general, $P_f < 1.0$, i.e., the probability of fixation of the best allele of each gene is increased by increasing N_e, but the approach to unity is asymptotic. It follows that $Y_{wi} - E(\ddot{Y}_{w\infty})$ is reduced as N_e is increased.[1] The effective population size of a long term recurrent selection program should obviously be such that $[E(\ddot{Y}_{w\infty}) - \ddot{Y}_{wo}]/(Y_{wi} - \ddot{Y}_{wo})$ is close to unity. To optimize N_e, it should be increased until the consequent increase in $E(\ddot{Y}_{w\infty})$ from any further increase of N_e is not sufficient to compensate for the cost involved.

Obviously, the optimum for N_e can only be approximated and, unfortunately, approximations arrived at may not be very close to the real optima. Reasons include

(1) lack of information on which to base assumptions concerning the minimum for sq_0 that deserves attention.

(2) erroneous estimates of the extent to which breeds or cultivars derived from the S-population(s) of the program will be employed in world agriculture, and

(3) the consequences of non-validity of assumptions involved in the method used for computing N_r.

The best that can be hoped for is that chosen values of N_e will not be grossly off the mark.

It is in this context that the inadequacies of 11.1, as a guide for decisions concerning N_e are properly considered. Are the biases of 11.1 due to non-validity of assumptions large relative to errors that may result for one or more other reasons? Consider the consequence of deciding that the probability of fixation should be equal to or larger than some specified value (say 0.96) for all alleles for which $sq_0 \geq 0.03$ instead of for all alleles for which $sq_0 \geq 0.06$. It is apparent from 11.2 that the N_r is in the first case 100 percent greater than in the second. Considering the limits of our knowledge with respect to u-values and allele frequencies, it is obvious that biases no larger than 50 or 60 percent can be viewed as unimportant.

The prime assumptions to which 11.1 is subject are (1) natural selection, (2) constancy of s, the selection coefficient, (3) an additive gene with two alleles and (4) N_e constant through time. By virtue of the third of these, $E(\Delta q) = sq(1-q)/2$. When employed in the context of artificial selection with s' substituted for s and s' defined to make $E(\Delta q) = s'q(1-q)/2$, the answers provided by 11.1 are essentially correct when both s' and N_e are constant through time, i.e., in successive cycles of selection. Because N_e is controllable and can therefore be held nearly constant in breeding programs, it need not be considered further. On the other hand, s' may be non-constant for any of a number of reasons, e.g., change due to change in allele frequency is indicated by expressions shown in Table 11.3 except in the case of additive genes ($a_x = 0.0$). But when 11.1 and 11.2 are employed, a single constant value of s' is assumed. The answer provided would be correct if that value persisted in all cycles of selection. The practice suggested in section 11.1.2 is that the appropriate expression from Table 11.3, but with q equated to 0.5, be employed. This will be done in all applications or discussions that follow in this and later chapters. What are the magnitudes of errors that may be involved when the actual value of s' is variable for one reason or another?

The expressions for s' shown in Table 11.3 were obtained assuming linkage equilibrium, no epistasis, no multiple alleles, and $-1.0 \leq a_x \leq 1.0$.[2] Subject to those restrictions and assuming near constancy of k, $\sigma(\hat{X})$, and f from cycle to cycle (because of constancy in the operation design of the selection program), significant variation in s' could occur only as the result of change in q (due to selection or drift) when $a_x \neq 0.0$, u_x (because of interaction with the environment in which criterion units are compared), or v in the case of progeny test selection when the S population is not the tester. These will be discussed in turn before turning to implications of overdominance, multiple alleles, linkage disequilibrium, and epistasis.

11.2.1. Change in q When $-1.0 \leq a_x \leq 1.0$

The s'-approximations of Table 11.3 are, with one exception, functions of a_x and q. The portions of the listed expressions that vary with those parameters are $[1+(1-2q)a_x]$ or $[1+(1-2q)a_x/2]$. We note that (1) both equal 1.0 when $a_x=0.0$, i.e., the value of s' does not change as q changes in the case of additive genes, (2) the effects of changes in q are greater when a_x is larger and (3) in the case of $[1+(1-2q)a_x/2]$ the effects of changes in q are only half as great. Some numerical values of $[1+(1-2q)a_x]$ are listed in Table 11.4. The fact of particular importance is that when frequency of the superior allele is low $[1+(1-2q)a_x]$, and hence s', is relatively small when $a = -1.0$, relatively large when $a=1.0$. The reasons why this is important are that the chance of loss of a useful allele <u>via</u> drift is greatest when its frequency is low, loss is less likely if s' is larger so that $E(\Delta q)$ is larger and probability of fixation increases as probability of loss decreases. It is apparent that when q_0 is small and other things

Table 11.4. Values of [1+(1−2q)a]

q	a = -1.0	a = -1/2	a = 0.0	a = 1/2	a = 1.0
.1	0.2	0.6	1.0	1.4	1.8
.3	0.6	0.8	1.0	1.2	1.4
.5	1.0	1.0	1.0	1.0	1.0
.7	1.4	1.2	1.0	0.8	0.6
.9	1.8	1.4	1.0	0.6	0.2

(u_x,k, etc.) are equal, probabilities of fixation will be larger when a_x is larger, i.e., smallest when the useful allele is completely recessive, largest when it is completely dominant.[3] The obvious corollary is that N_r, the effective population size required to make the probability of fixation equal to a specified large value, may be much larger for a favorable allele that is recessive than for one that is additive if the initial frequency of the allele is low. Conversely, it appears that N_r, when q_0 is small, should be smaller for a favorable allele that is dominant than for one that is additive. Confirming evidence from a computor simulation study of N_r for additive and recessive alleles are presented in Table 11.5.[4] The data were produced as follows.

Table 11.5. Comparison of recessive and additive favorable alleles with respect to N_e required for various probabilities of fixation

			N_e required		
s'	q_0	P_f	Recessive allele (a= -1)	Additive allele (a=0)	Ratio
.2	.2	.910	80	30.1	2.7
.2	.2	.965	100	41.9	2.4
.2	.1	.695	100	29.7	3.4
.2	.1	.815	150	42.2	3.6
.2	.1	.810	175	41.5	4.2
.2	.1	.920	300	63.1	4.8
.1	.2	.940	300	70.3	4.3

The computor was programmed to simulate the recurrent selection process using $a_x = -1.0$ with arbitrary values of s', q_0, and N_e. The result obtained was whether the favorable allele of a simulated gene was fixed or lost. The fraction in the third column of the table is the the fraction of 200 replications in which fixation was the result. This observed fraction is an estimate of P_f. Thus for recessive favorable alleles the estimate of P_f was 0.91 when s' = 0.2, q_0 = 0.2, and N_e = 80 were used. Then 11.2 was employed to obtain N_r= 30.1 for an additive allele, given s = 0.2, q_0 = 0.2 and P_f = 0.91. The significant general result was that, as anticipated, N_e required for a given probability of fixation and a small value of q_0 is much larger when the favorable allele is recessive. The number in the last column is the ratio of the number in the fourth column to that in the fifth.

On the basis of numerical values in Table 11.4 and the argument employed in the case of recessive alleles, we infer that fixation probabilities should be higher if the useful allele is dominant than if additive in effect when initial allele frequency is low. That this is so has been demonstrated in various computer simulation investigations [see, for example, Carr & Nassar (1970)].

11.2.2. Variation in s′ Not from Change in q

The value of s', as defined in section 11.1.2, is the quantity that makes $E(\Delta q) = s'q(1-q)/2$. It follows that the actual change in q during any one selection cycle can be viewed as

$$\Delta q = [s'q(1-q)/2] + d_c$$

where d_c is a random deviation arising from gamete sampling (see section 2.6.3) and

q is allele frequency in the genetic population in the cycle in question.

Making no assumption concerning constancy of s' and remembering 2.27, the variance of Δq is

$$\sigma^2(\Delta q) = [\sigma^2(s')q^2(1-q)^2/4] + q(1-q)/2N_e = \sigma^2(d_s) + \sigma^2(d_c) \qquad 11.11$$

where $\sigma^2(d_s)$ and $\sigma^2(d_c)$ are variances defined in section 2.6.3 and

$$\sigma^2(d_s) = \sigma^2(s')q^2(1-q)^2/4$$

Clearly, when $\sigma^2(s') = 0.0$, the variance of Δq reduces to $\sigma^2(d_c)$, the gamete sampling variance. In that special case,

$$\sigma^2(\Delta q) = \sigma^2(d_c) = q(1-q)/2N_e$$

Note (1) that $\sigma^2(s')$ measures variation in s' that is random relative to q and

(2) that if $\sigma^2(s') \neq 0.0$, the variance of Δq, and hence $(1-P_f)$, cannot be made as small as desired by increasing N_e.

The last of these facts poses the question of what the relative magnitudes of the two portions of the variance of Δq may be. I have been helped by rewriting 11.11 as follows.

$$\sigma^2(\Delta q) = [q(1-q)/2]\{(1/N_e) + [\sigma^2(s')q(1-q)/2]\} \qquad 11.12$$

This focuses attention on the relative magnitudes of $1/N_e$ and $\sigma^2(s')q(1-q)/2$. When $\sigma^2(s') = 0.0$, the variance of Δq is proportional to $1/N_e$ and approaches zero as N_e is increased. This was the situation assumed by Kimura (1957) and, consequently, the situation in which 11.1 and 11.2 are applicable. We see from 11.12 that in the more general situation in which $\sigma^2(s') \neq 0.0$, $\sigma^2(\Delta q)$ is proportional to $1/N_a$ where

$$1/N_a = (1/N_e) + \sigma^2(s')q(1-q)/2 \qquad 11.13$$

N_a will be referred to as N_e-adjusted, i.e., the effective population size adjusted for variance in s'. Note that the limit approached by $1/N_a$, as N_e is increased, is $q(1-q)\sigma^2(s')/2$ instead of zero. Though the difference between $1/N_a$ and $1/N_e$ approaches zero when q is very small or very large, it is obvious that because $\sigma^2(s')$ increases $\sigma^2(\Delta q)$ (unless one of the alleles of the gene is already fixed), the probability of fixation of the superior allele is reduced when s' is variable.

11.2.2.1. PROGENY TEST SELECTION—VARIATION IN S′ FROM VARIATION IN V

In RSSC with a pure line tester, v is not variable if the same tester is used cycle after cycle. If the tester were, as it conceivabley could be, a random mating population there would be modest random variation due to gamete sampling. The greatest variation in v would occur in the modification of RRS proposed by Russell and Eberhart (1975), outlined in section 9.3.3 and discussed by Comstock (1979). Because the object of RRS is to increase the combining ability of the S_1 and S_2 populations, the pure line testers must be recent derivatives from their respective populations if they are to be optimal substitutes as testers for the populations themselves. Accordingly, it will be assumed that new tester lines are employed in each successive cycle. It will suffice to consider selection in only one of the populations. Consider any gene and let q_2 be the frequency of the B allele in the S_2 population when formation of a pure line to be employed as tester for selection in the S_1 population is initiated. Then, assuming that the tester line is a random one, the probabilities that its genotype will be BB or bb are q_2 and $(1-q_2)$, respectively (section 2.11). Possible genotypes, their probabilities and the corresponding values of v and $[1+(1-2v)a_x]$ are shown below.

Genotype of the pure line tester	Probability	v	$Z=[1+(1-2v)a_x]$
BB	q_2	1	$1-a_x$
bb	$1-q_2$	0	$1+a_x$

From these quantities we find

$$\sigma^2(Z) = 4q_2(1-q_2)(a_x)^2 \qquad 11.14$$

From Table 11.3

$$s' = k(1+f)[1+(1-2v)a_x]u_x/\sigma(\hat{X}) \qquad 11.15$$

From 11.15 and 11.14, and remembering that $[1+(1-2v)a_x] = Z$, we see that if all parameters except v are constant

$$\sigma^2(s') = 4k^2(1+f)^2q_2(1-q_2)a_x^2u_x^2/\sigma^2(\hat{X}) \qquad 11.16$$

We are now concerned, as emphasized by 11.12 and 11.13 with how large $\sigma^2(s')q_1(1-q_1)$ may be relative to $1/N_e$. Clearly, its size depends on magnitudes of all quantities involved. As an example, suppose $k = 1.5$, $f = 0.5$, $a_x = 2.0$ and $u_x/\sigma_x = 0.1$. Then

$$\sigma^2(s')q_1(1-q_1) = 0.81q_1(1-q_1)q_2(1-q_2)$$

which could take any value between 0.0 and 0.05 depending on the values of q_1 and q_2. Compare to $1/N_e = 0.04$ or 0.025, given $N_e = 25$ or 40. It is apparent that $1/N_a$ could be increased substantially, relative to $1/N_e$ by variance of s' arising from variation in v. An interesting aspect is that $\sigma^2(s')q(1-q)$, being a function of a_x, u_x, and allele freqencies, is variable among genes while N_e is not.

• 11.2.2.2. VARIATION IN S′ FROM VARIATION IN TEST ENVIRONMENTS

We perceive intuitively that, as a consequence of genotype-environment interaction, the y-values associated with any set of single-locus genotypes will be different from one macro-environment (or set of macro-environments) to another. A more formal discussion in Appendix 5.1 established that because of this, the magnitude of $\sigma(\hat{q}_i, y_{xik})$ depends on the macro-environment set in which criterion units are observed (tested). In this context, y_{xik} pertains to the effect, in the k-th macro-environment set, of the i-th locus genotype on the selection criterion, as opposed to

y_{xi} which pertains to the average of y_{xik} over all macro-environments of the TPE.

The procedure outlined in section 11.1.2 yields

$$(s')_i = 2k\sigma(\hat{q}_i,\hat{X})/q_i(1-q_i)\sigma(\hat{X})$$

Assuming linkage equilibrium in the S population, $\sigma(\hat{q}_i,\hat{X}) = \sigma(\hat{q}_i,\hat{y}_{xi})$ as stated by 5.22. Thus in the linkage equilibrium situation

$$(s')_{ik} = 2k\sigma(\hat{q}_i,\hat{y}_{xik})/q_i(1-q_i)\sigma(\hat{X}) \qquad 11.17$$

Here the k subscript identifies the macro-environment set in which criterion units are observed. The variation among sets of y-values translates into variation of u's and a's among macro-environment sets. We can, therefore, employ familiar expressions. Consider the particular case of selection (type 1) among full-sib families. Then from Table 8.2

$$\sigma(\hat{q}_i,\hat{y}_{xik}) = q_i(1-q_i)[1+(1-2q_i)a_{xik}]u_{xik}/2$$

For illustrating principles it will be simpler, and will suffice, to examine the additive case in which all a_{xik} values are zero. Then

$$\sigma(\hat{q}_i,\hat{y}_{xik}) = q_i(1-q_i)u_{xik}/2$$

and, by substitution in 11.17

$$(s')_{ik} = ku_{xik}/\sigma(\hat{X})$$

from which

$$\sigma^2(s') = k^2\sigma^2(u)/\sigma^2(\hat{X}) \qquad 11.18$$

where $\sigma^2(u)$ = the variance of u_{xik} among macro-environment sets.

To facilitate numerical approximations 11.18 is modified by multiplying and dividing by u_{xi}^2 where u_{xi} is the average of u_{xik} over all macro-environments of the TPE. That yields

$$\sigma^2(s') = k^2[u_{xi}/\sigma(\hat{X})]^2[\sigma(u)/u_{xi}]^2 \qquad 11.19$$

Rationale for reasonable assumptions concerning the magnitude of $u_{xi}/\sigma(\hat{X})$ is provided in section 6.3.3 for mass selection and in Chapter XIV for family selection. Here, in an exploratory example, we will assume $u_x/\sigma(\hat{X}) = 0.12$, a value within the range projected as reasonable in Chapter XIV.

The ratio, $\sigma(u)/u_{xi}$, will be greatest in those instances where u_{xik} is sometimes positive and sometimes negative, which is possible though u_{xi} is by definition always positive. Consider variation in soil moisture. Crop scientists are well aware that at any specific location the moisture available to a growing and maturing crop is variable and may range from excessive in some years to the drouth or semi-drouth conditions of other years. Moreover, the pattern of annual variation is inevitably variable from location to location within any TPE not restricted to a very limited geographical area. In this circumstance, there must be genes of which the allele most favorable with respect to the entire TPE is not most favorable in every macro-environment. Stated in terms of variation (among macro-environments) in u_{xik}, we visualize that, though on average this quantity is positive, a portion of its distribution extends below zero. Suppose that u_{xik} is normally distributed and let p = the probability of a negative value in the case of random macro-environments from the TPE. Then corresponding to any particular value of p, the value of $\sigma(u)/u_{xi}$ is available from information in standard tables of the normal distribution. All such tables show the fractions of the distribution in which individual values of the variate are smaller or larger) than selected deviations (measured in standard deviations) from the distribution mean. For example, the fraction smaller than x = -0.8 is found to be approximately 0.212. Note (1) that if x = 0, the population mean must be 0.8 and (2) that in standard nomal distribution tables $\sigma = 1.0$. It follows that if p (as defined above) ≈ 0.21, the value of $\sigma(u)/u_{xi} \approx 1/0.8 = 1.25$. A few such values are shown in Table 11.6. The set [p=0.16, $\sigma(x) = 1.0$] is most familiar from elementary statistics where we learned that about 32 percent of the normal distribution lies beyond one standard deviation above or below the population mean

Table 11.6. Values of $\sigma_x/\ddot{u}_x$ when the distribution of x is "normal" and the probability of a negative value is p

p	0.05	0.1	0.16	0.2	0.3	0.4
$\sigma_x/\ddot{u}_x$	0.61	0.78	1.00	1.19	1.92	4.00

Now values of $\sigma^2(s')$ can be visualized. As an illustrative example, in the case of full-sib selection, $u_{xi}/\sigma(\hat{x}) = 0.12$ and p = 0.3, we find using 11.19 and Table 11.6 that

$$\sigma^2(s') = k^2(0.12)^2(1.92)^2 = 0.05308k^2$$

Next, using 11.13 , the values of N_a corresponding to particular values of N_e and q would be as follows where in each case the first value is based on k = 1.0 and the second on k = 1.5.

	N_e			
q	10	20	40	80
0.1	9.8, 9.5	19.1, 18.1	36.5, 32.9	67.2, 55.9
0.2	9.6, 9.1	18.4, 16.8	34.2, 28.9	59.7, 45.3
0.4	9.4, 8.7	17.7, 15.5	31.9, 25.4	53.0, 37.3

We note that, in accord with 11.13, the amount by which $N_a < N_e$ is increased by increases in N_e, k, or q. In this specific example the difference ranges from trivial to very substantial.

The issue is usefully examined in another way if we inquire how large N_e must be made to achieve a required value of N_a. Solving 11.13 for N_e, the following is obtained.

$$N_e = 1/[(1/N_a) - q(1-q)\sigma^2(s')/2]$$

Assuming, as before, full-sib selection, $u_{xi}/\sigma(\hat{x}) = 0.12$, p = 0.3, and k = 1.0 or 1.5, values of N_e corresponding to particular values of N_a and q would be as shown below. It is apparent

	N_a			
q	10	20	40	80
0.1	10.2, 10.6	21.0, 22.4	44.2, 51.0	98.9, 140.3
0.2	10.4, 11.1	21.9, 24.7	48.2, 64.7	121.2, 339.6
0.4	10.7, 11.7	22.9, 28.0	53.7, 93.7	163.1, *

*N_a = 80 not possible. In this case as N_e approaches infinity N_a approaches 69.8.

from the numbers shown that the effective population size required to achieve a high probability of fixation of the better allele can be increased substantially by the variation in s' arising from variation among macro-environments.

11.2.3. Overdominance

Dominance outside of the range from complete dominance to complete recessivity of the better allele was not considered by Kimura (1957) and 2.33 is not generally applicable in the case of overdominant genes (see Robertson, 1962). However, the critcal issue is not overdominance per se but whether the algebraic sign of $E(\Delta q)$ is affected by the magnitude of allele frequency in the S population. Thus, as emphasized in the introduction to this chapter, when selection is for combining ability by RRS or RSSC, 11.1 has about the same utility relative to overdominance genes as it does more generally in the case of non-overdominance genes.

Consider RSSC when the "tester" is a pure line. Then, given linkage equilibrium in the S population,

$$s' = k(1+f)[1+(1-2v_i)\,a_{xi}]u_{xi}/\sigma(\hat{X}) = k(1+f)[1 \pm a_{xi}]u_{xi}/N_e(\hat{X}) \qquad 11.20$$

(see Table 11.3 and remember that v_i = 1.0 or 0.0 when the tester is a pure line). It is apparent from 11.20 that the allele (in the S population) that is most useful with respect to performance in criterion units may be either B or b, depending on which is present in the pure line tester. However, in either case s' does not vary with q; hence 11.1 and 11.2 have validity as with additive genes.

If the tester is not homozygous, 11.20 tells us that s' is greater or less than zero depending on whether v is greater or less than (1+a)/2a but is, as before, not a function of q. Given a tester such that v is constant, s' would also be constant. Consider the single cross of two pure lines as tester. In that case v would be 0.0, 0.5, or 1.0 but constant. RSSC programs in which the tester is such that v is variable in time as the result of drift or drift and selection pose the same general situation as RRS.

Hill (1970) investigated response limits in RRS. Noting that the optimum end result for overdominant genes is fixation of different alleles in the S_1 and S_2 populations, he studied the joint probability of fixation of B in either of the populations and of b in the other. For our purpose, a less rigorous approach appears to me to be adequate. Consider first genes that are segregating in one of the populations but not in the other. For these the situation is exactly as in

RSSC when the tester is a pure line. Accordingly the results given by 11.1 or 11.2 are applicable. In the case of overdominant genes that are segregating in both foundation populations, remember (1) that whether B (or b) is eventually fixed in the S_1 or S_2 population is of no consequence and (2) that the selection system works, in every cycle, to increase the difference between q_{1i} and q_{2i} (see section 9.1). The first of these facts means that we need only be concerned with the probability that once one allele or the other is fixed in one population, the other will ultimately be fixed in the other population. With one of the alleles homozygous in one of the populations, The situation is once again that of RSSC with a pure line tester so that again 11.1 and 11.2 apply. Moreover, because the difference between q_{1i} and q_{2i} tends always to be increased by RRS, it is likely that if, for example, B is fixed first in one of the populations, the frequency of b in the other population will not be close to zero. From this it follows that the probability of the desired result must be higher for overdominance genes in RRS than for some non-overdominance genes in single-population selection systems.

- 11.2.4. **Independent Assortment**

It is apparent from 11.9 and Table 8.2 that, unless linkage equilibrium prevails, the s'-value for any favorable allele can be affected (increased or decreased) by the impacts of other genes. Moreover, linkage disequilibrium will always be present, and variable over time, in finite populations (Hill and Robertson, 1968, and section 2.4.3). Hill and Robertson (1966) and Robertson (1970) investigated effects on probability of fixation assuming (1) linkage equilibrium when recurrent selection is initiated, (2) additive gene effects and (3) artificial selection. Fixation probabilities were invariabley decreased (by amounts that depended on recombination frequencies, effective population size, selection intensity and magnitudes of effects of favorable alleles) but the decreases were not large.

Ho and Comstock (1980) investigated the effects of the extreme initial linkage disequilibria that are inevitable (section 2.4.3) in a population formed by crossing two pure lines when genes are closely linked. Their data indicate that substantially greater decreases in probabilities of fixation are then to be expected and that a three to four-fold increase in effective population size could be required to offset the impact of such extreme linkage disequilibria. A follow-up investigation by Cardellino and Comstock (unpublished) demonstrated that the problem is even greater in the case of favorable alleles with initial frequencies < 0.5 (as in a random mating population established from a back-cross generation after a cross of diverse populations).

Taken together the results of these several studies indicate (1) that the linkage disequilibrium that arises from finite population size alone does not have an important effect relative to fixation probabilities when effective population size is large enough so that P_f values would be large if there were no linkage disequilibrium but (2) the N_r values obtained using 11.2 may be much too small in cases where the foundation population is one in which extreme linkage disequilibrium is present and the frequencies of a portion of the useful alleles are small (say, ≤ 0.25). Of course, the impact of linkage disequilibrium is inversely proportional to frequency of recombination between adjacent genes and, hence, is more likely to be large if the number of genes bearing on the selection criterion is large.

It is intuitively obvious, and important to bear in mind, that when foundation stocks are formed by intercrossing diverse populations, a useful allele introduced from a phenotypically inferior population is likely to be linked with more inferior alleles than one from a superior population. As a result, the initial s' of that useful allele will be lower than it would be with linkage equilibrium, and may well be negative.[5] In these cases larger effective population size is of value because it increases the probability of the recombinations required to free useful alleles from their associations with inferior alleles at other loci, in addition to reducing random drift.

• 11.2.5. Epistasis and Multiple Alleles

Neither of these deviations from the two allele, no epistasis situation changes the general relationship between the probability of fixation of the best allele (or the best allele set in the case of epistasis) and effective population size. Given either epistasis or multiple alleles or both, it is still true that larger effective population size in a long term recurrent selection program increases the expected value of the genotypic end product of the program. However, as noted in section 3.6.4, in the event of multiple peak epistasis, the "peak" approached in response to recurrent selection is conditioned by the matrix of original frequencies of the alleles involved in the epistatic set of genes. Larger effective population size increases the probability that the "peak" favored by original allele frequencies will be the end product, not the probability that the best possible "peak" will be achieved.

It goes without saying that these deviations from the two allele, no epistasis model increase the imponderables involved in decision making with respect to effective population size in long term recurrent selection programs.

• 11.2.6. Summary

The explorations of the preceeding portions of this chapter have made it abundantly clear that the optimum effective population size in long-term recurrent selection programs cannot be determined with the precision that would be desirable. At best only rough approximations are possible. In the most favorable situation (gene effects exclusively additive, linkage equilibrium and no genotype x environment interaction) determinations cannot be as reliable as the breeder would like because we know so little concerning the distribution of u_x values and population frequencies of useful alleles. However, decisions (whether consciously made or not) concerning effective population size are embodied in every recurrent slection program -- they cannot be avoided -- and they are crucial relative to the long-run success of any such program.
An approach that I consider useful embodies the following components:

(1) An hypothesis, considered to be reasonable, concerning the smallest value of $s'q_0$ that would need to be considered to include enough useful alleles to account for a large portion (say, 95%) of all the genetic improvement that could be made if N_e were infinite.

(2) Use of 11.2 to determine N_r, the effective population size required to make P_f as large as desired for any additive allele having that value of $s'q_0$

(3) Identification of the deviations, from the situation in which 11.1 and 11.2 give good answers, that are or may be present and would increase the N_e requirement by 100 percent or more.

Such deviations have been shown to include

(a) Recessivity of useful alleles.

(b) Linkage disequilibrium such that the s'-values for a non-trivial portion of the useful alleles may be negative or much smaller than they would otherwise be.

(c) Heterogeneity of the TPE such that the random variation of s' is significant in magnitude.

(4) Employ N_e larger than the estimated N_r by an amount considered reasonable in the light of opinions concerning (a), (b), and (c).

Recessivity of favorable alleles would call for N_e as much as 3 to 5 times the estimate of N_r. However, the evidence provided by the extensive data on heterosis and inbreeding depression (see section 2.11) suggests that, in general, alleles favorable to value traits are not recessive

The types of populations in which linkage disequilibrium may be sufficient to have a significant impact have been identified in subsection 11.2.4.

This author has no basis for any general statements concerning the allowance to be made for heterogeneity among macro-environments. At the same time examples have been given (and more could be) of situations in which $\sigma^2(s')$ might be large enough to make N_a substantially smaller than N_e.

• 11.3. $E(\Delta\ddot{Y})$ as a Function of N_e

The direct impact of effective population size is on the magnitude of random variation in allele frequency change, Δq, from any one generation or cycle of selection to the next. That variation has so far, in this text, been ignored in derivations of expected responses to selection, e.g., N_e does not appear in 5.25, 5.26, 8.28, or 8.29. By the definition of N_e (see section 2.7)

$$\sigma^2(d_c) = \bar{q}(1-\bar{q})/2N_e \qquad 11.21$$

in which d_c symbolizes the random deviation of an allele frequency in selections from $\bar{q}$, its expectation. It follows from 2.25 that if the contributions of both mutation and variation in the effective selection coefficient are ignored, as trivial,

$$\sigma^2(\Delta q) = \sigma^2(d_c) = \bar{q}(1-\bar{q})/2N_e$$

On the other hand, when the variance of s' is taken into account (section 11.2.2) the following is obtained

$$\sigma^2(\Delta q) = \bar{q}(1-\bar{q})/2N_a$$

where

$$1/N_a = (1/N_e) + \bar{q}(1-\bar{q})\sigma^2(s')/2 \qquad 11.13$$

Because Δq exhibits random variation, the expectation (average value) of any function of q will be a function of $\sigma^2(\Delta q)$. Consequently, one of our tasks here is to express $\Delta\ddot{Y}$ in terms of changes in allele frequencies and then determine the expectation of each of the individual terms of which $\Delta\ddot{Y}$ is the sum. The results to be presented will pertain to the special case characterized by diploid meiosis, no epistasis, two alleles per gene and linkage equilibrium in S populations. In addition, it will be assumed that $\sigma^2(s') = 0.0$ so that $N_a = N_e$. In instances in which N_a is thought to be larger than N_e by a non-trivial amount, the estimate of N_a should be employed in place of N_e.

Consider the average values of genotypes in a T-population in successive cycles, identified by subscripts 1 and 2. then

$$\Delta\ddot{Y} = \ddot{Y}_2 - \ddot{Y}_1 = \Sigma_i(\ddot{y}_{2i} - \ddot{y}_{1i})$$

and
$$E(\Delta\ddot{Y}) = \Sigma_i E(\ddot{y}_{2i} - \ddot{y}_{1i})$$

Clearly, the question to be answered is this -- how is the expectation of $\Delta\ddot{y} = \ddot{y}_2 - \ddot{y}_1$ affected by N_e? In its general form, the answer will be the same for all genes.

Before going further, some statements concerning symbols are required. The numerals, 0,1,and 2, used as subscripts will identify variables with one or another of the first three cycles of a recurrent selection program. Thus, $\ddot{y}_{1i}$ and $\ddot{y}_{2i}$ will represent the average values of genotype at the i-th locus in the S populations of cycles 1 and 2 of the program. For the purpose of symbol construction each cycle is assumed to begin with the S population of that cycle and end with the selections to used as parents of the S population of the following cycle. Thus, for example,

q_{0i} = the frequency of the B allele of the i-th gene in the S population of cycle 0,

q_{1i} = the frequency of the B allele of the i-th gene in the S population of cycle 1,

q_{1si} = the frequency of B_i in the selections of cycle 1 that will be the parents of the S population of cycle 2, and

$(\Delta q_i)_0 = q_{1i} - q_{0i}$

Because concern herein is focused on the q's and $\ddot{y}$'s pertaining to a single gene (any single gene), the gene identifying subscript will, for the most part, not be required.

It will be crucial to keep in mind (1) that $\ddot{y}_2$ and $\ddot{y}_1$ are averages for genetic populations, as distinguished from averages for finite samples, (2) that the allele frequencies of a genetic population are the average allele frequencies of the parents from which that population was produced,[6] i.e., that $q_2 = q_{1s}$ and $q_1 = q_{0s}$, (3) that $q_{0s} - q_0 = (\Delta q)_0 = E(\Delta q)_0 + \bar{d}_0$ in which $\bar{d}_0$ symbolizes a random deviation of $q_{0s} - q_0$ from $E(\Delta q)_0$ that occurs because the number of parents is always finite, and (4) that $E(\Delta q)_0$ is a function of q_0 but $\bar{d}_0$ is not, it varies randomly.

Because Δq exhibits random variation whenever the number of selections is finite (as it always must be in practice) the expectation , average value, of any function of q will be a function of $\sigma^2(\Delta q)$.[7] Consider any genetic population (S_0) and selections from genetic samples drawn from that population. The expectation of q_{1i} can of course be stated to be

$$E(q_1) = q_0 + E(\Delta q)_0 \qquad 11.22$$

The consequences of effective population size are clarified by supposing the whole selection cycle process to be repeated an infinite number of times. In this context the "whole process" includes the drawing of a genetic sample from the genetic population (the S_0) followed by the selection process and random intermating of selections to produce the S_1 population. Assume the entire selection process to be identical in each replication. Obviously the process cannot actually be replicated an infinite number of times; like every infinite population it can only exist in the "mind's eye". The important thing to realize is that Δq in any single replication is not likely to be exactly equal to its expectation. Instead, in accord with the definition of expectation, the average of Δq over an infinity of replications and its expectation would be quantitatively identical, i.e., the average value of q_1 would be $q_0 + E(\Delta q)_0$ as indicated by 11.22. The variance, of q_1 around its expectation is by the definition of N_e (see section 2.7 and Appendix 11.1)

$$\sigma^2(\Delta q)_0 = \sigma^2(q_1) = [E(q_1)][1-E(q_1)]/2N_e. \qquad 11.22a$$

Here $\sigma^2(q_1)$ is the variance of q_1 among the S_1 populations of replications of the "whole selection process". The obvious prime consequence of this variance is the "drift" of allele frequencies. Significant secondary consequences are

(1) more homozygous (less heterozygous) genotypes when averages over replications of a selection process are considered,

(2) an inbreeding effect unless dominance is absent ($a = 0.0$), and

(3) because q_1 is variable among replications of the selection process, $E(\Delta q)_1$ is also variable and, in general, its average over replications is not $E(\Delta q)_1$ in the specific case in which $q_1 = q_0 + E(\Delta q)_0$.

A single recurrent selection program constitutes one replication of an infinite number that conceptually could be conducted. With this in mind, it is apparent that an extention of symbolism is required. In previous chapters $E(\Delta q)$ has pertained to the change in q to be anticipated as a function of the specific frequency of the B allele in the S population of a particular cycle of a selection program. Once the S_0 population has been determined, q_0 is the frequency of B in that population and $E(\Delta q)_0$ is a function of that single allele frequency. However, $E(\Delta q)_1$ in any particular recurrent selection program is a function, not of $q_0 + E(\Delta q)_0$, but rather of $q_1 = q_0 + (\Delta q)_0$ in which $(\Delta q)_0$ is the change in q that occurred in one replication (a specific selection program) out of the infinite number possible. To recognize this difference by symbols employed, let

$E(\Delta q/q)$ symbolize the expectation of Δq as a function of q in the S population,

$E(\Delta q/q = q_1)_1$ symbolize the expectation of Δq in cycle 1 as a function of q_1, the achieved frequency of the B allele in the S_1 population and

$E(\Delta q/q = q_0)_1$ symbolize the expectation of Δq in cycle 1 as a function of q_0.

The reality and logic of issues considered above can be demonstrated by numerical examples. Having gained understanding by dealing with examples, I judge that a selected small set will help some readers.

Tables 11.7 and 11.8 show single locus expectations in the case of two variations of recurrent selection among pure lines extracted without selection from S populations in which genotype frequencies are in both Hardy-Weinberg and linkage equilibrium. Computations were minimized by making the number of selections only three. While that would be unusual in a real breeding program, the important issues are nevertheless demonstrated. The basic quantities in each case are the probabilities of genotype combinations in the selections of cycle j, i.e., the pure lines chosen to be randomly intermated in formation of the S populations of cycle j+1. Given those probabilities, the expectations of q, $\ddot{y}$, and the proportion of heterozygosity are easily determined.

Cycle 0 is unique because the genetic samples of all replications are drawn from the same genetic population. Thus $E(q_1)$ is the same in all replications of the selection process. Because the only single locus genotypes possible in pure lines are those homozygous for one or the other of the two alleles, and the frequencies of B in the BB and bb genotypes are 1 and 0, respectively, $E(q_1)$ is the probability that the genotype of any single selection will be BB. It follows that the probabilities of the combinations of genotypes in a specific number of selections are given by the separate terms of a binomial expansion (Appendix 1). Let $\bar{q}_1 = E(q_1)$. Then in the case of M selections the probability that the number of BB lines will be x is given by the term in the expansion of $[q_1 + (1-\bar{q}_1)]^M$ in which the exponent of $\bar{q}_1$ is x. Thus, for example, when M=3, the probabilities of genotype combinations are as follows.

Genotypes of Selections	Probability
BB,BB,BB	$(\bar{q}_1)^3$
BB,BB,bb	$3(\bar{q}_1)^2(1-\bar{q}_1)$
BB,bb,bb	$3(\bar{q}_1)(1-\bar{q}_1)^2$
bb,bb,bb	$(1-\bar{q}_1)^3$

In all later cycles, q is variable among the S populations of different replications. When the number of selections is two, the frequency of B is constrained to one of three values (1,1/2, or 0); when M = 3, the frequency of B can only be 1, 2/3, 1/3, or 0; etc.

As indicated in the table itself, Table 11.7 shows probabilities and expectations when there are three selections taken randomly (k = 0) in each of three successive cycles and the frequency of the B allele in the S population of cycle 0 is $q_0 = 0.7$. Allele frequencies in the S populations of replicates in cycles after cycle 0 are determined by the genotype combinations in the three selected lines from which those S populations are generated. For example, given BB,BB,bb as the genotype combination, the allele frequency of the S population is two-thirds.[8] It is apparent that in cycles after cycle 0 there are four classes of replications of the program and that the frequencies of the B allele are 1, 2/3, 1/3, and 0 in these four classes. Consequently, the probabilities of a selected line that is BB (or bb) in genotype are different in

Table 11.7. Illustration–Heterozygosity, q, and ȳ in successive cycles of selection
Assumptions: (1) HW genotype frequencies and linkage equilibrium in S populations; (2) Selection units–random pure lines; (3) Selections–Three lines taken randomly from the genetic sample; thus k = 0 and N_e = 1.5; (4) u = 1/2 and q_0 = 0.7.

Cycle	x	Allele freq. in S popns	P	H	E(ȳ) = 2qu + Hau: a = 0	a = 1.0	E(Δq)	E(q)
0		0.700		.42	.700	.910	.00	.700
1	3	1.000	.343	.0000	1.000	1.000	.00	1.000
	2	.667	.441	.4444	.667	.889	.00	.667
	1	.333	.189	.4444	.333	.555	.00	.333
	0	.000	.027	.0000	.000	.000	.00	.000
	Totals	.700	1.000	.2800	.700	.840	.00	.700
2	3	1.000	.4807	.0000	1.000	1.000	.00	1.000
	2	.667	.2380	.4444	.667	.889	.00	.667
	1	.333	.1820	.4444	.333	.555	.00	.333
	0	.000	.0993	.0000	.000	.000	.00	.000
	Totals	.700	1.0000	.1866	.700	.793	.00	.700
3	3	1.000	.5580	.0000	1.000	1.000	.00	1.000
	2	.667	.1462	.4444	.667	.889	.00	.667
	1	.333	.1338	.4444	.333	.555	.00	.333
	0	.000	.1620	.0000	.000	.000	.00	.000
	Totals	.700	1.0000	.1244	.700	.762	.00	.700

Note: x = number of previous cycle selections with BB genotype, P = probability, H = fraction of S population genotypes that are heterozygous, and E(q) = the probability of the B allele in random gametes from the S population.

the replicate members of those classes. It follows that the probabilities of the four genotype combinations possible in three selections differ among the classes of replications and that a weighted average of those probabilities is required to obtain the overall probabilities of the next cycle. To illustrate, consider the probability of a replicate in cycle 2 in which the S population is homozygous for the B allele because its three parent lines, selected in cycle 1, were all BB in genotype. The probabilities of a BB line selection are 1, 2/3, 1/3, or 0, given that q = 1, 2/3, 1/3, or 0 in the S population from which they are selected. Taking account of the probabilities of these allele frequencies in the S populations of replicates in cycle 1, we obtain

$$(.343)(1) + (.441)(2/3)^3 + (.189)(1/3)^3 = .4807.$$

As another example, consider the probability of a replicate in cycle 3 in which the S population allele frequency is q = 2/3. When genetic samples are obtained from S populations with allele frequencies of 1, 2/3, 1/3, or 0,the corresponding probabilities of a selection set composed of two BB lines and one bb line are 0, $3(2/3)^2(1/3)$, $3(2/3)(1/3)^2$, and 0. Taking account of the probabilities of the four allele frequencies in S populations in the cycle 2 replications we find that the probability of q = 2/3 in S populations of cycle 3 is

$$(.2380)3(2/3)^2(1/3) + (.1820)3(2/3)(1/3)^2 = .1462$$

With one exception, the quantities shown as "averages" were obtained by taking the probabilities of classes into account. The exception was the <u>total</u> in the P column. The individual probabilities of all possible classes must obviously add to one and the fact that they do provides a check on computations. Noteworthy features of quantities shown in Table 11.7 are

(1) The average frequency of the more favorable allele is 0.7 in cyles 1-3 as it was in cycle 0. That, of course, reflects the fact of random selection so that $E(\Delta q) = 0$ as shown.

(2) The "drift" of allele frequencies among individual finite populations is evidenced by the cycle to cycle changes in the probabilities of replications in the four classes. Note in particular the increases in probabilities of the homozygous ($x = 3$ or 0) classes.

(3) On average, H (the fraction of heterozygous genotypes) decreases from cycle to cycle. Expressed as a fraction, the reduction is in each case $1/2N_e = 1/3$, i.e., H in cycle j+1 is $(1-1/2N_e)H_j$.

(4) $E(\ddot{y})$ when $a = 0$ is the same in successive cycles because, on average, q remains the same. Unless dominance is completely absent ($a=0$), the magnitude of the second term varies in proportion to the heterozygosity fraction. This change is, of course, the <u>inbreeding effect</u> which is negative if $a > 0$, positive if $a < 0$. Obviously, in the absence of epistasis, the observed inbreeding effect is the sum over all loci of the effect here illustrated in terms of just one locus.

By virtue of the assumption of selection for high performance some of the quantities shown in Table 11.8 differ from those of Table 11.7. The primary impact of that selection is to make $E(\Delta q) > 0$ in the case of all non-homozygous S populations. All other differences between quantities shown in the two tables flow from those non-zero values of $E(\Delta q)$. The following are noteworthy features of the content of Table 11.8.

(1) The average of allele frequencies of S populations increases in successive cycles -- though at a diminishing rate because the probability of replicates in which parents of the S population have different genotypes decreases in successive cycles.

(2) As intuitively expected, the decrease in the heterozygosity fraction from cycle to cycle is affected by both N_e and the change in q due to selection. In Table 11.7 the decrease from one cycle to the next was 1/3 in each case. Here the decrease from cycle 0 to cycle 1 is about 40% and in each cycle the heterozygosity fraction indicated is less in Table 11.8 than in Table 11.7.[9]

(3) Because the frequency of the favorable allele is increased by positive (rather than random) selection, $E(\Delta\ddot{y})$ is larger in the S populations of each successive cycle when $a=0$. On the other hand, when $a=1$ the negative inbreeding effect is larger in absolute magnitude than the positive effect of selection via increase in frequency of the favorable allele, i.e., $E(\Delta\ddot{y})$ decreases from cycle to cycle. This would not be the case if effective population size were enough larger. The allele frequency "drift", reflected by the cycle to cycle increases in the probabilities of homozgous ($x = 1$ or 0) S populations with their consequent reductions of the over all average of heterozygosity and $E(\Delta q)$, becomes smaller as N_e is increased.

Changes in $E(q)$ and $E(\ddot{y})$ are seen to depend on selection intensity, effective population size (N_e), allele frequency, level of dominance (a) and the magnitude of u. The first two of these can (within limits) be chosen by the breeder. In Chapters V, VI, VIII, and IX expressions that ignore N_e are derived for $E(\Delta q)$ and $E(\Delta\ddot{Y})$ as functions of q, u, a, and selection intensity. In

Table 11.8. Illustration—Heterozygosity, q, and $\bar{y}$ in successive cycles of selection

Assumptions: (1) HW genotype frequencies and linkage equilibrium in S populations; (2) Selection units— random pure lines; (3) Selections—the three lines with highest performance in a 100 line genetic sample, thus $N_e = 1.5$ and $k = 2.27$; (4) $q_0 = 0.7$, $u = 1/2$, $\sigma(X) = 10$, and $2uK = .227$.

$E(\Delta q/q) = 2q(1-q)uK$, $E(\Delta q/q=.7) = .0477$, and $E(\Delta q/q=2/3) = E(\Delta q/q=1/3) = .0504$.

Cycle	x	Allele freq. in S popns	P	H	$E(\bar{y}) = 2qu + Hau$, a = 0	a = 1.0	$E(\Delta q)$	$E(q)$
0		0.7		.42	.700	.910	.0477	.7477
1	3	1.000	.4180	.0000	1.000	1.000	.0000	1.0000
	2	.667	.4231	.4444	.667	.889	.0504	.7171
	1	.333	.1428	.4444	.333	.555	.0504	.3837
	0	.000	.0161	.0000	.000	.000	.0000	.0000
Totals		.748	1.0000	.2515	.748	.873	.0285	.7762
2	3	1.000	.5821	.0000	1.000	1.000	.0000	1.0000
	2	.667	.2235	.4444	.667	.889	.0504	.7171
	1	.333	.1353	.4444	.333	.555	.0504	.3837
	0	.000	.0591	.0000	.000	.000	.0000	.0000
Totals		.776	1.0000	.1595	.776	.850	.0181	.7943
3	3	1.000	.6722	.0000	1.000	1.000	.0000	1.0000
	2	.667	.1344	.4444	.667	.889	.0504	.7171
	1	.333	.0976	.4444	.333	.555	.0504	.3837
	0	.000	.0958	.0000	.000	.000	.0000	.0000
Totals		.794	1.0000	.1031	.794	.846	.0117	.8060

Note: x = number of previous cycle selections with BB genotype; P = probability; H = fraction of S population genotypes that are heterozygous; E(q) = the probability of the B allele in random gametes produced by cycle m selections, (i.e., the expectation of q in S_{m+1} populations) -- a function of the frequency of B in the S_m population.

effect the assumption involved is that magnitudes of N_e in practice are such that its impact can be considered trivial with respect to designs of breeding programs. That has been shown to be incorrect in the context of long term recurrent selection programs (sections 2.8 and 11.1) and the values of $E(\Delta\bar{y})$ in Table 11.8 demonstrate that N_e can be small enough to have a significant effect in short term recurrent selection programs. Theory, amenable to application by breeders, and that takes account of effective population size would obviously be useful. Expressions that yield $E(\Delta q)$ and $E(\Delta\bar{y})$ in cycle 1 as functions of N_e are derived in Appendix 11.3. They could be used to obtain either of those expectations in later cycles by successive applications but the computer and computer programs required would mitigate against general use for those purposes. Two expressions for $E(\Delta q/q=q_0)_1$ are shown below. While q_0 does not appear in either of the two, that both are functions of q_0 is implicit in the fact that $\bar{q}_1$ is a function of q_0.

(1) When the selection units are non-inbred individuals or families of non-inbred individuals

$$E(\Delta q/q{=}q_0)_1 = \bar{q}_1(1-\bar{q}_1)\{[1+(1-2\bar{q}_1)a][1-1/2N_e]$$
$$-(1-2\bar{q}_1)a/N_e\}ucK \qquad \text{(A11.3.9)} \quad 11.23$$

in which $\bar{q}_1 = E(q_1)$, $K = k/\sigma(\hat{X})$, and c = the coefficient shown in Table 8.2 that is variable among selection systems.

(2) When selection is among S_n families on the basis of their own performance

$$E(\Delta q/q{=}q_0)_1 = \bar{q}_1(1-\bar{q}_1)\{[(1+\hat{h})+(1-\hat{h})(1-2\bar{q}_1)a/2][1-1/2N_e]$$
$$-[(1-\hat{h})(1-2\bar{q}_1)a/2]/N_e\}uK \qquad \text{(A11.3.10)} \quad 11.24$$

in which h = the inbreeding coefficient of parents of the S_n families.

In the case of selection among pure lines (as assumed in Tables 11.7 and 11.8), h = 1.0 and 11.24 reduces to

$$E(\Delta q/q{=}q_0)_1 = 2\bar{q}_1(1-\bar{q}_1)(1-1/2N_e)uK. \qquad 11.25$$

The application and validity of 11.25 is demonstrated by checking against results presented in Table 11.8. In that example $\bar{q}_1 = q_0 + E(\Delta q)_0 = .7477$, $N_e = 1.5$, and $2uK = .227$. Making these substitutions in 11.25, we obtain .0285, the value indicated in the table. It would be convenient if, by analogy with 11.25, it were true that

$$E(\Delta q/q{=}q_0)_2 = 2\bar{q}_2(1-\bar{q}_2)(1-1/2N_e)uK.$$

Actually, employing the true value of $\bar{q}_2$ (.776 in this case) the answer given by this analog of 11.25 is 0.0263 instead of .794 - .776 = 0.0181, the correct value of $E(\Delta q)_2$. The obvious reasons are (a) that allele frequencies are variable among the S populations of cycle 1 replicates, (b) that because of that variation $E(\Delta q/q{=}q_1)_2$ is variable among replicates, and (c) that the average over replications is not the same as $E(\Delta q/q{=}q_1)_2$ would be if q_1 were the same in the S populations of all cycle 1 replicates.

Before going on it must be noted that exact answers are given by 11.23 when a=0 and by 11.24 when a=0 or h=1; otherwise answers provided are not exact. An illustration is provided by the cycle 1 expectations in the case of mass selection when a=1 that are shown in Table 11.9. As indicated, parts (a) and (b) of Table 11.9 deal with two selections and four selections, respectively. In the case of two selections, the true value of $E(\Delta q)_1$ obtained by averaging over replicate classes is

$$(.4219)(.00656)+(.2109)(.01750)+(.0469)(.01969) = .00738$$

while the value provided by 11.23 is

$$(.75)(.25)[2(.25)(1-1/4)-(1-1.5)/2](.07) = .00820$$

The answer provided by 11.23 is in error because (a) $E(q_1^3)$ appears in $E(\Delta q/q{=}q_0)_1$, (b) in the derivation outlined in Appendix 11.3, $E(q_1^3)$ was based on A11.3.2 which is not exact unless the distribution of X is symmetrical and (c) the distribution of q_1 is skewed unless $E(q_1)=1/2$ and becomes more skewed as $E(q_1)$ approaches 1 or 0 and sample size becomes small.

In the case of four selections, part (b) of the table, the correct value of $E(\Delta q)_1$ obtained by averaging over replicate classes is

Table 11.9. Illustration–Calculation of $E(\Delta q)_1$
Assumptions: (1) HW genotype frequencies and linkage equilibrium in S populations; (2) Selection units–non-inbred individuals; (3) a = 1, u = 1/2, c = 1, $E(q_1)$ = 0.75, and $\sigma(X)$ = 17.3; (4) $E(\Delta q) = q(1-q)[1+(1-2q)a]uck/\sigma(X) = 2q(1-q)^2uk/\sigma(X)$.

Selections - best <u>two</u> from 100 in genetic sample; N_e = 2, k=2.42, and $uk/\sigma(X)$ = 0.07

Cycle	Allele freq. in S popns.	Probability[a]	$E(\Delta q)$
1	1.00	.3164	.000
	.75	.4219	.00656
	.50	.2109	.01750
	.25	.0469	.01969
	.00	.0039	.00000
Averages	.75	1.0000	.00738

$E(\Delta q)_1$ = (.4219)(.00656) + (.2109)(.01750) + (.0469)(.01969) = .00738

Selections - best <u>four</u> from 100 in genetic sample; N_e = 4, k=2.16, and $uk/\sigma(X)$ = 0.0624

Cycle	Allele freq in S popns.	Probability[a]	$E(\Delta q)$
1	1.000	.1001	.00000
	.875	.2670	.00171
	.750	.3115	.00586
	.625	.2076	.01099
	.500	.0865	.01562
	.375	.0231	.01831
	.250	.0038	.01758
	.125	.0004	.01196
	.000	.0000	.00000
	.750	1.0000	.00641

$E(\Delta q)_1$ = (.2670)(.00171)+(.3115)(.00586) ...... +(.0004)(.01196) = .00641
<u>$E(\Delta q)_1$ = the probability of the B allele in random gametes produced by cycle 1 selections.</u>

[a] These are binomial probabilities. When $E(q_1)$ = .75, the probabilities of B and b containing gametes are .75 and .25, respectively. Then letting n= the number of selections and x = the number of B containing gametes among m = 2n random gametes, the probability of an S population of which the allele frequency is x/m is given by $[m!/x!(m-x)!](.75)^x(.25)^{m-x}$. Thus, for example, the probability that q =.625 in an S population in cycle 1 when the number of selections is four is $[8!/5!3!](.75)^5(.25)^3$ = .2076.

$$[(.2670)(.00171) + (.3115)(.00586) + \cdots\cdots + (.0004)(.01196)] = .00641$$

and the value provided by 11.23 is

$$(.75)(.25)[2(.25)(1-1/8)-(1- 1.5)]/4](.0624) = .00658.$$

Note that the upward bias of the answer provided by 11.23 is 11.1% in the case of two selections (N_e= 2) but only 2.7% in the case of four selections (N_e=4). Whatever the number of selections, the bias would be greater if $E(q_1)$ were farther from 0.5. However, it is apparent that the approximation provided by 11.23 is good unless $E(q_1)$ is close to 1.0 or 0.0 and the number of selections is very small.

- 11.3.1. **General Formulas**

The logic of derivations is outlined in Appendix 11.3. Formulas 11.23 - 11.25, having to do with expected changes in frequency of the B allele, are shown above. Expressions dealing with change in the average value of single locus genotypes are presented below.

- 11.3.1.1. $E(\Delta\bar{y})_1$ IN THE TARGET POPULATION WHEN T= S

When selection units are non-inbred individuals or families of non-inbred individuals

$$E(\Delta\ddot{y})_1 = 2\bar{q}_1(1-\bar{q}_1)\{[1+(1-2\bar{q}_1)a]^2(1-1/2N_e) - Z/2N_e\}u^2cK$$
$$- \bar{q}_1(1-\bar{q}_1)au/N_e \qquad (A11.3.17) \quad 11.26$$

in which $Z \approx 4a\{(1-2\bar{q}_1) + [1-5\bar{q}_1(1-\bar{q}_1)]a\}$, $K = k/\sigma(\hat{X})$, and c = the coefficient shown in Table 8.2 that is variable among selection systems.

When selection is among S_n families on the basis of their own performance

$$E(\Delta\ddot{y})_1 = 2\bar{q}_1(1-\bar{q}_1)\{[(1+\hat{f})+(1-\hat{f})(1-2\bar{q}_1)a/2][1+(1-2\bar{q}_1)a](1-1/2N_e)$$
$$- Z/2N_e\}u^2K - \bar{q}_1(1-\bar{q}_1)au/N_e \qquad (A11.3.19) \quad 11.27$$

in which $Z \approx a\{(3+\hat{f})(1-2\bar{q}_1) + 2(1-\hat{f})[1-5\bar{q}_1(1-\bar{q}_1)]a\}$.

- 11.3.1.2. $E(\Delta\bar{y})_1$ IN THE TARGET POPULATION WHEN T= S_1 X S_2

When selection and criterion units (in both S_1 and S_2) are composed of non-inbred individuals from the population itself

$$E(\Delta\ddot{y})_1 = \bar{p}_1(1-\bar{p}_1)\{[1+(1-2\bar{p}_1)a](1-1/2N_{e1}) - 2(1-2\bar{p}_1)a/2N_{e1}\}$$
$$\times [1+(1-2q)a]\,u^2c_1K_1$$
$$+ \bar{q}_1(1-\bar{q}_1)\{[1+(1-2\bar{q}_1)a](1-1/2N_{e2}) - 2(1-2\bar{q}_1)a/2N_{e2}\}$$
$$\times [1+(1-2p)a]\,u^2c_2K_2 \qquad (A11.3.24) \quad 11.28$$

Here p and q symbolize the frequencies of the B allele in S_1 and S_2, respectively, and the numerical subscripts to the N_e's, the c's, and the K's reflect the possibility of different selection systems in S_1 and S_2. Note that 11.28 deals with the system, RSP, considered in section 9.4 and that 11.28, when summed over all loci, converges as it should to 9.24 as N_e becomes large (and the compositions of K_1 and K_2 are taken into account).

When selection in both S populations is among S_n families on the basis of their own performance

$$E(\Delta\ddot{y})_1 = \bar{p}_1(1-\bar{p}_1)\{[(1+\hat{f})+(1-\hat{f})(1-2\bar{p}_1)a/2](1-1/2N_{e1})$$
$$-[(1-\hat{f})(1-2\bar{p}_1)a/2]/N_{e1}\}[1+(1-2q_1)a]u^2K_1$$
$$+ \bar{q}_1(1-\bar{q}_1)\{[(1+\hat{f})+(1-\hat{f})(1-2\bar{q}_1)a/2](1-1/2N_{e2})$$
$$-[(1-\hat{f})(1-2\bar{q}_1)a/2]/N_{e2}\}[1+(1-2p_1)a]\,u^2K_2 \qquad (A11.3.25) \quad 11.29$$

Again numerical subscripts recognize that values of N_e, k, and $\sigma(\hat{X})$ may be different for S_1 and S_2. The lack of subscripts to h in the two portions of 11.29 reflects assumption that n, the generations of self-fertilization to produce the S_n families, would be the same in both S populations.

When selection is based on performance of test cross families the general expression that covers the variations of selection based on test cross performance is the following.

$$E(\Delta\ddot{y})_1 = \bar{p}_1(1-\bar{p}_1)[1+(1-2v_1)a][1+(1-2q)a](1-1/2N_{e1})u^2c_1K_1$$
$$+\bar{q}_1(1-\bar{q}_1)[1+(1-2v_2)a][1+(1-2p)a](1-1/2N_{e2})u^2c_2K_2 \qquad (A11.3.26) \quad 11.30$$

in which v_1 and v_2 symbolize the frequencies of B in the testers employed in the S_1 and S_2 selection programs. The reasons why 11.30 deviates in structure from 11.28 and 11.29 and the variations of 11.30 that are specific to RSSC and RRS are covered in Appendix 11.3.

• 11.3.2. Discussion and Summary

The following aspects of the several formulas in sections 11.3 and 11.3.1 that show how $E(\Delta q)_1$ and $E(\Delta \ddot{y})_1$ are affected by effective population size are all worthy of attention.

(1) As emphasized in discussion of the content of Table 11.8, $E(\Delta q)$ in cycles beyond cycle 1 are not provided by simple analogs of 11.25. Considering the reasons why that is true, it is obvious that the same principles apply in the case of $E(\Delta \ddot{y})_1$ formulas. It follows that the expectation of $\Delta \ddot{y}$ in any cycle beyond cycle 1 is the weighted average of $E(\Delta \ddot{y})$ over all classes of replications of the selection program. Thus

$$E(\Delta \ddot{y})_m = \Sigma_c P_{cm} E(\Delta \ddot{y})_c$$

where m, as a subscript, identifies the cycle,
c, as a subscript, identifies the class of replications, and
P_{cm} = the probability in the m-th cycle of a replication of the c-th class.

In the same way, $E(\Delta q) = \Sigma_c P_{cm} E(\Delta q)_c$

(2) As N_e becomes large all expressions converge to the forms derived earlier (when effects of N_e were ignored) if $\bar{q}_1$ and $\bar{p}_1$ are replaced by q and p, respectively.

(3) When the selection population is the target population (T=S) the $E(\Delta \ddot{y})_1$ expressions have two components.

The first of these measures the positive effect of selection -- it is a linear function of k, hence its magnitude is directly proportional to the selection differential. In $E(\Delta \ddot{y})_1$ its magnitude is partially dependent on the magnitude of N_e. But in $E(\Delta \ddot{y})_0$ (see A11.3.15) it is independent of N_e because, given the specific S population of cycle 0, q_0 is a constant.

The second component of $E(\Delta \ddot{y})_1$ is inversely proportional to N_e and is not a function of the selection differential. When it is summed over all genes and f is substituted for $1/2N_e$, it is recognized as the <u>inbreeding effect</u> given by 2.40.

(4) In the case of RSP, $E(\Delta \ddot{y})_1$ is given by 11.28, 11.29, and 11.30. In these no inbreeding effect is indicated. Its absence reflects the fact that the random components of allele frequency change are independent from one of the S populations to the other. Symbolically, the frequency of the heterozygous genotype in the $S_1 \times S_2$ cross is $p(1-q) + q(1-p)$ which is not a function of the inbreeding coefficient of either of the S populations.

(5) When dominance is absent (a = 0.0) all formulas reduce to $\bar{q}_1(1-\bar{q}_1)(1-1/2N_e)$ multiplied by K and appropriate constants. For example, 11.23 becomes

$$E(\Delta q/q{=}q_0)_1 = \bar{q}_1(1-\bar{q}_1)(1-1/2N_e)ucK,$$

11.24 becomes $E(\Delta q/q{=}q_0)_1 = \bar{q}_1(1-\bar{q}_1)(1+\hat{f})(1-1/2N_e)uK,$

11.26 becomes $E(\Delta \ddot{y})_1 = 2\bar{q}_1(1-\bar{q}_1)(1-1/2N_e)u^2cK,$

and 11.28 becomes

$$E(\Delta \ddot{y})_1 = \bar{p}_1(1-\bar{p}_1)(1-1/2N_{e1})u^2c_1K_1 + \bar{q}_1(1-\bar{q}_1)(1-1/2N_{e2})u^2c_2K_2.$$

(6) In each of the prediction formulas except 11.30, there are two terms within a pair of

curly brackets. Consider, for example, 11.23. There the content of the curly brackets is

$$[1+(1-2\bar{q}_1)a](1-1/2N_e) - (1-2\bar{q}_1)a/N_e$$

If the second of the two terms were absent 11.23 would be changed to

$$E(\Delta q/q=q_0)_1 = \bar{q}_1(1-\bar{q}_1)[1+(1-2\bar{q}_1)a](1-1/2N_e)ucK.$$

At face value, this would indicate that, whatever the level of dominance, the impact of N_e being finite would be to make $E(\Delta q/q=q_0)_1$ a linear function of $(1-1/2N_e)$ as when dominance is absent (a = zero). Actually, the second term, $(1-2\bar{q}_1)a/N_e$, together with the first, is required to show the impact of finite effective population size when $a \neq 0.0$.

As a second example, consider 11.26 in which the content of the curly brackets is

$$[1+(1-2\bar{q}_1)a]^2(1-1/2N_e) - Z/2N_e$$

If the second of these terms were absent 11.26 would be

$$E(\Delta\ddot{y}) = 2\bar{q}_1(1-\bar{q}_1)[1+(1-2\bar{q}_1)a]^2(1-1/2N_e)u^2cK - \bar{q}_1(1-\bar{q}_1)au/N_e$$

Aside from the "inbreeding effect" term, this would indicate that, whatever the level of dominance, the impact of the magnitude of N_e on $E(\Delta\ddot{y})$ due to change in q in response to selection would be a linear function of $(1-1/2N_e)$. Again, the modification required when $a \neq 0.0$ is provided by the second term within the curly brackets -- in this case $Z/2N_e$.

These two examples characterize each of the six formulae specified above. In each instance, the quantity in question is a linear function of $(1-1/2N_e)$ when $a = 0.0$ but not when $a \neq 0.0$. The compositions of Z in 11.26 and 11.27 and its counterparts in 11.23, 11.24, 11.28, and 11.29 are such that they become negative when $a > 0.0$ and $\bar{q}_1$ (or $\bar{p}_1$) > 0.5 and then have the effect of increasing instead of decreasing the expected response to selection. However, unless N_e is small and both a and q are large that effect is too small to have practical importance.

• 11.3.3. **Applications—Choice and Design of Short Term Programs**

In preceding summaries of the effect of N_e on the expectation of change in average value of genotype, $K = kc/\sigma(\hat{X})$ has been treated as a constant. However, in practice N_e can hardly be changed without concurrent change in either k or $\sigma(\hat{X})$ or both unless program cost is altered. Consider any program in which the criterion unit is a family. N_e cannot be increased at the same time that k remains the same unless the size of the genetic sample, i.e., the number of selection units per cycle, is increased. To do that without increasing the total number of individuals or plots observed would require a decrease in family size and a consequent increase in $\sigma(\hat{X})$. Clearly, what may be gained or lost by a change in N_e should be considered in the context of the effect or cost of the concurrent changes in k, $\sigma(\hat{X})$ or program size by which the modification of N_e is achieved. These interacting issues will be considered and illustrated in greater detail in Chapter XIV.

The preceding paragraph draws attention to the fact that once the scope, i.e., the operation cost, of a recurrent selection program has been decided, there will inevitably be an optimum balance among variables of the program. This fact leads to the notion that comparisons of the potential effectiveness of different selection systems should be based on $E(\Delta\ddot{Y}) = \Sigma_i E(\Delta\ddot{y}_i)$ when for each of the sysems compared the "balance of its variables" is near-optimum. The potential advantage and inherent difficulties of doing so will also be explored in Chapter XIV.

12 Choice of the Selection Criterion

Data concerning more than one aspect of phenotype are invariably collected in selection programs. Assuming selection is to be based on a single criterion, how should the different items of data (measurements or scores) be combined in the formation of that criterion?[1] The objective of breeding is to increase the total value of the cultivars, breeds, varieties, or crossbreds available for use in agriculture. The following are obvious corollaries.

(1) It would be reasonable, if not optimum, to employ total value itself as the basis for selection. However, very frequently, the cost of making all measurements required to quantify total value would appear to be, or would be, prohibitive.

(2) The best selection criterion is the one that maximizes the genetic increase in total value that is likely to be achieved. In the context of recurrent selection, that equates to the expectation of change in genotype for total value during the entire period of the program.

• 12.1. Expected Change per Cycle in Genotype Values

Consider p different measurements of phenotype and a selection criterion, $\hat{X}$, that is a linear function of those measurements. Let $\hat{X}_m$ symbolize the m-th measurement for a criterion unit. Then

$$\hat{X} = b_1 \hat{X}_1 + b_2 \hat{X}_2 + \cdots\cdots b_p \hat{X}_p = \Sigma_m b_m \hat{X}_m \qquad 12.1$$

where b_m is the "weight" that the m-th measurement receives in the selection criterion. Since any b can be zero, 12.1 covers the cases where any number of measurements are not obtained or, for any reason, are not used in the selection criterion, including the extreme in which a single measurement is employed as the basis for selection. Let W symbolize total value. Then from 5.8 and 5.18

$$E(\Delta\ddot{Y}_w) = [k/\sigma(\hat{X})]\Sigma_i\,\sigma(\hat{q}_i,\hat{X})\,[\partial\ddot{Y}_w/\partial q_i] \qquad 12.2$$

and, in general

$$E(\Delta\ddot{Y}_m) = [k/\sigma(\hat{X})]\Sigma_i\,\sigma(\hat{q}_i,\hat{X})[\partial\ddot{Y}_m/\partial q_i] \qquad 12.3$$

where $\ddot{Y}_m$ symbolizes the mean value of genotype with respect to the measurement X_m. The meanings of other symbols should be apparent from previous use. For example,

$E(\Delta\ddot{Y}_w)$ symbolizes the expected change per cycle of $\ddot{Y}_w$ (the average value of genotype with respect to total value) in the target population and

$\partial\ddot{Y}_m/\partial q_i$ symbolizes the partial derivative of $\ddot{Y}_m$ in the target population with respect to the frequency of B_i in the selection population.

By virtue of 12.1,

$$\sigma(\hat{q}_i,\hat{X}) = b_1\,\sigma(\hat{q}_i,\hat{X}_1) + b_2\sigma(\hat{q}_i,\hat{X}_2) + \cdots\cdots\cdots\; b_p\,\sigma(\hat{q}_i,\hat{X}_p). \qquad 12.4$$

Then substitution in 12.2 yields

$$E(\Delta\ddot{Y}_w) = [k/\sigma(\hat{X})][b_1\Sigma_i\,\sigma(\hat{q}_i,\hat{X}_1) + b_2\Sigma_i\,\sigma(\hat{q}_i,\hat{X}_2) + \cdots\cdots b_p\,\sigma(\hat{q}_i,\hat{X}_p)][\,\partial\ddot{Y}_w/\partial q_i]$$

$$= [k/\sigma(\hat{X})][b_1G_{1w} + b_2G_{2w} + \cdots\cdots\cdots b_pG_{pw}] \qquad 12.5.$$

where

$$G_{mw} = \Sigma_i\,\sigma(\hat{q}_i,\hat{X}_m)\,[\partial\ddot{Y}_w/\partial q_i] \qquad 12..6$$

and substitution in 12.3 yields

$$E(\Delta\ddot{Y}_{m'}) = [k/\sigma(\hat{X})][b_1G_{1m} + b_2G_{2m} + \cdots\cdots\cdots b_pG_{2p}]$$

where

$$G_{mm'} = \Sigma_i\,\sigma(\hat{q}_i,\hat{X}_m)\,[\partial\ddot{Y}_{m'}/\partial q_i] \qquad 12.7$$

The quantity symbolized by G is not new or unfamiliar. The symbol, G, which has not been used previously, is introduced to allow some algebraic expressions to be written more compactly and to facilitate referrence to sums of the type indicated in 12.6 and 12.7. Note that the first subscript of G identifies the measurement involved in the $\sigma(\hat{q}_i,\hat{X})$ of the sum and the second subscript identifies the trait (measurement) to which the $\partial\ddot{Y}/\partial q_i$ of the sum pertains.

Because $\sigma(\hat{q}_i,\hat{X})$ varies with the selection unit and the criterion unit and $\partial\ddot{Y}/\partial q_i$ varies with the target population, the exact composition of any G depends on those details of the program in question. When the target population is the S population and the criterion units are composed of S population individuals, G_{mw} is equal or proportional to the additive genetic covariance between X_m and W, $G_{mm'}$ is equal or proportional to the additive genetic covariance between X_m and $X_{m'}$ and G_{mm} to the additive genetic variance of X_m. For illustration, consider progeny test selection with the S population itself employed as "tester" and assume the S population to be the target population. Then from Table 8.2

$$\sigma(\hat{q}_i,\hat{X}_m) = [(1+f)/2]\{q_i(1-q_i)[1+(1-2q_i)a_{mi}]u_{mi}$$
$$+\Sigma_{j\neq i}(pt-rs)_{ij}[1+(1-2q_j)a_{mj}]u_{mj}\}$$

and from 5.11

$$\partial\ddot{Y}_w/\partial q_i = 2[1+(1-2q_i)a_{wi}]u_{wi}$$

Substituting in 12.6 we obtain

$$G_{mw} = (1+f)\Sigma_i\{q_i(1-q_i)[1+(1-2q_i)a_{mi}][1+(1-2q_i)a_{wi}]u_{mi}u_{wi}$$
$$+\Sigma_i\Sigma_{j\neq i}(pt-rs)_{ij}[1+(1-2q_j)a_{mj}][1+(1-2q_i)a_{wi}]u_{mj}u_{wi}\} \quad 12.8$$

In these expressions the a-values for X_m are symbolized as a_{mi} and a_{mj} and the u-values as u_{mi} and u_{mj}. When 12.8 is compared with 6.20, we find that in this example $G_{mw} = [(1+f)/2]\,\sigma_{gmw}$ where σ_{gmw} is the additive genetic covariance between W and X_m. In cases where two populations are involved because the target population is not the S population, the G's identify quantities analogous to additive genetic covariance (or variance).

The nature of the G's is particularly important for the following reasons. Because additive genetic covariances and analogous quantities can be negative, $G_{mm'}$ can be negative (except when m = m'). It follows that, depending on the relative values of the weighting coefficients (the b's), either $E(\Delta\ddot{Y}_{m'})$ or $E(\Delta\ddot{Y}_w)$ can be negative even when the b's are such that $\hat{X}$ is increased by all favorable deviations in measurements contributing to $\hat{X}$. This, of course, is consistent with previous observations (Chapter VI). Unless X = W, $\Delta\ddot{Y}_w$ is a correlated response and the expectation of a correlated response can be either positive or negative. In the event of b's such that X = W, $\Delta\ddot{Y}_w$ would be a direct response and $E(\Delta\ddot{Y}_w)$ would be > 0.0.

• 12.2. The Smith-Hazel Selection Index (Criterion)That Maximizes Genetic Improvement per Cycle

Assuming that the selection criterion in a single-stage selection process will be a linear function of p measurements, see 12.1, how can the weighting coefficients (the b's) that maximize $E(\Delta\ddot{Y}_w)$ be determined? Theoretical treatments of this problem were presented by Smith (1936) and Hazel (1943). Their treatments were less general than the one being offered but the logic employed here was provided by those papers. Smith concerned himself with the genotypic value of selections without reference to population improvement in recurrent selection programs. Hazel focused on "breeding value", thus on the expected difference between successive generations in the average value of genotypes,[2] in the special case of mass selection with the selection population as the target population.

Standard calculus procedure provides (see Appendix 12.1) the following set of equations that can be solved simultaneously for optimum b-values, i.e., the b's that maximize $E(\Delta\ddot{Y}_w)$.

$$\begin{aligned}
\beta_1\sigma_{11} + \beta_2\sigma_{12} + \cdots\cdots\beta_p\sigma_{1p} &= G_{1w}\\
\beta_1\sigma_{21} + \beta_2\sigma_{22} + \cdots\cdots\beta_p\sigma_{2p} &= G_{2w}\\
\cdot \quad\quad \cdot \quad\quad \cdot \quad\quad & \cdot\\
\cdot \quad\quad \cdot \quad\quad \cdot \quad\quad & \cdot \qquad\qquad 12.9\\
\beta_1\sigma_{p1} + \beta_2\sigma_{p2} + \cdots\cdots\beta_p\sigma_{pp} &= G_{pw}
\end{aligned}$$

Here σ_{mm} = the phenotypic variance of $\hat{X}_m$ (m = 1,2,3,⋯p),

$\sigma_{mm'}$ = the phenotypic covariance of $\hat{X}_m$ with $\hat{X}_{m'}$ (m ≠ m'),

p = the number of measurements to be included in the selection criterion and, collectively, the β's constitute a set of b-values that maximizes $E(\Delta\ddot{Y}_w)$. As explained in Appendix 12.1, only the relative values of the β's are relevant. If a specific set maximizes $E(\Delta\ddot{Y}_w)$, any other set within which relative magnitudes are the same would serve as well. Changing absolute values while maintaining the same relative magnitudes would alter the absolute values of $\hat{X}$ but not their relative values for any set of criterion units. Thus selections made would be unaffected. The specific set of β's provided by equations 12.9 are the ones (see Appendix 12.1) that make $\sigma^2(\hat{X}) = \Sigma_m\beta_m G_{mw}$. As a result we find from 12.5 that the expectation of $\Delta\ddot{Y}_w$, if the σ's were known and used in $\hat{X}$, would be

$$E[\Delta\ddot{Y}_w/b\text{'s} = \beta\text{'s}] = k(\Sigma_m\beta_m G_{mw})^{1/2} \qquad 12.10$$

Unfortunately, the β's cannot be known. It is apparent from 12.9 that to know the β's we must first know the numerical values of p phenotypic variances, $p(p-1)/2$ phenotypic covariances, and p genotypic parameters (the G_{mw}'s). In the most favorable circumstances good estimates, not the exact values, of these quantities will be available. This means that, at best, only good approximations to the β's can be obtained. Let these be symbolized as B's. Because these, being other than the β's, are not optimum

$$E[\Delta\ddot{Y}_w/B\text{'s} \neq \beta\text{'s}] < k(\Sigma_m\beta_m G_{mw})^{1/2}$$

The actual loss of effectiveness resulting from the use of B's instead of β's is not readily apparent. However, our attention is drawn to the following practical issues.

(1) $\hat{X} = \Sigma_m B_m X_m$ will not be a good selection criterion unless adequate estimates of the σ_{mm}, the $\sigma_{mm'}$, and the G's are available.

(2) Estimation of the G's requires prior definition of total value. Otherwise the data needed would not be identified.

(3) One or more of the measurements required to quantifyl W may be very expensive.

(4) If selection is worthwhile, it is because allele frequencies change in response to selection. It follows that the G's, which are functions of allele frequencies, and the genetic fractions of the σ's will be changed by effective selection and hence that if B-values are to be kept reasonably close to optimum, re-estimation of parameters at appropriate time intervals will be required.

The problems involved in estimation of the σ's are, in general, considerably less troublesome than those encountered in estimation of the G's. There are various reasons for this but it will suffice to note that in any on-going program new data on the measurements being employed for selection unit evaluation are obtained in each cycle and these data, since they consist of $\hat{X}_m$ values for the criterion units of the specific selection system involved, are exactly the data needed for direct estimation of the σ's. In contrast, the data that flows from the on-going program is less likely to be all that is needed for estimation of the G's. The most obvious examples are those in which, for one reason or another, data are not regularly obtained on all aspects of total value. The reason may be the sheer cost of the measurement or its impact relative to cycle length. Consider longevity of dairy cows and fruit trees or the quality at harvest time of lumber from pine trees. Routine measurement of such traits would be very undesirable in terms of cycle length in breeding programs.

• 12.3. Definition of Total Value

This is a two step process. First, all of the value traits of the organism, i.e., the measurements required for quantification of total value, must be recognized. Second, total value is identified as an explicit mathematical function of those value trait measurements. Without this identification, numerical data on total value cannot be compiled.

The values of agricultural species lie in their products (grain, eggs, meat, fiber, lumber, etc.). It follows that value of product minus cost of production equals net value (total value).[3] The application of this rule requires knowledge of the organism, how it is managed in agriculture and how attributes of the organism and its product(s) relate to costs and value of product(s). Note, however, that quantities affecting total value but not the amount by which it is changed when genotypes are modified by selection can, for our purposes, be ignored in definitions of total value (see Appendix 12.2 for discussions of total value in the cases of maize and swine).

The concept that total value can be properly defined without reference to all traits that are significant with respect to the utility of the organism deserves comment. Consider lodging resistance in maize. Without it a large proportion of the plants in a field might be recumbent at harvest time with the result that realized yield to the producer (the grain gathered by conventional harvesting equipment) would be extremely low. Accordingly, lodging resistance is a significant attribute of maize in so far as agricultural utility is concerned. However, its impact is totally reflected by realized yield which is a value trait that must appear in the definition of total value. Lodging resistance is only one of many traits critical relative to total value but not required for its quantification, its definition. To define in a way that would recognize all such attributes would lead us on and on into all of the intricacies of the molecular biology of the organism. Fortunately, that is not required. However, omission from the definition of total value does not imply that a trait should not receive attention in the selection criterion. Whether it should or not is determined (aside from the practical consideration of cost of data) by the parameters of equations 12.9. Consider any measurement, X_m. Whether it appears or does not appear in the definition of total value, if it has an important bearing on W, the parameter G_{mW} may be of significant magnitude. Then, unless the heritability of X_m is too small, the solution of equations 12.9 would probably indicate a real advantage of including the trait as a component of the selection criterion.

In the end, any apppropriate definition of total value will be a function of a relatively small number of value trait measurements and a set of numerical coefficients. This function could be linear but is likely to be a non-linear function of the value trait measurements. For illustration, consider an over-simplification of the case in dairy cattle such that

$$W = d_1X_1 - d_2X_2 - d_0$$

where X_1 = annual milk production in kilograms, X_2 = the concomitant feed requirement in kilograms, d_1 = value of a kilogram of milk, d_2 = cost of a kilogram of feed, and d_0 = overhead cost per year per cow (assumed here to be unaffected by genotype). Here X_1 and X_2 are the value trait measurements and d_0, d_1, and d_2 symbolize the numerical coefficients of the definition.

When total value is defined as a linear function of value trait measurements, the general form of the definition is $W = d_1X_1 + d_2X_2 + d_3X_3 + \cdots d_0$ where any of the d's can be positive or negative. The change in W resulting from changes in the X's can be represented as

$$\Delta W = c_1(\Delta X_1) + c_2(\Delta X_2) + c_3(\Delta X_3) + \cdots \quad 12.11$$

where

$$c_m = \partial W/\partial X_m \quad 12.12$$

(= d_m when W <u>is</u> a linear function of value trait measurements).

The introduction of the c-series of symbols, which serves no purpose when W is a linear function of the X's, sets the stage for treatment in the cases of non-linear definitions. It follows from 12.11 that

$$\Delta Y_w = c_1(\Delta Y_1) + c_2(\Delta Y_2) + c_3(\Delta Y_3) + \cdots$$

and

$$\Delta \ddot{Y}_w = c_1(\Delta \ddot{Y}_1) + c_2(\Delta \ddot{Y}_2) + c_3(\Delta \ddot{Y}_3) + \cdots \quad 12.13$$

in which Y_m symbolizes genotypic value with respect to X_m. Then, when the changes in the $\ddot{Y}_m$'s that result from a change in q_i are considered, it is apparent from 12.13 that

$$\partial \ddot{Y}_w/\partial q_i = c_1(\partial \ddot{Y}_1/\partial q_i) + c_2(\partial \ddot{Y}_2/\partial q_i) + c_3(\partial \ddot{Y}_3/\partial q_i) + \cdots \quad 12.14$$

Using 12.14 for substitution in 12.6 we obtain

$$G_{mw} = c_1\Sigma_i\, \sigma(\hat{q}_i, \hat{X}_m)\, [\partial \ddot{Y}_1/\partial q_i] + c_2\Sigma_i\, \sigma(\hat{q}_i, \hat{X}_m)\, [\partial \ddot{Y}_2/\partial q_i] + \cdots$$

Then by the definition of $G_{mm'}$ given by 12.7

$$G_{mw} = c_1G_{m1} + c_2G_{m2} + c_3G_{m3} + \cdots \quad 12.15$$

Remember that $c_m = \partial W/\partial X_m$ as given by 12.12 and, below, by 12.17.

Now consider the more probable situation in which total value is defined as a non-linear function of value trait measurements. Symbolically

$$W = F(X_1, X_2, X_3; \cdots d_0, d_1, d_2, \cdots)$$

and, whatever the form of the function, the following is a close approximation when the ΔX's (the changes in the X's) are small.

$$\Delta W \approx (\Delta X_1)\partial W/\partial X_1 + (\Delta X_2)\partial W/\partial X_2 + (\Delta X_3)\partial W/\partial X_3 + \cdots$$
$$= c_1(\Delta X_1) + c_2(\Delta X_2) + c_3(\Delta X_3) + \cdots \quad 12.16$$

where

$$c_m = \partial W/\partial X_m \text{ (as in 12.12).} \quad 12.17$$

Note that this partial derivative, in non-linear cases, will always be a function of one or more of the X's. To obtain a single value for each c, all X's are replaced by the corresponding $\ddot{X}$'s in the partial derivatives. Then the c's apply most exactly to individuals that are average in all respects.

Accepting 12.16 as a sufficient approximation to ΔW and noting the concordance between 12.16 and 12.11, it is apparent that the logic employed in moving from 12.11 to 12.15 holds here as it did in the case of a linear function definition of total value. Thus, the relationship set out by 12.15 can be assumed in all cases.

There is, however, one very important difference between the cases where the total value definition is linear and those where it is non-linear. It is this. When the definition is linear in form the c's do not change as a consequence of genetic change in response to selection. On the other hand, when the definition is non-linear, $c_m = \partial W/\partial X_m$ is always a function of the mean value of one or more of the X's. It follows that if selection is effective, the c-values change as generation time goes on.

The quantities here symbolized by c_m were called "economic" values by Hazel (op. cit.).

• 12.4. The Selection Criterion That Maximizes the Total Genetic Improvement That Can Be Made via Recurrent Selection

Consider first the genetic improvement of a random mating selection population assuming (1) no overdominance, (2) no multiple peak epistasis and (3) no selection other than the artificial selection based on the criterion being discussed. In this case maximum genetic improvement requires fixation of the most favorable allele of each gene that has a bearing on total value. Probability of fixation is discussed in sections 2.8 and 11.1.

When we focus on a gene that segregates independently from all others with effects on the selection criterion, the following facts (attested in the two allele case by equations 2.33 and 2.34) are intuitively obvious.

(1) If the selection criterion is such that the selection coefficient (the s-value discussed in sections 2.6.1 and 6.3.6) is zero, the probability of fixation of the favorable allele is q_0, its original frequency in the population and

(2) If the probability of fixation is to be $>q_0$, s must be > 0.0 and, if the probability of fixation is to approach 1.0, effective population size must be large enough to insure that the force of selection is not likely to be negated by random drift.

It follows that for eventual genotypic value to be maximum, the requirement with respect to the selection criterion is that it makes s>0.0 for all alleles required in an ideal genotype. Of course, maximum improvement cannot be achieved without sufficient effective population size but, so far as the selection criterion is concerned, the requirement is the one stated above.

Now consider a gene that affects only one of the measurements taken into account by the selection criterion. Let the measurement affected by the gene be X_a. If larger values of X_a increase total value, i.e., if c_a >0.0, and the corresponding "weighting" coefficient (b-value) is positive, the s-value pertaining to effect of selection on frequency of the favorable allele will be >0.0. On the other hand if b_a is negative, the s-value would be negative and the probability of fixation of that favorable allele would be less than its original frequency in the population. Conversely, if $c_a < 0.0$, meaning that smaller values of X_a increase total value, $b_a < 0.0$ is required to make $s > 0.0$ for the allele favorable to total value. In summary, if s is to be positive for the favorable alleles of genes belonging to the class being considered, the "weighting" coefficient for the measurement involved must be positive if the c-value for that measurement is positive but negative if the c-value is negative.

The general case to be dealt with is that of pleiotropic genes, specifically those with effects on more than one of the measurements used in the selection criterion. It will suffice for our purpose to consider genes that affect two measurements for which the "weighting" coefficients are non-zero. The relevant quantities for a gene that affects X_1 and X_2 are shown symbolically in Table 12.1. It reminds us that value of genotype always pertains to an individual trait (measurement) and that contributions to total genotypic value (the y-values), in the case of pleiotropic genes, are specific to each trait affected by the gene. In accord with

Table 12.1. The y's of a pleiotropic gene that affects two traits (X_1 and X_2)

Genotype	y_1	y_2	y_w	y_x
BB	$2u_1$	$2u_2$	$2c_1u_1 + 2c_2u_2$	$2b_1u_1 + 2b_2u_2$
Bb	$u_1+a_1u_1$	$u_2+a_2u_2$	$c_1(u_1+a_1u_1)+c_2(u_2+a_2u_2)$	$b_1(u_1+a_1u_1)+b_2(u_2+a_2u_2)$
bb	0	0	0	0

Note: As elsewhere in this chapter, the numerical subscript that identifies single measurements of the X_m series is employed to identify the associated values of u, a, c, and b.

earlier usage herein, B symbolizes the more favorable allele in the sense that $u_w = c_1u_1 + c_2u_2$ is assumed to be > 0.0. However, it is not to be assumed that both u_1 and u_2 or both c_1u_1 and c_2u_2 are > 0.0. The reality of pleiotropy is that any single allele may be favorable with respect to one trait and unfavorable with respect to another.

Because $E(\Delta q) = k\sigma(\hat{q},\hat{X}) / \sigma(\hat{X})$ it will be positive when $\sigma(\hat{q},\hat{X})$ is positive.[4] Given the assumptions stated earlier in this section, we know from Table 8.2 that $\sigma(\hat{q},\hat{X})$ is proportional to $q(1-q)[1+(1-2q)a_x]u_x$ and hence that it is >0.0 when $u_x > 0.0$. Thus $s > 0.0$ when $u_x > 0.0$ and we, therefore, are concerned with the b-values that make $u_x = b_1u_1 + b_2u_2 > 0.0$. The requirements are that

$$b_1u_1 > -b_2u_2 \text{ or } b_2u_2 > -b_1u_1. \qquad 12.18$$

Note the following:

(1) In the special case of a gene that affects only one of the measurements used in the selection criterion, the conclusion reached earlier is apparent from 12.18. For example, if the gene affects only X_1, $u_2 = 0.0$ and the condition for $s > 0.0$ reduces to $b_1u_1 > 0.0$ which indicates that b_1 must be positive when u_1 is positive and negativewhen u_1 is negative. Clearly, this is guaranteed when $b_1 = c_1$.

(2) Whatever the values of u_1 and u_2, values of b_1 and b_2 that make u_X negative are possible. It will be shown in the following section that this possibility is not eliminated when $b_m = \beta_m$ for all values of m.

(3) Because $u_W = c_1u_1 + c_2u_2 > 0.0$, $u_X = b_1u_1 + b_2u_2 > 0.0$ when $b_1 = c_1$ and $b_2 = c_2$.

It is apparent from the foregoing that, for all favorable alleles, selection coefficients are positive when b = c for all measures of phenotype. Moreover, these are the only b-values that insure $s > 0.0$ for all best alleles. Note that setting b = c means that all value trait measurements and no others are used in the selection criterion and that, in effect, total value itself is the selection criterion.[5]

The specifications of the circumstance for which the foregoing conclusions were reached included (1) no selection other than that based on the selection criterion considered, (2) independent assortment of genes affecting total value, (3) the selection population itself is the target population, and (4) no overdominance. The first of these identifies the selection process to which the conclusions apply as single stage selection in which all of the phenotypic information employed is integrated into just one measure per selection unit. It requires no further comment.

Independent assortment: More exactly, this specification should have called for linkage equilibrium. Our conclusion rests on the facts that (a) if $u_X = u_W$, because c's are used as the weighting coefficients of the selection criterion, u_X will by definition always be positive but (b) if $u_X \neq u_W$ it may be negative because of one or more negative b-values and/or because a particular allele is favorable to one or more value traits and unfavorable to one or more other value traits. However, it is apparent from 6.15 that, whatever the composition of $\hat{X}$, the covariance between q_i and $\hat{X}$ can be negative as a consequence of linkage disequilibrium though it would be positive otherwise. Thus, in the absence of linkage equilibrium, using the c's as weighting coefficients does not guarantee $E(\Delta q) > 0.0$ for every useful allele. Nevertheless, if the assumption (section 11.2.4) that linkage disequilibrium will be trivial whenever effective population size is sufficient by the probability of fixation criterion (Chapter XI) is correct, $E(\Delta q) > 0.0$ for useful alleles can be assumed in the case of well designed recurrent selection programs in which total value is the selection criterion.

No overdominance and the S population = the T population: For reasons that will almost immediately become obvious, these are properly discussed together. It is emphasized in sections 6.3.3 and 9.5 and by the content of Table 8.2 that in the case of genes that are overdominant and criterion units that belong to the S population, $E(\Delta q)$ can be positive or negative depending on the frequency of the B allele. This means, if the S population is the T population, that which allele is the most useful is subject to change depending on allele frequency and that insuring $u_X > 0.0$ because it is equal to u_W, does not make $E(\Delta q) > 0.0$ at all times. However, if $u_X = u_W$ and the gene is overdominant relative to total value, the equilibrium frequency that tends to be maintained maximizes the contribution of the gene to $\ddot{Y}_W$. Further, assuming fixation as the result of drift, the probability that the allele fixed is the one most favorable relative to total value is maximized if the selection criterion is such that $u_X = u_W$. The argument of this paragraph can be extended to those systems in which selection in each of two or more S populations uses criterion units that belong to the population but the target population is a cross (two-way, three-way or rotation) of the populations involved. As an example, consider the RSP system of section 9.4.

When the target population is the cross of two populations and the criterion units are composed of members of the cross (RSSC or RRS) the allele of an overdominant gene that is "most useful" in the S population of RSSC or either S population of RRS depends on allele frequency in the other population of the system. Thus in RRS, for example, if the frequency of the B-allele is higher in population 1 than in population2, the most useful allele is B in population 1 and b in population 2. Making $u_X = u_W$ by using total value as the selection criterion insures that selection will work to fix B in one population and b in the other <u>if the gene is overdominant relative to total value (W)</u>. This desired result <u>is not insured</u> if any other selection criterion is employed, i.e., if $\hat{X} \neq W$ the gene may be overdominant relative to W but not relative to $\hat{X}$.

In summary, we see that, in various systems, the argument for using total value as the selection criterion (when the goal is to maximize total genetic improvement by recurrent selection) holds for overdominant as well as non-overdominant genes.

• 12.5. The Decrease in $E(\Delta\ddot{Y}_w)$ Associated with Using Total Value as the Selection Criterion

Consider, as alternatives, selection on the basis of (1) total value or (2) the criterion that uses data on all value traits and maximizes $E(\Delta\ddot{Y}_W)$. The former has the advantage that it insures $s > 0.0$ for all best alleles so that total improvement in long term recurrent selection programs may be substantially greater, but may well suffer by comparison in the context of short term programs. Problems posed by the nature of data required for estimation of ß's will be discussed in the following section but ignored here.

• 12.5.1. When $\sigma_{mm} = AG_{mm}$ and $\sigma_{mm'} = AG_{mm'}$ for All Values of m and m'

Consider equations (12.9). Substituting for G_{1W}, G_{2W}, etc. on the basis of 12.15 and taking account of the postulated relations between σ_{mm} and G_{mm} and between $\sigma_{mm'}$ and $G_{mm'}$, we obtain

$$\begin{aligned} \beta_1 AG_{11} + \beta_2 AG_{12} + \cdots \beta_p AG_{1p} &= c_1 G_{11} + c_2 G_{12} + \cdots c_p G_{1p} \\ \beta_1 AG_{21} + \beta_2 AG_{22} + \cdots \beta_p AG_{2p} &= c_1 G_{21} + c_2 G_{22} + \cdots c_p G_{2p} \\ &\;\;\vdots \\ \beta_1 AG_{p1} + \beta_2 AG_{p2} + \cdots \beta_p AG_{pp} &= c_1 G_{p1} + c_2 G_{p2} + \cdots c_p G_{pp} \end{aligned} \qquad 12.18$$

The obvious solution to these equations is $\beta_m = c_m/A$, i.e., $\beta_1 = c_1/A$, $\beta_2 = c_2/A$, etc. It is apparent, therefore, that in this very special situation the weighting coefficients that maximize $E(\Delta\ddot{Y}_w)$ are proportional to the c's. They could as well be the c's themselves because the absolute magnitudes of the weighting coefficients are irrelevant to $E(\Delta\ddot{Y})$ so long as relative magnitudes are not disturbed. It follows that in this situation there is no decrease in $E(\Delta\ddot{Y}_w)$ because selection is based on total value.

The significance of this result will be pointed up by examining special cases. Consider mass selection where the S population itself is the target population and the criterion (and selection) units are single animals or plants. Since the criterion units are individuals, the σ_{mm}'s and $(\sigma_{mm'})$'s are phenotypic variances and covariances, respectively, among individuals. The composition of any G can be obtained in terms of 12.7 using 5.11 and 6.15 or Table 6.2 as sources of the composition of $\partial\ddot{Y}_m/\partial q_i$ and $\sigma(\hat{q}_i,\hat{X}_m)$. When this is done we learn that G_{11} = the additive genetic variance of X_1 as given by 6.19 and G_{12} = the additive genetic covariance between X_1 and X_2 as given by 6.20. In general G_{mm} is the additive genetic variance of X_m and $G_{mm'}$, the additive genetic covariance between X_m and $X_{m'}$. Clearly, $\sigma_{mm} = G_{mm}$ when the heritability of X_m is 1.0 and, in addition, if the heritabilities of all value trait measurements are 1.0, $\sigma_{mm'} = G_{mm'}$ for all pairs of such measurements. The overall result would be $\sigma_{mm} = AG_{mm}$ and $\sigma_{mm'} = AG_{mm'}$ (in which A = 1). Hence the c_m's would constitute an optimum set of "weighting" coefficients. This would have been obvious without demonstration because with heritability of one in the case of all value trait components of W, the heritability of total value $(W = c_1X_1 + c_2X_2 + \cdots\cdots c_pX_p)$ itself would be 1.0 and direct response to selection for W would be superior to any possible correlated response. Two predictions appear justified by the foregoing.

(1) While the published evidence indicates that the heritabilities of most value traits are closer to zero than to one, the penalty, measured by $E(\Delta\ddot{Y}_w)$, for using c's as the weighting coefficients of the selection criterion should be reduced if heritabilities could, by some means, be increased.

(2) If heritabilities were the same for all measurements, it would mean that $\sigma_{mm} = AG_{mm}$ for all values of m. While this would not constrain the ratios of $G_{mm'}$ to $\sigma_{mm'}$ to equality for the $m \neq m'$ parameters, it appears that in general, the penalty, once more measured by $E(\Delta\ddot{Y}_w)$, for using the c's rather than β's should be less when the heritabilities of value traits are similar than when they differ by substantial amounts.

Now consider organisms that can be propagated asexually. This enables selection among individual genotypes on the basis of the average performance of several or many individuals having the same genotype. Each criterion unit is then a family of genetically identical individuals. Obviously, comparisons of such families can be made as precise as practical by increasing family size, replications, and the number of macro-environments in which the comparison is conducted. The result is an approach to the $\sigma_{mm} = AG_{mm}$ and $\sigma_{mm'} = AG_{mm'}$ situation in which the c's become optimum as weighting coefficients.

Errors in comparisons of clonal families are entirely non-genetic. In contrast, when there is genetic variation within criterion unit families, comparison error variance has a genetic as well as a non-genetic component. However, as with clonal families, phenotypic variance among families can be made to approach its inter-family genetic component by increasing the scope of test comparisons. Though less obvious than in the clonal family case, the result is the same - an approach to the $\sigma_{mm} = AG_{mm}$ and $\sigma_{mm'} = AG_{mm'}$ situation. The remaining question is whether the precision required to make this approach such that the c's are near optimum as selection criterion weighting coefficients is practical by other criteria (see Chapter XIV.

12.5.2. Other Situations

From 12.5, it is apparent that when selection is based on total value, because b = c for all value trait measurements,

$$E(\Delta\ddot{Y}_w)_V = E(\Delta\ddot{Y}_w/b\text{'s} = c\text{'s}) = [k/\sigma(\hat{X})][c_1G_{1w} + c_2G_{2w} + \cdots] \qquad 12.19$$

where, in accord with A12.1.2,

$$\sigma^2(\hat{X}) = \Sigma_m c_m^2 \sigma_{mm} + \Sigma_m \Sigma_{m' \neq m} c_m c_{m'} \sigma_{mm'} \qquad 12.20$$

Assuming that all parameters called for by equations 12.9 could be known so that the ß's, the b-values that maximize $E(\Delta\ddot{Y}_w)$, could also be known, we have from 12.10

$$E(\Delta\ddot{Y}_w)_M = E(\Delta\ddot{Y}_w/b\text{'s} = \beta\text{'s}) = k[\beta_1 G_{1w} + \beta_2 G_{2w} + \cdots\cdots]^{1/2} \qquad 12.21$$

Equations 12.19, 12.20, and 12.21 together with equation 12.9 provide the basis for assessing $E(\Delta\ddot{Y}_w)_V/E(\Delta\ddot{Y}_w)_M$ but, except in the special case discussed in section 12.5.1, answers can only be obtained by the indicated numerical processes.

Table 12.2. The consequences of selection based on total value instead of the criterion that maximizes the expected genetic increase in average phenotypic value in the current generation ($G_{11} = G_{22} = 10$, $c_1 = 1.0$ and $c_2 = 2.0$ in all parameter combinations)

Example	σ_{11}	σ_{12}	σ_{22}	G_{12}	β_1	β_2	E_M/k	E_V/k	E_V/E_M
1	40	0	40	0	.25	.50	3.54	3.54	1.0
2	40	20	40	-5	-.25	.50	2.74	1.79	.65
3	40	20	40	5	.25	.50	4.18	4.18	1.0
4	40	-20	40	-5	.25	.50	2.74	2.74	1.0
5	40	-20	40	5	1.08	1.17	7.13	6.39	.89
6	40	0	160	0	.25	.125	2.34	1.92	.82
7	40	40	160	-5	-.125	.125	1.37	1.04	.76
8	40	40	160	5	.45	.04	3.17	2.42	.76
9	40	-40	160	-5	.125	.125	1.37	1.32	.96
10	40	-40	160	5	.875	.375	5.18	3.07	.59
11	16	-4	40	5	1.42	.75	6.87	5.53	.80
12	13	0.5	25	5	1.50	.97	7.37	6.53	.89

Note: E_m and E_V are abbreviations of $E(\Delta Y_w)_m$ and $E(\Delta Y_w)_V$ and other symbols are employed as defined for the purposes of this chapter.

Numerical results are shown in Table 12.2 for a series of examples involving two measurements, both of value traits. It is assumed that these are the only value traits and that no other measurements are to be used in the selection criterion. The magnitudes of G_{11}, G_{22}, c_1, and c_2 are the same in each of the twelve examples. G_{1w} and G_{2w}, computed using equation 12.15, vary among the examples because G_{12} varies. In examples 1 -5, $G_{11}/\sigma_{11} = G_{22}/\sigma_{22} = 1/4$. Thus, depending on the magnitudes of G_{12} and σ_{12}, the condition, $\sigma_{mm'} = AG_{mm'}$ for all values of m and m', is possible.[6] That condition prevails in examples 1, 3, and 4 and, consequently $E_V/E_m = E(\Delta\ddot{Y}_w)_V/E(\Delta\ddot{Y}_w)_M = 1.0$ in those cases. In examples 6-10, G_{11}/σ_{11} is four times as large as G_{22}/σ_{22} which means that non-genetic variation among criterion units is a much larger portion of the total in the case of the second measurement. Finally, examples 11 and 12 are modifications of 10 included to provide insight concerning the effect, discussed in section 12.5.1, of reducing the phenotypic variance among criterion units and causing thereby an approach to the $\sigma_{mm} = AG_{mm}$ and $\sigma_{mm'} = AG_{mm'}$ condition. Consider selection in plants and criterion units that are a type of family permitting as much replication as desired. It is apparent from 10.10 that in such a case variance other than the genetic variance among families can be reduced at will by increases in v, t, and r. The reductions assumed (relative to example 10) are 80 and 90 percent, respectively, in examples 11 and 12. Note the following aspects of information provided in Table 12.2.

(1) Where $\sigma_{mm'} \neq AG_{mm'}$ for all m and m', the difference between E_V and E_M is substantial in two instances (examples 2 and 10). Contrary to the second prediction of section 12.5.1, G_{11}/σ_{11} and G_{22}/σ_{22} were equal in one of those cases (example 2).

(2) The magnitudes of σ_{12} and G_{12} are responsible for great variation in E_M. Within the sets comprised by examples 1-5 and 6-10, respectively, where σ_{12} and G_{12} are the only variable parameters, E_M ranges from 2.74 to 7.13 and 1.37 to 5.18. The smaller values for the second set reflect the fact that G_{22}/σ_{22} is only one-quarter as large in examples 6-10.

(3) Though E_V/E_M is variable, the correlation between E_V and E_M is strong. In general E_V is large when E_M is large.

(4) Examples 10, 11, and 12 illustrate the validity of the first prediction offered in section 12.5.1. As σ_{11}, σ_{12}, and σ_{22} are reduced by replication and/or increases in number of individuals per criterion unit, E_V approaches E_M. Note, however that example 11 assumes a five-fold and example 12 a ten-fold increase over example 10 in numbers of individuals (or plots) per criterion unit.

(5) Examples 2 and 7 constitute straightforward demonstrations that the weighting coefficients (the β's) that maximize $E(\Delta\ddot{Y}_w)$ relative to the current cycle of selection do not insure $s > 0.0$ for all "best" alleles. Obviously, when $\beta_1 < 0.0$, as in these examples, $s < 0.0$ for any allele that is favorable to X_1 and without effect on any other measurement employed in the selection criterion.

These observations provoke questions concerning situations in which more than two measurements are available and possibly meaningful. For example, (1) should smaller values of E_V/E_M be anticipated when there are more value traits and (2) are smaller values of E_V/E_M likely when there are more than two measurements of which only a portion pertain to value traits? Initial insights relative to these questions are provided by the three measurement examples detailed in Table 12.3. The magnitudes of parameters that are the same in all twelve examples are indicated in the table title.

Table 12.3. The consequences of selection based on phenotypic value—three measurement examples ($\sigma 11 = 200$, $\sigma 22 = \sigma 33 = 40$, $G_{11} = G_{22} = G_{33} = 10$ and $c_1 = c_2 = 1.0$)

Example	c_3	σ_{12}	σ_{13}	σ_{23}	G_{12}	G_{13}	G_{23}	β_1	β_2	β_3	E_M/k	E_V/k	E_V/E_M
1	1	0	0	0	0	0	0	.05	.25	.25	2.35	1.79	.76
2	1	0	0	0	0	5	5	.075	.375	.50	4.09	2.99	.73
3	1	0	0	0	0	-5	-5	.025	.188	.00	1.03	.60	.58
4	1	0	40	20	0	0	0	.023	.18	.14	1.85	1.50	.81
5	1	0	40	20	0	5	5	.011	.39	.43	3.82	2.50	.65
6	1	0	40	20	0	-5	-5	.06	.20	-.16	1.14	.50	.44
7	0	0	0	0	0	0	0	.05	.25	.00	1.73	1.29	.75
8	0	0	0	0	0	5	5	.05	.25	.25	2.35	1.29	.55
9	0	0	0	0	0	-5	-5	.05	.25	-.25	2.35	1.29	.55
10	0	0	40	20	0	0	0	.09	.41	-.32	2.24	1.29	.57
11	0	0	40	20	0	5	5	.023	.18	.14	1.85	1.29	.70
12	0	0	40	20	0	-5	-5	.20	.64	-.77	4.01	1.29	.31

Note: E_m and E_v are abbreviations of $E(\Delta Y_w)_m$ and $E(\Delta Y_w)_v$ and other symbols are employed as defined for the purposes of this chapter.

Note that $G_{22}/\sigma_{22} = G_{33}/\sigma_{33}$ but that G_{11}/σ_{11} is only one-fifth as large. In examples 1-6, $c_3 = 1.0$ but in examples 7-12, $c_3 = 0.0$, i.e., X_3 is a value-trait in the first six examples but is not in the second six. Within these two sets of examples the parameter variation is in the quantities, σ_{13}, σ_{23}, G_{13}, and G_{23}. The pattern of variation in the first set is repeated in the second. Thus examples 1 and 7, 2 and 8, etc. are identical except that $c_3 = 1.0$ in the first member of the pair and zero in the second. Notable aspects of the information contained in the table include the following.

(1) In examples 1 and 7 (and 1 and 6 of Table 12.2), in which all covariance parameters are zero, $\beta_m = c_m G_{mm}/\sigma_{mm}$ in accord with Hazel and Lush (1942).[7] Except in example 1 of Table 12.2, this means that the β's are not proportional to the c's and accounts for the observation that $E_V < E_M$.

(2) The lowest values of E_V/E_M are smaller than the lowest shown in Table 12.2. Not unexpectedly, the answers to our two questions (posed above) appear to be in the affirmative.

(3) As expected, the non-value trait X_3 in examples 7-12 makes no positive

contribution to the effectiveness of the selection criterion unless one or more of its covariance parameters is non-zero (see example 7). It should be noted that such a trait can affect E_M through a phenotypic covariance even when all G's relating to the trait are zero (compare examples 7 and 10).

(4) Again, the magnitudes of E_M and E_V/E_M are strongly affected by the covariance parameters. It is apparent that the interactions of the various parameters in their effects on E_M and E_V are complex. However, it does appear that E_V/E_M tends to be smallest when the deviations from zero of phenotypic covariances and the corresponding G's are in opposite directions (compare example 5 with 6 and 11 with 12).

Not shown in Table 12.3 is the effect of increasing the information obtained per criterion unit so that phenotypic parameters are made to approach their irreducible (genetic) minimums.

• 12.6. Phenotypic Information from Two or More Sources

In connection with the definition of criterion unit (Chapter III) it is noted that the information on which selection is based is sometimes provided from more than one source, i.e., more than one criterion unit may be involved in the selection system. In the simplest of such systems, only one trait would be measured. For example, the criterion for selection among individuals in the case of dairy cattle might involve milk production of the individual (if a female), its dam, full-sibs and half-sibs but no data on any other trait. In general, however, the selection criterion can be a function of data on several traits for which the sources of data are not the same. Consider mass selection in swine. Growth rate, carcass value and litter size are all value trailts. However, measurement of actual carcass value requires slaughter of the animal and hence cannot be obtained on candidates for selection.[8] Litter size is not expressed in the male and to use observed litter size of female selection units would entail an increase in cycle length beyond that required when selection is practiced as soon as data on growth rate (to market weight) and carcass value are available. One way to resolve these problems is to employ a selection criterion based on (1) growth rate of the individual, (2) carcass value of a small number (one, two, or three) of full-sibs, and (3) size of one or more litters farrowed and weaned by the dam.

The discussion that has accompanied theory so far presented has included no explicit reference to programs in which the selection criterion data comes from two or more sources. Thus it is in order to emphasize here that the generalizations embodied in equations 5.19 and 5.20 are not subject to any limitations concerning the observations used in computing $\hat{X}$. It follows that 12.6, 12.7, and the 12.9 set of equations are as applicable when $\hat{X}$ is a function of data from two or more criterion units as when the data are all obtained from just one criterion unit.

The special case involving measurements on only one trait, the S population as the T population, and all selection criterion data obtained from criterion units that are parts of the S population was treated first by Lush (1947) and later by various authors [including Lerner

(1950), Robertson (1955), and Henderson (1963)].[9] The treatment by Henderson is the most general of those cited and corresponds more closely than the others with that presented here. The theory sketched here yields the results recorded by other authors and appears to the author to handle the general problem in a broader array of cases by virtue of its defining $G_{mm'}$ as $\Sigma_i \sigma(\hat{q}_i,\hat{X}_m)\ (\partial\ddot{Y}_{m'}/\partial q_i)$.

In section 12.4, where it is tacitly assumed that all selection criterion information is obtained from a single criterion unit, it is shown that maximum total genetic improvement via recurrent selection can only be assured by using the c-values of traits as their weighting coefficients in the selection criterion. The supporting argument can be summarized as follows:

(1) Maximum genetic improvement requires fixation of the most favorable allele of each gene that has a bearing on total value.

(2) The probability of fixation of any best allele is not greater than its original frequency unless the expected change (per cycle of selection) in the frequency of that allele is positive, i.e., unless $E(\Delta q) > 0.0$.

(3) $E(\Delta q_i) > 0.0$ when $\sigma(\hat{q}_i,\hat{X}) > 0.0$ but the latter is only guaranteed when the selection criterion is $\hat{X} = \Sigma_m c_m \hat{X}_m$.

We now inquire whether this result is correct when more than one criterion unit is employed. Two special cases will be examined.

Example 1 Consider selection among individuals in the case of swine and assume (1) that growth rate of the indivdual ($\hat{X}_1$) and the average size of litters weaned by the dam ($\hat{X}_2$) are among the measures employed in the selection criterion, (2) that $\hat{X}_1$ and $\hat{X}_2$ are the only measures of growth and litter size employed and (3) linkage equilibrium in the S population. Then by 12.4, if the i-th gene has effects on both growth rate and litter size but not on any other trait in the selection criterion

$$\sigma(\hat{q}_i,\hat{X}) = b_1\sigma(\hat{q}_i,\hat{X}_1) + b_2\sigma(\hat{q}_i,\hat{X}_2) \qquad 12.22$$

where $\hat{X} = \Sigma_m b_m \hat{X}_m$. Given linkage equilibrium and the definitions of X_1 and X_2 in this example, we know from 6.3 that

$$\sigma(\hat{q}_i,\hat{X}_1) = q_i(1-q_i)[1+(1-2q_i)a_{1i}]u_{1i} \qquad 12.23$$

and it is easily shown that

$$\sigma(\hat{q}_i,\hat{X}_2) = [q_i(1-q_i)/2][1+(1-2q_i)a_{2i}]u_{2i} \qquad 12.24$$

It follows, by substitution in 12.22, that

$$\sigma(\hat{q}_i,\hat{X}) = q_i(1-q_i)[(b_1u_{1i} + b_2u_{2i}/2)$$
$$+ (1-2q_i)(b_1a_{1i}u_{1i} + b_2a_{2i}u_{2i}/2)]$$

In this case, then, if $b_1 = c_1$ and $b_2 = 2c_2$, we have

$$\sigma(\hat{q}_i,\hat{X}) = q_i(1-q_i)[1+(1-2q_i)a_{wi}]u_{wi}$$

where $u_{wi} = c_1u_{1i} + c_2u_{2i}$ and $a_{wi} = c_1a_{1i}u_{1i} + c_2a_{2i}u_{2i}$ in accord with principles illustrated in Table 12.1. By definition, $u_{wi} = c_1u_{1i} + c_2u_{2i} > 0.0$ when X_1 and X_2 are the only value traits affected by the i-th gene. Hence, in this instance, $b_1 = c_1$ and $b_2 = 2c_2$ assure that $\sigma(\hat{q}_i,\hat{X})$ and $E(\Delta q_i)$ are both positive.[10] On the other hand, $E(\Delta q_i) > 0.0$ is not insured by any values of b_1 and b_2 that are not proportional to c_1 and $2c_2$, respectively.

Example 2: Consider example 1 modified by including the average growth rate ($\hat{X}_3$) of n full-sibs as well as the growth rate ($\hat{X}_1$) of the individual and average litter size of the dam ($\hat{X}_2$)among the measurements employed in the selection criterion. In this case, if the i-th gene has effects on growth rate and litter size but not on any other trait of $\hat{X}$,

$$\sigma(\hat{q}_i,\hat{X}) = b_1\sigma(\hat{q}_i,\hat{X}_1) + b_2\sigma(\hat{q}_i,\hat{X}_2) + b_3\sigma(\hat{q}_i,\hat{X}_3) \quad 12.25$$

Given linkage equilibrium, standard procedure yields

$$\sigma(\hat{q}_i,\hat{X}_3) = [q_i(1-q_i)/2][1+(1-2q_i)a_{1i}]u_{1i} \quad 12.26$$

where a_{1i} and u_{1i} relate to growth rate as they do in the discussion of example 1. Substitution in 12.25 in terms of 12.23, 12.24, and 12.26 yields

$$\sigma(\hat{q}_i,\hat{X}) = q_i(1-q_i)\{(b_1+b_3/2)u_{1i}+(b_2/2)u_{2i}$$
$$+(1-2q_i)[(b_1+b_3/2)a_{1i}u_{1i}+(b_2/2)a_{2i}u_{2i}]\}$$

This becomes $q_i(1-q_i)[1+(1-2q_i)a_{wi}]u_{wi}$ so that $E(\Delta q_i) > 0.0$ if $(b_1+b_3/2) = c_1$ and $b_2 = 2c_2$. Among the values of b_1 and b_3 such that $(b_1 + b_3/2) = c_1$ there must be a set that is optimum relative to the expected genetic improvement of growth rate and it is intuitively obvious that the optimum value of b_3 relative to b_1 would increase as the number of full-sibs per individual increases.

It is apparent from these examples that when selection criterion information is obtained from more than one criterion unit, the weighting coefficients that guarantee $E(\Delta q_i) > 0.0$, and hence $s > 0.0$, for all useful alleles are functions of both the c-values of the traits and the genetic relationships of criterion units to the selection unit instead of being equal (or proportional) to the c-values as they are when all information is obtained from one criterion unit. The method for finding this special set of weighting coefficients is demonstrated in the discussion of the examples.

12.7. Summary and Practical Problems

This chapter is devoted to the structure of the selection criterion in breeding programs involving single-stage selection. Relevant facts include the following.

(1) Total value is ordinarily a function of more than one trait. The questions posed are (a) what measurements should be employed and (b) what weight should each receive in the selection criterion?

(2) When the S population is in linkage equilibrium and all selection criterion information is obtained from one criterion unit, using the c-values (defined in section 12.3) as weighting coeffficients insures $E(\Delta q) > 0.0$ for all useful alleles.[11] This is not accomplished in any other way. Thus, if the goal is to maximize the level of total value ultimately achieved via recurrent selection, the selection criterion weighting coefficients should be proportional to the c-values, in which case only value traits would be employed in the selection criterion.

(3) Given the exact values of all relevant parameters, the σ's and G's of 12.9, weighting coefficients that maximize $E(\Delta\bar{Y}_w)$ in the current cycle of selection could be

determined. These have been designated as the ß's and $E(\Delta\ddot{Y}_w)$ given the ß's as the weighting coefficients has been symbolized as E_M.

(4) In general the ß's are not proportional to the c's. Consequently, if $E(\Delta\ddot{Y}_w)$ given the c's as weighting coefficients is symbolized as E_V, $E_V/E_M \leq 1.0$ and, depending on the relevant parameters, may be substantially less than 1.0 (see tables 12.2 and 12.3). However, as the sampling variance of criterion unit information is reduced (the nature of the units permitting), E_V/E_M is increased.

(5) The parameters (σ's and G's) that determine the ß's are never known exactly, i.e., estimates must be employed in equations 12.9. Hence only estimates of the ß's can be available and when these are used as weighting coefficients, $E(\Delta\ddot{Y}_w)$ is less than if the ß's themselves could be used. It follows that the penalty, measured in terms of $E(\Delta\ddot{Y}_w)$, for using c's as weighting coefficients will in all cases be less than suggested by the value of E_V/E_M.

Unfortunately, not much information is available concerning the decrease in $E(\Delta\ddot{Y}_w)$ that results from the inescapable differences between the estimated and true values of the ß's. A very specific two-trait example was thoroughly investigated by Harris (1964). His results indicate that rather large, but not impossible, amounts of data are required if these decreases are to be reasonably small. However, there are no obvious guidelines for accurate extrapolation from his results to more complex situations.

(6) The problem of estimating the σ's and G's is aggravated by the fact that magnitudes will be changed by selection that is effective. Thus, if estimates employed are to be continuously valid, new estimates must be obtained at regular and appropriate intervals. In many breeding programs the criterion unit data are, or can be, such that parameter estimates can be obtained from them. Then if the data from one cycle are not sufficient, that from the most recent cycles can be pooled to provide estimates for which expected values are close, though not equal, to the correct true values and for which sampling errors are less serious.

(7) The c-values may change as time goes on because (a) total value is a non-linear function of the value trait measurements or (b) the "constants" of the total value definition are altered. Consider, in the case of swine, number of offspring weaned per female per unit of time. To increase this number is to decrease the overhead cost for female parents. As the number increases the value of still another unit increase becomes less, i.e., the c-value decreases. In the case of sheep, both mutton and wool are produced. The development and extensive use of synthetic fibers resulted in reduced demand for wool and a consequent change in the relative values of the two products. An unfortunate implication, in the case of genes that affect more than one value trait is that the rank order of alleles, with respect to their effects on total value, may be altered by temporal change, for whatever reason, in c-values.

(8) In many programs it will not be (or seem to be) feasible to measure all value traits. This is least likely in the case of annual plants and becomes more likely as the productive lifetime of the organism becomes longer. Consider any annual cereal. The value of the product depends on the yield and quality of grain. If, as is most common, the criterion unit is a family

with no severe restriction on size, enough grain can be obtained to measure most quality attributes as well as yield. But even here there are exceptions, e.g., because of cost it probably would not be considered feasible in work with bread wheat to obtain measures of baking quality for all criterion units.

The situation is different with most longer-lived organisms. In dairy cattle, fruit tree species and the forest species it is obvious that major components of total value are lifetime attributes. However, if these are measured as part of the basis for selection, cycle length is necessarily longer than when selection is based on measures available at an earlier date. In dairy cattle, despite the apparent importance of longevity and lifetime production, selection is based in large part on the yield and quality of milk produced in the first first one or two lactations. The facts are parallel for virtually all breeding programs involving long-lived species.

Collectively, these facts provide context for the following generalizations regarding selection criteria.

(1) When the primary goal is maximum eventual total value of the target population and all traits can be measured without undue cost, the c-values should be employed as weighting coefficients.

The problem posed when these change substantially as time passes would be no less a problem if ß estimates were used in place of the c's. The best that can be done is to foresee change as much as possible and modify the selection criterion accordingly.[12] Certainly, the weighting coefficients employed should not be based on c-values already out of date.

(2) When maximum improvement within a shorter time frame is the goal, all value traits can be measured, and estimates of all the σ's and G's required for the estimation of ß's are available, the best decision may vary from one instance to another because the penalty for using c-values may or may not exceed the corresponding effect of errors in ß estimates. In this situation, the breeder can obtain, and be guided by, an approximation to E_V/E_M. A value near 1.0 would indicate that, at worst, little would be lost by using c-values. Conversely, a value substantially less than 1.0 would provide reason to believe that the use of ß estimates would result in more improvement of phenotypic value during a restricted time span.

(3) Frequently a portion of the value traits cannot be measured in each cycle of the selection program because (a) costs, by the nature of those measurements, are considered to be prohibitive or (b) traits in question have not been expressed by the time selection occurs.

Before going further, note that using c-values as the weighting coefficients for the portion of measurements that are available will not insure $E(\Delta q) > 0.0$ for all useful alleles because one of the necessary conditions for that outcome is that the selection criterion be based on all value traits.

Now consider programs of which the prime objective is maximum total value regardless of time required. These are the programs in which a selection criterion based on all value traits, and using c's as weighting coefficients, is optimum. In these cases the impact of selection systems deserves attention. What is possible or feasible may differ substantially from one system to another. In the case of egg production poultry, commercial practice is such that value trait measurements not available when cycle length is one year become available when the selection cycle is extended to two years. In maize breeding using reciprocal recurrent selection, if the inbreeding coefficient (f) of selection unit individuals is increased from zero to 0.94, cycle length would be increased but, depending on program procedures affecting cycle length, the associated negative effect on expected improvement per year might be offset by the favorable effect of inbreeding (see eq. 9.13) but the number of criterion units to be evaluated per year would be reduced (quite possibly by 50 percent) with the result that measurements thought too expensive if more were required might then appear worthwhile.

The most troublesome situations are those in which the data available on value traits is insufficient to support adequately precise estimates of the G_{mW}'s. It is apparent from 12.15 that estimation of each G_{mW} requires data on all value traits. When those data are not available no truly satisfactory selection criterion can be constructed. The apparent alternatives are (a) to employ the measurements of value traits that can be made each cycle, using c-values as weighting coefficients or (b) to employ both value trait measurements (those available each cycle) and measurements of other traits believed useful because of covariances with value traits, using as weighting coefficients estimates of ß's that maximize $E(\Delta\ddot{Y}_{W'})$ where $W' = \Sigma_m c_m X_m$ in which summation is over value traits for which data are available.

Selection based on criteria constructed in either of these ways may or may not improve the value traits that could not be taken into account in the search for weighting coefficients. It follows that temporal changes in those traits should be monitored as closely as possible (or needed) so that if unfortunate trends are detected, suitable adjustments in the selection criterion can be sought. These cannot be guaranteed to be efficient procedures; there can be no substitute for complete data.

13 The Design of Breeding Programs

• 13.1. Introduction

Issues properly considered in the comparison of alternative breeding programs were discussed in a preliminary way in Chapter IV. The purpose in this chapter is (a) to focus more explicitly on the decisions involved in shaping a breeding program and (b) to introduce the following chapter in which bases for decisions are treated in much greater detail.

Comstock(1977) stated that the decisions required pertain to (1) the goal, including the phenotypic goal and the target population of environments (TPE), (2) the foundation population or populations, (3) the selection system,(4) the selection criterion, and (5) effective population size.

In his 1977 paper the author focused on long term recurrent selection programs as defined in Chapter XI. To encompass short term as well as long term programs, the goal should include a time dimension. Clearly, any short term program is concerned with the production of a target population in which total value at the end of a specific period of time is maximum. The projected program length would condition decisions concerning the foundation population(s), the selection criterion , effective population size and the selection system.

Selection systems are considered in Chapters VI, VIII, and IX without being explicitly defined. However, the implicit definition by the content of those chapters is that systems are distinguished by the selection unit, the criterion unit, whether a single selection population (or more than one) is involved, and the target population. The expressions for $E(\Delta q)$ and $E(\Delta \ddot{Y})$ that have been presented together with considerations of effective population size in section 2.7, section 2.8, and Chapter XI and of the selection criterion in Chapters X and XII make it clear that within each system there is a set of variables to be jointly optimized if the effectiveness of the system is to be maximum.

The phenotypic goal is identified by definition of total value (see sections 1.4 and 12.3). Though the immediate focus of a program may be on only one (or some subset) of the value traits that determine total value (W), the obvious underlying goal is an increase in $\ddot{W}$ through improvement in one or more (not necessarily all) of the value traits. It follows that definition of W is always required (a) as part of the basis for structuring an approach to the optimum selection criterion and (b) as the basis for judging accomplishment.

13.2. Criteria and Procedure for the Comparative Evaluation of Recurrent Selection Programs

Decisions concerning the program goal and the foundation population(s) ordinarily precede and are independent of those concerning selection procedures. Thus our discussion of criteria for comparison of selection systems can appropriately assume prior decisions with respect to those aspects of programs.

Because the objectives of long and short term programs are not the same (see Chapter XI), the appropriate primary criteria are not the same for both.

13.2.1. Primary Criteria

The goal in long term programs is synthesis of ideal genotypes, i.e., maximum eventual value of genotype ($\ddot{Y}_W$), without special emphasis on time required. It follows that primary criteria in procedure comparisons should be (a) probability of fixation of superior alleles, (b) whether the criterion unit is such that the selection criterion, $\hat{X}$, can be total value, W, (so that $s > 0.0$ is guaranteed for all favorable alleles -- Chapter XII), and (c) whether the structure of the program is such that the best genotype can be achieved at each locus in the target population.[1]

In short term programs the goal is maximum value of $\ddot{Y}_W$ at the end of some pre-determined period of time. This, in turn, is a function of $\ddot{Y}_W$ at the outset and the sum of $\Delta\ddot{Y}_W$ over the cycles of selection that are possible within that time period. Obviously, then, $\ddot{Y}_{W0}$ (= $\ddot{Y}_W$ in the target population at time zero) and $E(\Delta\ddot{Y}_W)$ are the appropriate primary criteria.

13.2.2. Secondary Criteria

Rapid genetic change, though not primary, is obviously desirable in long term as in short term programs. It follows that among long term programs equal in other respects, choice would be based, appropriately, on $E(\Delta q)$ or $E(\Delta\ddot{Y}_W)$.

A criterion unit such that all value traits can be measured is not as crucial in short term as in long term programs but is desirable (a) because it improves the chances that the optimum selection criterion can be employed and (b) as a source of data on which the construction of a desirable selection criterion can be based.

The target population in the recurrent selection phase of a breeding program may or may not be the derivative of the program to be recommended for agricultural use. Consider, for example, recurrent selection in which the T population is the S population. In the case of livestock, the derivative to be used in agriculture is ordinarily the T population itself.[2] In contrast, in an asexual plant species (strawberries, apples, etc.) the cultivar is ordinarily a clone (a single genotype); the target population of recurrent selection is a population of potential cultivars and selection among those potential cultivars is obviously useful. In these cases any breeding program that employs recurrent selection is necessarily two-phased and consists of (1) recurrent selection and (2) selection among the potential cultivars that are the units of the T population. The latter has been designated the cultivar search phase of such programs (see sections 9.6.1.2 and 9.6.4). There is an obvious advantage when potential cultivars are the criterion units in the recurrent selection phase. Though the testing that is optimum for recurrent selection will never be sufficient to assure selection of the best or near-best unit in the cultivar search phase, it can serve to identify promising units worthy of further testing in the search for superior cultivars.

• 13.2.3. **Procedures**

The obvious first requirement is a list of the systems worth considering. These, of course, will vary with the organism in question. For example, selection among S_n families on the basis of their own performance is most practical in operational terms in the case of naturally self-fertilizing species but is not possible in animal species. As another example, reciprocal recurrent selection (RRS) would not be among the listed systems unless there was reason to believe that overdominance of a significant portion of genes affecting total value of the organism was possible.

Actual evidence concerning the relative effectiveness of systems would then be based on expected responses to selection provided by prediction equations, the crucial quantities being probabilities of fixation in long term programs and expectations of change per unit of time of allele frequencies or of average genotype value in the target population in both short and long term programs. Results obtained will of course vary with assumptions concernng

(a) operational details of selection systems, e.g., size of the genetic sample, selection intensity, size of families employed as criterion units, and the structure of criterion unit comparisons, and

(b) attributes of the selection populations, e.g., genetic and non-genetic variances, the number of segregating genes with effects on traits of interest, magnitudes of gene effects and dominance levels.

How many assumptions must be made? Consider the relatively simple problem of optimizing procedure in full-sib family selection for grain yield of maize. From 11.25 we note that the expectation of $\Delta\ddot{y}$ per cycle is a function of selection intensity (which determines k); $\sigma(\hat{X})$; effective population size; and the frequency (q), level of dominance (a), and the effect (u) of the more favorable allele of the i-th gene--and from 10.11 that $\sigma^2(\hat{X})$ is a function of the genetic variance among families [$\sigma^2(\bar{z})$], four non-genetic variances, and the number of locations (v), years (t), and replications (r) in each year-location in the comparison of criterion units. It follows that if $E(\Delta\ddot{y}_i)$ is to be quantified, twelve or more assumptions are required.[3] Some, in particular those pertaining to the variance components of $\sigma^2(\hat{X})$, can be made with a degree of confidence on the basis of accumulated evidence concerning their magnitudes. Others impinge on each other and are constrained by total program cost. For example, if v, t, and r are to remain the same, increasing the number of selection units in the genetic sample would increase the program cost per cycle but would allow (a) greater selection intensity without decrease of effective population size or (b) greater effective population size without sacrifice in selection intensity. Alternatively, without much change in cost, genetic sample size could be increased at the expense of $\sigma(\hat{X})$ by reduction of one or more of the parameters (v, t, or r) in the field comparison of criterion units. General information and logic provide a degree of guidance toward reasonable assumptions concerning the genetic parameters (q, a, and u). For example, given pronounced inbreeding depression, as in the case of grain yield of maize or litter size of swine, to assume that most a-values are greater than zero is in order. On the other hand, information regarding u-values is, at best, minimal. Whatever the basis for limiting the range of reasonable assumptions, the investigator will certainly try, at the very least, two or three values of each of these parameters in order to obtain minimal information concerning direction and magnitude of effects.

In view of the number of variables that affect, or may affect, responses to selection and the potential interactions among them, it is apparent that expectations associated with numerous assumption sets may be required before all information desired is obtained when evaluating procedure variations and/or comparing two or more systems of selection. Fortunately, given assess to a modern high speed computer, the larger portion of time and cost of information based on theory could be that involved in writing programs required.

As emphasized earlier (section 1.5) the alternative is actual comparison of systems using real populations--a network of trials that realistically will not be conducted with any economic species because of costs in manpower, facilities, and time that would be involved. The time cost is particularly significant because breeding program decisions must be made now, not postponed for the collection of convincing results of empirical studies--studies that even with an annual crop could very well take a minimum of 15, 20, or more years.

For the purpose of system comparisons, specification of the selection criterion is not critical when it is to be the same in all systems to be compared. X = W is optimum in the sense that it makes $s > 0.0$ for all useful alleles (see section 12.4). However, if s is negative for some desirable alleles, because for some reason $X \neq W$, we will still want (1) N_es in long term programs to be large enough for all alleles for which $s > 0.0$ and (2) $E(\Delta q)$ in either short or long term programs as large as possible for those same alleles.

13.2.3.1. LONG TERM PROGRAMS

The choice of selection procedures to be compared will vary with the species in question but clearly whether the criterion unit is such that all value traits can be measured is an attribute deserving special attention. Such a system that is desirable in other respects is most likely in the case of annual plants and least in the case of species in which longevity is a value trait of significance (as in dairy cattle and blueberries) or age at harvest is great relative to the age when reproduction is first possible (as in forest species).

The primary objective in long term programs is fixation of superior alleles. The probability of fixation of any desirable allele is increased as N_es' is increased (if, as will be assumed herein, N_es' is constant in successive cycles of selection) and is greater when the original frequency (q_0) of the allele is greater. However, when systems are being compared, it is proper to suppose that the foundation population will be the same whatever the selection system to be employed. Thus the values of q_0 do not depend on the selection system and have no bearing on differences among systems in fixation probabilities. Effective population size (N_e), not likely to be the same for different selection systems, remains constant in successive cycles of selection by any single system if operational details of the the program are the same in all cycles. On the other hand, s', being a function of allele frequency (q) when there is any level of dominance (see Table 11.3), is only constant in successive cycles of selection in the case of additive (a = 0.0) genes. However, because s' is a multiple of the gene effect (u) it is variable among additive genes as well as among others.

The best procedure to follow in comparison of selection systems is not beyond question. Rather, it depends on the relative weights to be given to (a) probability of fixation and (b) expected rates of increase in the frequencies of superior alleles. If maximum weight is placed on probability of fixation, selection systems (to be conducted in ways such that costs are about equal and N_es' maximized for each system) would be ranked in terms of the value of N_es' achieved for a specified category of genes.[4] The alternative would be to adjust selection in the systems being compared to the intensities that would achieve the same N_es' value for a specified class of additive genes and then rank the systems in terms of the expected rates of increase in allele frequencies. Again, the comparison should be predicated on conduct of systems compared such that operation costs would be approximately equal. It is worth noting that regardless of the choice made between these procedures a decision is in the end required concerning the value of N_es' judged to be desirable. In the case of the first procedure, having identified the "winning" system, the breeder must decide on the value of N_es' desired in order to identify the size of program required. In the case of the second procedure the desired N_es' value is built into the comparison procedure. The illustrative comparisons presented in Chapter XIV follow the second procedure outlined above.[5]

The procedure given by 11.10 is a straightforward way to arrive at a value of N_es' to employ in the second procedure described above for selection system comparisons. It gives the

$$(N_es')_r \approx -\ln(1-P_a)/2q_0 \qquad 11.10$$

value required if the probability of fixation is to be P_a in the case of the superior allele of an additive gene of which the frequency at the beginning of the selection program is q_0. For example, if the probability of fixation of an additive gene for which q_0 = .08 is to be .96 the required value of N_es' is $-\ln(.04)/(.16) = 20.12$.

- 13.2.3.2. **SHORT TERM PROGRAMS**

The goal of any short term program is maximum performance at the end of some limited period of time, e.g., 15 or 20 years. Thus the comparison of systems will logically focus on the expectation of total genetic gain whenever the initial performance of target populations can be assumed to be the same, as when the foundation population(s) would be the same no matter which selection system is used. If, as noted elsewhere, this were not so, the necessary focus would be on initial performance of the target population plus the expectation of total genetic gain.

Because $E(\Delta\ddot{Y}_x) \approx \Sigma_i E(\Delta\ddot{y}_i)$ the comparative effectiveness of selection systems in short term programs can be explored in terms of $E(\Delta\ddot{y}_i)$. Procedure is outlined and illustrated in sections 11.3, 14.2, and 14.3.

Ideally, expected genotypic value at the end of the projected time period of selection would be in terms of total value. Realistically, all of the genetic parameter information required (see Chapter XII) will rarely be available. The alternatives are to measure expected value in terms of (a) the selection criterion to be employed or (b) some better approximation to total value for which the necessary genetic parameter estimates are available.

13.3. Foundation Populations

If an ideal genotype is to be obtained via long term recurrent selection the best allele of every gene must either be present in the foundation population or arise by mutation during the course of the program. This raises the fundamental question of how populations differ. Consider a single gene. Is the allele of that gene that is best, with respect to performance in a specific TPE, present at some frequency in every subpopulation of the species? The answer is obviously "No" because in populations for which N_e has been small for a series of generations there is a non-trivial probability that such a "best" allele (if not homozygous from the beginning) will have been lost as the result of drift. Let us rephrase the question in terms of single subpopulations of which the effective population size has always been large and ask whether it is reasonable to assume (or argue) that the "best" allele of every gene is present in some non-zero frequency in each of them. As noted in Section 11.1.3, (a) the arguments for and very considerable investment in germplasm collections reflect the belief of many geneticists and breeders that a significant number of "best" alleles are absent from the genotypes of even the best of subpopulations and (b) there are valid justifications for that belief.

To insure the presence in a foundation population of a high proportion of the "best" alleles of a species, inputs from a number of subpopulations are clearly in order. Logically, the subpopulations employed should be divergent with respect to their origins and desirable as measured by their own performance. However, emphasis on origin diversity will almost certainly mean sacrifice with respect to frequencies of many of the superior alleles actually present in the foundation population. That, in turn, would mean more time required to bring those alleles to fixation but, assuming more of them, the genotypic value ultimately achieved would be greater.

Short term programs by definition encompass a limited number of selection cycles. Let

$\ddot{Y}_{w0}$ and $\ddot{Y}_{wn}$ symbolize average genetic value, with respect to total value, in the target population at the beginning and the end (after n cycles of selection). Then

$$\ddot{Y}_{wn} = \ddot{Y}_{w0} + \Sigma(\Delta\ddot{Y}_w)$$

in which summation is over the n cycles of the selection program; and it is obvious that the choice of the foundation population(s) should be based on estimates of (a) the initial genotype value of the target population and the genetic variance within the foundation population (or populations).

• 13.4. The Target Population of Environments

Ideally, the entire population of environments in which a given species should be employed in agriculture would be divided into TPEs in such a way that (a) the rank order values of genotypes would be the same throughout each TPE though variable from one TPE to another and (b) the number of TPE's would be as small as possible. Realistically, the cost of obtaining the data for such a division would be too great for the effort to be justified. Fortunately, the accumulated experience of breeders and agriculturists provides the basis for a delineation of TPE's that may approach the ideal rather closely. Another way of saying this is that many of the factors responsible for substantial variation in rank order values of genotypes are known. For example, in the case of crops, latitude, soil moisture and fertilization are among the significant variables and sites chosen to be representative of a relatively large area with respect to such variables are logical as locations for comparison of the criterion units of a recurrent selection program.

If each TPE were sufficiently homogeneous so that the same genotype would be ideal for every environment of any single TPE, the potential for genetic improvement over the whole production area of the species would be maximum. But, if for no other reason than the unpredictable temporal variation in climate, such perfect delineation of TPE's would be impossible, even if practical. Whether practical or not would depend on number of TPE's required since for each an addition to the total breeding program would be required, i.e., any gain in the potential for increase in $\ddot{Y}_w$ that is achieved <u>via</u> greater homogeneity within TPE's by increasing their numbers must be balanced against the consequent increment in the total cost of breeding programs. The other benefit of greater uniformity within TPE's (probably secondary though it has received more attention in the literature) is the related decrease in variance, $\sigma^2(\hat{X})$, of the selection criterion due to decrease in the genotype-environment component of that variance.

14 Comparison of Selection Systems

The purpose in this chapter is to illustrate how theory can be employed in evaluations of (a) alternative selection systems or (b) procedural variations of specific systems. Essential aspects of all systems are (1) the selection units, the criterion units, the target population and (2) procedural details , e.g., size of the genetic sample, the structure of criterion unit evaluations on which selection is based, and selection intensity. It will be shown, though it should come as no surprise, that there are interactions among these program components and hence that optimum decisions with respect to them require their joint consideration Both long and short term programs are considered. Comparisons presented are ones that, for professional breeders, should have value going beyond the technics illustrated.

Problems and procedural possibilities vary from one organism to another. However, applications framed in terms of maize serve to illustrate the general details involved in obtaining guidance from theory and to provide background for thinking about the special problems presented by other species. The variety of possible selection systems is great in the case of maize. Because it is naturally cross-fertilizing, close approximation to random mating within defined populations is possible by wind-pollination with sufficient isolation of each population from all others or by manual transport of pollen from massed collections to receptive silks that have been shielded from wind pollination. On the other hand, specific matings (including self-fertilizations) are easily accomplished by manual transport of pollen from the tassel of one plant to its own silks or to those of any one or more other plants.[1] In addition to the ease with which specific hand-pollinations can be made, the number of seeds resulting from a single pollen transfer is ordinarily large enough to allow observation of the resulting progeny in a significant number of test plots.

Expressions that will be utilized in the evaluation of long term programs are 11.10, 11.9, and 11.7. For convenience, they are renumbered and shown below.

$$(N_e s')_r \approx -\ln(1-P_a)/2q_0 \qquad (11.10)\ 14.1$$

$$s' \approx 2k\sigma(\hat{q},\hat{X})/q(1-q)\sigma(\hat{X}) \qquad (11.9)\ 14.2$$

$$E(\Delta q) \approx k\sigma(\hat{q},\hat{X})/\sigma(\hat{X}) \qquad (11.7)\ 14.3$$

Equation 14.3 pertains to Δq per cycle. Thus, for comparison of procedures among which cycle length is variable change per unit of time, obtained by dividing by cycle length, is required. The appropriate expression for $\sigma^2(\hat{X}) = [\sigma(\hat{X})]^2$ is not the same for all criterion units and, hence, must be considered as an aspect of individual programs.

When 14.1 is used to obtain a value for $(N_e s')_r$ values of P_a and q_0 must be supplied, but these cannot be expected to be optimum. The issue, of course, is the proportion of the potential of target populations that will be forfeited by setting P_a too low or q_0 too high. Generalizations concerning the distribution of initial allele frequencies are recorded in section 11.1.3.

When theory is used herein to compare plant breeding systems, family criterion units will be assumed to involve r replications at each of v locations in each of t years and, hence, that $\sigma^2(\hat{X})$, as given by 10.11, is

$$\sigma^2(\hat{X}) = \sigma^2(\bar{z}) + \sigma^2(\bar{z}a)/v + \sigma^2(\bar{z}b)/t + \sigma^2(\bar{z}ab)/vt + \sigma^2(e)/rvt \qquad (10.11)\ 14.4$$

Equations 11.22 to 11.29 and A11.3.27, having to do with $E(\Delta q)$ and $E(\Delta \ddot{y})$, are employed in the comparative evaluation of short term programs.

• 14.1. Long Term Programs

Because the goal in long term programs is the synthesis of near-ideal genotypes, the evaluation criteria to be employed in this section are (1) is the program one with which the required value of $N_e s'$ can be achieved and (2) if so, then how does it compare to other possible programs in terms of the expected magnitude of Δq? It is apparent from equations 14.1 - 14.4 that the answers to these questions will always depend on the situation and hence that the situation with respect to which one or more systems are evaluated must be specified in all its relevant detail. Specifications required include details of the program(s) in question and are otherwise all apparent from equations 14.1, 14.2, 14.3 and 14.4.

To obtain $(N_e s')_r$ using 14.1 requires numerical values for P_a and q_0. But note that many sets of P_a and q_0 values yield the same answer. For example, $-\ln(1-P_a)/2q_0 = 12.63$ when P_a = 0.92 and q_0 = 0.10 and also whenever P_a and q_0 are such that $-\ln(1-P_a) = 25.26q_0$ (for example, when P_a = 0.96 and q_0 = 0.127). To evaluate s' and $E(\Delta q)$ using 14.2 and 14.3 requires specification of (a) the fraction, p, of the genetic sample to be selected, as the basis for evaluating k, (b) the value of u_x to substitute in the appropriate expression for $\sigma(\hat{q},\hat{X})$, and (c) values for the variances and the parameters (v, t, and r) of the field comparison of criterion units that are required for evaluation of $\sigma(\hat{X}) = [\sigma^2(\hat{X})]^{1/2}$ using 14.4 Finally, the size (T) of the genetic sample is required; it along with the fraction selected yields the number of selections (M), which is needed for computation of the effective population size.

14.1.1. An Example

The specifications of this example together with significant consequences are summarized in Table 14.1.

The numerical magnitudes of u_X, and of each of the variance components of $\sigma^2(\hat{X})$, are functions of the scale of measurement. Let the latter be such that $\ddot{Y}_X = 100$ and let u_X,

Table 14.1. Specifications, some consequences, and symbols employed

SPECIFICATIONS

$P_a = 0.92$ and $q_0 = 0.1$ or $P_a = .96$ and $q_0 = .064$

Additive genes ($a_X = 0.0$) $2u_X = 0.01(\ddot{Y}_X)$

Scale of measurement such that $\ddot{Y}_X$ (of the S population) = 100 at the outset, making $u_X = 0.5$[a]

HW frequencies of genotypes and linkage equilibrium in the S(= target) population

The selection units which are also the criterion units are S_6 families (developed without selection) of which the single plant parents are S_5 individuals.

Three plant generations per year possible

Lines divided into sets of 50 for randomized block field comparison, followed by intra-set selection, $t = r = 1$, $v = 1, 2,$ or 3 in the field comparisons

$\sigma^2(\bar{z}) = 25(1+h) = 49.22$, $\sigma^2(\bar{z}a) = \sigma^2(\bar{z}b) = \sigma^2(z)/4 = 12.3$,

$\sigma^2(\bar{z}ab) = \sigma^2(\bar{z})/2 = 24.61$, and $\sigma^2(e) = 144$

CONSEQUENCES

$h = 0.96875$ $L = 3$

(N_es') required = 12.63 or 25.15

$N_e \approx M/(1+h)$	(from Appendix 11.1)
$s' = k(1+h)/\sigma(\hat{X})$	(from 14.2 and $u_X = 0.5$)
$N_es' = kM/\sigma(\hat{X})$	(given $u_X = 0.5$)
$\sigma(\hat{X}) = 15.6, 12.33,$ or 11.04	(when $v = 1, 2,$ or 3)
$\sigma(\hat{q},\hat{X}) = q(1-q)(1+h)u_X$	(from Table 8.2, $a = 0.0$, and linkage equilibrium)
$E(\Delta q)/q(1-q)L = k(1+h)u_X/\sigma(\hat{X})L$	(from 14.3)

SYMBOLS

T = number of selection units in the genetic sample

v, t, and r = numbers of locations, years, and replications per location-year in which criterion units are grown for comparison

M = number of selections

p = M/T = selection intensity (fraction of units selected)

X = the selection criterion

$\hat{X}$ = average of X for a single criterion unit in the field comparison specified

$\sigma^2(\hat{X}) = [\sigma(\hat{X})]^2$ = variance of $\hat{X}$ among selection units

k = selection differential in $\sigma(\hat{X})$ units (see Table 6.1 or section 6.3.2)

L = cycle length in years

$h \approx 1-(1/2)^{n-1}$ = coefficient of inbreeding of the parents of S_n families

[a] The specification that $u_X = 0.5$ pertains only to the genes for which we wish the probability of fixation of B to be 0.92 if their initial frequencies are 0.1 (or .96 if their initial frequencies are 0.064)

measured on that scale, be 0.5. Note (a) that u_X = 0.5 means that the substitution of the BB for the bb genotype at the locus in question would mean an increase of one percent in $\ddot{Y}_X$, the average value of genotype of the population, and (b) that the specification, u_X = 0.5, pertains only to the hypothetical genes for which we wish the probability of fixation of B to be P_a -- no constraint is being placed on the u-values at other loci.

The attributes of the selection system considered are (a) the target population is the S population, (b) HW frequencies of genotypes and linkage equilibrium in the S population, and (c) S_6 families are both the criterion and the selection units. The single plants from which seed is multiplied for field comparisons are to be S_5 plants produced without selection by five successive generations of self-fertilization starting from random S_0 plants.

The plan for obtaining performance data on criterion units is the following. First the T units of the genetic sample would be divided into sets of 50. Then the units of each set would be compared in randomized block trials---a separate trial for each set --- and from each set the family or families that performed best would be selected. The obvious reason for intra-set comparison and selection is limitation of within block variation among plots. Choice of 50 as set size in our example was arbitrary and does not imply that 50 is (or is not) near-optimum.

The value specified for $\sigma^2(\bar{z})$ assumes 25 as the additive genetic variance among individual plants in the foundation S population. Twenty-five is the value that would result if there were 200 genes for each of which a was 0.0 and q and u were both 0.5. It could of course result from an infinite number of combinations of a, u, and q (that are all expected to vary among genes) and the number of genes with effects on the selection criterion (X) but it is the numerical value of $\sigma^2(\bar{z})$, not how it originates, that is of prime importance here.

The value specified for $\sigma^2(e)$ is premised on a coefficient of plot error variance equal to 12 percent in a randomized block field trial in which the number of entries is 50. The relations among the genetic variance, $\sigma^2(\bar{z})$, and the genotype-environment interaction variances conform reasonably with estimates recorded by Sprague and Federer (1951) and Comstock and Moll (1963). Hopefully, breeders would be in position to employ estimates obtained from their own data that might therefore be more appropriate relative to variances of their own material and the conditions in which it would be grown.

Table 14.2 shows the kind of answers that can be obtained for long term programs by processes that have been described. It is focused on the impacts of variations in three variables (selection intensity, size of the genetic sample, and the structure of field comparisons of criterion units). You will note that it is $E(\Delta q)/q(1-q)L$, not $E(\Delta q)/L$, that receives attention. The reason is that the $q(1-q)$ that appears in all expressions for $\sigma(\hat{q},\hat{X})$ and therefore in all expressions for $E(\Delta q)$ is cancelled out in $E(\Delta q)/q(1-q)L$. Because $q(1-q)$ is not a function of variables of the selection program the same information concerning the <u>relative</u> impacts of variations in the selection program is obtained as would be by $E(\Delta q)/L$ (which only rarely can be obtained because q can only rarely be known).

Examine each portion of Table 14.2 separately and note that when size of the genetic sample (T) and the structure of the field comparison are constant: (1) as p increases (selection intensity

Table 14.2. Effects of variation in p (percent selected)

% selected	$\sigma(\hat{X})$	M	k	$N_e s'$	$E(\Delta q)u_x/q(1-q)L$
(a) T = 1200, v = 1					
.08	15.6	96	1.86	11.45	.039
.09		108	1.81	12.53[a]	.038
.10		120	1.76	13.54	.037
.24		288	1.30	24.00	.028
.26		312	1.25	25.00[b]	.026
.28		336	1.20	25.85	.025
(b) T = 600, v=2					
.16	12.33	96	1.52	11.83	.040
.18		108	1.46	12.78[a]	.039
.20		120	1.40	13.63	.038
.50		300	.80	19.46	.022
.60		360	.64	18.69	.017
(c) T = 2400, v = 1					
.035	15.6	84	2.21	11.90	.046
.04		96	2.16	13.29[a]	.046
.045		108	2.10	14.54	.044
.08		192	1.86	22.89	.039
.09		216	1.81	25.06[b]	.038
.10		240	1.76	27.08	.037
(d) T = 1200, v = 2					
.06	12.33	72	2.00	11.68	.053
.07		84	1.92	13.08[a]	.051
.08		96	1.86	14.48	.050
.16		192	1.52	23.67	.040
.18		216	1.46	25.58[b]	.039
.20		240	1.40	27.25	.038
(e) T = 800, v=3					
.09	11.04	72	1.81	11.80	.054
.10		80	1.76	12.75[a]	.052
.11		88	1.71	13.63	.051
.28		224	1.20	24.35	.036
.30		240	1.16	25.22[b]	.034
.32		256	1.12	25.97	.034

[a] closest to the required 12.63.
[b] closest to the required 25.15.

decreases), M increases and k decreases but their rates of change are such that their product and, consequently, $N_e s'$ increases as proportion selected increases until p > .50 (see part b of the table) and (2) as expected, $E(\Delta q)$ decreases as selection intensity decreases. When parts of the table are compared, we find the following impacts of increasing v from 1 to 2 or 3 while the total of plots, Tv, is held constant.

(1) Because $\sigma(\hat{\bar{X}})$ is smaller when v is larger, E(Δq) is increased and this effect is great enough so that comparing points at which Tv is the same and required values of $N_e s'$ are reached, the value for E(Δq) is slightly larger when v =2 than when v = 1. However, when increasing v makes the genetic sample too small, the required $N_e s'$ may not be possible (see part b) and if v is increased too much at the expense of T, E(Δq) when the required value of $N_e s'$ is reached may not be increased (compare parts e and d of the table).

(2) When T is increased without change in structure of the field comparison (illustrated when parts a and c of the table are compared) required values of $N_e s'$ are reached with greater selection intensity and a consequent advantage in E(Δq).

The effect of increasing t was not checked because to have made it 2 instead of 1 would not have decreased $\sigma(\hat{\bar{X}})$ enough to offset the resulting increase of L from 3 to 4 (remember that criterion units must be compared in the season in which the target population is grown commercially). Increasing v instead of r was investigated because as evident from 14.4 its effect on $\sigma(\hat{\bar{X}})$ is greater whenever $\sigma^2(\bar{z}a) > 0.0$.

14.1.2. Comparison of Programs

Problems are encountered in the making of program comparisons. Two will be cited at the outset but without intention to imply that others may not be important.

Obviously, cost deserves attention. It is tempting to insist that comparisons should be made on a cost-equivalent basis. But that would, at best, be a formidable task because there are a number of elements of total cost that may vary from one system to another. Consider the production of selection and/or criterion units. Hand-pollinations are required in some cases, but not all, and the number required varies among systems. Given sufficient isolation, half-sib families can be obtained without hand-pollination. On the other hand, S_n families and full-sib families cannot, but between them the numbers of hand-pollinations required differ. As another example, consider the structure of field comparisons and remember that with total plot number constant $\sigma^2(\hat{\bar{X}})$ is always reduced more by increasing locations than by increasing replications (equation 14.4). However, the cost of two replications at one location must be assumed to be less than that of one replication at each of two locations.

A second significant issue is the sensitivity of comparisons to specifications on which they are based. To illustrate, it is widely recognized and has been amply demonstrated (see, for example, Falconer, 1981) that between mass selection and family selection, the former is superior when heritability is high but may not be when heritability is low.

In view of such problems it may be most productive to make a first comparison using specifications that take obvious issues into account but without pretense that they are adequate in all respects. Systems to be considered are listed in Table 14.3. These can all be viewed as single population systems. On the other hand, each of them can be viewed as the system of selection in one or both of the populations of an RSP program.

Specifications pertaining to the eight selection systems to be compared are shown in Table 14.4 and the compositions of N_e, s', $N_e s'$, and E(Δq)u_x/q(1-q)L are listed in Table 14.5.

Table 14.3. Selection systems considered in Chapter 14

System 1	selection among S_6 families based on their own performance
System 2	selection among S_4 families based on their own performance
System 3	selection among S_1 families based on their own performance
System 4	selection among S_6 families based on the performance of test-cross progenies of the S_5 parents of those families, the tester being the population itself
System 5	selection among S_1 families based on the performance of test-cross progenies of the S_0 parents of those families, the tester being the population itself
System 6	selection (type 1) among full-sib families
System 7	selection (type 1) among half-sib families
System 8	selection among half-sib families based on performance of seed parents, commonly designated as "mass selection" (of seed parents)

Specifications should logically be the same in some respects for all of the selection systems but just as logically must vary in other respects. Expressions for $\sigma(\hat{q},\hat{X})$ are from Table 8.2 with the assumptions of a = 0.0 and linkage equilibrium in the S population taken into account;[2] those for N_e are from Appendix 11.1; and those for s' were obtained using 14.2. Values of $\sigma(\hat{X})$, computed using 14.4, are shown in Tables 14.8 and 14.9.

The relative magnitudes of $\sigma^2(\bar{z})$ for the several systems are those indicated to be correct by genetic variance component theory presented in Chapter XV and the ratios among assumed values of $\sigma^2(\bar{z})$ and the three GE interaction variances are the same for all systems. Values assumed for $\sigma^2(e)$ are proportional to the inbreeding of criterion unit plants; largest in the case of system 1, smallest for systems 4 through 7, and intermediate for systems 2 and 3.[3] With respect to System 8, only $\sigma^2(\bar{z})$ and the total phenotypic variance, $\sigma^2(p)$, had to be quantitized. By definition, $\sigma^2(p)$ = (additive genetic variance)/(heritability). It was approximated for the purpose of the comparison being presented assuming heritability equal to 0.084 as reported for grain yield in the Iowa Stiff Stalk Synthetic by Hallauer and Miranda (1981) and additive variance equal to 25. $\sigma^2(p) = \sigma^2(\hat{X}) = 298$ was obtained.

Values specified for m in the case of systems 3 and 5 and for F in the case of system 7 were arbitrary. In the case of 3 and 5, N_e increases as $2m/(2m+1)$ increases but the latter is close to its maximum when m=10 which seems reasonable with respect to labor involved. In the case of system 7, N_e increases as $F/(F+1)$ increases but the latter is at 80 percent of its maximum F=4 and there is some increase in cost as F is increased. There is of course a slight, but only slight, decrease in $\sigma(\hat{X})$ when F is increased beyond 4.

The expressions for $N_e s'$ and $E(\Delta q)/q(1-q)/L$ in Table 14.5 enable qualitative predictions of the relative merits of the eight systems and provide the basis for understanding

Table 14.4. Background for system comparisons

Specifications that apply for all systems

Linkage equilibrium and zero inbreeding in the S population

Scale of measurement such that $\ddot{Y}_X$ of the S population = 100 at the outset

Additive genes (a_X = 0.0) $(N_e s')_r$ = 12.63 or 25.15, based on u_X = 0.5

$\sigma^2(\bar{z}a) = \sigma^2(\bar{z}b) = \sigma^2(\bar{z})/4$ $\sigma^2(\bar{z}ab) = \sigma^2(\bar{z})/2$

Variable specifications and consequences (all specifications considered)

SYSTEM 1 $\sigma^2(\bar{z}) = 49.22$ $\sigma^2(e) = 144$ $N_e \approx M/(1+h)$

$\sigma(\hat{q}_i,\hat{X}) = (1+h)q(1-q)u_X$ h =.96875

SYSTEM 2 $\sigma^2(\bar{z}) = 46.88$ $\sigma^2(e) = 138$ $N_e \approx M/(1+h)$

$\sigma(\hat{q}_i,\hat{X}) = (1+h)q(1-q)u_X$ h = .875

SYSTEM 3 $\sigma^2(\bar{z}) = 25$ $\sigma^2(e) = 121$ $N_e = M[2m/(2m+1)]$

$\sigma(\hat{q}_i,\hat{X}) = q(1-q)u_X$ m = 10

SYSTEM 4 $\sigma^2(\bar{z}) = 12.30$ $\sigma^2(e) = 100$ $N_e \approx M/(1+h)$

$\sigma(\hat{q}_i,\hat{X}) = (1+h)q(1-q)u_X/2$ h = .96875

SYSTEM 5 $\sigma^2(\bar{z}) = 6.25$ $\sigma^2(e) = 100$ $N_e = M[2m/(2m+1)]$

$\sigma(\hat{q}_i,\hat{X}) = q(1-q)u_X/2$ m = 10

SYSTEM 6 $\sigma^2(\bar{z}) = 12.5$ $\sigma^2(e) = 100$ $N_e = 4M/(2-p)$

$\sigma(\hat{q}_i,\hat{X}) = q(1-q)u_X/2$

SYSTEM 7 $\sigma^2(\bar{z}) = 6.25$ $\sigma^2(e) = 100$ $N_e \approx 8M[F/(F+1)]/(2-p)$

$\sigma(\hat{q}_i,\hat{X}) = q(1-q)u_X/4$ F = 4

SYSTEM 8 Selection units are half sib families; criterion units are their common parents

$\sigma^2(\bar{z}) = 25$ $\sigma^2(\hat{X}) = 298$ $\sigma(\hat{X}) = 17.3$ $N_e = 8M/(1+p)(2-p)$

$\sigma(\hat{q}_i,\hat{X}) = q(1-q)u_X/2$

Criterion Unit Comparisons

SYSTEMS 1-7: Units grouped into random sets of 50 for separate randomized block trials followed by intra set selection. v = 1 or 2, t = r = 1 in all field trials

SYSTEM 8: Units are individual plants, hence no replication is possible. Selection within sets of 200. The plants of each set, uniformly spaced, constitute a separate field block, as many blocks as sets. This is the grid system described by Gardner (1961).

the more quantitative information shown in Tables 14.8 and 14.9. Note, for example, that kM appears in each of the $N_e s'$ expressions. It follows, for example, that if selection intensity (p) and the size of genetic sample (T) were the same for all systems, $N_e s'$ would probably be much smaller for Systems 4 and 5 than for any of the others. The expressions for N_e and s' are included in the table because together they explain the variation among the expressions for $N_e s'$.

Table 14.5. Critical formulas (when a = 0)

System	N_e	s'	$(N_e s')$	$E(\Delta q)/q(1-q)L$
1	$M/(1+h)$	$2k(1+h)u_x/\sigma(\hat{X})$	$2kMu_x/\sigma(\hat{X})$	$k(1+h)u_x/\sigma(\hat{X})L$
2	$M/(1+h)$	$2k(1+h)u_x/\sigma(\hat{X})$	$2kMu_x/\sigma(\hat{X})$	$k(1+h)u_x/\sigma(\hat{X})L$
3	$2Mm/(2m+1)$	$2ku_x/\sigma(\hat{X})$	$4kMmu_x/(2m+1)\sigma(\hat{X})$	$ku_x/\sigma(\hat{X})L$
4	$M/(1+h)$	$k(1+h)u_x/\sigma(\hat{X})$	$kMu_x/\sigma(\hat{X})$	$k(1+h)u_x/2\sigma(\hat{X})L$
5	$2Mm/(2m+1)$	$ku_x/\sigma(\hat{X})$	$2kMmu_x/(2m+1)\sigma(\hat{X})$	$ku_x/2\sigma(\hat{X})L$
6	$4M/(2-p)$	$ku_x/\sigma(\hat{X})$	$4kMu_x/(2-p)\sigma(\hat{X})$	$ku_x/2\sigma(\hat{X})L$
7	$8MF/(F+1)(2-p)$	$ku_x/2\sigma(\hat{X})$	$4kMFu_x/(F+1)(2-p)\sigma(\hat{X})$	$ku_x/4\sigma(\hat{X})L$
8	$8M/(1+p)(2-p)$[a]	$ku_x/\sigma(\hat{X})$	$8kMu_x/(1+p)(2-p)\sigma(\hat{X})$	$ku_x/2\sigma(\hat{X})L$

[a] Because the selection units of both system 8 and system 7 are half-sib families this expression can be derived from that for N_e in the case of system 7. The bridging assumptions are (a) that the S population of cycle k+1 is composed of n members of each of the M half-sib family selections of cycle k, (b) that all S population individuals after the initial cycle are constant parents of half-sib families in the genetic sample (T – N) and (c) that all individual members of the selection unit families of system 8 have different pollen parents, i.e., that F = n. It follows that N = Mn, N = M/p because M = pN, and hence that F = n = 1/p. Substituting 1/p for F in the N_e expression shown for system 7 yields the approximation shown for system 8.

Appropriate values for L (see Table 14.6) depend, in accord with Hallauer and Miranda (op. cit.), on number of plant generations required per cycle; the number that can, or will, be grown per year; and the fact that ordinarily the criterion unit performance trials must be grown in a particular season of the year, i.e., the season in which the crop is grown commercially.

Table 14.7 deals with selection intensity in the case of mass selection (system 8). Tables 14.8 and 14.9 focus on the relative effectiveness of the eight systems and whether relative effectiveness depends much on such things as the structure of the field comparison of criterion units, the size of the genetic sample, the $N_e s'$ requirement chosen, or the number of plant generations that can be grown per year. Primarily because the criterion units of mass selection are single plants and hence cannot be replicated or grown at more than one location but also because with observations being required on each plant it might be unreasonable to assume that the same total number of plants should be grown in the case of mass selection as in the others where obervations are made on plots, mass selection is considered separately (Table 14.7) before going to comparison of all eight systems.

By rearrangement of the expression for $N_e s'$ in the case of system 8 that is provided in Table 14.5

$$M = (N_e s')(1+p)(2-p)\sigma(\hat{X})/4k \qquad \text{(when } u_X = 0.5 \text{ as specified)}$$

Then using one of the specified values of $(N_e s')_r$ together with the specified value of $\sigma(\hat{X})$ we

Table 14.6. L(years per cycle) as a function of plant generations per year and cycle

System	Plant generations per cycle	Plant generations per year Three	Two	One
1	8	3	4	8
2	6	2	3	6
3	3	1	2	3
4	8	3	4	8
5	3	1	2	3
6	2	1	1	2
7	2	1	1	2
8	1	1	1	1

Table 14.7. Mass selection–effects of variation in percent selected (p)

p	M	k	N_es'	$E(\Delta q)/q(1-q)$	Required Size of Genetic Sample
.005	38	2.90	12.71	.042	7600
.01	42	2.66	12.85	.038	4200
.03	49	2.27	12.67	.033	1633
.05	55	2.06	12.79	.030	1100
.10	65	1.76	12.66	.026	650
.30	105	1.16	12.74	.017	350
.50	154	0.80	12.66	.012	308
.70	241	0.50	12.61	.010	344
.90	635	0.18	12.64	.002	706
.005	76	2.90	25.42	.042	15200
.01	83	2.66	25.40	.038	8300
.03	98	2.27	25.35	.033	3267
.05	109	2.06	25.35	.030	2180
.10	130	1.76	25.31	.026	1300
.30	208	1.16	25.24	.017	693
.50	306	.80	25.15	.012	612

obtain an expression for M that is a function of only p and k. Thus using $(N_es')_r = 12.63$ yields

$$M = 54.62(1+p)(2-p)/4k$$

The M's obtained using this are, in general, not whole numbers so those shown in Table 14.7 are the smallest whole number larger than the answer obtained. For example, given p = .03 and k=2.27, the M obtained is 48.82. This "rounds up" to 49, the value listed in the table. Using such "rounded up" values, the $N_e s'$ values shown were obtained. M and p, taken together, supplied the size of genetic sample required.

The facts that stand out in Table 14.7 are (a) that in the case of mass selection $(N_e s')_r$ can be achieved with any selection intensity, (b) as p is increased $E(\Delta q)u_x/q(1-q)$ per cycle decreases, and (c) as p is increased the size of genetic sample needed if $(N_e s')$ is to be achieved is minimized when p ≈ .5 but then $E(\Delta q)$ is less than one-third its value when p = .01.

Tables 14.8 and 14.9 provide comparative information concerning the eight selection systems listed in Table 14.3 and for which background information is shown in Tables 14.4 and

Table 14.8. Comparison of systems for long term programs, $(N_e s')_r = 12.63$

						$E(\Delta q)/q(1-q)L$[a]		
System	$\sigma(\hat{X})$	p	M	k	$N_e s'$	L(3)	L(2)	L(1)
(a) T = 1200, v = 1								
1	15.6	.09	108	1.81	12.53	.038	.028	.014
2	15.2	.09	108	1.81	12.86	.056	.037	.018
3	13.1	.08	96	1.86	12.98	.071	.036	.024
4	11.2	.16	192	1.52	13.03	.022	.016	.008
5	10.6	.15	180	1.56	12.61	.037	.018	.012
6	11.2	.06	72	1.99	13.19	.044	.044	.022
7	10.8	.07	84	1.92	12.38	.022	.022	.011
(b) T = 600, v = 2								
1	12.33	.18	108	1.46	12.78	.039	.029	.014
2	12.05	.17	102	1.49	12.61	.058	.038	.020
3	10.06	.14	84	1.59	12.64	.079	.040	.026
4	8.37	.32	192	1.12	12.85	.022	.016	.008
5	7.74	.30	180	1.16	12.85	.038	.018	.012
6	8.39	.10	60	1.76	13.25	.052	.052	.026
7	7.92	.12	72	1.67	12.92	.026	.026	.013
(c) T = 2400, v = 1								
1	15.6	.04	96	2.16	13.29	.046	.034	.017
2	15.2	.035	84	2.21	12.21	.068	.046	.022
3	13.1	.035	84	2.21	13.86	.084	.042	.028
4	11.2	.06	144	2.00	12.86	.030	.022	.011
5	10.6	.06	144	2.00	12.94	.047	.024	.016
6	11.2	.025	60	2.34	12.69	.052	.052	.026
7	10.8	.03	72	2.27	12.29	.026	.026	.013
(d) T = 1200, v = 2								
1	12.33	.07	84	1.92	13.08	.051	.038	.019
2	12.05	.065	78	1.95	12.62	.076	.050	.026
3	10.06	.055	66	2.02	12.62	.102	.050	.034
4	8.39	.10	120	1.76	12.62	.034	.026	.013
5	7.74	.10	120	1.76	13.00	.057	.028	.019
6	8.39	.04	48	2.16	12.61	.064	.064	.032
7	7.92	.05	60	2.06	12.80	.032	.032	.016
(e) T = 4200								
8	17.3	.01	42	2.66	12.85	.038	.038	.038

[a] Numbers in columns headed L(3), L(2), and L(1) are those obtained using the L that is appropriate when plant generations per year are 3, 2, or 1, respectively (see Table 14.6).

Table 14.9. Comparison of systems for long term programs, $(N_e s')_r$=25.15

						E(Δq)/q(1-q)/L[a]		
System	$\sigma(\hat{X})$	p	M	k	$N_e s'$	L(3)	L(2)	L(1)
(a) T= 1200, v = 1								
1	15.6	.24	288	1.30	24.96	.028	.020	.010
2	15.2	.25	300	1.27	25.06	.039	.026	.013
3	13.1	.22	264	1.35	25.91	.052	.026	.017
4	11.2	.50	600	.80	21.24	---	---	---
5	10.6	.50	600	.80	21.56	---	---	---
6	11.2	.14	168	1.59	25.64	.036	.036	.018
7	10.8	.18	216	1.46	25.67	.017	.017	.008
(b) T = 600, v = 2								
1	12.33	.50	300	.80	19.46	---	---	---
2	12.05	.50	300	.80	19.92	---	---	---
3	10.06	.50	300	.80	22.72	---	---	---
4	8.37	.50	300	.80	14.34	---	---	---
5	7.74	.50	300	.80	14.77	---	---	---
6	8.39	.25	150	1.27	25.95	.038	.038	.019
7	7.92	.32	192	1.12	25.86	.018	.018	.009
(c) T = 2400, v = 1								
1	15.6	.09	216	1.81	25.06	.038	.028	.014
2	15.2	.09	216	1.81	25.72	.056	.037	.018
3	13.1	.08	192	1.86	25.96	.071	.036	.024
4	11.2	.15	360	1.56	25.07	.023	.017	.008
5	10.6	.15	360	1.56	25.23	.037	.018	.012
6	11.2	.06	144	1.99	26.38	.044	.044	.022
7	10.8	.075	180	1.89	26.18	.022	.022	.011
(d) T = 1200, v = 2								
1	12.33	.18	216	1.46	25.57	.039	.029	.014
2	12.05	.17	204	1.49	25.22	.058	.038	.020
3	10.06	.14	168	1.59	25.29	.079	.039	.026
4	8.39	.31	372	1.14	25.27	.022	.016	.008
5	7.74	.30	360	1.16	25.69	.038	.018	.012
6	8.39	.10	120	1.76	26.50	.052	.052	.026
7	7.92	.12	144	1.67	25.84	.026	.026	.013
(e) T = 8300								
8	17.30	.01	83	2.66	25.40	.038	.038	.038

[a] Numbers in columns headed L(3), L(2), and L(1) are those obtained using the L that is appropriate when plant generations per year are 3, 2, or 1, respectively (see Table 14.5).

14.5. Together they illuminate how the comparative effectiveness of the systems is affected by variations in (a) the required value of $N_e s'$, (b) size of the genetic sample, (c) change in structure of the field comparison of criterion units, and (d) the number of plant generations that can be grown per year. In parts (a) - (d) of both tables the information shown is that which relates to the most intense selection with which the required value of $N_e s'$ can be achieved. For example, in part (b) of Table 14.8 which deals with T (size of genetic sample) = 600 and a single replication comparison of criterion units at two locations (v = 2) in the same year, the near minimum value of p that would generate the required value of $N_e s'$ is shown to be 0.10 in the case of system 6. Going to the last columns of the table, we see that E(Δq)/q(1-q) per year should then be .052 if two (or three) plant generations could be grown per year. Information regarding system 8 (mass selection) when p = 0.01 is shown separately (part e) in both tables. The choice of this selection intensity for inclusion in these tables was arbitrary except for the fact that the size of genetic sample required (4200 or 8300 as seen in Table 14.7) was not excessive by comparison with the number required for the 1200 or 2400 plots assumed in the case of the other seven systems.

Focusing now on the last three columns of the tables, because when the $(N_es')_r$ is achieved the system of choice would logically be the one for which $E(\Delta q)$ is largest,[4] note the following facts.

(1) No values are shown on some lines of Table 14.9. These are the cases in which $N_es' = 25.15$ cannot be achieved by the system in question. Values of N_es' shown are the largest possible, i.e., those achieved with p = .50.

(2) $E(\Delta q)$ values shown in the case of system 8 (mass selection) are all the same. This is because (a) it requires only one plant generation per cycle with the result that L for this system is not affected by number of plant generations that can be grown per year and (b) the greater N_es' of Table 14.9 is accomplished by doubling the genetic sample instead of increasing M by increasing p as with the other systems.

(3) Conversely, with only two exceptions, values of $E(\Delta q)$ for the other seven systems decrease as the plant generations that can be grown per year decrease from 3 to 2 to 1. In the case of systems 6 and 7 there is no decrease from the L(3) to the L(2) column because for those systems (as described) the fact that the criterion unit comparison must be conducted in the commercial growing season means that cycle length must be one year whether three plant generations per year or only two are possible.

(4) When T = 600 and v = 2 (Table 14.9, part b), $N_es' = 25.15$ could only be reached with systems 6, 7, and 8. The ratios of N_e to M (see Table 14.5) were primarily responsible.

(5) The importance of generations per cycle and generations that can be grown per year are emphasized when rankings in the L(3), L(2), and L(1) columns are compared. Note, for example, that system 6, consistently third in the L(3) column, is as consistently the leader in the L(2) column. Equally striking is that system 8, not a standout in the L(3) column, is the leader in each case when only one generation can be grown per year.

14.1.3. Discussion

Because $\sigma(\hat{q}_i,\hat{X})$ is proportional to u_{xi} and therefore zero if u_{xi} is zero, it is apparent from 14.2 that s' is greater than zero in the case of alleles having positive effects on X, the selection criterion. On the other hand, s' can only be zero in the case of any allele without effect on the selection criterion. Thus the social value of long term recurrent selection programs will be greatest when X = W (selection criterion = total value) or when X is a dominant component of total value.

Quantitative information of the sort presented in tables 14.8 and 14.9 rests on several assumptions that cannot be made with anything approaching great certainty. It is natural therefore to question the utility of the time-consuming intellectual exercise involved. However,

(1) in view of probable growth of world population during the decades immediately ahead and the evidence of on-going decline in the world land resource for food production (Brown, 1990), the importance of capitalizing on the full potential of the germplasm of agricultural species cannot be doubted,

(2) this can only be accomplished via one or another form of recurrent selection, and

(3) all such programs require decisions concerning effective population size despite the fact that good information concerning all variables that affect $N_e s'$ is not now available and not likely to be available in the foreseeable future. Obviously, there can be no escape from just doing the best that can be done given available information.

Despite uncertainties concerning the bases (listed in table 14.4) for quantities presented in Tables 14.8 and 14.9, the following conclusions of major significance appear justified.

(1) Within the range of variations in genetic sample size, structure of criterion unit comparisons, and $N_e s'$ required that were investigated the relative effectiveness of selection systems was not much affected. Significant changes in rank that were exhibited were those resulting from changes in the number of plant generations grown per cycle of selection.

(2) Judged by $E(\Delta q)/q(1-q)$ per annum given the required value of $(N_e s')$, there are substantial differences among the selection systems considered. System 4 trails all other systems (except 7) in all comparisons involving the same combination of T,v, and $(N_e s')_r$. When S_1 families are the selection units (systems 3 and 5), $E(\Delta q)/q(1-q)$ is in every comparison much larger in the case of system 3 as would have been predicted from expressions in the final column of Table 14.5. It is in every case about twice as large for system 6 as for system 7, again predictable from expressions in Table 14.5. Each of the top three positions is occupied by system 2,3, or 6 in all instances when two or three generations can be grown per year. If only one plant generation can be raised per year system 8, mass selection conducted as described, appears superior to any of the other systems (this assumes the selection criterion could and would be the same as in other systems).

(3) Rather large progams are required to achieve reasonable probability of fixation goals together with near maximum increases per annum in frequencies of favorable alleles. Consider systems 2, 3, and 6 and their values of $E(\Delta q)/q(1-q)$ in the L(3) and L(2) columns. In all instances those values were larger when the size of the genetic sample was doubled -- from 1200 to 2400, given v = 1, or from 600 to 1200, given v = 2.. The increases ranged from 18 to 31 percent when the required value of $N_e s'$ was 12.63 and from 25 to 44 percent when $(N_e s')_r$ was 25.15. These results raise questions concerning (a) the values of still larger genetic samples and of more than two plots per unit (v > 2) in criterion unit comparisons and (b) the program size required to capitalize on whatever potential is inherent in new favorable alleles arising by on-going mutation (see section 11.1.3). These issues receive further attention in section 14.4.

Table 14.10 provides quantitative information pertaining to selection among S_1 families (system 3) of the type contained in Table 14.9 but for more T,v combinations. As expected,

Table 14.10. Variation in T and v when selection is among S_1 families

T	v	$\sigma(\hat{X})$	p	k	M	$N_e s'$	E(Δq/q(1-q)L. L(3)	E(Δq/q(1-q)L. L(2)	Tv
1200	1	13.10	.22	1.35	264	25.9	.052	.026	1200
	2	10.06	.14	1.59	168	25.3	.079	.040	2400
	4	8.14	.10	1.76	120	24.6	.108	.054	4800
	6	7.38	.09	1.81	108	25.2	.122	.061	7200
	8	6.98	.085	1.83	102	25.4	.131	.066	9600
2400	1	13.10	.08	1.86	192	26.0	.071	.035	2400
	2	10.06	.055	2.02	132	25.2	.100	.050	4800
	3	8.82	.045	2.10	108	24.4	.119	.055	7200
	4	8.14	.0425	2.13	102	25.4	.131	.065	9600
4800	1	13.10	.032	2.24	156	25.3	.085	.043	4800
	2	10.06	.0225	2.38	108	24.3	.118	.059	9600

Note: T = number of selection units in the genetic sample

v = number of locations in the criterion unit comparisons

when T is constant E(Δq)/q(1-q)L increases as v is increased and when v is constant it increases as T is increased. Of more practical interest are the E(Δq)/q(1-q)L values when Tv (the total number of plots in criterion unit comparisons) is constant. The E(Δq) data from the L(3) column of Table 14.10 arranged according to values of Tv are shown below. In each case the highest value is obtained with T = 1200 and v = 2, 4, or 8. The increase from doubling the total number of plots from 2400 to 4800 is 36 per cent and from doubling again to 4800 to 9600 is 21 percent.

Tv = 2400 T	v	E(Δq/q(1-q)L	Tv = 4800 T	v	E(Δq/q(1-q)L	Tv = 9600 T	v	E(Δq/q(1-q)L
1200	2	.079	1200	4	.108	1200	8	.131
2400	1	.071	2400	2	.100	2400	4	.131
			4800	1	.085	4800	2	.118

Whether program sizes considered are adequate with respect to chances of eventual fixation of new favorable alleles depends on the magnitudes of their beneficial effects and the probabilities of the mutations giving rise to such alleles (see Appendix 11.2). Whatever the selection system, the size of the genetic sample, or the structure of the criterion unit comparison, effective population size can be increased by reduction of the selection intensity. As shown by A11.2.3 the result is to increase the genetic gain to be realized in the long term via fixation of new favorable alleles arising from mutation. However, as noted earlier this is accomplished at the cost of some reduction in E(Δq) per cycle. This becomes apparent when we compare quantities shown in Tables 14.8 and 14.9. In the latter values of M and $N_e s'$ are greater than in Table 14.8 and values of E(Δq) are all lower.

Brief comments concerning three of the assumptions that have been employed are in order at this point.

(1) The value of $(N_es')_r$ employed was obtained assuming $a_X = 0.0$. As noted in section 11.2.1, when other things are equal the probability of fixation is higher for favorable alleles that are partially or completely dominant ($0.0 < a \leq 1.0$) and lower for favorable alleles that are in some degree recessive. When a value trait is heterotic there is reason to believe that the majority of alleles with positive effects on the trait are at least partially dominant.

(2) Absolute linkage equilibrium in the S population is unlikely but a close approach to that condition is judged probable (by the author) when effective population size is as great as required to make fixation probabilities high. Of course, if the foundation population were for any reason in disequilibrium, a close approach to the equilibrium situation assumed would require a number of cycles of the selection program.

(3) Whether the magnitude of u_X has a bearing on the relative effectiveness, as judged by $E(\Delta q)/q(1-q)L$, of different selection systems is not revealed by the contents of Tables 14.8 and 14.9. From the expressions for $E(\Delta q)/q(1-q)L$ shown in Table 14.5, we note (a) that u_X appears in each and (b) that, aside from quantities that are constant for each selection system though variable among systems, the only variable among the eight expressions is k. Thus whether the relationships among the $E(\Delta q)/q(1-q)L$ values of different systems are affected by the magnitude of u_X depends on whether the ratios among k's depend on the magnitude of u_X.

Consider, for example, systems 2 and 6 and the compositions of N_es' (shown in Table 14.5) for those systems. In the case of system 2, the quantity $2/\sigma(\hat{X})$ is an attribute of the system, unchanged by the value of u_X. Thus whether an arbitrary value of N_es' is to be achieved will depend on the magnitude of kMu_X -- if u_X is smaller, kM must be larger. In the same way, it is apparent that in the case of system 6, the magnitude of N_es' varies directly with that of $kMu_X/(2-p)$ so that to achieve a specific value of N_es', if u_X is smaller (or larger), $kM/(2-p)$ must be larger (or smaller). The issue is whether the ratios of k's involved in the required values of kM and $kM/(2-p)$ remain the same without regard to the value of u_X. They do not, as demonstrated in Table 14.11. However, changes are small and, in general, not sufficient to cause changes in the $E(\Delta q)/q(1-q)L$ ranks of selection systems.

• 14.2. Short Term, Single Population Programs

The goal in short term programs is maximum increase in the average value of genotypes of target populations during the contemplated finite periods of time. Thus, aside from considerations of cost and assuming the same base population, the selection system chosen should be the one for which the sum over cycles of $E(\Delta\ddot{Y})$ is greatest. This sum can be approximated, possibly well enough for selection system comparisons, by extrapolation from the expected response per year to the selection of a single cycle. The general expression for this single cycle response (Chapers V, VI, and VIII) in the case of single population programs in which a random mating selection population is itself the target population is[5]

$$E(\Delta\ddot{Y}_c) = [k/\sigma(\hat{X})]\Sigma_i\,\sigma(\hat{q}_i,\hat{X})2[1+(1-2q_i)a_{ci}]u_{ci}. \qquad (8.28)\ \ 14.5$$

Table 14.11. Magnitude of u_x and comparison of systems for long-term programs

u_X	System	$\sigma(\hat{X})$	p	M	k	$N_e s'$	$E(\Delta q)/q(1-q)/L$ L(3)	L(2)	L(1)
(a) T = 1200, v = 2, $(N_e s')_r$ = 12.63									
.5	2	12.05	.065	78	1.95	12.62	.076	.050	.026
	3	10.06	.055	66	2.02	12.62	.102	.050	.034
	6	8.39	.04	48	2.16	12.61	.064	.064	.032
	7	7.92	.05	60	2.06	12.80	.032	.032	.016
.25	2	12.05	.17	204	1.49	12.61	.029	.019	.010
	3	10.06	.14	168	1.59	12.64	.040	.020	.013
	6	8.39	.10	120	1.76	12.80	.026	.026	.013
	7	7.92	.12	144	1.67	12.78	.013	.013	.007
(b) T = 1200, v = 2, $(N_e s')_r$ = 25.15									
.5	2	12.05	.17	204	1.49	25.22	.058	.038	.020
	3	10.06	.14	168	1.59	25.29	.079	.039	.026
	6	8.39	.10	120	1.76	26.50	.052	.052	.026
	7	7.92	.12	144	1.67	25.84	.026	.026	.013
.25	2	12.05	.50	600	.80	19.91	.---	.---	.---
	3	10.06	.50	600	.80	22.72	.---	.---	.---
	6	8.39	.24	288	1.30	25.35	.019	.019	.009
	7	7.92	.31	372	1.14	25.35	.009	.009	.005
(c) T = 2400, v = 2, $(N_e s')_r$ = 12.63									
.5	2	12.05	.03	72	2.27	13.56	.088	.059	.029
	3	10.06	.025	60	2.34	13.29	.116	.058	.039
	6	8.39	.02	48	2.42	13.98	.072	.072	.036
	7	7.92	.02	48	2.42	11.85	.036	.036	.018
.25	2	12.05	.065	156	1.95	12.62	.038	.025	.013
	3	10.06	.055	132	2.02	12.62	.050	.025	.017
	6	8.39	.04	96	2.16	12.61	.032	.032	.016
	7	7.92	.05	120	2.06	12.80	.016	.016	.008
(d) T = 2400, v = 2, $(N_e s')_r$ = 25.15									
.5	2	12.05	.065	156	1.95	25.24	.076	.050	.025
	3	10.06	.055	132	2.02	25.24	.100	.050	.033
	6	8.39	.04	96	2.16	25.22	.064	.064	.032
	7	7.92	.05	120	2.06	25.61	.033	.033	.017
.25	2	12.05	.17	408	1.49	25.22	.029	.019	.010
	3	10.06	.14	336	1.59	25.29	.040	.020	.013
	6	8.39	.10	240	1.76	26.50	.026	.026	.013
	7	7.92	.12	288	1.67	25.84	.013	.013	.007

When the selection criterion is slso the response criterion, 14.5 becomes

$$E(\Delta\ddot{Y}_X) = [k/\sigma(\hat{X})]\Sigma_i\,\sigma(\hat{q}_i,\hat{X})2[1+(1-2q_i)a_{Xi}]u_{Xi}. \qquad 14.6$$

In each of these Σ_i indicates summation over loci. That sum is a genetic covariance between traits X and C in the case of 14.5 and a genetic variance of trait X in 14.6. The first of the expressions is the expectation of a correlated response and the second is the expectation of a direct response. The latter is the usual basis for ranking the effectiveness of selection systems (see, for example, Empig et al, 1972; Sprague and Eberhart, 1977; and Hallauer and Miranda, 1981) and it is the variation among systems with respect to the expectation of $\Delta\ddot{Y}_X$ that is examined in what follows. Doing so is obviously justified whenever total value is to be the selection criterion or whenever increasing genotype value with respect to the measure (X) to be employed as selection criterion is the real prime purpose of a breeding program.

Because $\Sigma_i \sigma(\hat{q}_i,\hat{X})2[1+(1-2q_i)a_{Xi}]u_{Xi}$ is a quantity that can be estimated (see Chapter XV) extrapolation based on 14.6 provides a preview of the genetic improvement that may be achieved as well as a basis for system comparisons. However, it ignores (a) impacts of effective population size, which are not the same for all selection systems, and (b) the fact that changes in allele frequencies resulting from selection may change the relative effectiveness of different forms of selection by significant amounts.

Change in $\ddot{Y}_X$, the average value of total genotypes, as depicted in 14.6, is the sum over all genes of changes in the values of single locus genotypes. Consequently, the relative effectiveness of selection systems can be appraised using $E(\Delta\ddot{y}_i)$ instead of $E(\Delta\ddot{Y}) = E\Sigma_i(\Delta\ddot{y}_i)$. Moreover, as will be demonstrated, focus on $E(\Delta\ddot{y})$ clarifies the interactions in recurrent selection among population size (N_e), level of dominance (a), selection intensity, and allele frequencies with respect to the relative effectiveness of selection systems.

From 14.6 it is apparent that in the case of single population programs

$$E(\Delta\ddot{y}_{Xi}) = [k/\sigma(\hat{X})]\sigma(\hat{q}_i,\hat{X})2[1+(1-2q_i)a_{Xi}]u_{Xi}. \qquad 14.7$$

The composition of $\sigma(\hat{q}_i,\hat{X})$ is shown in Table 8.2 for various selection systems. Consider those listed in Table 14.3 and identified by the numbers 1 - 8. In the case of systems 4 - 8, the criterion units are non-inbred individuals (in system 8) or families composed of non-inbred individuals and for this group

$$\sigma(\hat{q}_i,\hat{X}) = cq_i(1-q_i)[1+(1-2q_i)a_{Xi}]u_{Xi}$$

when the S population is in linkage equilibrium, i.e., when pt-rs = 0 for all pairs of genes. The values of c (shown in Table 8.2) are variable among the five systems. In systems 1-3 the criterion units are families composed of inbred individuals and for this group

$$\sigma(\hat{q}_i,\hat{X}) = q_i(1-q_i)[(1+h + (1-h)(1-2q_i)a_{Xi}/2]u_{Xi}$$

in which h = the coefficient of inbreeding of the parents of the S_n families. Substitutions in 14.7 yield

$$E(\Delta\ddot{y}_X) = 2q(1-q)[1+(1-2q)a_X]^2u_X^2c[k/\sigma(\hat{X})] \qquad 14.8$$

in the case of systems 4-8 and

$$E(\Delta\ddot{y}_X) = 2q(1-q)[(1+h)+(1-h)((1-2q)a_X/2][1+(1-2q)a_X]u_X^2[k/\sigma(\hat{X})] \qquad 14.9$$

in the case of systems 1-3. Because 14.8 and 14.9 are the same for all genes, the gene identifying subscript is not required. In addition x, the trait identifying subscript will be omitted in the remainder of this chapter because only direct responses to selection are to be considered. Of course, 14.8 and 14.9 like 14.6 ignore effects of finite population size.

Table 14.12 lists numerical values of 14.8 or 14.9 that do not take account of the magnitudes of the $[k/\sigma(\hat{X})]$'s. Thus quantities shown in the final three columns are values of $E(\Delta\ddot{y})/[k/\sigma(\hat{X})]$ when N_e is infinitely large. To adjust them for comparison of the eight selection systems listed in Table 14.3, assumptions must be made concerning number of plant

Table 14.12. Expectation of $\Delta\ddot{y}$ per cycle, given that u = 0.5

Selection system	Years/cycle L(3)	L(2)	L(1)	$E(\Delta\ddot{y})/[k/\sigma(\hat{X}_u)]$ q = 0.5	q = 0.8, a = 0.5	q = 0.8, a = 0.8
1	3	4	8	.2461	.1100	.0816
2	2	3	6	.2344	.1040	.0768
3	1	2	3	.1250	.0476	.0316
4	3	4	8	.1230	.0386	.0213
5	1	2	3	.0625	.0196	.0108
6	1	1	2	.0625	.0196	.0108
7	1	1	2	.0312	.0098	.0054
8	1	1	1	.0625	.0196	.0108

generations that would be grown per year and the magnitude of $\sigma(\hat{X})$ in the case of each system. However, the unadjusted values illustrate effects of variation in allele frequency (q) and the level of dominance (a).

(1) In the case of each selection system the magnitude of $E(\Delta\ddot{y})/[k/\sigma(\hat{X})]$ is a function of $q(1-q)$ and of $(1-2q)a$. Thus, unless $a<0.0$ and sufficiently large in absolute magnitude, $E(\Delta\ddot{y})$ must be smaller (as shown) when q is greater than 1/2. And because, whatever the value of a, $(1-2q)a = 0.0$ when $q = 1/2$, no specific value of a is indicated in the heading of the first of the of the last three columns of the table.

(2) When $q>1/2$, increases in a cause decreases in $(1-2q)a$. This is illustrated by the differences between values shown in the last two columns of the table.

(3) The ratios between quantities in the last three columns of the table are the same for systems 4 - 8 but differ from those for systems 1 - 3. This reflects the difference between $[1+(1-2q)a]$ and $[(1+h)+(1-h)(1-2q)/2]$.

The values shown in Table 14.13 were derived from those in Table 14.12 using L's listed in Table 14.12 and the $\sigma(\hat{X})$ values shown in Table 14.13 itself. For example, in the case of selection system 1, the expectation of $\Delta\ddot{y}$ when q = 0.5 and plant generations grown per year are two is $0.246/(4)(11.04) = .00557$. The $\sigma(\hat{X})$'s of Table 14.13 were obtained using 14.4 and the variance specifications presented in Table 14.4 (assuming v = 3 and t = r = 1). Because no account has been taken of selection intensity, the values shown in the table must be viewed as multiples of k or as the values to be expected given k = 1.[6]

Considered together the contents of Tables 14.12 and 14.13 refocus attention on the importance of L, $\sigma(\hat{X})$, a, and q. We note the following.

(1) It comes as no surprise that depending on number of generations that can be grown per year the relative utilities of systems, as judged by $E(\Delta\ddot{y})$, may change significantly. Note, for example, that under L(3) in Table 14.13 system 5 is indicated to be slightly superior to system 6, but under L(2) the indication is that system 6 is much better than system 5. The reason for this switch is, of course, that system 5 requires three generations per cycle compared to the two required by system 6.

Table 14.13. Expectation of $\Delta\ddot{y}$ per year, given that u = 0.5 and k = 1

Selection system	$\sigma(\hat{X})$*	L(3) q=.5	L(3) q=.8, a=.5	L(3) q=.8, a=.8	L(2) q=.5	L(2) q=.8, a=.5	L(2) q=.8, a=.8
1	11.04	.0074	.0033	.0025	.0056	.0025	.0018
2	10.80	.0109	.0048	.0036	.0072	.0032	.0024
3	8.82	.0142	.0054	.0036	.0071	.0027	.0018
4	7.20	.0057	.0018	.0010	.0043	.0013	.0007
5	6.54	.0096	.0030	.0016	.0048	.0015	.0008
6	7.22	.0086	.0027	.0015	.0086	.0027	.0015
7	6.53	.0048	.0015	.0008	.0048	.0015	.0008
8	17.30	.0036	.0011	.0006	.0036	.0011	.0006

* the $\sigma(\hat{X})$'s were obtained using 14.4 and the variance specifications presented in Table 14.4 (assuming v = 3 and t=r=1)

(2) Differences among systems change as level of dominance and frequency of the more favorable allele are increased. Compare, for example, the numbers in Table 14.13 for systems 1 and 5. Under L(3), $E(\Delta\ddot{y})$ for system 5 is 29 percent larger when q = 0.5 but 33 percent smaller when q = a = 0.8. In general, if there is a degree of dominance of better alleles, gains from selection among S_n families can be expected to hold up better, as allele frequencies are increased by selection, than gains by any of the other systems considered.

(3) Differences among systems in $\sigma(\hat{X})$ can be large enough to have considerable effects on $E(\Delta\ddot{y})$.

(4) The absolute, but not the relative, magnitudes of $E(\Delta\ddot{y})$ values shown in the table would have been different if the assumption about the magnitude of u had been different.

(5) Judged on the basis of $E(\Delta\ddot{y})$ values shown in Table 14.13, systems 2 and 3 are the best of the three in which selection is among S_n families and 5 and 6 are the best of the other five systems.

Clearly, any choice of selection system based on the kind of information presented in Table 14.13 should take account of (a) number of generations that can and would be grown per year (or other unit of time) and (b) available information pertaining to allele frequencies and levels of dominance in the populations in which selection would be practiced. Of less, but considerable, importance are reasonably good estimates of $\sigma(\hat{X})$ for the several systems.

14.2.1. Impact of Effective Population Size

Effects of N_e are considered in Chapter XI. Equation 11.26 pertains to single population systems in which the selection units are non- inbred individuals or families of non-inbred individuals, e.g., systems 4 - 8 of those listed in Table 14.3. In the case of systems 1 - 3 of that list in which the selection unit individuals are inbred, the applicable equation is 11.27. These two equations pertain to $E(\Delta\ddot{y})$, the expected change, in response to a single cycle of selection, in the average effect of genotypes at a single locus with respect to the selection criterion (X).

For convenience, 11.26 and 11.27, with minor rearrangements, are renumbered and shown below.[7]

$$E(\Delta\ddot{y})_1 = 2\bar{q}_1(1-\bar{q}_1)(1-1/2N_e)[1+(1-2\bar{q}_1)a]^2u^2cK - 2\bar{q}_1(1-\bar{q}_1)(Z/2N_e)u^2cK - \bar{q}_1(1-\bar{q}_1)au/N_e \qquad (11.26)14.10$$

in which

$Z \approx 4a\{(1-2\bar{q}_1) + [1-5\bar{q}_1(1-\bar{q}_1)]a\}$,

$\bar{q}_1$ = the expectation of q in cycle 1,

c is a quantity listed in Table 8.2 that is variable among selection systems, and

$K = k/\sigma(\hat{X})$.

$$E(\Delta\ddot{y})_1 = 2\bar{q}_1(1-\bar{q}_1)\{[(1+h)+(1-h)(1-2\bar{q}_1)a/2][1+(1-2\bar{q}_1)a](1-1/2N_e) - Z/2N_e\}u^2K - \bar{q}_1(1-\bar{q}_1)au/N_e \qquad (11.27)14.11$$

in which

$Z \approx a\{(3+h)(1-2\bar{q}_1) + 2(1-h)[1-5\bar{q}_1(1-\bar{q}_1)]a\}$

and $h = f_{n-1}$ = the inbreeding of the parents of the S_n families. Note the following.

(1) As N_e becomes large 14.10 and 14.11 approach the forms of 14.8 and 14.9, respectively. Stated differently, if N_e were infinite there would be no random sampling deviation of q_1 from $\bar{q}_1 = q_0 + E(\Delta q)_0$. Consequently, there would be no inbreeding effect and the difference between $E(\Delta\ddot{y})_0$ and $E(\Delta\ddot{y})_1$ would depend solely on the differencee between q_0 and $q_1 = q_0 + E(\Delta q)_0$.

(2) When N_e is finite the first two terms on the right sides of 14.10 and 14.11 are both directly proportional to k, the standardized selection differential. Taken together they constitute the expectation of response due to selection.

(3) The final term on the right in both 14.10 and 14.11 is <u>not</u> a function of the selection practiced. It is the contribution of the single locus to the total inbreeding effect.[8]

(4) When a = 0, the second and third right hand terms go to zero. Then since the first term varies directly with $(1-1/2N_e)k$, it is obvious that there must be an optimum selection intensity. Whatever the size of the genetic sample, k is increased and N_e decreased when selection intensity is increased (fewer units selected). The reverse results when a larger fraction of the genetic sample is selected.

When a≠ 0, the necessary consequences of increasing or decreasing selection intensity are not as obvious as when a = 0. However, barring overdominance (a>1), it is reasonable to suppose that the sum of the first two right hand terms must always be positive. Then, given a>0, the situation is qualitively the same as when a = 0 and hence there must be an optimum selection intensity.

The net effect of change in selection intensity is more obscure when a < 0 so that the inbreeding effect is positive and larger when N_e is smaller. However, this is not a very significant issue because inbreeding effects on value traits are in general negative, not positive.

(5) It is apparent from 14.10 and 14.11 that the optimum selection intensity is a function of q, a, and u. Not quite as obvious, it is also a function of T (the size of the genetic sample) because N_e varies with both T and p.

Though an explicit algebraic expression for optimum selection intensity cannot be obtained from either 14.10 or 14.11, approximations can be produced by iteration.[9] To facilitate computations (without the aid of a high speed computer) and the understanding of results, numerical values of two components of 14.10 and 14.11 corresponding to specific values of $\bar{q}_1$ and a are presented in Table 14.14.

Among the questions that come to attention when effective population size is considered are (a) what are optimum selection intensities for specific selection systems, (b) would choice

Table 14.14. Components of $E(\Delta\bar{y})$ that are functions of both a and $\bar{q}_1$

(a) When T = S and selection is among S_4 families on the basis of their own performance

Values of $[1+h+(1-h)(1-2\bar{q}_1)a/2][1+(1-2\bar{q}_1)a]$

$\bar{q}_1$	a = 0	a = 0.2	a = 0.5	a = 0.8	a = 1
.2	1.875	2.11	2.46	2.82	3.10
.5	1.875	1.88	1.88	1.88	1.88
.65	1.875	1.76	1.58	1.41	1.30
.8	1.875	1.64	1.30	0.96	0.73

Values of $Z \approx a\{(3.875)(1-2\bar{q}_1) + 2(.125)[1-5\bar{q}_1(1-\bar{q}_1)]a\}$

.2	0.00	0.47	1.18	1.89	2.38
.5	0.00	-0.01	-0.02	-0.04	-0.06
.65	0.00	-0.23	-0.59	-0.95	-1.12
.8	0.00	-0.46	-1.15	-1.83	-2.28

(b) When T = S and selection is among S_1 families on the basis of their own performance.

Values of $[1+h+(1-h)(1-2\bar{q}_1)a/2][1+(1-2\bar{q}_1)a\}$

$\bar{q}_1$	a = 0	a = 0.2	a = 0.5	a = 0.8	a = 1
.2	1.00	1.19	1.50	1.84	2.08
.5	1.00	1.00	1.00	1.00	1.00
.8	1.00	0.83	0.60	0.40	0.28

Values of $Z \approx a\{3(1-2\bar{q}_1) + 2[1-5\bar{q}_1(1-\bar{q}_1)]a\}$

.2	0.00	0.37	1.00	1.69	2.20
.5	0.00	-0.02	-0.13	-0.32	-0.50
.8	0.00	-0.34	-0.80	-1.18	-1.40

(c) When T = S and criterion units are composed of non-inbred S population individuals.

Values of $[1+(1-2\bar{q}_1)a]^2$

$\bar{q}_1$	a = 0	a = 0.2	a = 0.5	a = 0.8	a = 1
.2	1.00	1.25	1.69	2.19	2.56
.5	1.00	1.00	1.00	1.00	1.00
.65	1.00	0.88	0.72	0.58	0.49
.8	1.00	0.77	0.49	0.27	0.16

Values of $Z \approx 4a\{(1-2\bar{q}_1)+[1-5\bar{q}_1(1-\bar{q}_1)]a\}$

.2	0.00	0.51	1.40	2.43	3.20
.5	0.00	-0.04	-0.25	-0.64	-1.00
.65	0.00	-0.26	-0.74	-1.31	-1.75
.8	0.00	-0.45	-1.00	-1.41	-1.60

of selection system be different if based on $E(\Delta\ddot{y})$'s that take proper account of effective population size (when selection intensities are near optimum for each system considered) than if based on $E(\Delta\ddot{y})$'s given by 14.8 and 14.9, and (c) what are the magnitudes of upward bias in values for $E(\ddot{y})$ when obtained without attention to effective population size? The first of these questions is explored via 14.10 and 14.11 in terms of numerical values of $E(\Delta\ddot{y})_1$ shown in Tables 14.15 - 14.18 and the second and third in Table 14.19 in which values provided by 14.10 and 14.11 are compared with values provided by 14.8 and 14.9.

Tables 14.15 - 14.18 have to do with selection systems 2, 3, 5, and 6 (but are not designed for comparison of the systems - values of $E(\Delta\ddot{y})$ per cycle, not per year, are shown). On the evidence in Table 14.13 the first pair of these appears to be the most promising of the S_n family systems and the final pair the most promising of the systems in which criterion units consist of non-inbred individuals.

Table 14.15. $E(\Delta\ddot{y})_1$ per cycle and optimum selection intensity as functions of a and $\bar{q}_1$ (system 2—selection among S_4 families)

					a=1/2		a=0.8	
p	k	M	N_e	k/σ	$\bar{q}_1=.5$	$\bar{q}_1=.8$	$\bar{q}_1=.5$	$\bar{q}_1=.8$
(a) $T = 1200, v = 1, \sigma(\hat{X}) = 15.2, u = 1/2, c = 1$								
.01	2.66	12	6.4	.175	.0282	.0118	.0223	.0044
.02	2.42	24	12.8	.159	.0310[a]	.0133	.0281	.0076
.03	2.27	36	19.2	.149	.0309	.0134[a]	.0289[a]	.0084
.04	2.16	48	25.6	.142	.0303	.0132	.0288	.0086[a]
.06	2.00	72	38.4	.132	.0290	.0125	.0280	.0085
.08	1.86	96	51.2	.122	.0272	.0119	.0264	.0082
(b) $T = 400, v = 3, \sigma(\hat{X}) = 10.8, u = 1/2, c = 1$								
.02	2.42	8	4.3	.224	.0320	.0137	.0234	.0041
.04	2.16	16	8.5	.200	.0369[a]	.0160[a]	.0325	.0087
.06	2.00	24	12.8	.185	.0369	.0160	.0340[a]	.0097
.07	1.92	28	14.9	.178	.0362	.0158	.0338	.0098[a]
.08	1.86	32	17.1	.172	.0356	.0155	.0334	.0098
.14	1.59	56	29.9	.147	.0319	.0139	.0306	.0093
(c) $T = 200, v = 6, \sigma(\hat{X}) = 9.35, u = 1/2, c = 1$								
.04	2.16	8	4.3	.231	.0335	.0144	.0249	.0047
.08	1.86	16	8.5	.199	.0367[a]	.0158	.0327	.0086
.09	1.81	18	9.6	.194	.0367	.0159[a]	.0329[a]	.0089
.10	1.76	20	10.7	.188	.0363	.0157	.0328	.0091[a]
.16	1.52	32	17.1	.163	.0335	.0146	.0314	.0091
.20	1.40	40	21.3	.150	.0315	.0137	.0297	.0088
(d) $T = 100, v = 12, \sigma(\hat{X}) = 8.55, u = 1/2, c = 1$								
.07	1.92	7	3.7	.225	.0289	.0122	.0189	.0021
.08	1.86	8	4.3	.218	.0308	.0131	.0222	.0036
.13	1.63	13	6.9	.191	.0326[a]	.0139[a]	.0272	.0063
.15	1.56	15	8.0	.182	.0323	.0138	.0277[a]	.0068
.16	1.52	16	8.5	.178	.0320	.0137	.0277	.0069[a]
.28	1.20	28	14.9	.140	.0276	.0118	.0251	.0068
.32	1.12	32	17.1	.131	.0262	.0112	.0241	.0066

[a] Near optimum for the particular combination of a and $\bar{q}_1$.

Table 14.16. $E(\Delta\bar{y})_1$ per cycle and optimum selection intensity as functions of a and $\bar{q}_1$ (System 3—selection among S_1 families)

					a=1/2		a=0.8	
p	k	M	N_e	k/σ	$\bar{q}_1$=.5	$\bar{q}_1$=.8	$\bar{q}_1$=.5	$\bar{q}_1$=.8
(a) T = 1200, v = 1, $\sigma(\hat{X})$ = 13.1, u = 1/2, c = 1								
.01	2.66	12	11.4	.203	.0189	.0064	.0158	.0014
.02	2.42	24	22.8	.185	.0199[a]	.0072[a]	.0184	.0033
.03	2.27	36	34.3	.173	.0195	.0072	.0185[a]	.0038
.04	2.16	48	45.7	.165	.0191	.0071	.0183	.0040[a]
.06	2.00	72	68.5	.153	.0181	.0068	.0176	.0040
.08	1.86	96	91.4	.142	.0170	.0064	.0166	.0039
(b) T = 400, v = 3, $\sigma(\hat{X})$ = 8.82, u = 1/2, c = 1								
.02	2.42	8	7.6	.274	.0240	.0082	.0196	.0014
.04	2.16	16	15.2	.245	.0256[a]	.0093[a]	.0233	.0041
.06	2.00	24	22.8	.227	.0251	.0092	.0236[a]	.0047
.08	1.86	32	30.5	.211	.0239	.0089	.0228	.0049[a]
.12	1.67	48	45.7	.189	.0220	.0082	.0213	.0048
.16	1.52	64	60.9	.172	.0203	.0076	.0197	.0045
(c) T = 200, v = 6, $\sigma(\hat{X})$ = 7.38, u = 1/2, c = 1								
.03	2.27	6	5.7	.308	.0246	.0082	.0168	.0003
.06	2.00	12	11.4	.271	.0271[a]	.0097[a]	.0241	.0038
.07	1.92	14	13.3	.260	.0267	.0096	.0242[a]	.0041
.12	1.67	24	22.8	.226	.0250	.0092	.0235	.0047[a]
.16	1.52	32	30.5	.206	.0233	.0086	.0222	.0047
.24	1.30	48	45.7	.176	.0204	.0076	.0196	.0044
(d) T = 100, v = 12, $\sigma(\hat{X})$ = 6.55, u = 1/2, c = 1								
.05	2.06	5	4.8	.315	.0227	.0073	.0158	-.0012
.09	1.81	9	8.6	.276	.0255[a]	.0089	.0215	.0024
.11	1.71	11	10.7	.261	.0254	.0090[a]	.0222[a]	.0032
.24	1.30	24	22.8	.198	.0215	.0078	.0200	.0038[a]
.40	.97	40	38.1	.178	.0166	.0061	.0157	.0032

[a] Near optimum for the particular combination of a and $\bar{q}_1$

Selection intensities for which values of $E(\Delta\bar{y})_1$ are shown vary among the four tables. They were selected to include (1) a near optimum intensity, obtained by iteration, for each q,a combination and (2) those approximately half as large and twice as large for each q,a combination. The four sections of each table have to do with the effect of the magnitude of $\sigma(\hat{X})$, assuming the total number of plots involved in the criterion unit comparison to be constant. As in Table 14.13, values of $\sigma(\hat{X})$ were obtained using 14.4 and variance specifications shown in Table 14.4. Throughout, t= r= 1 was assumed, i.e., that the comparison of each cycle was to be completed in just one year with only one replication at each location. As indicated by the magnitudes of v assumed, $E(\Delta\bar{y})_1$ values are shown for comparisons involving 1, 3, 6, or 12 plots per criterion unit.

The quantities shown in these tables are conditioned by a specific set of assumptions and hence must be interpreted with care. However, the following merit attention.

(1) Because the relation between the number of selections (M) and effective population

Table 14.17. E($\Delta\bar{y}$), per cycle and optimum selection intensity as functions of a and $\bar{q}_1$ (system 5—selection among S_1 families based on the performance of test cross progenies of S_0 parents)

p	k	M	N_e	k/σ	a=1/2 $\bar{q}_1$=.5	a=1/2 $\bar{q}_1$=.8	a=0.8 $\bar{q}_1$=.5	a=0.8 $\bar{q}_1$=.8
(a) T = 1200, v = 1, $\sigma(\hat{X})$ = 10.6, u = 1/2, c = 1/2								
.015	2.53	18	17.1	.239	.0110	.0025	.0089	-.0008
.03	2.27	36	34.3	.214	.0114[a]	.0031	.0104	.0006
.04	2.16	48	45.7	.204	.0113	.0032[a]	.0105[a]	.0009
.06	1.99	72	68.5	.189	.0108	.0031	.0103	.0012
.08	1.86	96	91.4	.175	.0102	.0030	.0098	.0012
.12	1.67	144	137.1	.158	.0094	.0028	.0091	.0013[a]
.24	1.30	288	274.2	.123	.0074	.0021	.0073	.0011
(b) T = 400, v = 3, $\sigma(\hat{X})$ = 6.54, u = 1/2, c = 1/2								
.03	2.27	12	11.4	.347	.0155	.0036	.0126	-.0011
.06	2.00	24	22.8	.306	.0161[a]	.0044[a]	.0146[a]	.0008
.12	1.67	48	45.7	.255	.0144	.0042	.0137	.0015
.16	1.52	64	60.9	.232	.0134	.0039	.0128	.0015[a]
.32	1.12	128	121.9	.171	.0101	.0030	.0099	.0014
(c) T = 200, v = 6, $\sigma(\hat{X})$ = 5.03, u = 1/2, c = 1/2								
.03	2.27	6	5.7	.451	.0154	.0026	.0097	-.0045
.05	2.06	10	9.5	.410	.0180	.0043	.0147	-.0013
.06	2.00	12	11.4	.390	.0166[a]	.0046	.0157	-.0005
.10	1.76	20	19.0	.350	.0182	.0049[a]	.0164[a]	.0008
.12	1.67	24	22.8	.332	.0177	.0049	.0162	.0011
.24	1.30	48	45.7	.258	.0146	.0042	.0139	.0015[a]
.48	.83	96	91.4	.165	.0096	.0028	.0092	.0011
(d) T = 100, v = 12, $\sigma(\hat{X})$ = 4.07, u = 1/2, c = 1/2								
.05	2.06	5	4.8	.506	.0161	.0027	.0096	-.0055
.07	1.92	7	6.7	.472	.0185	.0040	.0138	-.0028
.10	1.76	10	9.5	.432	.0194[a]	.0047	.0160	-.0010
.14	1.59	14	13.3	.391	.0190	.0050[a]	.0166[a]	.0001
.32	1.12	32	30.5	.275	.0149	.0042	.0138	.0011[a]
.60	.64	60	57.1	.157	.0087	.0024	.0080	.0006

[a] Near optimum for the particular combination of a and $\bar{q}_1$.

size varies among the four selection systems (see Table 14.5) it comes as no surprise that optimum selection intensities vary among the selection systems. When, for example, we consider systems 2 and 6, we find (a) that when selection intensity is the same N_e is about four times as large in the case of system 6 and (b) that optimum selection intensity is greater (optimum p smaller) for system 6 except in the case of q = a = 0.8. That exception is explained when we note that the ratio of the $[1+(1-2\bar{q}_1)a]^2$ component of 14.10 to the $[(1+h)+(1-h)(1-2\bar{q}_1)a/2][1+(1-2\bar{q}_1)a]$ component of14.11 is smallest in the $\bar{q}_1$ = a= 0.8 situation (see Table 14.14).

(2) When decreases in $\sigma(\hat{X})$ are obtained at the expense of size of the genetic sample, T, there is (must be) an optimum in the balance between T and the number of plots per criterion unit. This fact is demonstrated but the evidence in these tables indicates that the expectations of

Table 14.18. $E(\Delta\ddot{y})_1$ and optimum selection intensity as functions of a and $\bar{q}_1$ (system 6—selection among full-sib families)

p	k	M	N_e	k/σ	a=1/2 $\bar{q}_1$=.5	a=1/2 $\bar{q}_1$=.8	a=0.8 $\bar{q}_1$=.5	a=0.8 $\bar{q}_1$=.8
(a) T = 1200, v = 1, $\sigma(\hat{X})$ = 11.20, u = 1/2, c = 1/2								
.01	2.66	12	24.1	.238	.0121	.0031	.0106	.0001
.015	2.53	18	36.3	.226	.0123*	.0034	.0112	.0008
.02	2.42	24	48.5	.216	.0121	.0035*	.0114*	.0011
.03	2.27	36	73.1	.203	.0118	.0035	.0113	.0014
.04	2.16	48	98	.193	.0114	.0034	.0110	.0015
.06	1.99	72	148.5	.178	.0108	.0032	.0101	.0015*
.12	1.67	144	306.4	.149	.0091	.0028	.0090	.0012
(b) T = 400, v = 3, $\sigma(\hat{X})$ = 7.22, u = 1/2, c = 1/2								
.015	2.53	6	12.1	.350	.0160	.0038	.0133	-.0008
.03	2.27	12	24.4	.314	.0168*	.0046	.0154	.0010
.04	2.16	16	32.7	.299	.0166	.0047	.0155*	.0015
.05	2.06	20	41.0	.285	.0161	.0047*	.0153	.0017
.06	1.99	24	49.5	.276	.0159	.0047	.0152	.0018
.08	1.86	32	66.7	.258	.0151	.0045	.0146	.0019
.10	1.76	40	84.2	.244	.0144	.0043	.0140	.0019*
.20	1.40	80	177.8	.194	.0117	.0036	.0121	.0018
(c) T = 200, v = 6, $\sigma(\hat{X})$ = 5.82, u = 1/2, c = 1/2								
.02	2.42	4	8.1	.416	.0171	.0037	.0131	-.0022
.04	2.16	8	16.3	.371	.0188*	.0050	.0168	.0006
.06	1.99	12	24.7	.344	.0186	.0053*	.0173*	.0014
.08	1.86	16	33.3	.320	.0179	.0052	.0169	.0018
.12	1.67	24	51.1	.287	.0166	.0049	.0159	.0020
.14	1.59	28	60.2	.273	.0159	.0047	.0154	.0020*
.28	1.20	56	130.2	.206	.0124	.0037	.0121	.0018
(d) T = 100, v = 12, $\sigma(\hat{X})$ = 4.97, u = 1/2, c = 1/2								
.03	2.27	3	6.1	.457	.0166	.0032	.0113	-.0038
.06	1.99	6	12.4	.400	.0192*	.0049	.0166	-.0001
.08	1.86	8	16.7	.374	.0191	.0052*	.0171	.0007
.09	1.81	9	18.8	.364	.0190	.0052	.0172*	.0010
.14	1.59	14	30.1	.320	.0177	.0051	.0166	.0016
.20	1.40	20	44.4	.282	.0161	.0047	.0153	.0017*
.40	.97	40	100.0	.195	.0115	.0034	.0112	.0015

*Near optimum for the particular combination of a and $\bar{q}_1$.

$\Delta\ddot{y}$ at or reasonably near to optimum selection intensities are not greatly affected by variation in the ratio of T to the total number of plots employed when the latter is held constant. However, $E(\Delta\ddot{y})$ at near optimum selection intensity is in almost all cases smallest when T= 1200 and v = 1.

(3) Both optimum selection intensity and $E(\Delta\ddot{y})$ are less when level of dominance and /or allele frequencies are greater. These differences are more pronounced in the case of systems 5 and 6.

(4) Optimum selection intensities in S_4 and S_1 family selection are less affected by variation of allele frequency and level of dominance than is the case for systems 5 and 6. Moreover, the impact of selection intensity on $E(\Delta\ddot{y})$ is less, particularly at higher values of q and a, in the case of S_4 and S_1 family selection. This is a definite advantage because in practice

insufficient information regarding the distributions of q's, a's, and u's means that only rough approximations to optimum selection intensities are possible.

(5) For the most part, decreases in $E(\Delta\ddot{y})$ values as selection intensities deviate from optimum are not drastic. The largest exceptions are to be seen in Tables 14.17 and 14.18 (selection systems 5 and 6) in the case of a = q = 0.8 and selection intensity greater than the optima for the other a,q combinations. For example, at the near optimum selection intensity when a = q = 0.5 we note that $E(\Delta\ddot{y})$ when a = q = 0.8 ranges from half its value at optimum selection intensity down to small negative values.

This observation seems important; it indicates that the selection intensity chosen in practice may deviate quite a bit from what would be optimum without major loss in response to selection. At the same time it must be remembered that optimum selection intensity is strongly affected by the magnitude of $\sigma(\hat{X})$ when the total number of plots in the criterion unit comparisons is constant (because when T is less N is less if p is the same)-- compare parts a, b, c, and d of any one of the four tables.

The appropriate conclusion is that for any specific value of $\sigma(\hat{X})$, the selection fraction (p) to be employed can deviate from optimum by as much as 50 percent on the downside or 100 percent on the upside without large loss in response to selection if the ranges of q, a, and u values that have been assumed are reasonable.[10]

Table 14.19 contrasts expectations of $(\Delta\ddot{y})_1$ obtained when effective population size is taken into account and when it is not. Quantities in part (a) are derived from the $E(\Delta\ddot{y})_1$ values at near optimum selection intensities shown in part (b) of Tables 14.15 - 14.18.[11] Those values pertained to response per cycle; hence, for system comparisons each had to be divided by the appropriate cycle length shown in Table 14.12. Thus the .0184 for system 2 in the first column under L(3) is the .0369 of part (b) in Table 14.15 divided by two and the four values under L(2) for system 5 are the four "near optimum" values in part (b) of Table 14.17 divided again by two. The quantities shown in part (b) of Table 14.19 were obtained from Table 14.13. Since those assumed k = 1, each was multiplied by the k expected given the selection intensity associated with the corresponding value in part (a) of Table 14.19. For example, in the case of system 3 (Table 14.16), the near optimum selection intensity for the a = .8, q = .8 situation was p = .08 with k = 1.86. Hence, the .0036 and .0018 shown in Table 14.13 were each multiplied by 1.86 to yield the .0067 and .0033 shown in part (b) of Table 14.19. Values of $E(\Delta\ddot{y})$ in part (b) were computed using 14.9 in the case of systems 2 and 3--14.8 in the case of systems 5 and 6. Values of q assumed in part (b) paralleled those of $\bar{q}_1$ in part (a).

Information is provided relative to questions posed earlier. First, comparing systems in terms of quantities shown, essentially the same answers are obtained whether part (a) or part (b) is employed. Stated differently, the evidence in this table provides no reason to believe that, for practical purposes, comparing selection systems for use in short term single population programs is improved by taking effective population size into account and comparing on the basis of expected responses given that each system would be conducted using its own optimum selection intensity.

Table 14.19. Inbreeding cost measured by E(Δÿ) per year

	L(3)				L(2)			
	a = .5		a = .8		a = .5		a = .8	
System	$\bar{q}_1$ = .5	$\bar{q}_1$ = .8	$\bar{q}_1$ = .5	$\bar{q}_1$ = .8	$\bar{q}_1$ = .5	$\bar{q}_1$ = .8	$\bar{q}_1$ = .5	$\bar{q}_1$ = .8
(a) N_e considered								
2	.0184	.0080	.0170	.0049	.0123	.0053	.0113	.0033
3	.0256	.0093	.0236	.0049	.0128	.0046	.0118	.0024
5	.0161	.0044	.0146	.0015	.0080	.0022	.0073	.0008
6	.0168	.0047	.0155	.0019	.0168	.0047	.0155	.0019
(b) N_e ignored								
	q = .5	q = .8	q = .5	q = .8	q = .5	q = .8	q = .5	q = .8
2	.0235	.0104	.0217	.0068	.0156	.0069	.0145	.0046
3	.0306	.0108	.0284	.0067	.0153	.0054	.0142	.0033
5	.0191	.0060	.0191	.0023	.0095	.0030	.0095	.0012
6	.0196	.0056	.0187	.0026	.0196	.0056	.0187	.0026
(c) Ratio of values obtained when N_e is ignored to those when N_e is considered								
2	1.28	1.30	1.28	1.39	1.27	1.30	1.28	1.39
3	1.20	1.16	1.20	1.37	1.20	1.17	1.20	1.38
5	1.19	1.36	1.31	1.53	1.19	1.36	1.30	1.50
6	1.17	1.19	1.21	1.37	1.17	1.19	1.21	1.37

Note: $\sigma(\hat{X})$ for each system was that shown in Table 14.13 and in Part b of Tables 14.15 - 14.18.

Second, when quantities in parts (a) and (b) that pertain to the same selection system and a,q combination are compared, those in part (a) are always smaller. This decrement is the consequence of finite effective population size. In Table 14.19 it ranges from 14 to 27 percent when q and/or a = 0.5 but is greater, ranging from 28 to 40 percent, when q = a = 0.8. Remember, however, that these quantities are specific to the set of magnitudes assumed for relevant parameters [T, a, u, and $\sigma(\hat{X})$] and to the actual selection intensity. It is apparent, for example, that if the fraction selected is less than optimum, k will be larger and E(Δÿ), computed without attention to effects of N_e, will be larger while the actual E(Δÿ) will be less than when selection intensity is optimum, i.e., the decrement due to finite N_e will be larger than when selection intensity is optimum.

- 14.2.2. **Variations of Parameter Magnitudes**

Size of program (Tv): When the number (T) of units in the genetic sample is greater any specific effective population size can be achieved with greater selection intensity. It is obvious therefore that optimum p's will be less and that larger values of E(Δÿ) should be possible when T is greater. Table 14.20 provides insights concerning quantitative relationships in the case of systems 2 and 6.

These two were chosen because in one case the selection unit families are produced by self-fertilizations and in the other by matings between random non-inbred individuals. As in the case of Tables 14.15 to 14.18, four levels of $\sigma(\hat{X})$ achieved by four assumptions concerning number of locations at which criterion units are to be compared were employed. Values of

Table 14.20. $E(\Delta\ddot{y})_1$ per cycle and optimum selection intensity as functions of a, q, and Tv (when u = 0.5)

		Tv = 1200					Tv = 2400				
					a = .5	a = .8				a = .5	a = .8
v	$\sigma(\hat{X})$	p	N_e	k/σ	q = .5	q = .8	p	N_e	k/σ	q = .5	q = .8
(a) System 2 (selection among S_4 families)											
1	15.2	.02	12.8	.159	.0310[a]	.0076	.01	12.8	.175	.0347[a]	.0089
		.04	25.6	.142	.0303	.0086[a]	.02	25.6	.159	.0342	.0099[a]
3	10.8	.04	8.5	.200	.0369[a]	.0087	.03	12.8	.210	.0426[a]	.0117
		.07	14.9	.178	.0362	.0098[a]	.04	17.1	.200	.0420	.0120[a]
6	9.35	.08	8.5	.199	.0367[a]	.0086	.04	8.5	.231	.0438[a]	.0112
		.10	10.7	.188	.0363	.0091[a]	.07	14.9	.205	.0424	.0119[a]
12	8.55	.13	6.9	.191	.0326[a]	.0063	.07	7.5	.225	.0411[a]	.0098
		.16	8.5	.178	.0320	.0069[a]	.10	10.7	.206	.0403	.0105[a]
(b) System 6 (selection among full-sib families)											
1	11.20	.015	36.3	.226	.0123[a]	.0008	.01	48.2	.238	.0135[a]	.0014
		.06	148.5	.178	.0108	.0015[a]	.03	146.2	.203	.0122	.0018[a]
3	7.22	.03	24.4	.314	.0168[a]	.0010	.02	32.3	.335	.0188[a]	.0019
		.10	84.2	.244	.0144	.0019[a]	.06	99.0	.276	.0166	.0024[a]
6	5.82	.04	16.3	.371	.0188[a]	.0006	.02	16.2	.416	.0215[a]	.0011
		.14	60.2	.273	.0159	.0020[a]	.08	66.7	.320	.0190	.0026[a]
12	4.97	.06	12.4	.400	.0192[a]	-.0001	.03	12.2	.451	.0226[a]	.0005
		.20	44.4	.282	.0161	.0017[a]	.12	51.1	.336	.0196	.0025

Note: v= number of locations at which criterion unit comparison is conducted.

[a]The near-maximum value of $E(\Delta\ddot{y})_1$ for the (a,q,v,Tv) combination in question.

$E(\Delta\ddot{y})$ are presented for two values of Tv (1200 and 2400) and for two a,q combinations (a = q = 0.5 and a = q = 0.8). The asterisks indicate the near-maximum values of $E(\Delta\ddot{y})_1$ for each (a,q,v,Tv) combination of parameters. The following aspects of the quantities shown in the table are all of some interest.

(1) As expected, when other things are the same, $E(\Delta\ddot{y})_1$ is always larger when Tv = 2400 instead of 1200. The differences range from 10 to 26 percent in the case of a=q=0.5 genes but are larger in the case of a=q=0.8 genes, reaching 52 percent when selection intensity

is near optimum for that class of genes and 90 percent at the selection intensity that is optimum for the a=q=0.5 class of genes.

(2) Again as expected, optimum selection intensities are always greater (p less) when Tv = 2400 instead of 1200, other things being equal. The reason of course is that with genetic sample size doubled, adequate effective population sizes are obtained with smaller values of p.

Note that the near-optimum p values shown tend to be half as large when Tv = 2400 instead of 1200 and that the associated values of N_e are more often than not nearly equal.

Magnitude of u: Qualitatively, the impact of the size of u is apparent from equations 14.10 and 14.11. In each of those expressions the first two terms (which together measure the positive effect of selection) are multiples of u^2 while the final term, which quantifies inbreeding depression, is a multiple of u. It follows that when u is larger, other things being equal, the inbreeding effect must be smaller relative to the positive effect of selection -- and *vice versa* when u is smaller. However, the net impact of the magnitude of u, being a function of other variables [q, a, N_e, and $k/\sigma(\hat{X})$], is not so readily apparent. Table 14.21 lists values of $E(\Delta\ddot{y})_1$ obtained for selection systems 2 and 6 using 14.10 and 14.11. Results are shown for two classes of genes (defined by the magnitudes of a and q), three genetic sample magnitudes (T = 100, 200, and 400) and three values of u (0.25, 0.50, 1.00). The values used for $\sigma(\hat{X})$ were those used in previous tables when the assumed parameters of the criterion unit comparisons were v = 6 and t = r = 1. For each selection system and each combination of T and u two sets of $E(\Delta\ddot{y})_1$ values are presented. The first in each case are the values associated with the near optimum selection intensity for the a = q = 0.5 genes and the second the values associated with the near optimum selection intensity for a = q = 0.8 genes. Asterisks identify values associated with the near optimum selection intensity for the indicated class of genes. In three of the 18 combinations of variables the near optimum value of p was the same for both of the two classes of genes. In all other cases the optimum selection intensity, in accord with content of Tables 14.15 to 14.18 was greater (p less) for a = q = 0.5 genes than for a= q = 0.8 genes. Aspects of the content of the table that merit comment are the following.

(1) As reasonable to expect, optimum selection intensity is greater when u is greater and when T is greater. Reasons are these: (a) both $E(\Delta q)$ and $\partial\ddot{y}/\partial q$ are directly proportional to u and (b) the inbreeding effect is inversely proportional to N_e which in turn is directly proportional to T when p is constant.

(2) Of special interest are the differences in $E(\Delta\ddot{y})$ associated with the different values of u. If there were no effect of N_e and p were the same throughout, it is apparent from the first terms of 14.10 and 14.11 that $E(\Delta\ddot{y})$ would change as the square of u, e.g., its value given u =1.0 would be 16 times its value given u = 0.25. However, because the inbreeding effect (proportional to u, not u^2) is negative (when a > 0), change in $E(\Delta\ddot{y})$ was in each case greater than the underlying change in u^2. For example, in the T = 100 and System 2 section of the table the ratio of $E(\Delta\ddot{y})$ values given near optimum selection intensities for a = q =0.5 in the cases of

Table 14.21. $E(\Delta\ddot{y})_1$ per cycle and optimum selection intensity as functions of u, T, a, and q

		System 2 (S_4 Families)					System 6 (full-sib families)				
					a = .5	a = .8				a = .5	a = .8
T	u	p	N_e	k/σ	q = .5	q = .8	p	N_e	k/σ	q = .5	q = .8
100	.25	.20	10.7	.150	.0055[a]	.0000	.12	25.5	.287	.0032[a]	-.0004
		.40	21.3	.104	.0045	.0005[a]	.40	100.0	.167	.0023	.0001[a]
	.50	.12	6.4	.179	.0290[a]	.0047	.08	16.7	.320	.0158[a]	.0001
		.24	12.8	.139	.0265	.0061[a]	.24	54.5	.223	.0127	.0013[a]
	1.00	.09	4.8	.194	.1375[a]	.0386	.04	8.4	.371	.0737[a]	.0249[a]
		.12	6.4	.179	.1357	.0389[a]					
200	.25	.12	12.8	.179	.0077[a]	.0012	.07	29.0	.330	.0040[a]	-.0002
		.24	25.6	.139	.0068	.0015[a]	.26	119.5	.215	.0031	.0003[a]
	.50	.08	8.5	.199	.0367[a]	.0086	.04	16.3	.371	.0188[a]	.0006
		.10	10.7	.188	.0363	.0091[a]	.14	60.2	.273	.0159	.0020[a]
	1.00	.06	6.4	.213	.1652[a]	.0501[a]	.025	10.1	.402	.0844[a]	.0083
							.07	29.0	.330	0771	.0109[a]
400	.25	.06	12.8	.213	0096[a]	.0018	.04	32.7	.371	.0048[a]	-.0001
		.12	25.6	.179	.0091	.0022[a]	.15	129.7	.268	.0039	.0005[a]
	.50	.04	8.5	.231	.0438[a]	.0112	.02	16.2	.416	.0215[a]	.0011
		.07	14.9	.205	.0424	.0119[a]	.08	66.7	.320	.0190	.0026[a]
	1.00	.03	6.4	.243	.1912[a]	.0599[a]	.015	12.1	.434	.0948[a]	.0114
							.04	32.7	.371	.0879	.0132[a]

Note: v = 6 and t = r = 1 were assumed in computation of all $E(\Delta\ddot{y})$ values.

$\sigma = \sigma(\hat{X})$ = 9.35 and 5.82 were used for systems 2 and 6, respectively (see part c of Tables 14.15 and 14.18).

[a]The near maximum value of $E(\Delta\ddot{y})_1$ for the (u,T, a, q) combination in question.

u = 1.0 and u = 0.25 is 0.1375/ 0.0055 = 25.0. Because, as noted earlier, the absolute value of the inbreeding effect relative to that of selection becomes larger when u is smaller, there are instances in which $E(\Delta\ddot{y})$ is negative. Logically, these are associated with the smallest value of u and non-optimum selection intensities.

(3) The inbreeding component of the expectation of change in $\ddot{y}$ constitutes one reason to expect that realized heritability in programs of recurrent selection for even slightly heterotic traits should in general be less than the ratio of additive genetic variance to phenotypic variance. Moreover, because, other things being equal, the magnitude of the inbreeding effect is larger relative to the expected response to selection when u is smaller, the difference between the two types of estimates would be greater when additive genetic variance arises from genes of small effect (small u values).

• 14.3. Short Term, Two Population Programs in Which $T = S_1 \times S_2$

The focus in this section is on the relative effectiveness of one-population and two-population programs. The latter have the advantage noted in section 11.3.2, that $E(\Delta\ddot{y})$, and hence $E(\Delta\ddot{Y})$, is not subject to an inbreeding effect. However, because they consist of two selection programs the genetic samples of those programs, if equal in size and tested in the same way, can only be half as large as in a single-population program if the total effort is to be the same. For example, if 200 selection units are to be observed in a single-population program or the same number in a two-population program and ten selections are to be employed in the single-population program and in each population of the two-population program, E(k) would be 2.06 in the single-population program and 1.76 in the case of each population of the two-population program. Assuming $E(\Delta\ddot{Y})$ proportional to k, as it would be if there were no impact of effective population size, the expected response to selection would be 2.06/1.76 = 1.136 times as large in the one-population program. How would the systems compare assuming other numbers of selection units and taking the consequences of effective population size into account?

The two-population systems that are considered are the RSP and RRS systems described and discussed in Chapter IX. The two specific issues to be addressed are these: (a) when and why is a T= S (one-population) program superior to an RSP program or vice versa, and(b) in the absence of overdominance is RSP superior to RRS?

The first of these questions arises by virtue of the genetic sample size advantage of T= S programs and the no inbreeding effect advantage of RSP programs. The second question is significant because RRS is attractive when the selection criterion is heterotic and hence overdominance of some of the genes that affect it is a possibility though not a certainty.

• 14.3.1. Single Population (T= S) vs. RSP Programs

This comparison requires assumptions concerning allele frequencies and size of genetic samples that have not been needed in previous sections of this chapter.
Because $E(\Delta\ddot{y})$ is always a function of allele frequency in the population or populations involved assumptions must obviously be made concerning source materials. One or the other of the following logical assumptions will be employed.

A. That the source populations of the two-population program consist of large random samples from the random mating source population employed in cycle one of the single- population program. This justifies the secondary assumption that at every locus initial allele frequencies will be the same in all three populations of the two programs.

B. That the source populations employed in the two-population program differ in their allele frequencies because they are divergent in their origins and that the source population of the single-population program was derived by crossing those of the two-population program. Accordingly allele frequency at each locus in the single-population program will be assumed to be the average of those in the separate populations of the two-population program.

Genetic sample sizes assumed are constrained by the conditions that (a) the total number of plots in criterion unit comparisons shall be equal in the two types of programs and (b) genetic samples in the two S populations of two-population programs shall be of equal size. Thus, in each comparison, the number of selection units in genetic samples of the two-population program are set at one-half the number in the genetic sample of the one-population program.

Finally, every comparison assumes the same selection system in the two S populations of RSP and the single S population of the T= S program.

14.3.1.1 WHEN THERE IS NO DOMINANCE (A = 0)

Let p and q symbolize the cycle 0 frequencies of the better allele in the two S populations of the RSP program and r the cycle 0 frequency of that allele in the S population of the T = S program. We know then, from 5.10 and 5.12 that, given a=0, values of $\ddot{y}$ are 2ru when T = S and (p+q)u when $T = S_1 \times S_2$. It follows that the expectations of $2(\Delta r)$ and $(\Delta p) + (\Delta q)$ in the case of the one and two population programs, respectively, constitute the key to their relative effectiveness. While $E(\Delta q)_1$ is not perfectly correlated with $E(\Delta q)$ in any following cycle, there is no obvious reason to doubt that comparison of the two program types on the basis of $E(\Delta q)_1$ would differ seriously from one based on the sum of expectations over any series of cycles. When a = 0, 11.22 reduces to

$$E(\Delta q/q=q_0)_1 = \bar{q}_1(1-\bar{q}_1)[1-1/2N_e]uck/\sigma(\hat{X}) \qquad 14.12$$

and 11.23 reduces to

$$E(\Delta q/q=q_0)_1 = \bar{q}_1(1-\bar{q}_1)(1+\hat{f})[1-1/2N_e]uk/\sigma(\hat{X}) \qquad 14.13$$

where for brevity, $E(\Delta q/q=q_0)_1$ is denoted also as $E(\Delta q)_1$.

If the selection system, the design for comparison of criterion units, and the initial allele frequencies are the same (r = p = q) in all three of the S populations the quantities of 14.12 (or 14.13) that could be different in the two program types are only k, N_e, and $\bar{q}_1(1-\bar{q}_1)$. Because the genetic sample would be twice as large in the single-population program, selection could more intense (and k larger) without making N_e smaller than in the S populations of the RSP program. To simplify, let us assume selection intensity twice as great in the single-population program so that N_e would be the same in the two program types. Then, because any possible difference between $\bar{r}_1(1-\bar{r}_1)$ and $\bar{p}_1(1-\bar{p}_1) = \bar{q}_1(1-\bar{q}_1)$ must be trivial (see Appendix 14.1)

$$2E(\Delta r)_1/[E(\Delta p)_1 + E(\Delta q)_1] \approx k/k' \qquad 14.14$$

in which k and k' symbolize the standardized selection differentials in the T = S program and the RSP program, respectively. Thus, as anticipated intuitively, 14.14 indicates that, by virtue of the greater selection intensity possible without making N_e smaller, a single-population program is superior [$E(\Delta\ddot{y})$ greater] to an RSP program when r= p= q and a = 0.

Now consider the other logical assumption, described and labelled B above, i.e., that r = (p+q)/2. Because of the greater selection intensity in the T = S program $\bar{r}_1 \neq (\bar{p}_1+\bar{q}_1)/2$. However, it is apparent from the demonstration in Appendix 14.1that because r= (p+q)/2 the deviation of $\bar{r}_1$ from $(\bar{p}_1+\bar{q}_1)/2$ must be trivial in consequence relative to the difference between k and k'. If the selection system and the design for comparison of criterion units are the same in all three of the S populations, the quantities of 14.12 (or 14.13) that could be different among the S populations are again k, N_e, or $\bar{q}_1(1-\bar{q}_1)$. In addition to $\bar{r}_1 \approx (\bar{p}_1+\bar{q}_1)/2$, assume as before that selection intensity is twice as great in the T = S program and hence that N_e is the same in all three S populations because genetic sample size is twice as great in the single-population program. Finally, because the a=0 situation is being considered our question is whether

$$2E(\Delta r)_1 > [E(\Delta p)_1 + E(\Delta q)_1]. \quad 14.15$$

By arguments and assumptions set out above

$E(\Delta r)_1$ is directly proportional to $(\bar{r}_1)(1-\bar{r}_1)k \approx [(\bar{p}_1+\bar{q}_1)/2][1-(\bar{p}_1+\bar{q}_1)/2]k$,

$E(\Delta p)_1$ is directly proportional to $\bar{p}_1(1-\bar{p}_1)k'$, and

$E(\Delta q)$ is directly proportional to $\bar{q}_1(1-\bar{q}_1)k'$.

It follows that our question reduces to whether

$$2[(\bar{p}_1+\bar{q}_1)/2][1-(\bar{p}_1+\bar{q}_1)/2]k > [\bar{p}_1(1-\bar{p}_1)+\bar{q}_1(1-\bar{q}_1)]k'. \quad 14.16$$

We know that k > k' and it can be shown (see Appendix 14.2) that

$$2[(\bar{p}_1+\bar{q}_1)/2][1-(\bar{p}_1+\bar{q}_1)/2] = [\bar{p}_1(1-\bar{p}_1)+\bar{q}_1(1-\bar{q}_1)+(\bar{p}_1-\bar{q}_1)^2].$$

It is therefore clear that 14.16 is correct, hence that 14.15 is correct, and finally that when dominance is absent (a = 0) a single-population program is superior [$E(\Delta\ddot{y})$ greater] to an RSP program whether r = p = q or r = (p+ q)/2.

- **14.3.1.2. WHEN THE FAVORABLE ALLELE IS IN SOME DEGREE DOMINANT (A>0)**

As noted earlier a single-population (T= S) program has an advantage and a disadvantage relative to an RSP program. Its advantage is that when total effort (measured by the total number of plots in the criterion unit comparisons) is the same in the two program types, the genetic sample is twice as large in the T = S program and therefore selection intensity can be greater without making effective population size smaller than in the RSP populations. The disadvantage is the inbreeding effect (when a> 0.0) suffered that reduces $E(\Delta\ddot{y})$ in T = S programs.

It is shown in the preceding section that when there is no inbreeding effect because a = 0 the T = S program is superior, as intuitively expected, whether r = p = q or r = (p +q)/2. Our question now is whether and in what circumstances the inbreeding effect disadvantage outweighs the greater selection intensity advantage of the T= S program. For three reasons, it makes sense to assume a = 1 at the outset: (1) we know that when a > 1 the RRS two population program is superior in the long run to either T = S or RSP programs (Chapter IX), (2) if RSP is not superior when a = 1, it will not be when a< 1, and (3) the formulas relevant to our question are simplified by setting a = 1. The formulas to be employed at the outset are modifications of 11.26 and 11.28. These pertain to one population (T = S) and RSP programs, respectively in the case of selection systems in which selection and criterion units are composed of non-inbred individuals. When a = 1 and r is substituted for q, 11.26 reduces to

$$E(\Delta\ddot{y})_1 = 8\bar{r}_1(1-\bar{r}_1)^2[(1-\bar{r}_1) - 3(1-2\bar{r}_1)/2N_e]u^2cK - \bar{r}_1(1-\bar{r}_1)u/N_e \quad 14.17$$

and when a = 1, 11.28 reduces to

$$E(\Delta\ddot{y})_1 = 4\bar{p}_1(1-\bar{p}_1)[(1-\bar{p}_1) - (2-3\bar{p}_1)/2N_e](1-q_0)u^2cK'$$
$$+ 4\bar{q}_1(1-\bar{q}_1)[(1-\bar{q}_1) - (2-3\bar{q}_1)/2N_e](1-p_0)u^2cK' \quad 14.18$$

In these expressions $K = k/\sigma(\hat{X})$ and $K' = k'/\sigma(\hat{X})$ Where k and k' are the standardized selection differentials in the T = S and RSP programs, respectively. Because the selection systems employed in the two programs are assumed (for the purpose of our comparison) to be the same, the standard deviation of the selection criterion, $\sigma(\hat{X})$, is appropriately assumed the same in the two programs. On the other hand, as noted earlier $k \neq k'$.

To use 14.17 and 14.18 for our purpose, assumptions must be made concerning the magnitudes of r_0, p_0, and q_0 (the frequencies of B in the three source populations), the size of the genetic samples in the two programs, and the selection intensities in the two programs. Values of the expected allele frequencies in cycle 1 are obtained from the assumed values in cycle 0. For example, $\bar{q}_1 = q_0 + E(\Delta q/q=q_0)$, where in accord with 8.27,

$$E(\Delta q/q=q_0) = q_0(1-q_0)[1+(1-2q_0)a]ucK$$

when the S population is in linkage equilibrium. In Table 14.22, part (b) shows results obtained when the r = p = q assumption is made and parts (c) and (d) when the r = (p + q)/2 assumption is made. We note that, when r = p = q, $E(\Delta\ddot{y})_1$ is larger in the case of the two-population system except when the initial allele frequencies are very low. While the difference between the program types is never as large in absolute magnitude as the inbreeding effect suffered in the case of the T = S program, the difference expressed as a percentage of $E(\Delta\ddot{y})$ values in the case of the single-population program becomes large when initial allele frequencies are over 0.5.

When r = p = q, the initial target population $\ddot{y}$ values of the one- and two-population programs are equal (random individuals of the two RSP source populations are in fact random individuals of the T = S source population). As a consequence, only the expectations of $\Delta\ddot{y}$ are required for qualitative comparison of genetic gains to be expected from the two programs. On the other hand, when r = (p+q)/2, the difference between the initial target population $\ddot{y}$ values of the programs together with the expected changes in succeeding cyles are required for comparison of the programs in terms of the expectations of their target population $\ddot{y}$ values in post-zero cycles.[12] We note in part (c) of the table that the pattern of program differences in $E(\Delta\ddot{y})$ is not as consistently favorable to RSP as when initial allele frequencies are equal (part b). However, it is demonstrated in part (d) that because initial $\ddot{y}$ values are higher in the case of RSP, the $\ddot{y}$ values of RSP are consistently greater in following cycles.

In summary, the content of Table 14.22 suggests that when a = 1, RSP is superior to the T = S program except when initial allele frequencies in the three source populations of the two programs are equal and less than 0.3. But because the quantities of Table 14.22 are conditioned by the assumptions listed at the top of the table, it is apparent that results would have been different if assumptions had been different. Which of the assumptions are most critical and how much would be the effects of varying them?

(1) We know that when a = 0 the single population program is superior to RSP. Thus it is obvious that with partial dominance (0 < a < 1) the range of allele frequencies within which RSP is better than the T = S program would not be as great as when a = 1. For example, instead of being better in the r = p = q situation unless initial allele frequencies were less than 0.3 the range would be less by an amount dependent upon the actual value of a -- slightly less given almost

Table 14.22. $E(\Delta \ddot{y})_1$ per cycle for T = S and RSP programs

Assumptions: (1) a = 1 and selection among full-sib families; (2) Selection of M = 10 from 200 in T = S and of M = 10 from 100 in RSP; (3) u = 1/2, c = 1/2, $\sigma(\bar{X})$ = 7.22

(a) Quantities affecting $E(\Delta \ddot{y})_1$ that differ because selection intensity differs in the two programs

Program	si[a]	k	k/σ	$N_e = 4M/(2-si)$
T = S	.05	2.06	.285	20.5
RSP	.10	1.76	.244	21.05

(b) $E(\Delta \ddot{y})_1$ as a function of initial allele frequencies when r = p = q.

r = p = q	T = S	RSP
.1	.0246 - .0024[b] = .0222	.0182
.3	.0275 - .0053[b] = .0222	.0244
.5	.0166 - .0061[b] = .0105	.0144
.7	.0055 - .0050[b] = .0005	.0045
.9	.0004 - .0022[b] = -.0018	.0003

(c) $E(\Delta \ddot{y})$ and $E(\ddot{y})$ as functions of initial allele frequencies when r = (p+q)/2.

			$E(\Delta \ddot{y})_0$		$E(\Delta \ddot{y})_1$	
r	p	q	T = S	RSP	T = S	RSP
.3	.1	.5	.0293 - .0051[b] = .0242	.0187	.0275 - .0053[b] = .0222	.0181
.5	.3	.7	.0178 - .0061[b] = .0117	.0108	.0166 - .0061[b] = .0105	.0104
.5	.1	.9	.0178 - .0061[b] = .0117	.0020	.0166 - .0061[b] = .0105	.0021
.7	.5	.9	.0054 - .0051[b] = .0003	.0021	.0055 - .0050[b] = .0005	.0021

(d) Expectations of $\ddot{y}$ in cycles 0, 1, and 2 when r = (p+q)/2.

			$E(\ddot{y}_0)$		$E(\ddot{y}_1)$		$E(\ddot{y}_2)$	
			T = S	RSP	T = S	RSP	T = S	RSP
.3	.1	.5	.51	.55	.5342	.5687	.5564	.5868
.5	.3	.7	.75	.79	.7617	.8008	.7722	.8112
.5	.1	.9	.75	.91	.7617	.9120	.7722	.9141
.7	.5	.9	.91	.95	.9103	.9521	.9108	.9542

[a] si = selection intensity (here p symbolizes frequency of the favorable allele in one of the S populations of RSP).

[b] inbreeding effect.

complete dominance, less by larger amounts given lower degrees of dominance. Just as obviously, because heterosis becomes less as the degree of dominance (a) decreases, the advantage of RSP when a = 1 and r = (p+q)/2 (but $p \neq q$) becomes smaller and at some point disappears as the degree of partial dominance decreases.

(2) As noted in section 14.2.2 the positive effect of selection is proportional to the magnitude of u^2 while the inbreeding effect is a multiple of u. It is apparent therefore that the inbreeding effect disadvantage of the single population program hinges in part on the actual magnitude of u. It becomes less relative to the positive effecct of selection as u increases but

greater as u decreases. The assumption that $u = 1/2$ (when the measurement unit is such that $\ddot{X}$, the phenotypic mean of the selection criterion in the S population, is 100) on which quantities in Table 14.22 were based was arbitrary. Unfortunately in real world situations most existing values of $u/\ddot{X}$ are unknown. However, as emphasized in section 6.3.3, the larger the value of u, the more likely that the favorable allele will have become fixed or brought to high frequency by past selection.

(3) Equations 11.26 and 11.27 reveal that in the case of single population programs the ratio between expected absolute magnitudes of (a) the positive response to selection and (b) the inbreeding effect depends in part on the magnitude of N_e. It follows that the relative effectivess of T = S and RSP programs is likely to be altered by change in either (or both) size of genetic sample or the selection system employed. Larger genetic samples enable more selections without reduction in selection intensity and the relation between number of selections (M) and N_e varies among selection systems.

Values of $E(\Delta\ddot{y})_1$ for the T = S and RSP programs, assuming selection among S_1 families, are shown in Table 14.23. The applicable formulas in this case are 11.27 and 11.29. When $a = 1$, $h = 0$, and r is substituted for q, 11.27 reduces to

$$E(\Delta\ddot{y})_1 = 2\bar{r}_1(1-\bar{r}_1)[(1-\bar{r}_1)(3-\bar{r}_1) - (8-21\bar{r}_1+12\bar{r}_1^2)/2N_e]u^2K$$
$$- \bar{r}_1(1-\bar{r}_1)u/N_e \qquad 14.19$$

and when $a - 1$ and $h - 0$, 11.29 reduces to

$$E(\Delta\ddot{y})_1 = 2\bar{p}_1(1-\bar{p}_1)(1-\bar{q}_1)[(1.5-\bar{p}_1)-(2.5-3\bar{p}_1)/2N_e]u^2K'$$
$$+ 2\bar{q}_1(1-\bar{q}_1)(1-\bar{p}_1)[(1.5-\bar{q}_1)-(2.5-3\bar{q}_1)/2N_e]u^2K' \qquad 14.20$$

Table 14.23. $E(\Delta\dot{y})_1$ per cycle for T = S and RSP programs (selection among S_1 families)

Assumptions: (1) $a = 1$; (2) Selection of $M = 10$ from 200 in T = S and of $M = 10$ from 100 in RSP (3) $u = 1/2$, $\sigma(\hat{X}) = 8.82$, $N_e = 2Mm/(2m-1) = 9.5$.

(a) Quantities affecting $E(\Delta\ddot{y})_1$ that differ because selection intensity differs in the two programs

Program	si[a]	k	$K = k/\sigma(\hat{X})$
T = S	.05	2.06	.233
RSP	.10	1.76	.200

(b) $E(\Delta\ddot{y})_1$ as a function of initial allele frequencies when $r = p = q$.

r = p = q	T = S	RSP
.1	.0254 - .0053[b] = .0201	.0226
.3	.0372 - .0116[b] = .0256	.0324
.5	.0262 - .0131[b] = .0131	.0219
.7	.0114 - .0106[b] = .0008	.0088
.9	.0017 - .0047[b] = -.0030	.0010

[a] si = selection intensity.

[b] inbreeding effect.

The assumption in our comparison of the T = S and RSP programs that selection practiced in the two populations of the RSP program would be the same in all respects justifies the single values of N_e and K' in 14.20 in place of the possibly different values allowed in 11.29.

When values of $E(\Delta\ddot{y})_1$ in parts (b) of Tables 14.22 and 14.23 are compared we note (a) that because N_e is smaller when the selecton units are S_1 families instead of full-sib families, the inbreeding effect in the case of the single population program is larger (compare to values shown in Table 14.22) and (b) consequently, the RSP program is superior to the T = S program at all levels of the initial allele frequencies and, in general, the advantage of the RSP program is greater than in the case of selection among full-sib families. Values of $E(\Delta\ddot{y})_1$ and $E(\ddot{y})$ when $r = (p+q)/2$ are shown in Table 14.22 but not in Table 14.23 because it is apparent that the initially larger values of $\ddot{y}$ in the case of the RSP program due to heterosis would be maintained in part in succeeding cycles as shown in Table 14.22 when selection is among full-sib familes.

Because Tables 14.22 and 14.23 show $E(\Delta\ddot{y})$ per cycle instead of per year adjustments for number of years per cycle would obviously be required if values shown were to be used as the basis for choice between the selection systems.

- 14.3.2. **RSP vs. RRS Programs When a ≤ 1.0**

When the selection criterion exhibits a considerable inbreeding depression, there is reason to believe that the level of dominance of favorable alleles is sufficient to make an RSP program superior to a single-population program. In that event the possibility of some overdominant genes exists. Given overdominance, reciprocal recurrent selection (RRS) is superior to RSP. This, in turn, raises the practical question of whether when dominance is sufficient to make RSP superior to a single population (T = S) program, is RSP always superior to RRS when there is no overdominance. Values of $E(\Delta\ddot{y})_1$ shown in Table 14.24 suggest that RSP is superior to RRS when overdominance is absent. The comparison made is between RSP with selection among S_1 families and the full-sib version of RRS (see section 9.3.2). In both systems the selection units are non-inbred individuals and, hence, with equal numbers of selections in the two programs effective population size would also be equal. Remember that in full-sib RRS the same sample of criterion units provides the basis for selection in both S populations and consequently the number of selections in the RRS and RSP programs can be the same even though selection intensity is twice as great in the RRS program. Another simplification of this comparison is that the number of plant generations required per cycle of selection is the same (three) in both programs. Equations employed to obtain $E(\Delta\ddot{y})_1$ values were 14.20 in the case of RSP and A11.3.27 in the case of RRS. The latter reduces when a = 1 to

$$E(\Delta\ddot{y})_1 = 4(1-\bar{p}_1)(1-\bar{q}_1)[\bar{p}_1(1-\bar{q}_1) + \bar{q}_1(1-\bar{p}_1) + \bar{p}_1\bar{q}_1/N_e](1-1/2N_e)cK. \quad 14.21$$

The $E(\Delta\ddot{y})_1$ values shown in the table indicate superiority of the RPS program even though complete dominance (a = 1) was included among the assumptions on which calculations were based. At face value, the content of this table indicates that RRS should not be chosen over RSP unless the

Table 14.24. $E(\Delta\bar{y})_1$ per cycle for RSP and RRSf (full sib RRS) programs (selection among S_1 families)

Assumptions: (1) Selection of M = 10 from 100 in RSP and of M = 10 from 200 in RRSf; (2) a = 1, u = 1/2, $N_e = 2Mm/(2m - 1) = 9.5$

(a) Quantities affecting $E(\Delta\ddot{y})_1$ that differ between the two programs

Program	si[a]	k	c	$\sigma(\hat{X})$	$K = k/\sigma(\hat{X})$
RSP	.10	1.76	1	8.82	.200
RRSf	.05	2.06	1/2	7.22	.285

(b) $E(\Delta\ddot{y})_1$ as a function of initial allele frequencies when p = q.

p = q	RSP	RRSf
.1	.0226	.0213
.3	.0324	.0278
.5	.0219	.0166
.7	.0088	.0053
.9	.0010	.0003

(c) $E(\Delta\ddot{y})_1$ as a function of initial allele frequencies when $p \neq q$.

p	q	RSP	RRSf
.1	.5	.0265	.0230
.1	.9	.0060	.0091
.3	.7	.0173	.0158
.5	.9	.0047	.0034

[a] si = selection intensity

breeder has sufficient reason to believe that overdominance is involved in the genetics of the selection criterion that he intends to employ.

14.4. Discussion and Summary

The focus of this chapter, as indicated by its title, is on application of theory in comparison of selection systems and variations of systems. Eight systems are identified (Table 14.3) and considered. All have been used, some more than others, in ccommercial and experimental maize breeding programs but not all of them have a place in work with other species. The most obvious reason is that self fertilization is not possible in some species and cross-fertilizations are, at best, time consuming and therefore costly in others.

Breeders must think in terms of genetic improvement per unit of time (years). Thus both response per cycle of selection and length of cycle must be considered when comparing procedures. The significance of cycle length is emphasized herein by tabulation of response expectations per year on the assumption that three, two, or one plant generation(s) could (and would) be grown per year. Noteworthy in this connection is the indication in Tables 14.8 and 14.9 that mass selection may be the system of choice for use in long term programs when only one plant generation is grown per year.[13] In Chapter XII, it is noted and emphasized that to ensure that the effective selection coefficient (s') is positive for all alleles favorable to total value the selection criterion (X) must be total value itself. This presents an obvious dilemma in the case of organisms, other than annual plants (blueberries, fruit trees, chickens, dairy cattle, etc.), for which assessment of total value is not possible as soon as one or more significant components of total value can be measured.[14] Then to employ total value as the selection criterion can lengthen cycle length substantially. The problem posed is discussed further in Chapter XVI.

Comparisons of selection systems and considerations of how response expectations are affected by genetic and operational variables are conducted at the level of single genes. The focus is on probability of fixation and expected change per year in the frequency of a favorable allele in the case of long term programs and on $E(\Delta\ddot{y}_i)$, the change in average value of genotype at a single locus, in the case of short term programs. The obvious rationale is that total genetic change is composed of the changes at all the separate loci of the organism and hence can only be understood in terms of the reasons for variations in change of the contributions of single-locus genotypes.

- 14.4.1. **Short Term Programs**

Issues investigated are (1) the relative effectiveness of selection systems, (2) optimum selection intensities when impacts of effective population size (N_e) are taken into account, (3) whether taking account of N_e effects can improve judgements concerning relative effectiveness of selection systems, and (4) the amounts by which $E(\Delta\ddot{y})$ may be over-estimated by theory that does not take account of the impacts of effective population size.

The eight systems (Table 14.3) are compared in Table 14.13 on the basis of $E(\Delta\ddot{y})$ per year with consequences of N_e ignored. Four of the eight systems are compared in Table 14.19 in terms of $E(\Delta\ddot{y})$ when selection intensity is near optimum in the light of impacts of N_e. The significance of years per cycle of selection, noted above and earlier, is demonstrated by the relative magnitudes of quantities in the L(3) and L(2) portions of Table 14.13. The other sources of variation in $E(\Delta\ddot{y})$ are apparent from examination of 14.8 and 14.9, the expressions from which values of $E(\Delta\ddot{y})$ per cycle were computed. Note (a) that the denominator of both is $\sigma(\hat{X})$ and from Table 14.13 that $\sigma(\hat{X})$ is variable among selection systems even when the structure of criterion unit comparisons is the same and (b) that the component, $[(1+h)+(1-h)(1-2q)a_x/2]$ of 14.9 is replaced by $[1+(1-2q)a_x]$ in 14.8. The variations in c (shown in Table 8.2), in $\sigma(\hat{X})$, and in the differences between $[(1+h)+(1-h)(1-2q)a_x/2]$ and $[1+(1-2q)a_x]$ as affected by h, q, and a_x explain the variations in $E(\Delta\ddot{y})$ not accounted for by differences among systems in cycle length. The increase of the $[(1+h)+(1-h)(1-2q)a_x/2]$ to $[1+(1-2q)a_x]$ ratio as q increases when $a_x > 0$ favors the S_n family selection systems when the selection criterion is one such as grain yield in maize that exhibits a significant negative effect of inbreeding.

From 14.10 and 14.11 it is apparent that (given a particular selection system) $E(\Delta\ddot{y})_1$ is a function of $\bar{q}_1$, the expected frequency of the B allele in the S population of cycle 1; a, the level of dominance of the B allele; u, the quantitative difference between the contributions of the BB and bb genotypes to the value of total genotype; k, the standardized selection differential of which the expected magnitude increases as the fraction (p) of the genetic sample selected for use as parents of the following generation is decreased ; $\sigma(\hat{X})$, the standard deviation of the selection criterion; N_e, the effective population size; and T, the number of selection units in the genetic sample. T does not appear in either 14.10 or 14.11 but is an obvious factor because if T is larger (or smaller) both k and N_e will be larger (or smaller) even though p (selection intensity) is unchanged. Further variables by which $E(\Delta\ddot{y})_1$ is affected are c and h, quantities that distinguish and may be variable among selection systems (though one or even both can be the same for certain pairs of systems). All else being constant, if the number of units selected is smaller (p smaller), k is larger and N_e is smaller. Conversely, if p is larger (number of units selected larger) k is smaller and N_e is larger. Noting from either 14.10 or 14.11 that $E(\Delta\ddot{y})_1$ increases as k increases but decreases as N_e decreases, it is apparent that, for any combination of values of the other variables of which $E(\Delta\ddot{y})_1$ is a function, there must be an optimum value of p, the selection intensity.

Variation in optimum selection intensity [the value of p that maximizes $E(\Delta\ddot{y})_1$] is explored in Tables 14.15 - 14.18. Four selection systems are treated, a different one in each table. Near-optimum selection intensities are shown for 16 combinations of q, a, T, and $\sigma(\hat{X})$ in each table. The total number of plots assumed in criterion unit comparisons was the same (1200) in all of these 16 combinations. The variation in T and $\sigma(\hat{X})$ resulted from assuming one plot per selection unit at each of v (= 1, 3, 6, or 12) locations so that T (= 1200, 400, 200, or 100) and Tv = 1200 in all cases. Throughout u = 0.5, as well as Tv, is constant. A common and significant feature of the four tables is the variation indicated in optimum selection intensities which range, for example, from p = 0.02 to 0.16 in Table 14.15 and from p = 0.03 to 0.32 in Table 14.17. Table 14.20 deals with the effect of doubling Tv, the total number of plots employed in the comparison of criterion units, from 1200 to 2400. Results are shown for selection systems 2 and 6 and for number of plots (v) per selection unit. The effect of doubling Tv is to increase optimum selection intensity as intuitively expected; the decrease shown in p when other things are constant approaches 50 percent on average. Effects of variation in the magnitude of u are shown in Table 14.21. Optimum selection intensity increases substantially as the assumed value of u is increased from 0.25 to 0.5 to 1.0. The meaning of the content of these tables is that to actually optimize selection intensity is effectively impossible. To even come close could only result from truly inspired guesses concerning actual distributions of the u, a, and q values of segregating favorable alleles together with adequate information regarding the magnitude of $\sigma(\hat{X})$.

The content of Table 14.19 provides information relative to two practical issues. The first of these is whether comparisons of selection systems would be improved if based on $E(\Delta\ddot{y})_1$ when impacts of effective population size (N_e) are taken into account and selection intensity is optimized for each system. The apparent answer is no. Magnitudes of $E(\Delta\ddot{y})$ are shown in the case of eight combinations of allele frequency, level of dominance, and plant generations grown per year. Within each of these the rank order magnitudes of $E(\Delta\ddot{y})$ for the different selection systems are essentially the same whether or not N_e is ignored. Beyond this is the fact that selection intensity which is variable among selection systems cannot, in practice, be optimized (see above) and hence assumptions about optima would inevitably be more favorable to one system than another. Part (c) of the table shows that depending on allele frequency, level of dominance, and selection system, $E(\Delta\ddot{y})_1$ is over-estimated by from 16 to 53 percent when the impacts of effective population size are ignored. The comparisons assume the same selection intensity whether impacts of N_e are considered or ignored, the intensity that is near optimum for each (q, a, and system) combination. However, in practice, (a) there can be only one selection intensity in any cycle of selection, not a different one for different genes, and (b) as emphasized above, the selection intensity that would maximize $\Sigma_i E(\Delta\ddot{y}_i)$ cannot be approximated because adequate information concerning the joint distributions of a, u, and q for genes with effects on the selection criterion is not now available and cannot be expected to become available in the foreseeable future. The solution that is available when there is no dominance (all a's equal to zero) provides some guidance. When a = zero 14.10 reduces to

$$E(\Delta\ddot{y})_1 = 2\bar{q}_1(1-\bar{q}_1)(1-1/2N_e)u^2ck/\sigma(\hat{X}) \qquad 14.10a$$

and 14.11 to

$$2\bar{q}_1(1-\bar{q}_1)(1-1/2N_e)(1+h)u^2k/\sigma(\hat{X}). \qquad 14.11a$$

In each of these the optimum selection intensity is obviously the one that maximizes $(1-1/2N_e)k$. This optimum differs among selection systems because the ratio of N_e to M, the number of selections, varies among systems (see Table 14.5). Consider system 2 (selection among S_4 families) and system 6 (selection among full-sib families) for which k, N_e, and their product are listed below as functions of selection intensity (p) when genetic sample size (T) is 400 or 100. Maximum values of $(1-1/2N_e)k$ are bold-faced.

				System 2		System 6	
T	p	M	k	$N_e = M/(1+h)$	$(1-1/2N_e)k$	$N_e = 4M/(2-p)$	$(1-1/2N_e)k$
100	.01	1	2.66	0.53	0.15	2.01	2.00
	.02	2	2.42	1.07	1.29	4.04	**2.12**
	.03	3	2.27	1.60	1.56	6.09	2.08
	.04	4	2.16	2.13	1.65	8.16	2.03
	.05	5	2.06	2.67	1.67	10.26	1.96
	.06	6	2.00	3.20	**1.69**	12.37	1.92
	.07	7	1.92	3.73	1.66	14.51	1.85
400	.01	4	2.66	2.13	2.06	8.04	**2.49**
	.02	8	2.42	4.27	**2.14**	16.16	2.35
	.03	12	2.27	6.40	2.09	24.36	2.22

The associated selection intensities are of course the optima. Note the following.

(1) At each selection intensity N_e is roughly four times as great when selection is among full-sib families as when selection is among S_4 families. Consequently optimum selection

intensiy is greater (p smaller) in the case of system 6 in both cases examined and would be at any genetic sample size.

(2) When the genetic sample is larger optimum selection intensity is greater because $M = pT$ so that the same N_e is possible with greater selection intensity. Consequently $\Sigma_i E(\Delta\ddot{y}_i)$ is greater at optimum selection intensity when the genetic sample size is larger. For example, in the case of system 2 it would be 2.14/1.69 = 1.27 times as large when T = 400 as when T = 100. Of course, the program cost if judged by size of the genetic sample would be four times as great and the 27 percent increase in $E(\Delta Y)$ might not be considered enough to justify the extra cost.

(3) It is apparent from 14.10 and 14.11 that when favorable alleles are in some degree dominant the decrease in $E(\Delta\ddot{y})$ resulting from smaller effective population size will be greater than when such dominance is absent. This is reflected in parts (d) and (b) of Tables 14.16 and 14.18 by the fact that the near optimum values of p are considerably greater than when dominance is absent (illustrated above). The lowest near-optimum values of p tabulated for cases in which $a > 0.0$ are those when $a = \bar{q}_1 = 0.5$. Comparisons of near-optimum values of p when dominance is absent ($a = 0$) with values from Tables 14.16 and 14.18 are shown below. These suggest amounts by which optimum p may be larger when $a > 0.0$, but note that the

System	T	$a = 0.0$	$a = \bar{q}_1 = 0.5$	$a = \bar{q}_1 = 0.8$
2	100	.06	.09	.11
	400	.02	.04	.06
4	100	.02	.06	.09
	400	$\leq$ .01	.03	.04

optima when $a > 0.0$ are functions of the values of u (Table 14.21), $\bar{q}_1$, and $\sigma(\hat{X})$ in contrast to the case when $a = 0.0$. The point of all this is that optimum p can be appproximated in terms of genetic sample size when there is reason to believe that a-values are near zero (as when the trait in question exhibits little effect of inbreeding) but is greater (probably considerably greater) when the trait in question exhibits substantial inbreeding depression.

14.4.2. **Long Term Programs**

The most challenging decisions required in the context of long term selection programs are those concerning (1) species to be employed, (2) foundation populations, and (3) the values of N_es' to be required. The first two of these will be discussed in Chapter XVI. The third, to be considered here, has two variations: (a) the choice when the focus is to be on favorable alleles existing in the species at the time the program is initiated and (b) the choice when the focus is on favorable alleles arising by mutation during the course of the program in addition to those present at the outset.

As indicated by 14.1, the magnitude of N_es' that makes the probability of fixation of a superior allele equal to a specified value (P_a) is a function of P_a and the initial frequency (q_0) of the allele. The alternative to fixation is loss of the allele. If the superior allele is fixed, the contribution of the gene to $\ddot{Y}$ (the value of the total genotype) becomes 2u. Let this quantity by recognized as the potential with respect to the gene in question. However, the probability of fixation of any superior allele is P_f, not 1.0, so the expected contribution of the gene to $\ddot{Y}$ is $P_f(2u)$ and the reduction from its full potential is $2u - P_f(2u) = (1-P_f)(2u)$. Now remember (a) that when the value of N_es' achieved using a specific variation of a specific selection system is calculated as a guide toward optimum selection intensity the magnitude of u has been assumed, (b) that s' is a linear function of u, (c) that the "required" value of N_es' is based on a selected value of q_0 and (d) that both u and q_0 are variable among genes. It follows that N_es' achieved and therefore probability of fixation and "reduction from potential" will be variable among genes. For example, it was found, for the purpose of Table 14.2, that in the case of a superior allele with $q_0 = 0.10$, $N_es' = 12.63$ would yield $P = 0.92$. However, all other things being the same, the value of N_es' with respect to the superior allele of another gene for which $u = 0.3$ (instead of 0.5) would be $12.63(0.3)/(0.5) = 7.578$ and by 11.1 the probability of fixation of that allele would be

$$P_f \approx 1 - e^{-2(7.578)(0.1)} = 0.78$$

if the initial frequency of this second allele were also 0.10. Then the reduction from potential of this allele would be $(1- P_f)(2u) = (0.22)(0.6) = 0.132$. The issue is further illustrated in Table 14.25. Parts (a) and (b) focus on alleles having the same initial frequncy (0.1) but that vary from 0.5 to 0.1 in the magnitude of effect (u)and parts (c) and (d) on alleles for which u is the same (0.5) for all but initial frequency varies from 0.02 to 0.10. Part (a) assumes that selection is such that $N_es' = 12.63$ in the case of alleles for which $u = 0.5$.

As outlined above, the achieved values of N_es' in the case of other alleles deviate from 12.63 in correspondence with the deviation of their u-values from 0.5. If a set of genes consisting of one with each of the five u-values shown in the table is considered the full potential of the set is $2\Sigma_i u_i = 3.0$ and total "reduction from potential" is $2\Sigma_i(1-P_{fi})u_i = 0.585$. Thus the "reduction from potential" for the five gene set is 19.5 percent and the reduction for any number of such sets would also be 19.5 percent. The u and q_0 values assumed in part (b) are the same as in part (a) but the N_es' requirement is higher, being based on that required to yield $P_f = 0.92$ in the case of an allele for which $q = 0.1$ and $u = 0.3$. The total potential of the five gene set is 3.0 as before but the "reduction from potential" is lower, 0.25 or 8.3 percent of the total. In parts (c) and (d) the desired values of N_es' are those that make $P_f = 0.92$ for the alleles with $q_0 = 0.10$ and 0.06, respectively. Note

(1) that the variation in probability of fixation in parts (a) and (b) arose from variation in u-values and in parts (c) and (d) from variation in initial allele frequencies,

(2) that the total for the five alleles in "reduction from potential" is in both instances decreased by the increase in N_es'; from 19.5 to 8.3 percent in the first and from 28.0 to 14.9 percent in the second,

(3) that the "prices" for those gains are reductions in expectations of $\Delta q/q(1-q)$ per year but those reductions are directly proportional to u_x, the effect of the gene on the selection criterion and inversely proportional to $q(1-q)$ where q is the frequency of the favorable allele.

Table 14.25. Probability of fixation and "reduction from potential" for alleles with different u values or different initial frequencies when achieved $N_e s'$ is that required to make $P_f = 0.92$ in the case of one of the alleles (selection among S_4 families, L = 2, and $\sigma(\hat{X}) = 12.05$)

Achieved $N_e s'$	u	q_0	P_f	Potential	RfP[b]	$E(\Delta q)/q(1-q)L$
Part (a)	p = 0.065	k = 1.95				
12.63	0.5	0.1	0.92	1.00	0.080	0.0759
10.10	0.4	0.1	0.867	0.80	0.106	0.0606
7.58	0.3	0.1	0.780	0.60	0.132	0.0455
5.05	0.2	0.1	0.636	0.40	0.146	0.0303
2.53	0.1	0.1	0.397	0.20	0.121	0.0152
				3.00	0.585 =19/5% of total potential	
Part (b)	p = 0.13	k = 1.63				
21.10	0.5	0.1	0.985	1.00	0.015	0.0634
16.88	0.4	0.1	0.966	0.80	0.027	0.0507
12.66	0.3	0.1	0.920	0.60	0.048	0.0380
8.44	0.2	0.1	0.814	0.40	0.074	0.0254
4.22	0.1	0.1	0.569	0.20	0.086	0.0127
				3.00	0.250 = 8.3% of total potential	
Part (c)	p = 0.065	k = 1.95				
12.62	0.5	0.10	0.920	1.00	0.080	0.0759
12.62	0.5	0.08	0.867	1.00	0.133	0.0759
12.62	0.5	0.06	0.780	1.00	0.220	0.0759
12.62	0.5	0.04	0.636	1.00	0.364	0.0759
12.62	0.5	0.02	0.397	1.00	0.603	0.0759
				5.00	1.40 = 28.0 % of total potential	
Part (d)	p =0.13	k = 1.63				
21.10	0.5	0.10	0.985	1.00	0.015	0.0634
21.10	0.5	0.08	0.966	1.00	0.034	0.0634
21.10	0.5	0.06	0.920	1.00	0.080	0.0634
21.10	0.5	0.04	0.815	1.00	0.185	0.0634
21.10	0.5	0.02	0.570	1.00	0.430	0.0634
				5.00	0.744 =14.9% of total potential	

[a] As in Table 14.8, part d.

[b] Reduction from potential.

(4) that quantities in this table, as those in Tables 14.8 compared with those in Table 14.9, suggest that smaller values of $E(\Delta q)/q(1-q)$ are the invariable price to be paid for larger values of $N_e s'$. This is so when differences in $N_e s'$ result entirely from differences in selection intensity. However, when genetic sample size (T) is increased without change in other

parameters N_e is changed without change in s'. Results are increases in N_es' and probability of fixation without change in $E(\Delta q)$. For that matter, when T is increased both N_es' and $E(\Delta q)$ can, within limits, be increased.

In the context of long term selection programs designed to capitalize on the potential of new useful alleles that may arise via on-going mutation, the basic fact is that the initial frequency of such alleles must be 1/2N (or very close thereto) where N is the number of individuals per cycle in the population. Then (see Appendix 11.2) the approximate probabilities in a recurrent selection program of (a) fixation of a new and superior mutant allele and (b) that such an allele will be formed in one selection cycle and subsequently fixed in the population are N_es'/N and $2\pi N_es'$, respectively.[15] The consequent improvement of the genetic population would be 2u and hence the expectation of ultimate improvement from the pair of events (the potentialy beneficial mautation and fixation of the new allele) is $4\pi N_es'u$. For illustration, consider mass selection, as described in Table 14.3. the selection units are half-sib families and the criterion units are the individual seed parents of those families. Assume operational procedure as follows.

(1) An equal number of kernels from each of the M selections of cycle k are bulked for random field positions when planted to produce the S population of cycle k+1.

(2) Each plant functions as (a) one of the criterion units of cycle k+1 and as the common parent of the corresponding half-sib family selection unit and (b) as a potential pollen parent of any member of each half sib-family selection unit.

(3) An approximation to random pollination by (a) wind pollination if the population is grown in a sufficiently isolated field or (b) by manual distributions from mass collections of pollen.

To simplify, assume that the S population in each cycle consists of N = Mn plants (n from each of the M selections) and that no two plants have the same pollen parent. Then since each of the N plants functions as the common parent of a selection unit, we see that N = T = Mn and M = pT = T/n so that p = 1/n. Because M = pN where N is number of seed parents in the genetic population,

$$N_es' = 8kpNu/(1+p)(2-p)\sigma(\hat{X}) \qquad \text{(from Table 14.5)}$$

and

$$P_f = N_es'/N = 8pku/(1+p)(2-p)\sigma(\hat{X}).$$

Quantities flowing from a few specific assumptions concerning selection intensity (p) and N are shown below. As earlier quantification of s' was based on $\sigma(\hat{X})$ = 17.3.

p	k	pk	N = M/p	M	N_e	N_es'	$P_f \approx N_es'/N$	$4\pi N_es'u$	$E(\Delta q)/q(1-q)L$
.10	1.76	.176	100	10	38	3.9u	.039u	$15.6\pi u^2$	.051u
.50	0.80	.400	100	50	178	8.2u	.082u	$32.8\pi u^2$	.023u
.10	1.76	.176	1000	100	383	38.9u	.039u	$155.8\pi u^2$	.051u
.50	0.80	.400	1000	500	1778	82.2u	.082u	$328.8\pi u^2$	.023u
.10	1.76	.176	10000	1000	3828	389.4u	.039u	$1557.7\pi u^2$	.051u
.50	.80	.400	10000	5000	17778	822.1u	.082u	$3288.4\pi u^2$	.023u

Note the following.

(1) As p (the proportion selected) increases k (the standardized selection differential) decreases but their product(pk) and the probability of fixation increase. These relations change when p exceeds .50. For example when p = 0.6, k ≈ .64, and pk≈ .384.

(2) As shown in Table 14.25, increases in probability of fixation come at the expense of decreases in E(Δq) per unit of time when size of the genetic sample is constant.

(3) $N_e s'$, P_f, and E(Δq)/q(1-q)L are linear functions of u.

(4) The probability of fixation (P_f) is not a function of population size.The explanation is that when population size is smaller the initial frequency (q_0 =1/2N) of any mutant allele is larger.

(5) The ultimate genetic improvement, $4\pi N_e s'u$, is a linear function of πu^2.

Perspective is achieved by considering specific values of u and π, e.g., u = 0.2 and π = 10^{-6}. Then given p = 0.5, k=0.8, and N = 10000 as in the last line of the table above, $4\pi N_e s'u$ = 0.000132. This would be the expectation of genetic improvement from mutation of one gene. Assuming n = 20,000 mutable genes, the total potential for genetic improvement from the mutations in one cycle of the recurrent selection considered might be (20,000)(0.000132) = 2.63. This would amount to a 2.63% improvement if the units of measurement were such that $\ddot{Y}$, the average value of total genotyes in the population, equals 100. (See Appendix 11.2 for discussion of possible values of n, π, and u).

Obviously, 2.63 per cycle may be too large (or too small) but summed over successive cycles it could make long term programs well worthwhile when viewed in terms of food production for future generations. Unfortunately, the associated time requirement is not apparent from quantities shown above. In the case of any single allele Δq, the actual change in its frequency in any selection cycle, is E(Δq) plus d, a random deviation from that expectation. The variance of such deviations is inversely proportional to the effective population size (N_e) and would be zero if N_e were infinite. Then deviations from E(Δq) would all be zero and the change in frequency of each allele would be exactly q(1-q)E(Δq)/q(1-q).[16] Given p = 0.5, the largest value of E(Δq)/cycle would result when q = 0.5 and would be 0.25(0.023u) = 0.0058u. Otherwise, it would be smaller, e.g., only (0.01)(0.99)(.023u) = 0.00023u, given q = 0.01. Obviously, the number of cycles required to fix the new favorable allele would be very great. But note, the probability of fixation would then be 1.0. As a consequence of random deviation of Δq from E(Δq) the number of positive deviations can be sufficient in the case of a few new favorable alleles to make the average of Δq much greater than what it would be if always equal to E(Δq). That this would not occur often is indicated by the low probabilities of fixation. It is apparent that in the case of most favorable alleles arising by mutation that are eventually fixed the averages of deviations from E(Δq) are probably greater than zero and hence that the averages of Δq are larger (perhaps much larger) than the corresponding averages of E(Δq), but this leaves unresolved the the question of how much larger. Kimura and Ohta (1969) derived a formula for the average number of generations required for a new mutant allele to reach fixation. For two reasons it is not presented here: (1) it is beyond the mathematical competence of this author and (2) what would be more to the point is how long it would take to increase the frequencies of the alleles that would eventually be fixed to some arbitrary level, say to $q = x$ where $0 < x < 1$. Then in the case of alleles with additive effects ($a = 0$) the improvement achieved would $x(2u)$ which in the explicit example considered above would be $0.2(2.63) = 0.52$ percent from favorable mutations arising during one selection cycle. Obviously, the information in question can be obtained by simulation using a modern high speed computer.

• 14.4.3. Comparison of Probabilities of Fixation and E(Δq) in Long Term and Short Term Programs

It was noted in discussion of Table 14.25 that when probabilities of fixation of low frequency favorable alleles are increased by making $N_e s'$ larger the expected increases in allele frequencies are decreased unless the increase in $N_e s'$ is obtained by using a larger genetic sample without change in other parameters. That this should be so is apparent from the formulas for $N_e s'$ and $E(\Delta q)/q(1-q)L$ shown in Table 14.5. Other parameters [e.g., u_X, $\sigma(\hat{X})$ and m] being constant and remembering that $M = pT$, where T is the number of selection units in the genetic sample, we note that $N_e s'$ is proportional to kpT.[17] Then because kp is increased and k decreased when p is increased (selection intensity decreased), $N_e s'$ and probability of fixation are increased but $E(\Delta q)/q(1-q)L$ is decreased, i.e., the effects of change in selection intensity on probabilities of fixation and expected changes in allele frequencies are opposite in direction when genetic sample size is constant. Just as obviously when T is increased without change in other parameters probabilities of fixation are increased without decreases in $E(\Delta q)$.

15 Methods in Quantitative Genetics

Previous chapters have identified quantities relevant to prediction and explanation of the outcomes of selection programs. This one focuses on sources of information concerning the magnitudes of those quantities.

• 15.1. Variances and Covariances

The variance (or its square root) of the selection criterion is an element in each of the expressions that have been presented for prediction of a selection response (Δq, $\Delta\ddot{Y}$, or $\Delta\ddot{y}$). In turn, $\sigma^2(\hat{X})$ is a sum of genetic and non-genetic variances in proportions that vary with the structure of criterion unit comparisons (see Chapter X). When the selection criterion is to be a linear function of two or more measures, estimates of variances and covariances of those measures are required for approximation of optimum weights to be accorded the individual measurements in the structure of that criterion unless all value traits, and no others, are to be employed (Chapter XII). Additive genetic variance or covariance or an analog of one of those is the genetic component of the approximations to the expectation of change in average value of genotypes ($\Delta\ddot{Y}$) in response to selection when (a) criterion units are non-inbred individuals or families of non-inbred individuals and the target population is a random mating population or is obtained by mating random individuals of two such populations and (b) impacts of effective population size are ignored. Thus, the

$$\Sigma_i 2q_i(1-q_i)[1+(1-2q_i)a_{xi}][1+(1-2q_i)a_{ci}]u_{xi}u_{ci}$$
$$+\Sigma_i\Sigma_{j\neq i}\, 2(pt-rs)_{ij}[1+(1-2q_j)a_{xj}][1+(1-2q_i)a_{ci}]u_{xj}u_{ci}$$

of 6.17 is the additive genetic covariance between the selection criterion (X) and the response criterion (C) orthe additive genetic variance of X when the selection criterion is also the response criterion (C = X). As another example, the

$$\Sigma_i q_{1i}(1-q_{1i})[1+(1-2q_{2i})a_{xi}][1+(1-2q_{2i})a_{ci}]u_{xi}u_{ci}$$

of 9.11 is an analog of additive genetic covariance between traits X and C when $C \neq X$ or an analog of the additive genetic variance of X, when the selection criterion (X) is also the response criterion.

The primary sources of estimates of these various parameters are (1) analyses of variance of data from family comparisons and (2) the covariance between phenotypes of parents and offspring.

15.1.1. Analyses of Variance and Covariance—Plant Family Comparison Data

Because real data are always measures of phenotype, the expectations of statistics that can be computed vary with the nature of the data employed. That this should be so follows from the discussion of models for phenotype measures in Chapter X. The steps leading from data to estimates are the following.

(1) Formulation of an appropriate model showing effects that contribute to variation among the elements of the data.

(2) Specification of the analysis to be computed.

(3) Derivation of the expectations of relevant mean squares or mean products.

(4) Symbolic statements of estimates as functions of the mean squares or mean products of the analysis specified.

(5) Computations in accord with (2) and (4).

15.1.1.1. ESTIMATION OF VARIANCES

Consider data on any quantitative trait from a randomized block comparison of n families in each of v locations in each of t years, let the locations be the same ones in all years of the comparison, let the number of replications (r) be the same in all year-locations, and assume a single measurement per family per replication. In accord with discussion and definitions in Chapter X, the model identified as 10.9 is appropriate for the data just described.

$$X_{ijkm} = \ddot{X} + \bar{z}_i + a_j + b_k + (ab)_{jk} + (\bar{z}a)_{ij} + (\bar{z}b)_{ik} + (\bar{z}ab)_{ijk} + B_{jkm} + e_{ijkm} \qquad (10.9)15.1$$

in which symbols have meanings assigned in Chapter X. The appropriate analysis of variance in the case of this data is shown in Table 15.1. Expectations are shown in the case of the mean squares (labelled M_1, M_2, etc.) from which estimates of the variance among families, $\sigma^2(\bar{z})$, and the variances of effects arising from interaction between families and macro-environments can be obtained. Assumptions on which the mean square expectations in Table 15.1 are based, and which are therefore an integral portion of the working model, include the following.[1]

(1) That the families on which data were obtained were a random sample from the genetic population of which the variance estimates obtained will be said to be descriptive.

(2) That the macro-environments in which data were obtained are random members from the environment population to which the variance estimates obtained will be said to apply. In this connection remember that none of the variances of interest can be expected to be the same in different environment populations.

Table 15.1. ANOVA, comparison of families in v locations in t years

Source of Variance	d.f.	Mean Square	Expectation of Mean Square
Years (Y)	t-1		
Locations (L)	v-1		
Y x L	(t-1)(v-1)		
Repls. within Y &L	tv(r-1)		
Families (F)	n-1	M_1	$\sigma^2(e) + r\sigma^2(\bar{z}ab) + rt\sigma^2(\bar{z}a) + rv\sigma^2(\bar{z}b) + rvt\sigma^2(\bar{z})$
F x Y	(n-1)(t-1)	M_2	$\sigma^2(e) + r\sigma^2(\bar{z}ab) + rv\sigma^2(\bar{z}b)$
F x L	(n-1)(v-1)	M_3	$\sigma^2(e) + r\sigma^2(\bar{z}ab) + rt\sigma^2(\bar{z}a)$
F x Y x L	(n-1)(t-1)(v-1)	M_4	$\sigma^2(e) + r\sigma^2(\bar{z}ab)$
F x repls. within Y& L	vt(n-1)(r-1)	M_5	$\sigma^2(e)$

(3) That the family x environment interaction variances and the plot error variance, are uniform among macro-environments.

The process by which such mean square expectations are derived is sketched in Appendix 1. The obvious estimators of the variances of variable elements in 15.1 are shown in Table 15.2. Estimates there, as elsewhere, are signified by a preceeding bullet (•).

Table 15.2. Estimates of components of Table 15.1 mean squares

•$\sigma^2(e) = M_5$

•$\sigma^2(\bar{z}ab) = (M_4 - M_5)/r$

•$\sigma^2(\bar{z}a) = (M_3 - M_4)/rt$

•$\sigma^2(\bar{z}b) = (M_2 - M_4)/rv$

•$\sigma^2(\bar{z}) = (M_1 - M_2 - M_3 + M_4)/rvt$

Frequently, available data are from comparisons made in more than one location but only during one year. The appropriate analysis of variance is then the collapsed version (Table 15.3) obtained when t is set equal to one in the analysis shown in Table 15.1. It is apparent from the

Table 15.3. ANOVA, comparison of families in v locations in one (t = 1) year

Source of variance	d.f.	Mean Square	Expectation of Mean Square
Locations (L)	v - 1		
Repls. within locations	v(r-1)		
Families (F)	n - 1	M_1	$\sigma^2(e) + r[\sigma^2(\bar{z}ab + \sigma^2(\bar{z}a)] + rv[\sigma^2(\bar{z}b) + \sigma^2(F)]$
F x L	(n-1)(v-1)	M_3	$\sigma^2(e) + r\,[\sigma^2(\bar{z}ab + \sigma^2(\bar{z}a)]$
F x Repls w. locations	(n-1)v(r-1)	M_5	$\sigma^2(e)$

mean square expectations that such data do not provide the basis for separate estimates of the family variance and each of the three family x environment interaction variances. Instead the possible estimates are those shown below.

Estimate of	Estimator
$\sigma^2(e)$	$= M_5$
$\sigma^2(\bar{z}a) + \sigma^2(\bar{z}ab)$	$= (M_3 - M_5)/r$
$\sigma^2(\bar{z}) + \sigma^2(\bar{z}b)$	$= (M_1 - M_3)/rv$

The still less informative analysis possible when the data are from a comparison of families in only one location and one year is shown in Table 15.4. It is the analysis of Table 15.1 with both v and t set equal to one. Obviously, in this instance, only the sum of the family variance and the three family x environment interaction variances can be estimated. But note that if $\sigma^2(\bar{z}ab) + \sigma^2(\bar{z}a) + \sigma^2(\bar{z}b) + \sigma^2(\bar{z})$ is > 0.0, so also is $\sigma^2(\bar{z})$ > 0.0. There can be no genotype-environment interaction variance when there is no genetic variance.

Table 15.4. ANOVA, comparison of families in one location and one year

Source of Variance	d.f.	Mean Square	Expectation of Mean Square
Replications	r - 1		
Families	n - 1	M_1	$\sigma^2(e) + r[\sigma^2(\bar{z}ab) + \sigma^2(\bar{z}a) + \sigma^2(\bar{z}b) + \sigma^2(\bar{z})]$
Families x Repls.	(n-1)(r-1)	M_2	$\sigma^2(e)$

For the sake of focus on essentials the forms shown in Tables 15.1, 15.3, and 15.4 assume that the numbers of plots per replication in the field comparisons were n, the number of families, with no restriction on that number. In practice, data for estimation can come from field comparisons conducted for the primary purpose of getting data for variance estimation but is more frequently obtained in the course of selection programs. In the latter case plots per block would often be restricted for the sake of restricting the magnitude of plot error variance (which is expected to be larger when block size is larger). In work conducted for the primary purpose of variance estimation block size might also be restricted, in that case to reduce the sampling errors of variance estimates to be obtained (see discussion of sampling errors later in this section). Let us suppose that F families are divided into sets of size n and that the families of each set are compared (separately) in r replications at each of v locations in each of t years. Then whether (1) all of the data are treated in one overall analysis or (2) separate analyses are conducted of the data from each family set followed by pooling of the relevant sums of squares, and their respective degrees of freedom, before computation of mean squares, the results of primary interest (shown in Table 15.5) are the same.

Each of the mean square expectations shown in Table 15.5 is the same as in Table 15.1; hence the estimates of interest are all made in the same ways, e.g.

$$\bullet\sigma^2(\bar{z}a) = (M_3 - M_4)/rt.$$

Table15.5. ANOVA, comparison of families within sets in v locations in each of t years

Source of Variance	d.f.	Mean square	Expectation of Mean Square
Families w. sets	(F/n)(n-1)	M_1	$\sigma^2(e) + r\sigma^2(\bar{z}ab) + rt\sigma^2(\bar{z}a) + rv\sigma^2(\bar{z}b) + rvt\sigma^2(\bar{z})$
F x Y within sets	(F/n)(n-1)(t-1)	M_2	$\sigma^2(e) + r\sigma^2(\bar{z}ab) + rv\sigma^2(\bar{z}b)$
F x L within sets	(F/n)(n-1)(v-1)	M_3	$\sigma^2(e) + r\sigma^2(\bar{z}ab) + rt\sigma^2(\bar{z}a)$
F x Y x L w. sets	(F/n)(n-1)(t-1)(v-1)	M_4	$\sigma^2(e) + r\sigma^2(\bar{z}ab)$
F x repls. within Y, L, and sets	(F/n)(n-1)vt(r-1)	M_5	$\sigma^2(e)$

15.1.1.2. ESTIMATION OF COVARIANCES

Consider the type of family comparison specified in the preceeding subsection and data consisting of a single measurement per plot on each of two quantitative traits. The model for measurements would logically be the same, 15.1, as for the analysis of variance and the form of the analysis of covariance between the two traits the same as that shown in Table 15.1 for the analysis of variance of a single trait. It follows that the appropriate analysis of covariance, shown in Table 15.6, is parallel to the analysis of variance shown in Table 15.1.
The covariance estimators are obviously parallel to the variance estimators using the mean squares of Table 15.1. For example, the estimate of $\sigma(\bar{z})_{ij}$ is

$$\bullet\sigma(\bar{z})_{ij} = (P_1 - P_2 - P_3 + P_4)/rvt$$

Just as obviously the parallels go on to the analyses and estimators in the cases of data from

Table 15.6. Analysis of covariance, family comparison at v locations in t years

Source of Covariance	d.f.	Mean Product	Expectation of Mean Product
Years	t-1		
Locations	v-1		
Y x L	(t-1)(v-1)		
Repls. within Y&L	tv(r-1)		
Families	(n-1)	P_1	$\sigma(e)_{ij} + r\sigma(\bar{z}ab)_{ij} + rt\sigma(\bar{z}a)_{ij} + rv\sigma(\bar{z}b)_{ij} + rvt\sigma(\bar{z})_{ij}$[a]
F x Y	(n-1)(t-1)	P_2	$\sigma(e)_{ij} + r\sigma(\bar{z}ab)_{ij} + rv\sigma(\bar{z}b)_{ij}$
F x L	(n-1)(v-1)	P_3	$\sigma(e)_{ij} + r\sigma(\bar{z}ab)_{ij} + rt\sigma(\bar{z}a)_{ij}$
F xY x L	(n-1)(t-1)(v-1)	P_4	$\sigma(e)_{ij} + r\sigma(\bar{z}ab)_{ij}$
F x repls. w. Y&L	vt(n-1)(r-1)	P_5	$\sigma(e)_{ij}$

[a] $\sigma(\bar{z})_{ij}$ symbolizes the covariance between the $\bar{z}$ (family) effects of traits i and j, $\sigma(\bar{z}a)_{ij}$ symbolizes the covariance between the ($\bar{z}a$) effects of traits i and j, etc.

family comparisons in v locations in one (t=1) year and from comparisons in one location in one year.

15.1.1.3. SAMPLING ERRORS OF ESTIMATES

It is known that if all contributing effects are normally distributed, the variance of a mean square is a function of its expectation as follows.[2]

$$\sigma^2(M) = 2[E(M)]^2/f \qquad 15.2$$

where

M symbolizes the mean square, associated with

E(M), its expectation, and

f, the degrees of freedom associated with the mean square

The counterpart to 15.2 with respect to the variance of a mean product is

$$\sigma^2(P_{12}) = (1+\rho^2)[E(M_1)E(M_2)]/f$$

in which

P_{12} symbolizes a mean product between traits 1 and 2,

M_1 and M_2 are the associated mean squares of traits 1 and 2,

ρ is the product moment correlation between replicate pairs of M_1 and M_2, and

f = the degrees of freedom associated with P_{12}, M_1, and M_2.

Note that in the limiting case of a mean product of a variable with itself $E(M_1) = E(M_2 = E(M)$ and $\rho = 1.0$ so that

$$\sigma^2(P_{12}) = \sigma^2(M) = 2[E(M)]^2/f$$

The variance component estimators listed in Table 15.2 are all linear functions of mean squares that are independent in their sampling errors. Consequently, as outlined in Appendix 1, the variance of each estimator can be expressed as a function of the expectations of the mean squares involved. Consider the estimate of $\sigma^2(\bar{z})$ (the family variance) listed in Table 15.2. Because that estimate is $\bullet\sigma^2(\bar{z}) = (M_1 - M_2 - M_3 + M_4)/rvt$, its variance is

$$\sigma^2[\bullet\sigma^2(\bar{z})] = \{2[E(M_1)]^2/f_1 + 2[E(M_2)]^2/f_2 + 2[E(M_3)]^2/f_3 + 2[E(M_4)]^2/f_4\}/(rvt)^2 \qquad 15.3$$

in which subscripts identify degrees of freedom with mean squares. An obvious problem is that the values of the E(M)'s are not known. It can be dealt with in either of two ways.

(1) In the first of these, observed mean square values are substituted for their expectations, dividing in each case by (f + 2) instead of f to correct for bias in the mean square as an estimate of its population value, its expectation. For example, $2(M_2)^2/(f_2 + 2)$ would be substituted for $2[E(M_2)]^2/f_2$. When data have been obtained and an analysis of variance completed, this is reasonable. However, estimates obtained in this way are too large or too small because of the inevitable deviations of mean squares employed from their expectations. As a consequence they do not provide a completely reliable basis for setting confidence limits.

(2) In a second procedure, arbitrary values are assigned to mean squares as a basis for exploring the possible magnitudes of error in estimates of variance components. This can be

useful in the design of work to be conducted as the basis for obtaining one or more variance component estimates. Consider work designed to obtain an estimate of a family variance from an analysis of the type set out in Table 15.1. Values assigned should be consistent with one or more of the objectves of the investigator. Wanting information on the magnitude of a family variance, the data to be collected should at minimum be sufficient to enable an estimate of a specified magnitude to be judged significantly larger than zero. From previous experience with the crop, by himself or others, the investigator would be able to assign a reasonable value for $\sigma^2(e)$. The next step would be to inquire how large an estimate could be if the actual variance among families were zero. In that event it would be entirely logical to assign zero as the magnitude of $\sigma^2(\bar{z}a)$, $\sigma^2(\bar{z}b)$ and $\sigma^2(\bar{z}ab)$ as well as $\sigma^2(\bar{z})$. Suppose the unit of measurement were such that $\sigma^2(e) = 12$ could logically be assumed. That set of assumptions would lead to the assumption that

$$E(M_1) = E(M_2) = E(M_3) = E(M_4) = 12.$$

In the event that 101 families were to be grown in a randomized block comparison at each of five locations in each of two years with two replications in each of the ten year-locations the variance of $\bullet\sigma^2(\bar{z})$ would, from 15.3 and degrees of freedom shown in Table 15.1, be

$$[2(144)/100 + 2(144)/100 + 2(144)/400 + 2(144)/400]/400 = 0.018$$

from which the standard deviation (s) would be 0.134 and it would appear that any estimate of $\sigma^2(\bar{z})$ as large or larger than 0.3 could be assumed quite confidently to be statistically different from zero (note that, given normol distribution, the probability of $\bullet\sigma^2(\bar{z}) \geq 2s = 0.268$ would be slightly less than 0.023). It is apparent, as well, that if n (the number of families) were 51 instead of 101, the above variance would be twice as large and the standard deviation would be .190 instead of 0.134.

Let us extend the example by supposing the investigator wished to test whether $\sigma^2(\bar{z})$, the family variance were larger than 6. In that case, in addition to assigning 12 as the value of $\sigma^2(e)$, the assigned value of $\sigma^2(\bar{z})$ would logically be 6. Appropriate values to be assumed for the relevant interaction variances are not so obvious. Let us suppose that their sum were equal to $\sigma^2(\bar{z})$ and that all three were equal in magnitude. To summarize, assigned values would as follows: $\sigma^2(\bar{z}) = 6$, $\sigma^2(\bar{z}a) = \sigma^2(\bar{z}b) = \sigma^2(\bar{z}ab) = 2$, and $\sigma^2(e) = 12$. If, as before, 101 families were to be grown in five locations in each of two years with two replications in each of the ten year-locations, the consequences with respect to mean square expectations would be

$$E(M_4) = 12 + 2(2) = 16,$$

$$E(M_3) = 12 + 2(2) + 4(2) = 24,$$

$$E(M_2) = 12 + 2(2) + 10(2) = 36, \text{ and}$$

$$E(M_1) = 12 + 2(2) + 4(2) + 10(2) + 20(6) = 164.$$

Substituting in 15.3 yields

$$[2(16)^2/400 + 2(24)^2/400 + 2(24)^2/100 + 2(164)^2/100]/(20)^2 = 1.384$$

so that the standard deviation would be 1.18. Then if anslysis of the data obtained yielded an estimate of $\sigma^2(\bar{z}) \geq 8.5$ it could reasonably be concluded that $\sigma^2(\bar{z}) \geq 6.0$ unless the data indicated that assumptions concerning variances other than $\sigma^2(\bar{z})$ had been seriously in error.

15.1.2. Genetic Content of Family Variances and Covariances

The variance among families, $\sigma^2(\bar{z})$, derives basically from differences among the average values of family member genotypes.[3] There are different kinds of families and it is intuitively obvious that the genetic component of family variance should vary in type and/or magnitude from one kind of family to another. Types of genetic variance to which the reader has been introduced are (1) total genetic variance (section 9.6.4), (2) additive genetic variance (Chapters VI and VIII), and (3) analogs of additive genetic variance (sections 8.3 and 9.2). Each of these has value for predicting response ($\Delta\ddot{Y}$) to selection in at least one sort of selection program. Response is proportional to additive genetic variance in all single population (T= S) recurrent selection programs in which the criterion units are composed of non-inbred S population individuals and the response criterion is the selection criterion (Chapers VI and VIII). In RRS and RSSC in which the target for improvement is a population cross, the expected response is a function of an analog (referred to above) of additive genetic variance (sections 9.2 and 9. 3). In the case of non-recurrent selection among single crosses between pure lines or clonal families (in species that can be propagated asexually) expected response is proportional to total genetic variance (section 9.6.4).

Before going on to consideration of the genetic nature of particular family variances and covariances, it will be useful to identify and define dominance and epistatic variance, the two that together with additive genetic variance sum to total genetic variance in random mating populations. Symbolically,

$$\sigma_Y^2 = \sigma_g^2 + \sigma_d^2 + \sigma_i^2$$

where σ_Y^2 is the total genetic variance among the individuals of any single random mating population and σ_g^2, σ_d^2, and σ_i^2 are the additive, dominance, and epistatic components respectively, of the total genetic variance of such populations. Of course when there is no epistasis σ_i^2 = zero and

$$\sigma_Y^2 = \sigma_g^2 + \sigma_d^2 \qquad 15.4$$

As explained in Chapter V, the following symbolic description of genotype value is appropriate when absence of epistasis is assumed.

$$Y = \alpha + y_1 + y_2 + \ldots.. y_i + \ldots\ldots. y_n = \Sigma_i y_i \qquad (5.7)\ 15.5$$

where

α is a constant, the value of the genotype that is homozygous for the less favorable allele at all loci,

y_i is a quantity that varies with genotype at the at the i-th locus, and

n is the total number of loci (genes).

It follows from 15.5 that

$$\sigma_Y^2 = \sigma_g^2 + \sigma_d^2 = \Sigma_i \sigma^2(y_i) + \Sigma_i\Sigma_{j\neq i}\sigma(y_i,y_j)$$

Assuming no epistasis along with HW frequencies of genotypes at all loci (but not linkage equilibrium) and proceeding in the usual ways, the following expression is obtained (see Appendix 15.1).

$$\sigma_Y^2 = \sigma_g^2 + \sigma_d^2 = \Sigma_i 2q_i(1-q_i)\{1+2(1-2q_i)a_i + [1-2q_i+2q_i^2]a_i^2\}u_i^2$$

$$+2\Sigma_i\Sigma_{j\neq i}\{(pt-rs)_{ij}u_iu_j + (pt-rs)_{ij}(1-2q_i)a_ia_ju_j + (pt-rs)_{ij}(1-2q_j)u_ia_ju_i$$
$$+[pt+rs-2(p+r)(s+t)(p+s)(r+t)]a_iu_ia_ju_j\} \qquad \text{A15.1.5}$$

Then subtracting σ_g^2, as given by A15.1.6, it is shown in the same appendix that

$$\sigma_d^2 = 4\Sigma_i q_i^2(1-q_i)^2 a_i^2 u_i^2 + 4\Sigma_i\Sigma_{j\neq i}[(pt-rs)_{ij}]^2 a_iu_ia_ju_j \qquad 15.6$$

15.1.2.1. GENETIC VARIANCES AMONG FAMILIES WHEN THERE IS NO EPISTASIS

The procedure for obtaining the structure of these variances (or covariances) is straight forward. It will be illustrated by considering half-sib test cross families. It is apparent from 15.5 that the appropriate symbolic representation of the genetic portion of any family mean (in the absence of epistasis) is

$$\bar{z} = \alpha + \bar{y}_1 + \bar{y}_2 + \cdots\cdots \bar{y}_i + \cdots\cdots \bar{y}_n$$

from which

$$\sigma^2(\bar{z}) = \Sigma_i \sigma^2(\bar{y}_i) + \Sigma_i\Sigma_{j\neq i}\sigma(\bar{y}_i,\bar{y}_j) \qquad 15.7$$

and $\sigma(\bar{z}_x,\bar{z}_c) = \Sigma_i \sigma(\bar{y}_{xi},\bar{y}_{ci}) + \Sigma_i\Sigma_{j\neq i}\sigma(\bar{y}_{xi},\bar{y}_{cj})$

Assuming a random mating population, that constant parents are random members of that population, and that the random parents of each family are random members of the tester population, the joint distribution of $\bar{y}_i$ and $\bar{y}_j$ is shown in Table 15.7 for two traits, X and C. Note that the coded vaues of $\bar{y}_{xi}$ and $\bar{y}_{xj}$ are the same as those in Table 8.1 and that those of $\bar{y}_{ci}$ and $\bar{y}_{cj}$ are structurally the same in accord with the section 8.1.2 demonstration and statement that they would be for every locus (gene) and any trait. Coded values of y_{xi} in the case of constant parents are listed so that covariance between parents and offspring can be derived from the content of this table (section 15.1.3). Working from the table the following are obtained.

$$E^*(\bar{y}_i) = (p^2+2pr+r^2-s^2-2st-t^2)Z_i = [(p+r)^2-(s+t)^2]Z_i = (p+r-s-t)Z_i$$

$$E^*(\bar{y}_j) = (p^2+2ps+s^2-r^2-2rt-t^2)Z_j = (p+s-r-t)Z_j$$

$$\sigma^2(\bar{y}_i) = [p^2+2pr+r^2+s^2+2st+t^2-(p+r-s-t)^2](Z_i)^2$$
$$= 2(p+r)(s+t)(Z_i)^2 = 2q_i(1-q_i)[1+(1-v_i)a_i]^2u_i^2/4 \qquad 15.8$$

and $\sigma(\bar{y}_i,\bar{y}_j) = [p^2-r^2-s^2+t^2-(p+r-s-t)(p+s-r-t)]Z_iZ_j$

$$= (pt-rs)_{ij}[1+(1-v_i)a_i][1+(1-2v_j)a_j]u_iu_j/2 \qquad 15.9$$

Substituting in 15.7 in terms of 15.8 and 15.9 yields

Table 15.7. Joint distribution of y_{xj} in constant parent and of $\bar{y}_{xi}$, $\bar{y}_{xj}$, $\bar{y}_{ci}$, and $\bar{y}_{cj}$ in half-sib test cross families

Constant Parent			Family			
Genotype	Frequency	$y_{xi} - u_{xi}$	$*(\bar{y}_{xi})$	$*(\bar{y}_{xj})$	$*(\bar{y}_{ci})$	$*(\bar{y}_{cj})$
B_iB_j/B_iB_j	p^2	u_i	Z_{xi}	Z_{xj}	Z_{ci}	Z_{cj}
B_iB_j/B_ib_j	$2pr$	u_i	Z_{xi}	0	Z_{ci}	0
B_iB_j/b_ib_j	r^2	u_i	Z_{xi}	$-Z_{xj}$	Z_{ci}	$-Z_{cj}$
B_iB_j/b_iB_j	$2ps$	a_iu_i	0	Z_{xj}	0	Z_{cj}
B_iB_j/b_ib_j	$2pt$	a_iu_i	0	0	0	0
B_ib_j/b_iB_j	$2rs$	a_iu_i	0	0	0	0
B_ib_j/b_ib_j	$2rt$	a_iu_i	0	$-Z_{xj}$	0	$-Z_{cj}$
b_iB_j/b_iB_j	s^2	$-u_i$	$-Z_{xi}$	Z_{xj}	$-Z_{ci}$	Z_{cj}
b_iB_j/b_ib_j	$2st$	$-u_i$	$-Z_{xi}$	0	$-Z_{ci}$	0
b_ib_j/b_ib_j	t^2	$-u_i$	$-Z_{xi}$	$-Z_{xj}$	$-Z_{ci}$	$-Z_{cj}$

$*(\bar{y}) = (\bar{y}) - [(1+2v)u + au]/2$ (for explanation see section 8.1.2)

$Z_{xi} = [1+(1-2v_{xi})a_{xi}]u_{xi}/2$ $\quad$ $Z_{xj} = [1+(1-2v_{xj})a_{xj}]u_{xj}/2$

$Z_{ci} = [1+(1-2v_{ci})a_{ci}]u_{ci}/2$ $\quad$ $Z_{cj} = [1+(1-2v_{cj})a_{cj}]u_{cj}/2$

$$\sigma^2(\bar{z}) = \Sigma_i q_i(1-q_i)[1+(1-2v_i)a_{xi}]^2u_i^2/2$$

$$+ \Sigma_i\Sigma_{j\neq i}(pt-rs)_{ij}[1+(1-2v_i)a_i][1+(1-2v_j)a_j]u_{xi}u_{xj}/2 = \sigma^2_{ga}/4 \qquad 15.10$$

Comparing with 6.19 we see that 15.10 is the same as one-fourth of additive genetic variance with the exception that the v's of 15.10 take the place of q's in 6.19. It is because of this identity in form that σ^2_{ga} is referred to as an analog of additive genetic variance. When the constant and random parents are members of the same population the v's of 15.10 are appropriately replaced by q's. Hence $\sigma^2(\bar{z}) = \sigma^2_g/4$ in the case of within population half sib families, as indicated in Table 15.8.

Working again from Table 15.7 and following procedure parallel to that used to obtain $\sigma^2(\bar{z})$, yields

$$\sigma(\bar{z}_x,\bar{z}_c) = \Sigma_i q_i(1-q_i)[1+(1-2v_i)a_{xi}][1+(1-2v_i)a_{ci}]u_{xi}u_{ci}/2$$

$$+ \Sigma_i\Sigma_{j\neq i}(pt-rs)_{ij}[1+(1-2v_i)a_{xi}][1+(1-2v_j)a_{cj}]u_{xi}u_{cj}/2 = \sigma_{gcxa}/4 \qquad 15.11$$

in which σ_{gcxa} symbolizes an analog of additive genetic covariance between traits X and C. Again, if the random and constant parents are members of the same random mating population, the v's of 15.11 are logically replaced by q's. When that is done 15.11 is seen, when compared with 6.20, to be one-quarter of the additive genetic covariance between X and C.

The family types on which data are most frequently available are listed in Table 15.8 along with the nature of among family variance for all except S_n families. Genetically

Table 15.8. Genetic composition of family variances

Type of Families	Composition of variance
Genetically homogeneous families	σ^2_Y
Half sib test cross families	$\sigma^2_{ga}/4$[a]
Half sib families	$\sigma^2_g/4$
Full sib families	$\sigma^2_g/2 + \sigma^2_d/4$
Full sib within half sib families	$\sigma^2_g/4 + \sigma^2_d/4$
S_n families	(see section 15.1.2.2)

[a] σ^2_{ga} symbolizes the analog of additive genetic variance relevant to the direct response to selection when the target population is the cross between the S population and the tester population.

homogeneous families are those in which all members have the same genotype as in single cross hybrids between pure lines, clonal families in species that that can be asexually reproduced, or identical twins (or more than two siblings) in cattle, humans, or other species.[4]

15.1.2.2. GENETIC VARIANCE AMONG S_n FAMILIES WHEN THERE IS NO EPISTASIS

Consider the special case in which the non-inbred ancestors were random individuals from a random mating, linkage equilibrium population and parents of the S_n families were derived without selection by n-1 successive generations of self fertilization. In this special situation the probabilities of parents of the S_n families and the expectations of $\hat{y}_{xi} - u_{xi}$ and $\hat{y}_{xj} - u_{xj}$ are those shown below.

Parent Genotype	Frequency	$\hat{y}_{xi} - u_{xi}$	$\hat{y}_{xj} - u_{xj}$
B_iB_j/B_iB_j	$p^2 + (A+pr+ps)h + A(B-1)/2$	u_{xi}	u_{xj}
B_iB_j/B_ib_j	$(A + 2pr)(1-h) - AB$	u_{xi}	$a_{xj}u_{xj}/2$
B_ib_j/B_ib_j	$r^2 + (A+pr+rt)h + A(B-1)/2$	u_{xi}	$-u_{xj}$
B_iB_j/b_iB_j	$(A+2ps)(1-h) - AB$	$a_{xi}u_{xi}/2$	u_{xj}
B_iB_j/b_ib_j	AB	$a_{xi}u_{xi}/2$	$a_{xj}u_{xj}/2$
B_ib_j/b_iB_j	AB	$a_{xi}u_{xi}/2$	$a_{xj}u_{xj}/2$
B_ib_j/b_ib_j	$(A+2rt)(1-h) - AB$	$a_{xi}u_{xi}/2$	$-u_{xj}$
b_iB_j/b_iB_j	$s^2 + (A+ps+st)h + A(B-1)/2$	$-u_{xi}$	u_{xj}
b_iB_j/b_ib_j	$(A + 2st)(1-h) - AB$	$-u_{xi}$	$a_{xj}u_{xj}/2$
b_ib_j/b_ib_j	$t^2 + (A+st+rt)h + A(B-1)/2$	$-u_{xi}$	$-u_{xj}$

Here $A = pt + rs$, $B = [(1-2\gamma+2\gamma^2)/2]^{n-1}$,

γ = the probability of recombination between the i-th and j-th loci, and

$h = 1-(1/2)^{n-1}$ = the coefficient of inbreeding of the family parents.

The genotype frequencies shown here are those listed in Appendix 8.1 with h substituted for f because we are concerned here with the genotypes of the parents of S_n families instead of those of S_n individuals and recognition that $(pt-rs) = 0.0$ by virtue of linkage equilibrium in the random mating source population. Working from the above bivariate distribution we obtain

$$\sigma^2(\bar{z}) = 2\Sigma_i q_i(1-q_i)\{(1+h)+(1-h)(1-2q_i)a_{xi} + (1-h)[1-2q_i(1-q_i)(1-h)]a_{xi}^2/4\}u_{xi}^2$$

$$+2\,\Sigma_i\Sigma_{j\neq i} q_i(1-q_i)q_j(1-q_j)[B - (1-h)^2]a_iu_ia_ju_j \qquad 15.12$$

Does this expression have value in relation to unresolved questions of quantitative genetics and its implications for plant and animal breeding? This question draws immediate attention to the fact that despite linkage equilibrium in the random-mating source population, linkages among genes may have a significant bearing on the genetic variance among S_n families. It is apparent that before going further the possible magnitudes of $[B - (1-h)^2]$ must be explored.

Note first that when $n-1 = 0$, $[B - (1-h)^2] = 0.0$ because then both B and $(1-h)^2$ are equal to one. Whatever the value of γ, $[(1-2\gamma+2\gamma^2)/2]^0 = 1.0$ and, in like manner, $(1-h)^2 = [1-(1/2)^{n-1}]^2 = 1.0$. Thus linkages have no bearing on the genetic variance among S_1 family means. This is particularly fortunate because frequencies of genotypes are least likely to differ from their theoretical probabilities by reason of natural selection in the case of the non-inbred parents of S families. Values of B, $(1-h)^2$, and $[B - (1-h)^2]$ for some other combinations of n and γ are presented below. It comes as no surprise that the linkage contribution to the genetic variance is largest when the recombination probability is small or

n	$(1-h)^2$	$\gamma = 0.1$		$\gamma = 0.3$		$\gamma = 1/2$	
		B	$B-(1-h)^2$	B	$B-(1-h))^2$	B	$B-(1-h)^2$
2	0.250	.410	.160	.290	.040	.250	0
3	0.063	.168	.105	.084	.021	.063	0
4	0.016	.069	.053	.024	.008	.016	0

that it is zero when genes recombine independently ($\gamma = 1/2$). Finally, we see that at most the largest value of $[B - (1-\hat{f})^2]$ shown above is 0.16. This together with the fact that the single gene contribution to $\sigma^2(\bar{z})$ is a function of $q_i(1-q_i)$ whereas that of the linkage contribution of a pair of genes is a function of $q_i(1-q_i)q_j(1-q_j)$ indicates that the total linkage contribution cannot be the source of a significant portion of $\sigma^2(\bar{z})$ unless the a's are uniformly large and the number of genes affecting the trait is great.

The most obvious questions concerning the utility of estimates of the genetic variance among S_n families are: (1) would serious bias be involved if they were used as substitute sources of information concerning the magnitude of additive genetic variance and (2) how does the magnitude of 15.12 compare to that of $\Sigma_i\sigma(\hat{q}_i,\hat{X})\partial\ddot{Y}/\partial q_i$, the genetic component of $E(\Delta\ddot{Y}_x)$

when S_n families are both the selection units and the criterion units in selection for improvement of the selection population? These questions will be explored in the case of the genetic variance among S_1 families, the variation of 15.12 least likely to be seriously affected by natural selection among the genotypes of potential parents or the genotypes of family members.

Ignoring the final term of 15.12 [because $B - (1-h)^2$ equals zero in the case of S_1 families] and setting $h = 0$, the following is obtained as the i-th locus contribution to the genetic variance among S_1 families.

$$\sigma^2(\bar{z})_{iS1} = 2q_i(1-q_i)Z_i u_{xi}^2 \qquad 15.13$$

where $Z_i = 1+(1-2q_i)a_{xi} + [1-2q_i + 2q_i^2]a_{xi}^2/4$

For comparison, the i-th locus contribution to the additive genetic variance in a random mating population at linkage equilibrium is, from 6.8,

$$\sigma_{gxi}^2 = 2q_i(1-q_i)Z_{gi} u_{xi}^2 \qquad 15.14$$

where $Z_{gi} = [1+(1-2q)a]^2$

It is apparent that the ratio of the single locus contributions to the variances symbolized in 15.13 and 15.14 is $\sigma^2(\bar{z})_{iS1}/\sigma_{gxi}^2 = Z_i/Z_{gi}$

Assuming specific values of q and a this ratio is easily computed. A set of values is shown below.

	Z_i/Z_{gi}			
q	a=0	a = 0.3	a = 0.7	a=1.0
0.3	1.0	0.90	0.82	0.79
0.5	1.0	1.01	1.06	1.12
0.7	1.0	1.15	1.53	2.07
0.9	1.0	1.35	2.78	10.12

Remembering (1) that both allele frequency (q) and level of dominance (a) cannot be expected to be uniform from one locus to another and (2) that the above ratios pertain to (a) the genetic variance among S_1 families when the source population to which family parents belonged is in linkage equilibrium and (b) the additive genetic variance of that linkage equilibrium population, the following are worth noting.

(1) When a = 0, the Z_i/Z_{gi} ratio is 1.0, i.e., the magnitudes of $\sigma^2(\bar{z})_{iS1}$ and σ_{gxi}^2 are the same. Thus, when there is little or no inbreeding effect (presumably because the distribution of a-values is centered near zero) as indicated by little difference between average phenotypes of S_1 families and the source population, there should be little bias in the genetic variance among S_1 families as an estimate of the additive genetic variance in the source population. Moreover, because $\sigma^2(\bar{z})_{S1} \approx \sigma_g^2$ where as the estimates of the genetic variance among half-sib

families must be multiplied by four to obtain estimates of σ_g^2, the variance of estimates obtained using S_1 families should be much less.

(2) When a-values are greater than zero and allele frequencies are greater tha 1/2 the Z_i/Z_{gi} ratio is greater than 1.0 and as both a and q approach 1.0, the ratio becomes large. Thus in the case of traits that exhibit negative inbreeding effects additive genetic variance may be seriously overestimated by estimates of the genetic variance among S_1 families. Conversely, given a-values less than zero, the inbreeding effect would be expected to be positive and additive genetic variance would be underestimated by $\sigma^2(\bar{z})_{S1}$.

The covariance of $\hat{q}_i$ with $\hat{X}$ when S_n families are both the selection and criterion units and the S population is in linkage equilibrium is (see Table 8.2)

$$\sigma(\hat{q}_i,\hat{X}) = q_i(1-q_i)[(1+h)+(1-h)(1-2q_i)a_{xi}/2]u_{xi}$$

and in accord with 5.11,

$$\partial\ddot{y}_{xi}/\partial q_i = 2[1+(1-2q)a_{xi}]u_{xi}$$

Then

$$\sigma(\hat{q}_i,\hat{X})\partial\ddot{y}_{xi}/\partial q_i = 2q_i(1-q_i)[(1+h)+(1-h)(1-2q_i)a_{xi}/2][1+(1-2q)a_{xi}]u_{xi}^2$$

and setting h=0 we obtain in the case of S_1 families

$$\sigma(\hat{q}_i,\hat{X})\partial\ddot{y}_{xi}/\partial q_i = 2q_i(1-q_i)[1+(1-2q_i)a_{xi}/2][1+(1-2q_i)a_{xi}]u_{xi}^2$$

$$= 2q_i(1-q_i)[1+3(1-2q_i)a_{xi}/2+(1-4q_i+4q_i^2)a_{xi}^2/2]u_{xi}^2 = 2q_i(1-q_i)Z_{i\Delta}u_{xi}^2 \qquad 15.15$$

in which $Z_{i\Delta} = [1+3(1-2q_i)a_{xi}/2+(1-4q_i+4q_i^2)a_{xi}^2/2]$

Finally, from 15.13 and 15.15

$$\sigma^2(\bar{z})_{iS1}/\sigma(\hat{q}_i,\hat{X})\partial\ddot{y}_{xi}/\partial q_i = Z_i/Z_{i\Delta}.$$

Values of this ratio associated with specific values of q and a are shown below.

	$Z_i/Z_{i\Delta}$			
q	a = 0	a = 0.3	a = 0.7	a = 1.0
0.3	1.0	0.95	0.93	0.92
0.5	1.0	1.00	1.00	1.00
0.7	1.0	1.08	1.28	1.55
0.9	1.0	1.16	1.70	3.38

On the basis of these ratio values it is apparent that unless a-values and frequencies of alleles with positive effects are high predictions of response to selection among S_1 families based on the genetic variance among them will not be seriously biased. The cautious nature of this statement reflects the assumption of linkage equilibrium in the source population.

15.1.2.3. EPISTATIC VARIANCE WITHIN FAMILY GENETIC VARIANCES

Cockerham (1954) and Kempthorne (1954) defined sub-divisions of the epistatic component of genetic variance and showed how they contribute to the genetic variance of certain family variances. Their papers dealt with the same general issues but differed in symbolism and to some extent in models employed. Cockerham's model and symbolism is followed here. The assumptions of his basic model are (a) diploid meiosis, (b) no multiple alleles, and (c) linkage equilibrium. His subdivision of the total genetic variance parallels that of the analysis of variance of the data from a factorial experiment (involving three levels of n variables) in which the total interaction mean square is separated into portions associated with each single degree of freedom. Consider a 3 x 3 plant nutrition experiment comparing the nine treatments obtained by combining each of three levels of nitrogen with each of three levels of phosphorus. The total mean square for treatments with its eight degrees of freedom can be subdivided into eight portions as shown below.

Variance Source		d.f.
Levels of nitrogen (N)		2
Linear portion	1	
Quadratic portion	1	
Levels of phosphorus (P)		2
Linear portion	1	
Quadratic portion	1	
Interaction (N x P)		4
Linear N x Linear P	1	
Linear N x Quadratic P	1	
Quadratic N x Linear P	1	
Quadratic N x Quadratic P	1	
Total		8

The parallel in the case of genetic variance arising from two segregating genes with the BB, Bb, and bb genotypes viewed as levels is obvious but verbal descriptions of variance sources are more specific. The "linear" component is referred to as the "additive" component and the "quadratic" as the "dominance" component. Note that if dominance is absent the quadratic (= deviation from linear) component is zero. In the genetic variance case the four components of interaction variance are denoted as additive x additive (a x a); additive x dominance (a x d) and dominance x additive (d x a) denoted collectively as additive x dominance; and dominance x dominance (dxd). The extention to the genetic variance of n genes is suggested by the three gene example in which case the 3 x 3 x 3 = 27 single degrees of freedom in a factorial analysis would be composed as follows. Remember that with three genes there are three pairs (each gene being a member of two pairs).

Types of variance		d.f.
Additive		3
Dominance		3
a x a		3
a x d		6
d x d		3
Other epistatic variances		8
a x a x a	1	
a x a x d		
a x d x a	3	
d x a x a		
a x d x d		
d x a x d	3	
d x d x a		
d x d x d	1	
Total		26

The following symbolism was employed by Cockerham to distinguish the categories of variance.

σ^2_{AD} = the sum of the single degree of freedom components of the total variance of which a and d appear in the name of the category A and D times, respectively. For example, if A = 1 and D = 0, we have

σ^2_{10} = additive genetic variance

and, given A = 3 and D = 1, we have

σ^2_{31} = additive x additive x additive x dominance variance.

It follows that

$$\sigma^2_Y = \Sigma^n_{AD}\,\sigma^2_{AD}$$

in which Σ^n_{AD} indicates summation over all combinations of A and D such that $1 \le A + D \le n$.

Assuming HW frequencies of genotypes, and linkage equilibrium, we find then that

Genetic variance among half-sib families $= \Sigma^n_{A0}(1/4)^A\sigma^2_{A0}$, 15.16

Genetic variance among full-sib families $= \Sigma^n_{AD}(1/2)^A(1/4)^D\sigma^2_{AD}$, and

Genetic variance among full-sib families within half-sib families

$$= \Sigma^n_{AD}(1/2)^A(1/4)^D\sigma^2_{AD} - \Sigma^n_{A0}(1/4)^A\sigma^2_{A0} \qquad 15.17$$

These expressions reduce to those shown in Table 15.8 because only σ^2_{10} (additive genetic variance) and σ^2_{01} (dominance variance) are non-zero when epistasis is absent.

• 15.1.3. COVARIANCE BETWEEN PARENT AND OFFSPRING

It is intuitively obvious that both genotye and environment may be sources of covariance between parent and offspring phenotypes. Let trait X be symbolized as

$$X_p = Y_p + e_p$$

and

$$X_o = Y_o + e_o$$

in which Y is the value of total genotye, the e's are effects of environment, and the subscripts, p and o, identify parent and offspring. Assuming no covariance between either Y_p and e_o or Y_o and e_p, the covariance between X_p and X_o is seen to be

$$\sigma(X_p,X_o) = \sigma(Y_p,Y_o) + \sigma(e_p,e_o)$$

Thus, when there is reason to believe that the covariance between environments of parent and offspring, $\sigma(e_p,e_o)$, is close to zero, $\sigma(X_p,X_o)$ is an obvious source of information concerning the genetic variance in a population. Conversely, if the environments of parent and offspring are more alike than those of random individuals, the quantitative relation between their phenotypic covariance and the genetic variation in their population cannot be known. For illustration, consider the following examples.

(1) Dairy Herd Improvement Association records constitute data from which covariance between milk production of dams and daughters has been computed. For the most part, dams and daughters have made their records on the same farm and there is inevitable variation in overall environment from farm to farm. To use terminology from section 2.12, each farm must be viewed as a different macro-environment. It follows that covariances computed ignoring farm effects contain a component due to covariance between environment effects on production by dam and daughter. However, if pairs of which dam and daughter produced on different farms were identified, the data on those pairs were not used, and the covariance was computed on an intra-farm basis, environment covariance due to farm effects would not have been a part of the answer obtained. There are other ways in which this problem was or could have been handled but the point is that any potential for covariance from a non-genetic source should be recognized and dealt with in some way (if possible) if the parent-offspring covariance is to be interpreted with respect to the magnitude of genetic variance.

(2) In humans the magnitude of genetic variance of IQ (Intelligence Quotient) has been of special interest and the object of considerable research via covariances among relatives. With almost no exceptions, that research has been plagued by covariance between environments of the relatives involved. Consider the parent-offspring covariance. Within what was once the normal family structure in this country (if not in all others) absence of covariance between the environments of parent and offspring could never be reasonably assumed. Verbalized differently, the general environment of each family, like farms in the case of milk production data, must be considered a different macro-environment. But in monogamous cultures the rather easy alternative of computing a within family (macroenvironment) parent-offspring covariance is not available because within each family there would be only one parent per sex and hence no degrees of freedom for an estimate of covariance between parent and offspring.[5]

Table 15.7 contains the information required for derivation of the genetic portion of this covariance. The effects of genotype at the i-th locus in the case of parents have been coded by subtraction of u_i and the expected genetic effects of offspring genotypes at the i-th and j-th loci have been coded as indicated in the table. In accord with 5.4, which assumes no epistasis, the value of parent genotypes is considered to be the sum of effects of single locus genotypes, i.e.,

$$Y_p = \alpha + y_{p1} + y_{p2} + \cdots\cdots y_{pn} = \Sigma_i y_{pi}$$

in which p as subscript indentifies genotype values in parent individuals. Averaging over the genotypes possible in offspring given the genotype of an individual parent mated to random individuals of the population yields

$$\bar{Y}_o = \alpha + \bar{y}_{o1} + \bar{y}_{o2} + \cdots\cdots \bar{y}_{on} = \Sigma_i \bar{y}_{oi}$$

in which o as subscript identifies genotype values in offspring. From these expressions it is apparent that

$$\sigma(Y_p,\bar{Y}_o) = \Sigma_i \sigma(y_{pi},\bar{y}_{oi}) + \Sigma_i\Sigma_{j\neq i}\sigma(y_{pi},\bar{y}_{oj}) \quad 15.18$$

Working from table 15.7 and remembering that $(p+r) = q_i$, $(p+s) = q_j$, and $(p+r+s+t) = 1$, the following are obtained.

$$E(y_{xi}-u_{xi}) = [p+r)^2 + (s+t)^2]u_{xi} + 2(p+r)(s+t)a_{xi}u_{xi}$$
$$=(p+r-s-t)u_{xi} + 2(p+r)(s+t)a_{xi}u_{xi}$$
$$E^*(\bar{y}_{oxi}) = [(p+r)^2 - (s+t)^2]Z_i = (p+r-s-t)Z_i$$
$$E^*(\bar{y}_{oxj}) = [(p+s)^2 - (r+t)^2]Z_j = (p+s-r-t)Z_i$$
$$\sigma(y_{pxi},y_{oxi}) = [(p+r)^2+(s+t)^2 - (p+r-s-t)^2]u_{xi}Z_{xi}$$
$$- 2(p+r)(s+t)(p+r-s-t)\, a_{xi}u_{xi}Z_{xi} = 2q_i(1-q_i)[1+(1-2q_i)a_{xi}]u_{xi}Z_{xi} \quad 15.19$$

Substituting the value of Z_{xi} shown in Table 15.7, we obtain

$$\sigma(y_{pxi},y_{oxi}) = q_i(1-q_i)[1+(1-2q_i)a_{xi}][1+(1-2v_i)a_{xi}]u_{xi}^2 \quad 15.20$$

Next $\sigma(y_{pxi},y_{oxj}) = [p^2 - r^2 - s^2 + t^2 - (p+r-s-t)(p+s-r-t)\, u_{xi}Z_{xi}$

$$+2[ps - rt - (p+r)(s+t)(p+s-r-t)\, a_{xi}u_{xi}Z_{xj}$$
$$=2(pt-rs)u_{xi}Z_{xj} + 2(pt-rs)(s+t-p-r)\, a_{xi}u_{xi}Z_{xj} \quad 15.21$$

Substituting the value of Z_{xj} from Table 15.7 and remembering the relations between the two locus gamete frequencies (p,r,s, and t) and allele frequencies, we obtain

$$\sigma(y_{pxi},y_{oxj}) = (pt-rs)_{ij}[1+(1-2q_i)a_{xi}][1+(1-2v_j)a_{xj}]u_{xi}u_{xj} \quad 15.22$$

Finally, substituting in 15.18 in terms of 15.20 and 15.22,

$$\sigma(Y_{xp},Y_{xo}) = \Sigma_i\, q_i(1-q_i)[1+(1-2q_i)a_{xi}][1+(1-2v_i)a_{xi}]u_{xi}^2$$
$$+ \Sigma_i\Sigma_{j\neq i}(pt-rs)_{ij}[1+(1-2q_i)a_{xi}][1+(1-2v_j)a_{xj}]u_{xi}u_{xj} \quad 15.23$$

in which x as subscript specifies whatever trait is identified as trait X.

If the random parents belong to the same population as the constant parents, the v's of 15.23 are properly changed to q's. Then comparing with 6.19, we see that for any trait the genetic covariance between parent and offspring is equal to one-half of the additive genetic variance of the trait.[6] On the other hand, when the constant and random parents are drawn from different populations15.23, by virtue of similarity in form with 6.19, qualifies as an analog of $\sigma_g^2/2$.

Now consider the covariance between trait X in the constant parent with trait C (C≠X) in offspring. Referring to Table 15.7 we find that the distributions of $*(\bar{y}_{xi})$ and $*(\bar{y}_{ci})$ are identical as are those of $*(\bar{y}_{xj})$ and $*(\bar{y}_{cj})$. It is apparent, therefore, that $\sigma(y_{pxi},y_{oci})$ is obtained when Z_{ci} is substituted for Z_{xi} in 15.19. In like manner, substitution of Z_{cj} for Z_{xj} in 15.21 yields $\sigma(y_{pxi},y_{ocj})$. Finally, substitution in 15.18 of $\sigma(y_{pxi},y_{oci})$ for $\sigma(y_{pi},\bar{y}_{oi})$ and $\sigma(y_{pxi},y_{ocj})$ for $\sigma(y_{pi},\bar{y}_{oj})$ yields

$$\sigma(Y_{px},Y_{oc}) = \Sigma_i\, q_i(1-q_i)[1+(1-2q_i)a_{xi}][1+(1-2v_i)a_{ci}]u_{xi}u_{ci}$$
$$+ \Sigma_i\Sigma_{j\neq i}(pt-rs)_{ij}[1+(1-2q_i)a_{xi}][1+(1-2v_j)a_{cj}]u_{xi}u_{cj} \quad 15.24$$

The structural similarity between 15.24 and 6.20 justifies reference to $2\sigma(Y_{px},Y_{oc})$ as an analog of the additive genetic covariance between any traits identified as X and C. If the random and constant parents are drawn from the same population, v = q for all genes and $2\sigma(Y_{px},Y_{oc})$ is then the additive genetic covariance between X and C in the population to which the constant and random parents belong.

15.1.3.1. **REGRESSION OF OFFSPRING ON PARENT**

Symbolically, this regression is $\sigma(X_p,X_o)/\sigma^2(X_p)$. When offspring result from random matings of individuals belonging to the same population and parents and offspring are distributed over environments in such a way that there is no covariance between their environments, it was deduced above from 15.23 that $2\sigma(X_p,X_o) = \sigma^2_{gX}$. It follows that, given those conditions,

$$2\beta(X_o,X_p) = \sigma^2_{gX}/\sigma^2(X_p) = \text{the heritability of X.}$$

Estimates of this regression have been the source of many estimates of the heritability of quantitative traits in animal species. The data employed have usually been collected for primary purposes other than heritability estimation. Such estimates are not so common in the case of plant species because operationally individuals are less important in plant than in animal breeding and consequently less data on individuals is collected in plant species.

15.13.2. **REALIZED HERITABILITY**

The expected direct response to selection based on the phenotypes of individuals has commonly been considered equal to the product of heritability and the selection differential as depicted in the following from Chapter VI.

$$E(\Delta\ddot{Y}) = (\bar{X}_s - \bar{X})h^2_X \qquad 6.10$$

Let R (= observed response) be substituted for $E(\Delta\ddot{Y})$. Then by rearrangement

$$h^2_X \approx R/(\bar{X}_s - \bar{X})$$

Because actual response has a sampling error, it is rarely, if ever, exactly equal to $E(\Delta\ddot{Y})$. It follows that $R/(\bar{X}_s - \bar{X})$, the ratio of actual response to the selection differential, provides an estimate of heritability though it is not consistently equal to heritability. Estimates obtained in this way are designated as "realized heritability" (Falconer, 1960). On average, such estimates have been somewhat smaller than those obtained using (a) estimates of additive genetic variance based on estimates of family genetic variances or parent-offspring covariances and (b) observed phenotypic variances among individuals. Rahnefeld et al. (1963) in their paper on 17 generations of selection for growth in mice reported such a difference. They demonstrated that imperfect genetic correlation between growth in males and females would have contributed to the type of difference observed. Other reasons for downward bias in "realized heritability" are (a) lack of linearity of the regression of allele frequency on the selection criterion that is implicitly assumed in 6.10 (see section 7.2) and (b) the impacts of finite population size (see sections 11.3, 14.2, and 14.3).

Now consider selection in opposite directions as the basis for measuring realized heritability. For illustration consider growth in mice and suppose that the M fastest growing and the M slowest growing males are selected from contemporaries of the same generation of a population and simultaneously F fast growing and F slow growing females are selected from that population.[7] Let the "fast" males and females be mated randomly and, in the same way let there be random matings of the "slow" males and females. Growth will then be measured on the mice of all litters produced; F "fast" litters and F "slow" litters, all distributed randomly within the same general environment. Let

P_{fm} = the average growth of the "fast" male selections,

P_{ff} = the average growth of the "fast" female selections,

P_{sm} = the average growth of the "slow" male selections,

P_{sf} = the average growth of the "slow" female selections,

O_{fm} and O_{ff} symbolize the average growth of male and female offspring of the "fast" selections, and

O_{sm} and O_{sf} symbolize the average growth of male and female offspring of the "slow" selections.

Then $S = (P_{fm} + P_{ff})/2 - (P_{sm} + P_{sf})/2$ = the total selection differential,

$R = (O_{fm} + O_{ff})/2 - (O_{sm} + O_{sf})/2$ = the total response, and

R/S = realized heritability.

An advantage of this two-way selection procedure is that a control population is not required because the offspring of "fast" selections and "slow" selections are grown in the same general environment, whereas in one-way selection a control is required as a basis for estimation of the difference in the effects of environment on averages of the performance of the base population and of the offspring (one generation later) of selections from the base population. However, the performance of a control population would enable checking whether heritability is different depending on the direction of selection.

• 15.2. Estimation of Additive Genetic Variances and Covariances

The common and most important sources of information regarding the magnitudes of additive genetic variances and covariances are (1) the variation among half-sib family means and (2) the covariation of phenotypes of parents and their offspring.

It is shown in section 15.1.2.1 that when epistasis is absent the expectation of genetic variance among half-sib families, of which both the common and random parents are members of the same population, is one-quarter of the addititive genetic variance in the population. See 15.10 and the brief comments (four sentences) that follow its presentation. It is also shown that when all parents of half-sib families are members of the same population and epistasis is absent the expectation of genetic covariannce (of any two traits) among half-sib families is one-quarter of the population additive genetic covariance between the two traits (see 15.11).

In section 15.1.3 (equation 15.23 and related comments) it is shown that when matings are random among individuals of the same random mating population and there is no epistasis the expectation of genetic covariance between parent (either single parent) and offspring is one-half of the population additive genetic variance. In the case of covariance between different traits in parent and offspring the expectation of the genetic covariance is one-half of the additive genetic covariance of those traits in the population (see 15.24 and following comments).

Obviously, estimates of additive genetic variances or covariances can be obained using either half-sib family or parent-offspring data. The significant practical issues are (1) are the available data such that the observable variance (or covariance) is an unbiased estimate of the additive genetic parameter about which information is desired and (2) are the available data sufficient to provide estimates reliable enough to be useful?

In the case of estimates based on half-sib family variances, the problem of obtaining estimates unbiased by one or more components of genotype-environment interaction variance is demonstrated in section 15.1.1.1. It should also be noted that if the common parents of half-sib families are females variance due to maternal or cytoplasmic effects can be significant sources of upward bias.

In section 15.1.3 attention is drawn to the fact that a parent-offspring covariance may contain both genetic and environmental components but that the covariance between the environments of parent and offspring can be circumvented by computing the parent-offspring covariance on a within macro-environment basis when that is possible. Not noted in that connection was that even that procedure does not eliminate bias due to covariance of effects due to interaction of genotype with macro-environment. However, there are reasons to doubt that genotype x environment interaction effects are as large relative to genetic effects in the expression of animal traits as known to be true in plant traits.

See section 15.1.1.3 for basic information regarding the sampling errors of estimates of variances and covariances.

15.3. The Effect of Inbreeding

Estimates of the magnitudes of inbreeding effects are provided by comparison of phenotypic means of groups in which allele frequencies are expected to be the same though their average coefficients of inbreeding are different. It goes without saying that both groups should be grown in the same macro-environment (or set of macro-environments) and that members of the two groups should be distributed at random relative to micro-environments.

As an example, consider the estimation of inbreeding effects on quantitative traits of maize by comparison of the phenotypic means of individuals produced by random matings with those of individuals produced by self-fertilization of random individuals. Because the parents of both groups are random individuals from the same population the expectations of allele frequencies of the non-inbreds and the inbreds (the S_1 individuals) are obviously the same. Given no environment effect difference between means of the two groups, the estimate of I, the inbreeding effect per unit of the inbreeding coefficients is, according to 2.40,

$$I = [\ddot{Y}_I - \ddot{Y}_n]/f = -\Sigma_i 2q_i(1-q_i)a_i u_i \qquad 2.40$$

in which the subscripts, I and n, refer to the "inbred" and "non inbred" groups. In this case the inbreeding coefficient of the inbreds (the S_1's) is 0.5. It is worth noting that if the time pattern of development is not the same for the two groups it cannot be assumed with certainty that observed phenotypic mean differences in the case of other traits have not resulted in part from environment effect differences [see Sentz et al (1954)]. Reliable estimates can only be insured by repetitions of the comparison that adequately sample time patterns of climatic variation within the growing season.

As a second example, consider the estimation of inbreeding effects in a random mating population of swine. Families of inbred individuals ($f = 0.25$) can be obtained by full-sib matings (using random brother-sister pairs) for comparison with full-sib families from matings between unrelated individuals selected at random from the population. As in our first example, the composition of the inbreeding effect is given by 2.40.

In each of these examples, and in others of the same general nature, variation among families within groups arises in part from sources that do not contribute to variation within families (genetic variation among parents or pairs of parents; cytoplasmic effects if any; and, in the case of the swine example, maternal and other effects unique to individual litters). It follows that the design of the group comparison and the data gathered should be such that an error variance can be computed that takes proper account of the variance among families within groups.

Within herds or flocks of animals involved in breeding research projects the inbreeding coefficient has frequently been one of the items of data recorded for each individual. The regression of any measure of phenotype on the inbreeding coefficient is then an estimate of the inbreeding effect of that trait. Numerous estimates have been obtained in this way. Obviously, such estimates are potentially biased unless the variation of inbreeding is random relative to other variables that affect performance. Thus when individuals belong to groups that differ in mean performance for reasons other than average amount of inbreeding the regression estimates should obviously be on intra-group bases. In general, this has been standard procedure.

Finally, as emphasized in section 2.11and by 2.40, inbreeding effects on any specific trait, being functions of allele frequencies, must be variable among populations, e.g., among breeds of livestock or varieties of chickens though the magnitude of variation may not be great. In the case of the S populations of recurrent selection programs if there are responses to selection as expected it is because the frequencies of favorable alleles are increased. An obvious result is that $\Sigma_i 2q_i(1-q_i)$ and hence inbreeding effects on traits for which a-values are non-zero will be changed though slowly so that total change may only be discernable after several cycles of selection.

• 15.4. Level of Dominance

The ideal genotype, locus by locus, is BB when $a \leq 1.0$ but Bb when $a > 1.0$. The recurrent selection systems, RRS and RSSC, described and discussed in Chapter IX, are systems of choice when a large portion of the genes affecting total value are over-dominant. However, these selection systems are not as effective as some others when $a \leq 1.0$ (see section 14.3.2).

Clearly, the extent of overdominance has meaning relative to choice of selection system in the case of species in which cross-fertilizations are normal (or not too costly).

• 15.4.1. Heterosis and the Effect of Inbreeding as Sources of Information

The inbreeding depression or heterosis exhibited by a trait is an obvious clue to whether a significant portion of the genes that affect the trait may be overdominant with respect to their effects on that trait. Equations 2.40 and 2.44 tell us that locus by locus heterosis and inbreeding effects are proportional to a-values (of genes with effects on the trait) and hence that neither will be small unless a-values are small or small on average because large positive values are counter balanced by about as many that are negative and equally large in absolute value.

• 15.4.2. Response to Recurrent Selection Applied to an F_2 Source Population

A second clue is provided by the performance of the F_1 and F_2 of the cross between two inbred lines that are homozygous (or nearly so). Both the performance of the F_1 relative to the mean performance of its parent lines and the difference in performance between the F_1 and the F_2 tell us whether a-values are predominantly positive but neither indicates whether overdominant genes are involved. However, some definite information can be obtained by the response to positive recurrent selection applied to the F_2 population. Assuming homozygous parent lines the F_1 must be heterozygous at some loci (those homozygous for different alleles in the two parent lines) and homozygous at all other loci (those homozygous for the same allele in both parent lines). Moreover, the F_2 can segregate at only those loci that are heterozygous in the F_1. It follows that (1) if all genes heterozygous in the F_1 are overdominant ($a > 1$) the average of the selection criterion in the population derived by recurrent selection in the F_2 cannot be brought to the level of the F_1,(2) if all favorable alleles heterozygous in the F_1 are completely dominant ($a = 1$), the population derived by recurrent selection in the F_2 can theoretically be brought to the mean of the selection criterion in the F_1, but no higher, and (3) if recurrent selection in the S population which was initially the F_2 brought the performance of the S population to a level higher than that of the F_1, the existence of additive or partially dominant genes would have been demonstrated though absence of overdominance at all loci heterozygous in the F_1 would not have been proven.

Obvious limitations of this procedure are (1) near homozygous lines are required (not available, for example in the major economic species of livestock), (2) the amount of time and the cost in terms of effective population size that may be required to learn whether response to recurrent selection can move the S population initiated from the F_2 to or beyond the level of the F_1, and (3) the fact that information obtained pertains only to one trait, i.e., the selection criterion employed. However, if the selected trait is a value trait and data are obtained on other value traits useful information regarding pleiotropy would be obtained.

• 15.4.3. Comparison of Phenotypes Associated with Single-Locus Genotypes

When single locus genotypes can be identified by phenotypic effects the individuals of a population can be classified by genotype at the locus in question. In rare instances the distinguishing phenotypes are visibly different as in the red, roan, white coat colors of Shorthorn cattle determined by the RR, RW, and WW single locus genotypes. In other instances the possible genotypes can be identified by indirect methods. Examples are genotypes of the ABO blood type locus in humans identifiable by agglutination tests and the single locus isozyme genotypes distinguishable by the electrophoretic technique. In general, given alleles B and b and some way of classifying individuals by genotype (BB, Bb, or bb) the phenotype averages of the three groups provide estimates of the u and a parameters of the locus. Let $\bar{X}_2$, $\bar{X}_1$, and $\bar{X}_0$ symbolize the phenotypic means of the groups, the subscripts indicating the number of B alleles in the genotype. Then, given linkage equilibrium and remembering that the y's (single locus genotype contributions to total genotype value) are 2u, u + au, and 0 for the BB, Bb, and bb genotypes, respectively,

u is estimated by $(\bar{X}_2 - \bar{X}_0)/2$,

au is estmated by $(2\bar{X}_1 - \bar{X}_2 - \bar{X}_0)/2$,

a is estimated by $(2\bar{X}_1 - \bar{X}_2 - \bar{X}_0)/(\bar{X}_2 - \bar{X}_0)$, and

au - u is estimated by $\bar{X}_1 - \bar{X}_2$.

However, despite the appeal of of estimating a, whatever its magnitude, the practical question of the breeder is whether a > 1. Thus the critical issue is whether (au - u) > 0; when greater than zero there is overdominance, otherwise not. Being a simpler function of the $\bar{X}$'s , the estimate of (au - u) has a smaller error variance than the estimate of a. Even so, the number of individuals on which data are obtained must be large if decisive estimates are to be obtained. Any estimate of $\bar{X}_1 - \bar{X}_2 > 0.0$ suggests that a may really be greater than 1.0 but the estimate may also be a random deviation from $\bar{X}_1 - \bar{X}_2 \leq 0.0$. How large a sample must be examined to make the probability reasonably large that an estimate of $\bar{X}_1 - \bar{X}_2$ will be judged significantly greater than zero? The answer will vary with the magnitudes of a, u, and the phenotypic variance among individuals.

For illustration, consider an hypothetical trait for which the unit of measurement is such that the population mean of the trait is 100 and the coefficient of variation of the trait magnitude is 15, i.e., its variance is 225. Let n be the total number of individuals examined in which the numbers possessing the BB and Bb genotypes are 0.36n and 0.48n, respectively. Then the sampling variance of an estimate of $\bar{X}_1 - \bar{X}_2$ becomes 225/0.36n + 225/0.48n = 1093.75/n and for indicated values of n the standard errors (s) would be as follows.

n	s	1.65s
1000	1.046	1.73
3000	0.604	1.00
6000	0.427	0.70

We now ask--how large must $\bar{X}_1 - \bar{X}_2$ be to be considered greater than zero by a significant (at the 5% level) amount? Going to a "normal" distribution table we find that the probability of an observation as large or larger than 1.65 standard deviations greater than the null hypothesis expectation has a probability of 5%. The corollary in our example is that if $\bar{X}_1 - \bar{X}_2 \geq 1.65s$, the estimate of au - u is significant at the 5% level, i.e., the probability of a > 1.0 is 95% or greater. The remaining question concerns the probability of an observation as large or larger than is required to be significant. This depends obviously on the actual magnitudes of u and a, i.e., the expectation of $\bar{X}_1 - \bar{X}_2$. The larger this expectation (the real value of au - u), the greater the probability of an observation as large or larger than that required for statistical significance. If a = 1.6 and u = 1.0, $E(\bar{X}_1 - \bar{X}_2) = 0.60$, s = 0.427 if n = 6000 (see above), and the probability of an observed value of $\bar{X}_1 - \bar{X}_2$ large enough to be considered significantly greater than zero is the probability of $(\bar{X}_1 - \bar{X}_2) \geq [1.65s - (au - u)] = (0.70 - 0.60) = 0.10$. Divided by s = 0.427, this becomes a standardized normal deviate of 0.234 for which the probability of a value as large or larger (again from a "normal" distribution table) is about 0.407. Table 15.9 shows how the probabilities of estimates large enough to be statistically significant at the 5 percent level vary with n, a, and u if phenotypic variance of the trait is 225.

Table 15.9. Probability of $X_1 - X_2$ large enough to be statistically significant when a > 1.0 as function of u, a, and sample size (n)

	u = 1.0			u = 0.6		
n	a = 1.3	a = 1.6	a = 1.9	a = 1.3	a = 1.6	a = 1.9
1000	0.09	0.14	0.21	0.07	0.09	0.11
3000	0.13	0.25	0.44	0.08	0.12	0.18
6000	0.18	0.40	0.58	0.10	0.18	0.28

Note: Assumed phenotypic variance among individuals = 225.

As expected the probabilities of estimates indicating a > 1.0 that are statistically significant ($P \leq .05$) are lower when a is lower, when u is lower, and when n is lower. The largest value shown is 0.58 in the a = 0.9, u = 1.0, and n (the number of individuals in the sample) = 6000 situation. Though the magnitudes of u assumed can be questioned they serve the purpose of the example.

Limitations of the approach considered in this section include (1) the requirement for a linkage equilibrium population (obvious though not specified earlier), (2) the considerable amount of data likely to be required to make the probability of demonstrating a-values greater than 1.0 even when they actually exist, and (3) the number of genes of which single locus genotypes can in some way be distinguished without effects on the performance of the plant or animal. A significant advantage is that as many traits as desired and convenient can be studied using the same sample of individuals.

• 15.4.4. Effects of Genotypes at Marked and Linked Loci

Estimations of the u, au, and a parameters of single genes of which all zygotic genotypes can be distinguished are discussed in the preceeding section. They require a linkage equilibrium population as the source of individuals from which data are obtained. In this case the requirement for loci of which all genotyes can be distinguished (the marked loci) is the same but a source population in which linkage disequilibrium is maximum is required. The F_2 of a single cross between two highly inbred lines meets this requirement. Data collected on each individual are (1) genotype at the marked locus and (2) phenotypic measurements of each trait concerning which information is desired.

Following Edwards et al (1987), who describe procedure and theory in detail, let the marked locus genotypes and the phenotypic averages of any measure on individuals of those three genotypic classes be symbolized as MM, Mm, and mm, respectively. It is obvious that each of the three averages will reflect the effects of all single locus genotypes plus the total of non-genetic effects. However, the expectations of effects of genes that segregate independently of the marked locus (recombination rate = $r = 0.5$) plus non-genetic effects will be the same for each of the three genotype classes (if individuals have been distributed at random relative to environments) and hence, aside from sampling errors, the differences among the MM, Mm, and mm averages will reflect the sum of the effects of the genotypes at the marked locus plus those of genotypes at any loci that do not segregate independently of the marked locus. However, in the absence of other data, there would be no way of knowing whether the observed differences among the MM, Mm, and mm averages were due to effects of genotype at the marked locus itself or to effects of genotype at any number of loci linked ($r < 0.5$) with the marked locus. As a consequence, information obtained is ambiguous with respect to level of dominance. [see Appendix 15.2 for theory following Edwards et al (op cit)].

15.4.5. Functions of Genetic Variances

In the absence of both epistasis and multiple alleles the genetic variances of diploid organisms are known to be functions of (1) the effect parameters (the u's and a's) of genes, (2) allele frequencies, and (3) the levels of linkage disequilibrium in the distribution of genotypes at pairs of loci. All genetic variances are functions of both the u's and a's of all genes with effects on any trait of interest. It follows that any pair of variances that are different because they are different functions of the u's and a's constitutes a base for inference concerning levels of dominance if biasing effects of allele frequencies and linkage disequilibrium can in some way be avoided or taken adequately into account.

15.4.5.1. INFERENCE FROM ESTIMATES OF ADDITIVE GENETIC AND DOMINANCE VARIANCES

It is apparent from the content of Table 15.8 that, in absence of epistasis, an estimate of additive genetic variance is provided by variance among half-sib families and an estimate of dominance variance by the difference between the variance of half-sib families and the variance among full-sib families within half-sib families. This assumes that all parents of families on which data are obtained are random individuals from the same random mating population. By mating each of m males (pollen parents in the case of plants) with f females m half-sib families, each composed of f full-sib families, are obtained. If these mf full-sib families are grown in r replicate blocks (within which they are distributed randomly) at v locations in each of t years the appropriate variance analysis of data obtained would be a logical extension of the analysis shown symbolically in Table 15.1. The form of this analysis, mean square expectations, and the estimates of genetic variances among half-sib families and among full-sib families within half-sib families are set out in Table 15.10. To obtain more data without making the block size larger than desirable, any number of sets of half-sib and full-sib families can be produced and employed. The appropriate variance analysis would then be an extension of that shown in Table 15.10 in the same way that the form shown in Table 15.5 is an extension of that shown in Table 15.1.

Table 15.10. ANOVA, half- and full-sib families at v locations in each of t years

Variance Source	d.f.	M.S.	Expectation of Mean Square
Years (Y)	$t-1$		
Locations (L)	$v-1$		
Y x L	$(t-1)(v-1)$		
Repls. within Y&L	$tv(r-1)$		
Half-sib families (H)	$m-1$	M_1	$\sigma^2(e)+r\sigma^2(\bar{z}_2ab)+vr\sigma^2(\bar{z}_2b)+tr\sigma^2(\bar{z}_2a)$ $+vtr\sigma^2(\bar{z}_2)+rf\sigma^2(\bar{z}_1ab)+vrf\sigma^2(\bar{z}_1b)+$ $+trf\sigma^2(\bar{z}_1a)+vtrf\sigma^2(\bar{z}_1)$
H x Y	$(m-1)(t-1)$	M_2	$\sigma^2(e)+r\sigma^2(\bar{z}_2ab)+rv\sigma^2(\bar{z}_2b)+rf\sigma^2(\bar{z}_1ab)$ $+vrf\sigma^2(\bar{z}_1b)$
H x L	$(m-1)(v-1)$	M_3	$\sigma^2(e)+r\sigma^2(\bar{z}_2ab)+rt\sigma^2(\bar{z}_2a)+rf\sigma^2(\bar{z}_1ab)$ $+trf\sigma^2(\bar{z}_1a)$
H x L x Y	$(m-1)(v-1)(t-1)$	M_4	$\sigma^2(e)+r\sigma^2(\bar{z}_2ab)+rf\sigma^2(\bar{z}_1ab)$
Full-sib families within half-sib families (F)	$m(f-1)$	M_5	$\sigma^2(e)+r\sigma^2(\bar{z}_2ab)+vr\sigma^2(\bar{z}_2b)+tr\sigma^2(\bar{z}_2a)$ $+vtr\sigma^2(\bar{z}_2)$
F x Y	$m(f-1)(t-1)$	M_6	$\sigma^2(e)+r\sigma^2(\bar{z}_2ab)+vr\sigma^2(\bar{z}_2b)$
F x L	$m(f-1)(v-1)$	M_7	$\sigma^2(e)+r\sigma^2(\bar{z}_2ab)+tr\sigma^2(\bar{z}_2a)$
F x Y x L	$m(f-1)(t-1)(v-1)$	M_8	$\sigma^2(e)+r\sigma^2(\bar{z}_2ab)$
Families x repls. within Y&L	$vt(mf-1)(r-1)$	M_9	$\sigma^2(e)$

Note: $\bar{z}_1$ = effect of k-th half-sib family $\quad\bar{z}_2$ = effect of m-th full-sib within half-sib family.

$$[(M_1-M_2-M_3+M_4) - (M_5-M_6-M_7+M_8)]/vtrf = \bullet\sigma^2(\bar{z}_1) = \bullet\sigma_g^2/4$$

$$(M_5-M_6-M_7+M_8)/vtr = \bullet\sigma^2(\bar{z}_2) = \bullet(\sigma_g^2 + \sigma_d^2)/4$$

$$(M_5-M_6-M_7+M_8)/vtr - [(M_1-M_2-M_3+M_4) - (M_5-M_6-M_7+M_8)]/vtrf = \bullet\sigma_d^2/4$$

Subject to the standard assumptions herein, the composition of these two components of the genetic variance, given by 6.19 and 15.6, are

$$\sigma_g^2 = \Sigma_i 2q_i(1-q_i)[1+(1-2q_i)a_i]^2u_i^2$$

$$+ \Sigma_i\Sigma_{j\neq i}2(pt-rs)_{ij}[1+(1-2q_j)a_j][1+(1-2q_i)a_i]u_iu_j \qquad 6.19$$

and $$\sigma_d^2 = 4\Sigma_i q_i^2(1-q_i)^2a_i^2u_i^2 + 4\Sigma_i\Sigma_{j\neq i}[(pt-rs)_{ij}]^2a_iu_ia_ju_j \qquad 15.6$$

These expressions pertain to genetic variation in populations in which HW frequencies of single locus genotypes can reasonably be assumed, i.e., in populations produced by random matings among individuals belonging to the same population.

Because a-values enter 6.19 and 15.6 in different ways it is appparent that the relative magnitudes of additive genetic and dominance variances are influenced by levels of dominance but it is equally apparent that the average magnitude of the a's cannot be deduced from the dominance-additive genetic variance ratio because of unknown impacts on that ratio of allele frequencies and levels of linkage disequilibrium. To clarify, consider an hypothetical F_2 in which there is no linkage disequilibrium because the spacing of genes affecting the same trait is in every case such that all of those genes segregate independently (recombination rate $= \Upsilon = 0.5$). In addition, by the nature of F_2 populations and assuming no gametic selection or meiotic drive, all allele frequencies would be 1/2. In the case of such a population 6.19 and 15.6 would reduce to $\sigma_g^2 = \Sigma_i(1/2)u_i^2$ and $\sigma_d^2 = \Sigma_i(1/4)a_i^2u_i^2$.

Then letting the ratio of dominance variance to additive genetic variance be symbolized as D/A,

$$2D/A = \Sigma_i a_i^2u_i^2/\Sigma_i u_i^2,$$

a weighted average of a^2 in which each u_i^2 is the weight given a_i^2. This result, which depends on (pt-rs) being zero for all pairs of loci, was sketched by Comstock and Robinson (1948). However, thinking that the number of genes affecting quantitative traits are likely to be great enough to make some linkage disequilibrium inevitable, they indicated that 2D/A could be biased upward relative to $\Sigma_i a_i^2u_i^2/\Sigma_i u_i^2$. When q's are set equal to 1/2, as justified in the case of an F_2 population, 6.19 and 15.6 reduce to

$$D/A = \sigma_d^2/\sigma_g^2 = 2q(1-q)a^2/[1+(1-2q)a]^2. \qquad 15.25$$

From 15.25 it is apparent that when estimates of dominance and additive genetic variances are obtained from old random breeding populations the D/A ratio may be almost free of any linkage disequilibrium effect but in exchange is a function of allele frequencies as well as levels of dominance. Some values of the 15.25 are shown in Table 15.11. Note that when a =0.5 dominance variance is small relative to additive genetic variance whatever the frequency of the favorable allele. While the ratio is larger when a = 0.8, the largest value shown (when q = 0.9) is still a little less than 1.0. Values of the ratio in excess of 1.0 occur when dominance is

Table 15.11. Dominance-additive genetic variance ratio

q	a = 0.5	a = 0.8	a = 1.0	a = 1.2	a = 1.6
.4	.10	.23	.33	.45	.71
.5	.13	.32	.50	.72	1.28
.6	.15	.44	.75	1.20	2.66
.7	.16	.58	1.16	2.23	8.30
.8	.16	.76	2.00	5.88	512.00[a]
.9	.13	.89	4.50	162.00[a]	5.87

[a]These ratios are large because q is very close to (1+a)/2a, the value at which additive genetic variance is zero.

complete (a = 1) and q ≥ 0.7 and in the cases of overdominance (a > 1) and q > 0.5. Of course the variances that can be estimated are the totals of the contributions from all segregating genes; thus the ratio for which an estimate is available is $\Sigma_i \beta_{d_i}^2 / \Sigma_i \beta_{g_i}^2$ around which the ratios of the two variances arising from segregation at single loci are expected to vary. It follows that an estimate in the neighborhood of 1.0 would suggest a mix of partial, complete, and overdominance in the action of genes affecting the trait in question. It is obvious that the dominance-additive genetic ratio could be much larger than 1.0 if a large portion of the effective genes were overdominant.

It is apparent from 15.16 and 15.17 that if epistasis and consequently epistatic variance were present, a much greater portion of the epistatic variance would be contained in the variance among full-sib within half-sib families than in the variance among half-sib families. As a result the upward bias in estimates made assuming no epistasis would be greater in the estimate of dominance variance than in the estimate of additive genetic variance and bias in the dominance-additive genetic variance ratio would be upward. Thus an estimate of the ratio high enough to indicate overdominance could conceivably be the consequence of epistasis. On the other hand, while low estimates could result from low allele frequencies, they cannot be caused by presence of epistasis. The potential bias due to epistasis was recognized by Robinson et al (1955). Their emphasis was on the fact that their estimates of dominance-additive genetic variance ratio in three open-pollinated varieties of maize, the largest of which was 0.52, could not have been low as a consequence of epistasis even despite the presence of considerable overdominance.

This method is obviously limited to naturally cross fertilizing species in which cost of the specific matings required to create numerous full-sib families in half-sib family groups is not prohibitive. The linkage equilibrium requirement means that the individual parents of those families must come from a population in which random mating has been the norm for a sufficient number of generations in which effective population size has not been small. Traits studied cannot be subject to maternal effects (including those of cytoplasm and X-chromosome genes in dioecious species) because variance among full-sib families would contain maternal effect variance with the result that an unbiased estimate of dominance variance would not be available.[10] Finally, since the dominance-additive genetic variance ratio is a function of both q and a (see Table 15.11), selection history information is a part of the required basis for inference concerning dominance levels. Unless it is reasonable to postulate that frequencies of alleles favorable to the trait are in the range above 0.6, low values of the ratio cannot viewed as evidence for absence of overdominance. On the plus side, the data for dominance level information can be obtained from the same experimental material for as many traits as desired.

15.4.5.2. DESIGN III

The research procedure now widely known as Design III was proposed and described by Comstock and Robinson (1952). Their statement concerning the structure of experimental material to be employed in work with monoecious species is quoted below.

> The experimental material is produced from backcross matings of F_2 plants to the two inbred lines from which the F_2 was derived; the F_2 plants are used as pollen parents. A set of progenies is made up of the 2n progenies obtained from backcrossing n random F_2 plants to each of the parent inbreds. The number of inbred plants used in production of each backcross progeny is important only with respect to insuring suffient seed.
>
> The total experimental material consists of s sets of n pairs of progenies. The members of each pair have the same F_2 (male) parent but different inbred parents. the two inbred parents are, of course, the same for all pairs of progenies.

Given data on n progeny pairs grown in a randomized block field arrangement in a single macro-environment, the analysis of variance shown in Table 15.12 would be appropriate.[11] The

Table 15.12. ANOVA, Design III data, n progeny pairs grown in a single macro-environment

Variance source	d.f.	Mean Square	Expectation of Mean Square
Replications	r - 1		
Inbred lines	1		
Progeny pairs	n - 1	M_1	$\sigma^2(e) + 2r\sigma^2(m)$
Lines x progeny pairs	n - 1	M_2	$\sigma^2(e) + r\sigma^2(ml)$
Remainder among plots	(2n-1)(r-1)	M_3	$\sigma^2(e)$

variances from which inferences concerning levels of dominance are made are $\sigma^2(m)$, the variance arising from genetic variation among the F_2 parents of progeny pairs, and $\sigma^2(ml)$, the variance arising from interaction between the genotypes of F_2 parents and inbred lines. However, on the basis of Tables 15.1, 15.5, and 15.10 and associated discussion it is apparent that each of these variances is the sum of a genotypic portion and a genotype x environment portion, i.e.,

$$\sigma^2(m) = \sigma^2(mG) + \sigma^2(mE) \quad 15.26$$

and $\sigma^2(ml) = \sigma^2(mlG) + \sigma^2(mlE)$

in which the genotypic portions are symbolized by $\sigma^2(mG)$ and $\sigma^2(mlG)$. The questions posed by the genotype x environment variance content of $\sigma^2(m)$ and $\sigma^2(ml)$ are considered later in this section. It is demonstrated in Appendix 15.3 that

$$\sigma^2(mG) = (1/2)\Sigma_i q_i(1-q_i)u_i^2 + \Sigma_{ij}(pt-rs)_{ij}u_iu_j \quad 15.27$$

and $\sigma^2(mlG) = \Sigma_i q_i(1-q_i)a_i^2 u_i^2 + 2\Sigma^c_{ij}(pt-rs)_{ij}a_iu_ia_ju_j - 2\Sigma^r_{ij}(pt-rs)_{ij}a_iu_ia_ju_j$ 15.28

in which Σ^c_{ij} and Σ^r_{ij} indicate summation over pairs of genes for which the F_1 linkage phase was coupling or repulsion, respectively, and q_i symbolizes the frequency in the F_2 of B, the more favorable allele, of the i-th gene. Because, barring deviations resulting from selection or meiotic drive, the frequencies of alleles at segregating loci in an infinite F_2 population should be one-half, 15.27 and 15.28 can reasonably be written as follows

$$\sigma^2(mG) = (1/8)\Sigma_i u_i^2 + \Sigma_{ij}(pt-rs)_{ij}u_iu_j \quad 15.29$$

and $\sigma^2(mlG) = (1/4)\Sigma_i a_i^2 u_i^2 + 2\Sigma^c_{ij}(pt-rs)_{ij}a_iu_ia_ju_j - 2\Sigma^r_{ij}(pt-rs)_{ij}a_iu_ia_ju_j$.

In the absence of any linkage disequilibrium, the value of $\sigma^2(mlG)/2\sigma^2(mG)$ would reduce to $\Sigma_i a_i^2 u_i^2/\Sigma_i u_i^2$, a weighted average of a^2, and if greater than 1.0 would require overdominance ($a>1$) of at least a portion of the genes having effects on the trait. But we know that linkage disequilibrium is inevitable in an F_2 population if among the genes with effects on the trait being investigated there are pairs that do not segregate independently. As noted earlier, (pt-rs) in the F_2 generation is a positive quantity when the F_1 double heterozygote of a gene pair is in the coupling phase but conversely is a negative quantity when the F_1 double heterozygote is in the repulsion phase. Thus it is reasonable to suppose that among the non-zero values of (pt-rs) some will be positive and some negative. Then, by the definition of u, all non-zero values of u_iu_j must be positive and consequently there will be both positive and negative values of $(pt-rs)_{ij}u_iu_j$. These will partially cancel each other; hence the total linkage disequilibrium portion of $\sigma^2(mG)$ may be positive or negative but, in any case, not excessive.[12] Conversely, <u>because</u> (pt-rs) must be positive in the case of coupling but negative in the case of repulsion in F_1 double heterozygotes all linkage disequilibrium contributions to bias in $\sigma^2(mlG)$ are positive unless $a_iu_ia_ju_j$ is negative (only possible when a_i and a_j differ in algebraic sign). Because dominance of favorable alleles appears to play a major role in the phenomena of heterosis and inbreeding depression it is reasonable to believe that a majority of favorable alleles are in some degree dominant ($a > 0$) in the case of such heterotic traits as grain yield of maize and litter size of swine and, consequently, that $a_ia_j >$ zero is more frequent than $a_ia_j <$ zero.

In view of the possible linkage disequilibrium content of $\sigma^2(mlG)$ and $\sigma^2(mG)$, it is apparent that when the segregating populations employed in Design III research projects are F_2's, that estimates of $\sigma^2(mlG)/2\sigma^2(mG)$ in excess of 1.0 would not require the existence of any overdominance, i.e. they would suggest, but not prove, that overdominance is present.

Let us now suppose that the constant parents of backcross pairs are not drawn from the F_2 but, rather, from a population in which the segregating genes are exactly those represented by different alleles in the two pure lines that were parents, once removed, of the F_2. To meet this specification concerning the identity of segregating genes, that population would have to have been derived from the F_2 itself. Consider a derivative of the F_2 obtained by successive generations of random mating without selection other than whatever natural selection is inescapable and with effective population size sufficient to insure little "drift" of allele frequencies. In each generation obtained in that way the expectations of values of (pt-rs) would be smaller than in the preceding generation (see Appendix 6.3) and linkage equilibrium would be approached as the number of generations of random mating increased. However, in recognition of effects of natural selection and drift, allele frequencies equal to one-half could not be assumed in this derived population. Taking random individuals from such an F_n population to use as parents of the backcross progeny pairs of a Design III investigation, data obtained would again be appropriately analyzed as in Table 15.12. Then, assuming the remaining linkage disequilibrium to be trivial in amount, the associated terms in 15.27 and 15.28 could be deleted leaving

$$\sigma^2(mG) = (1/2)\Sigma_i q_i(1-q_i)u_i^2 \text{ and}$$

$$\sigma^2(mlG) = \Sigma_i q_i(1-q_i)a_i^2u_i^2 \text{ as the genetic portions of } \sigma^2(m) \text{ and } \sigma^2(ml).$$

Remember now (1) that the quantity of which an unbiased estimate is sought is $\Sigma_i a_i^2u_i^2/\Sigma_i u_i^2$, (2) that the estimate available when the F_2 is the source of parents of backcross progeny pairs can be biased upward (amount uncertain) by linkage disequilibrium, and (3) that when the parents of backcross pairs are from an F_n population (derived from the F_2 as described above), the quantity of which an estimate is available approaches

$$\sigma^2(mlG)/2\sigma^2(mG) = \Sigma_i q_i(1-q_i)a_i^2u_i^2/\Sigma_i q_i(1-q_i)u_i^2$$

as n is increased. To escape the bias arising from linkage disequilibrium, an estimate of the average of a_i^2 in which individual values are weighted by $q_i(1-q_i)u_i^2$ instead of u_i^2 values must be accepted. Weighting by u_i^2 has an obvious appeal; the a^2's of genes with larger effects on the trait are given greater weight. But whether the $q_i(1-q_i)u_i^2$ weighted estimate is likely to be very different or less meaningful is questionable. Magnitudes of $\Sigma_i q_i(1-q_i)a_i^2u_i^2/\Sigma_i q_i(1-q_i)u_i^2$ and $\Sigma_i a_i^2u_i^2/\Sigma_i u_i^2$, relative to 1.0, evoke the same inferences. If greater than one, either indicates overdominance of some portion of the effective genes; if less than one either casts doubt on the presence of any overdominant genes. However, comparison of estimates obtained when F_2

individuals are employed as backcross progeny parents with those obtained when F_n individuals are employed can obviously be informative. Let the estimates of $\sigma^2(mIG)/2\sigma^2(mG)$ obtained using F_2 and F_n populations be symbolized as E_2 and E_n, respectively. Then the following outcomes are all possible and, assuming small standard errors, logical interpretations would be as indicated.

(1) Difference between E_2 and E_n not significant, both larger than zero but substantially smaller than 1.0. Interpretation--primarily partial dominance, i.e., distribution of a^2's centered above zero but below one; little if any overdominance; and little close linkage of effective genes.

(2) Difference between E_2 and E_n not significant, both larger than 0.5 and smaller than 1.0. Interpretation--primarily partial dominance; distribution of a^2's centered between 0.5 and 1.0; some overdominat genes possible; little close linkage of effective genes.

(3) E_n significantly smaller than E_2; $E_2 > 1$ and $E_n < 1$. Interpretation--E_2 biased upward by linkage disequilibrium resulting from close linkage between some pairs of effective genes; some overdominant genes possible or probable depending on the magnitude of $1-E_n$, the probability of overdominant genes becoming less as $1-E_n$ becomes greater.

(4) E_n significantly smaller than E_2; both larger than 1.0. Interpretation--E_2 biased upward by linkage disequilibrium resulting from close linkage between some pairs of effective genes; a substantial portion of overdominant genes.

Obviously, Design III investigations that include a portion involving the F_2 as the source of parents of backcross progeny pairs and one or more portions involving F_n populations as the source of parents of backcross progeny pairs can provide important information concerning (a) levels of dominance of the genes that have effects on quantitative traits and (b) whether among the genes having effects on a particular trait a non-trivial number of pairs are closely linked. Unfortunately, utilization of the approach is limited by its requirement for at least two pure (i.e., highly inbred) lines and the numerous specific matings required. In this connection, (1) highly inbred lines are not available in large animal species and (2) the level of dominance issue is irrelevant in the case of the self-fertilizing species in which cultivars are pure lines or mixtures of pure lines. On the plus side, are the facts that (1) validity of the no multiple allele assumption is virtually guaranteed by derivation of the genetic material from two pure lines plus the low probability of mutations and (2) numerous traits can be studied using the same experimental material.

15.4.5.3. GENOTYPE X ENVIRONMENT INTERACTION VARIANCE AND THE ESTIMATION OF GENETIC VARIANCES

As demonstrated in section 15.1.1.1, estimates of the genetic variance among families derived from analyses of variance are inevitably biased (upward) by variance among effects of interaction between family and macro-environments when data are collected in only one macro-environment. This is clearly illustrated by comparison of mean square expectations in Tables 15.1, 15.3, and 15.4. Bias of this sort is an obvious problem when the magnitude of additive genetic variance is deduced from estimates of the genetic variance among half-sib families. When selection systems are such that expected response to selection is a linear function of additive genetic variance (or one of its analogs) estimates of that parameter that are too large lead to unrealistic expectations of the genetic improvement possible via breeding programs. There can be no doubt that bias in an estimate of genetic variance is a serious matter when it is to be used as the basis for deciding whether a breeding program should or should not be mounted.

Are "clean" estimates equally important when the critical issue is the relative magnitude of two genetic variances? This question was discussed by Comstock (1955). He pointed out that if our concern is with the relative magnitudes of two genetic variances, e.g., V_1 and V_2, and esimates obtained are of $V_1 + B_1$ and $V_2 + B_2$, where B_1 and B_2 are genotype x environment portions of the expectations, $(V_1+B_1)/(V_2+B_2) = V_1/V_2$ if the ratio of B_1 to V_1 is equal to the ratio of B_2 to V_2. It follows that an argument can be made for collection of data in only one macro-environment (or for averaging estimates of the ratio in question obtained from data collected in different single macro-environments). The argument would include (1) the hypothesis that, in symbolism employed above, $(V_1+B_1)/(V_2+B_2)$ should, at worst, be nearly equal to V_1/V_2 and (2) that given the same total effort the variance of the $(V_1+B_1)/(V_2+B_2)$ ratio estimate obtained from one macro-environment data would be substantially less than that of the V_1/V_2 ratio estimate obtainable when data on one sample of progenies or families are collected in two or more macro-environments. Support for the second part of the argument is provided by the numerical examples in Appendix 15.4 that deals with the $\sigma^2(mIG)/2\sigma^2(mG)$ estimates of Design III.

• 15.4.6. Opinions Concerning Heterosis and Levels of Dominance

Early papers dealing with inbreeding and heterosis in maize were by Shull (1908 and 1911) and East (1908 and 1910). According to Hayes(1963) "At this time neither East nor Shull accepted a Mendelian explanation of hybrid vigor. Both related heterosis to the heterozygous condition due to physiological stimulus from hybridity."

In the sense that they believed the heterozygous state superior to the homozygous the essence of their explanation was overdominance. At about the same time Bruce (1910) pointed out that with superior alleles partly or completely dominant to inferiors the hybrid obtained by crossing two populations would be superior to the average of the parental populations. This explanation was considered inadequate by many in view of the fact that inbred lines derived from the hybrid were never found to be as good as the hybrid itself. Jones (1917) invoked genetic linkages among genes having effects on performance as the explanation and Collins (1921) pointed out that, if performance were controlled by many genes, derived lines as good as the parental hybrid itself would be very improbable. In the light of these theoretical arguments by Bruce, Jones, and Collins, dominance of favorable alleles was accepted by most geneticists as a sufficient explanation until Hull (1945) argued that the rate of improvement of maize hybrids was too slow relative to the difference in performance between first round hybrids and their open-pollinated source varieties to be explained without resource to overdominance. As the rationale for overdominance he cited the argument by East (1936) that as genes mutate new alleles are formed that do good things that are slightly different from what is done by the alleles from which they arose and that to get the benefits of both the original allele and another that may have diverged by virtue of one, two, or more mutations both the original allele and its descendant by mutation would have to be present. In effect, overdominance was explained by pleiotropy. A few years later, Brieger (1950) postulated overdominance by pleiotropy proposing that different alleles of the same gene might be superior with respect to vigor during successive stages of the plant's development. These proposals were fortified by Fisher (1949) and Crow (1952) who pointed out that if selection in open pollinated varieties had brought allele frequencies of superior and dominant alleles into equilibrium between the forces of selection and mutation the difference between hybrids of pure lines and the random mating populations from which pure lines were derived could not be as great as had actually been observed.

The consequence of all this was that overdominance was once more considered distinctly possible--an hypothesis deserving investigation as the basis for choice between the selection systems that did not recognize overdominance as a phenomenon to be reckoned with and those, on the other hand, proposed by Hull (1945) and by Comstock et al (1949).

Evidence from quantitative genetic studies of the sorts described in section 15.4.5 soon provided the base for present convictions among maize breeders and geneticists that the total involvement of overdominant genes in the inheritance of grain yield is minimal at most.[13] [See Robinson et al (1955), Gardner and Lonnquist (1959), Gardner (1963) and Moll et al (1964)]. Most convincing were the data from Design III experiments conducted at the Nebraska and North Carolina Experiment Stations. In those studies both (a) F_2 plants and (b) F_n, advanced generation, plants were used as pollen parents of backcross progeny pairs. The estimates of the $\sigma^2(ml)/2\sigma^2(m)$ ratio when F_2 plants were used were too high to be explained on the basis of partial or complete dominance without linkage disequilibrium but much lower estimates when F_n pollen parent plants were used indicated little, if any, overdominance and showed that linkage disequilibrium was probably responsible for the high ratios observed in the F_2 phase of the studies.

Because grain yield in maize (1) typifies the heterotic traits of cross-fertilizing diploid species and (2) is a quantitative trait of major importance in agriculture, the evidence indicating that overdominance is no more than a minor aspect of its inheritance must be considered a significant contribution of quantitative genetics research. It strongly suggests, though obviously doesn't prove, that overdominance is not a significant factor in the case of other highly heterotic traits. In particular, is it important in some species in the genetics of total value when the latter is a function of several value traits? In accord with East (1936) and Brieger (1950) that is theoretically possible as a consequence of pleiotropy given that different alleles of the same gene are superior and in some degree dominant in their effects on different value traits. In this connection two portions of extensive investigations of pupa weight genetics in _Tribolium castaneum_ by F. D. Enfield and various associates are relevant and thought provoking.

The results of Design III experiments were reported by Enfield et al (1969). Segregating populations of which male parents of backcross progeny pairs were members were (a) the F_2 from the cross of long inbred lines, (b) F_7 and F_{17} populations generated without artificial selection, and (c) F_{10} and F_{20} populations in which selection for greater pupa weight had been employed. Results, reported separately for males and females, were qualitatively similar to those in the D-III work on maize grain yield: (1) linkage disequilibrium in the F_2 was indicated by estimates of the $\sigma^2(ml)/2\sigma^2(m)$ ratio that were greater in the F_2 than in the advanced generations and (2) no suggestion of more than a minor role of overdominance. The estimates of the $\sigma^2(ml)/2\sigma^2(m)$ ratio in the case of the advanced generations were on average a little above 0.40, indicating partial dominance ($\bar{a} > 0.5$) of alleles favoring large pupa weight. The related observation in the context of possible overdominance in the case of total value, reported by Enfield (1980), was that when selection for large pupa weight was relaxed after 50 generations of selection its average decreased in 68 generations by slightly over 50 percent of what the response to selection had been during the 50 generations prior to the relaxation of selection. Apparently, there is pleiotropy involving pupa weight and one or more component of reproduction, i.e.,in the case of some genes alleles favorable relative to pupa weight are unfavorable to reproduction and those favorable to reproduction are unfavorable relative to pupa weight. Given then some level of dominance of favorable effects, whether on reproduction or on pupa weight, overdominance with respect to total weight of pupa produced (the product of number of pupa per female and average weight of pupa) is a possibility.

• 15.5. Numbers of Genes and the Magnitudes of Their Effects

The significance of the number of genes that affect a specific trait relative to total potential response of that trait to recurrent selection is established in section 6.3.5. The significance of $u/\sigma(\hat{X})$ relative to the probability of fixation is explained in section 11.1. And the interrelated impacts of $u/\sigma(\hat{X})$ and effective population size on the comparative utilities of selection programs and variations of them are illustrated in Chapter XI. Most, if not all, of what is presently known concerning these parameters has been learned (a) by comparing groups of individuals classified according to single locus genotype or (b) by inference from response to recurrent selection.

• 15.5.1. Comparison of Groups with Different Single-Locus Genotypes

This procedure is discussed in sections 15.4.3 and 15.4.4 in the context of information pertaining to levels of dominance. It is apparent from those sections that what can be learned differs depending on the history of the genetic population from which data are obtained.

If linkage equilibrium can be assumed (as in section 15.4.3) and the phenotypic means of two or more of the genotype groups differ significantly, it can be concluded that the gene is <u>one</u> that affects the trait in question and (assuming only two alleles) information is provided on the u value of the gene.[14] Unfortunately, there are few genes with phenotypic effects such that their single locus genotypes can be separately identified. Moreover, those that qualify on that score may not constitute a random sample of all genes that affect quantitative traits of interest.

When data are obtained from a genetic population (typically an F_2 from the cross of two highly inbred lines) such that linkage equilibrium is inevitable in the case of all pairs of genes located on the same chromosome and close enough to make their recombination probability less than one-half, what can be learned is quite different. First, a significant difference between at least two of the three groups distinguished by the marked locus genotype indicates the existence of at least one, but very possibly more than one, gene that affects the measured trait. Second, the difference between phenotypic averages of the two groups with homozygous genotypes at the marked locus cannot be interpreted in terms of the u-value of the marker gene, the u-value of any single gene located near to the marker gene, or the sum of the u-values of two or more genes that affect the trait and do not segregate independently from the marked locus (see Appendix 15.2).

In summary, there are reasons to believe that reliable information concerning u-values is not provided by data from either type of genetic population. In the case of linkage equilibrium populations the "marked" loci available (or that can be employed without excess cost) are likely to be a very small fraction of all loci and may constitute a non-representative sample. In the case of data from linkage disequilibrium populations an explicit basis for inference concerning magnitues of u-values is lacking. On the other hand, the data from either type of population provides some information relative to gene numbers.

Consider the data available from linkage equilibrium populations. Definitive evidence concerning genes that affect any single trait is limited to the small number for which actual data are obtained. However, justified inference goes beyond that. Suppose that the data supplied by n "marked" genes demonstrate that of those n genes, x affect a specific trait. Assuming that the genes in question are a random sample (in all respects except that phenotypes evoked enable identification of single-locus genotypes), the fraction $p = x/n$ constitutes an estimate of the proportion of all genes that have effects on the trait. For illustration, consider experimental data in which $n=20$ and $x = 3$ so that $p = 0.15$. Do the results justify the conclusion that 15 percent of all genes affect the trait? Both sampling error and bias must be considered. Bias is downward because the amount of data collected per gene will be insufficient to demonstrate beyond doubt that genes having the smallest effects are in fact genes that do affect the trait. On the other hand, as a consequence of sampling error the observed $p = 0.15$ could be either above or below the fraction of genes that bear on the trait in question. Let P be the fraction of all genes of the organism that affect the trait. Then if $P = 0.03$, the expectation of the number to be observed in a sample of size 20 would be 0.6 and the probability that three or more of twenty would be identified as having effects can be shown without difficulty to be less than 0.021.[15] The comparable probabilities given $P = 0.04$ or 0.05 would be about 0.044 or 0.075, respectively. Considering the downward bias noted above, the probability of $x = 3$ in a random sample of 20 would have to be viewed as quite improbable (< 0.021), given $P = 0.03$ and pretty small (< 0.044) given $P = 0.04$. It would appear quite safe to believe that $P \geq 0.05$ in which case if the total number of genes of the organism were 16,000 there would be at least 800 genes with effects on the trait.[16]

Now consider what is demonstrated by data from an F_2 (a linkage disequilibrium population). Suppose in this instance that effects associated with x loci are shown to be statistically significant . While the absolute minimum number of genes demonstrated to have effects on the trait would then be x, the real number safe to infer would be greater. In the first place, the probability that any single gene segregating in a random mating population will also be segregating in an F_2 from the cross of two pure lines derived from that population is less than one. Consider pure lines derived randomly from the same non-inbred population and let q_{ij} be the frequency of the j-th allele of the i-th gene in that source population. We know that the probability that any of those pure lines will possess the allele in question is q_{ij} and that the probability that any pair of lines will both possess the j-th allele is q_{ij}^2. It follows that in an F_2 populatation the probability of homozygosity at the i-th locus is $\Sigma_j q_{ij}^2$ and the probability of segregation in the F_2 is $1 - \Sigma_j q_{ij}^2$. If the number of alleles is two the probability of segregation is $1 - q^2 - (1-q)^2 = 2q(1-q) \leq 1/2$. Thus, assuming two alleles in the source population, if differential effects of genotypes are discovered at x loci, it is reasonable to infer that the number of effective genes is at least 2x. Of course, if number of alleles of a marker gene in the source population is more than two, $1 - \Sigma_j q_{ij}^2$ can be greater than 1/2 and thus to infer 2x effective genes might entail a little upward bias. The second point to be taken into account is that while a statistically significant difference between groups distinguished by genotype at a single marked locus may result from the effect of only one gene, it may very well reflect non-zero effects by two or more genes. Taking account of both of the relevant facts noted above, it is obvious that real phenotypic differences between genotype groups associated with x marked loci provide the basis for justifiably inferring the existence of 2x or more genes that affect the trait in question. Clearly, the number may be much greater because significant differences associated with the genotypes of any single marker gene may reflect the sum of effects of numerous genes located on the same chromosome arm. While the data pertaining to a single marker provides no evidence concerning the number of genes contibuting to the observed difference(s), information concerning the genomic locations of each of the marker genes for which data have been obtained could enable more explicit inference. For example, knowledge that each marker was located on a different chromosome arm and that a considerable portion of the total number of arms carried a marker would suggest that the number of genes with effects on the trait measured was much greater than the number of markers and hence that the number of genes involved was much greater than 2x.

- 15.5.2. **Response to Recurrent Selection**

Total possible response to recurrent selection was considered in section 6.3.5 and the significance of number of genes with effects on the selected trait was emphasized. In general, the more such genes, the greater the possible response.

As a basis for understanding the expressions that together enable inferences concerning number of effective genes and the magnitudes of their effects on a quantitative trait, consider an F_2

population tracing to a single cross of two homozygous lines. Next suppose that the F_2 is sub-divided, that upward recurrent selection is initiated in one of the sub-populations and downward recurrent selection in the other and that the two way selection is continued until no further changes in the averages of the selected trait are being made because genotypes have become homozygous at the loci of all genes with effects on the selected trait.

Whether value is increased by the upward or downward selection will be seen to have no bearing on the logic to be presented; only the meaning of symbols is affected. However, in what follows alleles favored by upward selection are considered the more favorable ones. Let

Y = total genotype value when all of the more favorable alleles of genes that were segregating in the F_2 have become homozygous,

Y' = total genotype value when all of the less favorable alleles of those genes have become homozygous, and

$D = Y - Y'$.

Then in accord with 5.4,

$$Y = \alpha' + 2\Sigma_i u_i,$$

$$Y' = \alpha',$$

and $$D = Y - Y' = 2\Sigma_i u_i = 2n\bar{u} \qquad 15.30$$

where the indicated summation is over loci (genes) that were segregating in the actual F_2 source population, $\bar{u}$ = the average of the u's of those same genes, and n is the number of genes with effects on the trait that were segregating in the F_2. Note the following.

(1) α' instead of α appears in the Y and Y' expressions. As used throughout beginning with 2.36 and 5.4, α would symbolize the value of the genotype homozygous for the less favorable allele of all genes segregating in the population from which the two pure line progenitors of the F_2 were derived. Here we must remember that an F_2 population will almost certainly be homozygous for the more favorable allele at some loci and hence that the more favorable allele of some genes will be homozygous in the end product of downward as well as of upward selection. In the same way, a portion of the loci homozygous for the less favorable allele in the end product of downward selection will also be homozygous in the end product of upward selection. To summarize, α' symbolizes the value of the portion of the total genotype that consists of genes not segregating in the F_2 source population.

(2) Equation 15.30, like 5.4, assumes no epistasis. There can be no general expression for D because, as demonstrated in section 7.3, values of u vary both with allele frquencies at two or more loci and with the way or ways in which sets of genes interact whenever epistasis is present. Thus relaxation of the no epistasis assumption would require assumptions concerning the details of epistasis present in the genetic system.

From 15.30 it is apparent that D, the total difference (response) theoretically achievable by two way, recurrent selection is proportional to the product of n (the number of genes with effects on the trait that were segregating at the outset) and $\bar{u}$ (the average magnitude of those effects). We have one equation and two unknowns (n and $\bar{u}$). Two equations involving those two unknowns are required as a basis for determining their separate values. Another quantity that can be estimated from the data of a two way selection project is σ^2_{gF} (the additive genetic variance in the F_2). Because from two equations only two unknowns can be estimated, the following simplifing assumptions with respect to the genes segregating in the F_2 that affect the trait subjected to selection will now be made: (1) that the effects of those genes on the trait (u values) are all equal in in their magnitudes, (2) that all of those genes segregate independently, and for completeness, (3) no epistasis (noted above). With these assumptions

$$D = 2nu \tag{15.31}$$

and $$\sigma^2_{gF} = nu^2/2. \qquad (6.26)\ 15.32$$

From 15.31 and 15.32

$$D^2/8\sigma^2_{gF} = n \tag{15.33}$$

This expression together with its enabling assumptions has been an element of quantitative genetic theory for decades [e.g., Castle(1921), Wright(1934 and 1952), and Falconer(1960 and 1981)]. Letting an estimate be indicated by a preceeding bullet (•), we obtain from 15.33

$$\bullet n = (\bullet D)^2/8(\bullet\sigma^2_{gF}) \tag{15.34}$$

and from 15.31 and 15.34

$$\bullet u = \bullet D/2(\bullet n) = 4(\bullet\sigma^2_{gF})/(\bullet D) \tag{15.35}$$

These expressions assume (1) that $E(\bullet D) = D$, i.e., that all alleles having positive effects are fixed by upward selection and all having negative effects are fixed by downward selection, (2) that genes with effects on the trait subjected to selection segregate independently, and (3) no variation among u-values. Do they yield estimates that may be seriously biased in other instances in which epistasis is absent? The answer to this question depends on the deviations of $E(\bullet D)$ and $E[\bullet\sigma^2_{gF}]$ from 2nu and $nu^2/2$, respectively. Accepting the first of the listed assumptions but not the third, $E(\bullet D) = D = 2n\bar{u}$. Can $E[\bullet\sigma^2_{gF}]$ be realistically expressed as a function of n and $\bar{u}$ so that a basis may be available for separate estimates of n and $\bar{u}$. (It is obvious that individual u-values cannot be estimated using only two items of quantitative information.)

In an F_2 from the cross of two pure lines absence of multiple alleles is guaranteed and (barring gamete selection and meiotic drive) allele frequencies at segregating loci must be 0.5.[17] Then , in the absence of epistasis, we know from 6.19 that

$$\sigma^2_{gF} = (1/2)\Sigma_i u_i^2 + 2\Sigma_i\Sigma_{j\neq i}(pt-rs)_{ij}u_iu_j \tag{15.36}$$

This variance is not a function of levels of dominance because when all q's are 1/2 all (1-2q)'s are zero. Questions are posed concerning (1) the sign and magnitude of the linkage disequilibrium term and (2) the deviation of $\Sigma_i u_i^2$ from nu^2 when u's are variable.

The linkage disequilibrium term, $\Sigma_i\Sigma_{j\neq i}(pt-rs)_{ij}u_iu_j$, appears in 15.29 as in 15.36. In the discussion of 15.29 it is noted (a) that if all genes affecting the trait segregate independently (pt-rs) = zero for all gene pairs but (b) that if, as expected, there are non-zero values of (pt-rs) in an F_2 population the expectation of $\Sigma_i\Sigma_{j\neq i}(pt-rs)_{ij}u_iu_j$ should ordinarily be quite close to zero.[18] Ignoring the linkage disequilibrium term leaves

$$\sigma^2_{gF} \approx (1/2)\Sigma_i u_i^2$$

but in accord with A1.4, $E(u_i^2) = \bar{u}^2 + \sigma^2(u)$ so that $E(\Sigma_i u_i^2) = n[\bar{u}^2 + \sigma^2(u)]$. It follows that

$$E(\bullet\sigma^2_{gF}) \approx n[\bar{u}^2 + \sigma^2(u)]/2 \qquad 15.37$$

Hence, by 15.30 and 15.37

$$D^2/8E(\bullet\sigma^2_{gF}) \approx 4n^2\bar{u}^2/4n[\bar{u}^2 + \sigma^2(u)] = n/[1 + \sigma^2(u)/\bar{u}^2] \qquad 15.38$$

The assumptions pertaining to D as defined by 15.30 are (1) absence of epistasis--already discussed--and (2) that $E(\bullet D) = D$ (see above) which requires very large, if not infinite effective population size and absence of pleiotropic effects that would prevent homozygosity of some alleles that affect the trait. Because N_e can never be infinite in practice, $E(\bullet D)$ will inevitably be less than $D = 2n\bar{u}$ by an amount determined by the probability of fixation of alleles favored by selection. Making assumptions indicated in section 11.1, the magnitude of bias can be explored using a specific variation of the following expression.

$$P_f = (1 - e^{-2N_e s q_0})/(1 - e^{-2N_e s}) \qquad (2.34)\ 11.1$$

in which s = the selective value of the allele and
q_0 = the population frequency of the allele when recurrent selection is initiated.

Because allele frequencies at segregating loci in an F_2 population are 1/2 and because our concern is with artificial selection instead of the natural selection assumed in 2.34, the above expression becomes, for our purpose,

$$P_f \approx (1 - e^{-N_e s'})/(1 - e^{-2N_e s'}) \qquad 15.39$$

and is easily calculated.

The impact of finite population size is illustrated here (see Table 15.13) assuming selection among individuals in which selection intensity is the same for male and female (or pollen and seed) parents. Then $N_e = N$ where N is total number of parents (see section 2.7) and $s' \approx 2ku/\sigma(X)$ (see sections 6.3.6 and 11.1.2). Values of k are taken from Table 6.1. In the case of the unknowns, (u,n, and σ) there was of course no alternative to the use of arbitrary values.

Let $\bar{X}$ symbolize the average of the trait in question in the F_2 population. Then note that the absoute values of $\bar{X}$, u, and $\sigma(X)$ are all proportional to the unit of measurement employed in collection of data and, therefore, ratios among them are not affected by the measurement used. If the unit of measurement is such that $\bar{X} = 100$, a reasonable value for $\sigma(X)$ is supplied by the coefficient of variation of the trait. For the purposes of Table 15.13, $\bar{X} = 100$ and $\sigma(X) = 15$ were employed. Because 2u has been defined as the effect of substitution of the BB genotype for the bb genotype, values of u were chosen that made $2u/\bar{X}$ seem likely or possible. For example, if $u = 0.5$ and $\bar{X} = 100$, $2u/\bar{X} = .01$, i.e., the BB for bb substitution in all individuals increases the population mean by one percent. The content of Table 15.13 illustrates the process by which the effective population size bias can be explored and provides a basis for some conclusions. As quantities shown are examined, remember that each line of the table deals with the special

Table 15.13. Two way selection and effective population size

($\bar{X} = 100$, $\sigma = 15$, $D = 2nu = 100$, $p = .05$, and $k = 2.06$)

						$N_e = 10$				$N_e = 40$			
u	$2u/\bar{X}$	u/σ	s'	n	σ^2_{gF}	P[a]	E(•D)	E(•n)	E(•u)	P[a]	E(•D)	E(•n)	E(•u)
.50	.010	.0333	.137	100	12.50	.797	59.4	35.3	.84	.996	99.2	98.4	.504
.25	.005	.0167	.069	200	6.25	.666	33.2	22.0	.75	.938	87.6	153.5	.285
.125	.0025	.0083	.034	400	3.125	.584	16.8	11.3	.74	.796	59.2	140.2	.211

[a] $P = P_f$ = Probability of fixation.

(unreal) situation in which the u-values of all genes segregating in the F_2 are the same.

Consider the first line of the table in which the assumed values of u and n are 0.50 and 100 so that $D = 2nu = 100$, the limits of upward and downward selection are 150 and and 50, and $\sigma^2_{gF} = nu^2/2 = 12.5$. By the expression for s' (above), $s' = 2(2.06)(0.5)/15 = 0.137$.

P, provided by 15.39, is the probability of fixation of both the more favorable allele in the case of upward selection and the less favorable allele in the case of downward selection. It follows that (1-P) is the probability of fixation of the more favorable allele in the case of downward selection and that

$$(\bullet D) = nP(2u) - n(1-P)(2u) = 2[P - (1-P]un = 2(2P-1)un. \qquad 15.40$$

Then •n and •u are obtained using 15.34 and 15.35. Noteworthy facts emphasized by the content of the table are

(1) Probability of fixation (P) increases as $N_e s'$ increases, being a strong function of both u and N_e. It is largest and very close to 1.0 when $u = 0.5$ and $N_e = 40$. In the other five cases P is smaller than 1.0 by amounts sufficient to bias E(•n) downward by significant amounts that range from 23 percent when $u = 0.25$ and $N_e = 40$ to 97 percent when $u = 0.125$ and $N_e = 10$.

(2) Quantities shown suggest that serious underestimation of gene number (n) is probable if $N_e \leq 10$ in both populations of a two way selection experiment and that substantial downward bias is entirely possible unless $N_e \geq 40$.

(3) The bias in E(•u) is upward and largest when the downward bias in E(•n) is largest, as apparent from 15.34 and 15.35.

Unfortunately, because probability of fixation is dependent on both u and N_e, whether the N_e of any single experiment was large enough to make biases asociated with probability of fixation small cannot be inferred from the results of that experiment.

The content of Table 15.13 is conditioned by specific assumptions concerning number of genes, u-values, and selection intensity and does not deal directly with the fact that in the real world the u-values of the genes that affect any single trait are unknowns and must certainly be variable. Questions deserving attention are posed.

The assumptions concerning n and u were such that D = 2nu = 100 in all three instances. What would have been the impact if D, the total potential response to two-way selection, had been supposed to be 160 without change in u values considered? That would have required n = 160, 320, and 640 instead of 100, 200, and 400. Probabilities of fixation would have remained the same since N_es' is not a function of the number of segregating genes that affect the trait. Consequently, because σ^2_{gF}, D and E(•D) are all proportional to n when a single value of u is involved, the biases would have been exactly as indicated by the content of Table 15.13.

What would have been the result of a different assumption concerning selection intensitly? The effect on probability of fixation of change in selection intensity would result from change in N_es'. Suppose the number of individuals in the genetic sample is symbolized as T. Then with the selection system assumed (selection intensity the same for males and females) N_e= pT and N_es'= 2pTku/σ. It follows that with no change in T as p is changed, the variable portion of N_es' is pk. Consider a few values of p and k (from Table 6.1).

p	k	pk .
.05	2.06	.103
.1	1.76	.176
.3	1.16	.348
.5	.80	.400
.6	.64	.384

Obviously, with no change in size of the genetic sample, N_es' increases as selection intensity decreases until p reaches 0.5. However, as p increases k decreases. Thus E(Δq) decreases and more generations of selection would be required for response to reach its maximum. That is a serious defect because recurrent selection until there is no further response is at best a very long enterprise. What can be done about it? Consider the two selection intensities, p = .05 and p = .30 and let T_{05} be the size of the genetic sample when p = .05 and T_{30} be the size of the genetic sample when p = .30. Then if T_{05} is made (.348/.103)T_{30}, pTk and N_es' will be made the same for the two selection intensities and the advantage of p = .05 with respect to E(Δq) would be preserved. It should of course be noted that to obtain N_e = 40 when p = .30 requires T_{30} = 133 = size of genetic sample while to make N_es' the same for the two selection intensities requires T_{05} = 133(.348/.103) = 450.

A sense of how variation in u-values may be expected to affect the results of two way selection projects can be obtained using quantities shown in Table 15.13. For the purposes of illustration, consider an F_2 in which the u-values of segregating genes are 0.50, 0.25, or 0.125

and that the numbers of genes in the three classes are as shown below.

u	n
.50	20
.25	40
.125	80

Whether the skewness of those relative frequencies is realistic can obviously be questioned. However, I am imagining that if the trait to be investigated is one for which recurrent selection has been practiced in the past, alleles with the largest favorable effects will have been brought to homozygosity or higher frequencies and hence that fewer of them will be segregating in an F_2 employed in a two-way selection project. Assumed magnitudes of $\bar{X}$, σ, p, and k will be 100, 15, 0.05, and 2.06 respectively, as indicated in Table 15.13. Using the assumptions listed

$$\sigma^2_{gF} = [20(0.5)^2 + 40(0.25)^2 + 80(0.125)^2]/2 = 4.375$$

and $$D = 2(20)(0.5) + 2(40)(0.25) + 2(80)(0.125) = 60.$$

It will suffice for our purposes to obtain quantities expected given $N_e = 40$ for comparison with assumed real quantities and those that would be obtained if N_e in the selection project were infinite. In the case of $N_e = 40$ the contribution to E(•D) by each class of genes is 2nu(2P - 1) as explained earlier (see 15.40). Thus, using values of P shown in Table 15.13

$$E(\bullet D) = 2(20)(0.50)(.992) + 2(40)(0.25)(.876) + 2(80)(0.125)(.596) = 49.2$$

and by 15.34 and 15.35

$$\bullet n = (49.2)^2/8(4.375) = 69.2$$

and $$\bullet \bar{u} = 49.2/2(69.2) = 0.355$$

In the case of infinite effective population size, not possible in practice, but useful as a reference point when values of n and u have been postulated, we know that $\sigma^2_{gF} = 4.375$ and that, barring interference by pleiotropy, E(•D) = D = 60 . Then

$$\bullet n = (60)^2/8(4.375) = 102.86$$

and $$\bullet \bar{u} = (60)/2(102.86) = 0.292.$$

The quantities of special interest are shown in Table 15.14, part (b), for easy comparison. Qualitatively, the biases apparent in the table are as expected. E(•D) < D when N_e is not infinite. Of course, that bias becomes trivial as N_e becomes large. E(•n) < n both when N = 40 and when N is infinite and consequently E(•ū) > ū in both cases. That E(•n) should be less than n even when E(•D) = D is apparent from 15.38.

The lessons to be learned from the contents of Tables 15.13 and 15.14 are (1) estimates of n are ordinarily biased downward, (2) when that is the case the asociated biases in estimates of ū will be upward, and (3) there seems no way of knowing how large N_e must be to avoid substantial biases because E(•D) < D unless one takes the position that only genes with u-values above a certain magnitude are of concern. An alternative would be to conduct parallel experiments using different effective population sizes with the objective of finding whether the differences in

Table 15.14. Effects of variation among u-values—an illustration

($\bar{X}$ = 100, σ = 15, p = 15, and k = 2.06)

(a) Frequency distribution of u-values (assumption)

Number of Genes	u
20	0.500
40	0.250
80	0.125

(b) Expectations of estimates and population values

N_e = 40	N_e Infinite	Population Values
$E(\bullet n)$ = 69.2	$E(\bullet n)$ = 102.8	n = 140
$E(\bullet \bar{u})$ = 0.355	$E(\bullet \bar{u})$ = 0.292	$\bar{u}$ = 0.214
$E(\bullet D)$ = 49.2	$E(\bullet D)$ = 60	D = 60
		$\sigma^2(u)$ = .0166

magnitudes of estimates obtained were large enough to suggest that even the larger N_e might be too small.

15.5.2.1. ONE WAY SELECTION

Obviously, response to recurrent selection in one direction can also provide information concerning number of genes and the average ($\bar{u}$) of their effects on the selected trait (see Comstock, 1969). Assume upward selection for any single trait, that the starting population is an F_2 from a cross between two pure lines, and that epistasis is absent. Then, taking account of dominance, let

$Y_2 = \alpha' + \Sigma_i u_i + (1/2)\Sigma_i a_i u_i$ = the average value of genotypes in the F_2 and, as before,

$Y = \alpha' + 2\Sigma_i u_i$ = the value of genotypes when the more favorable alleles of all genes that were segregating in the F_2 have become homozygous. Then

$$2(Y - Y_2) + \Sigma_i a_i u_i = D = 2\Sigma_i u_i \qquad 15.41$$

in which D is the quantity defined by 15.30. The required estimate of $\Sigma_i a_i u_i$ is conveniently provided by the difference between the phenotypic means of the F_1 and F_2 generations from the pure line cross. As shown in Appendix 15.5,

$$\ddot{Y}_{F1} - \ddot{Y}_{F2} = (1/2)\Sigma_i a_i u_i$$

when epistasis is absent.

Clearly, aside from procedure leading to estimates of D, relevant logic is the same for one way and two way selection.

Whether two way selection or one way selection is the procedure of choice hinges on a variety of considerations that may depend on the organism, the trait, other issues that may be served by a long term selection project, etc. that cannot appropriately be discussed in useful detail here. However, this author inclines to the view that there are more significant issues to be illuminated by upward than by downward long term recurrent selection projects.

• 15.5.2.2. **CONSEQUENCES OF EPISTASIS**

The quantifications of D by 15.30 and 15.41 are predicated on absence of epistasis and it was pointed out in the discussion of 15.30 that an alternative expression that would be justified in the presence of epistasis would require specifications concerning the nature of that epistasis.

Comstock and Enfield (1981) reported that data provided by long-term selection can cast doubt on the assumption of no epistasis. In data from long-term selection for greater pupa weight in flour beetles and post-weaning weight gain of mice they observed (1) population mean increases greater than the means of the F_2 (from the cross of highly inbred lines) populations in which recurrent selection had been initiated and (2) that estimates of additive genetic variance increased "through one-half or more of the time required to reach plateaus". They found that, in their data, both of these phenomena were consistent with multiplicative gene action (a form of epistasis) and that the observed increase in additive genetic variance, at least, could not be rationalized without recourse to some form of epistasis. They developed and reported a theoretical base for estimation of gene number assuming multiplicative gene action. The estimates obtained were no more than half as large as those obtained assuming no epistasis.

While the evidence inconsistent with the absence of epistasis that is described in the above paragraph came from one way upward selection projects, there is a significant form of epistasis that would not be discerned by one way selection results--the type in which the positive effects of favorable alleles diminish in magnitudes as the number of such alleles in the total genotype increases. Suppose that there are n genes that affect the trait that were not homozygous in the F_2 population. Then the maximum genotype value might result when all favorable alleles are present or when some lesser number, $n < x < 2n$, is present. The maximum genotype value, achieved when enough (not necessarily all) favorable alleles are present in the genotype, has sometimes been referred to as a "physiological limit". Jain (1989) argued that genetic increases in the grain yield of various economic species (he mentioned sorghum, maize, rice, millet, and wheat) are slowing as they approach limits imposed by photosynthetic rates. Clearly, if the positive effects of favorable alleles decrease as their numbers in the genotype increase and, conversely, increase as their numbers in the genotype decrease, the response to recurrent selection must be less in the case of upward selection than in the case of downward selection. Moreover, it is reasonable to postulate that the direction of asymmetry in two-direction selection projects would depend on the history of selection for the trait in the species in question and, hence, that two-way selection would be more illuminating than one way selection with reference, particularly, to magnitudes of $\bar{u}$. In this connection, note that the epistasis would not be signalled by an increase in additive genetic variance as allele frequencies increased from the 0.5 levels of the F_2 starting population as in the case of multiplicative gene action. Instead, additive genetic variance in the population subjected to downward selection would probably increase for a number of generations. The additive genetic variance contributed by a single locus at each point in time would be

$$\sigma^2_{gi} = 2q_i(1-q_i)[1+(1-2q_i)a_i]^2u_i^2.$$

As q's were decreased by selection u's would be increased (by the definition of the type of epistasis being considered) and it is probable that σ^2_{gi} would increase until the downward selection had resulted in a substantial reduction (from 1/2) in allele frequencies.

• 15.6. Comments

Breeders know that selection is <u>the</u> means by which genetic improvement can be realized but must choose among selection systems and procedural variations of them. A theoretical base for those choices is required (see section 1.5). Their other major questions have to do with the amounts of genetic improvement that can properly be expected (a) currently and (b) eventually.

In its interactions with the breeding enterprises, quantitative genetics has addressed itself to these problems and questions and has produced a body of theory of which directly relevant portions have been presented in previous chapters.[19] As stated in its opening sentences this chapter deals with quantification of parameters required in practical applications of the theory.

Equations 8.28 and 8.29 are master formulas for prediction of direct or correlated responses to single stage selection in the case of a broad class of breeding programs. Taken together they explain why variances and covariances have received so much attention in the literature of quantitative genetics. First, except when selection units have been obtained by inbreeding, both

$$\Sigma_i \sigma(\hat{q}_i,\hat{X})2[1+(1-2q_i)a_{ci}]u_{ci} \text{ and } \Sigma_i \sigma(\hat{q}_i,\hat{X})[1+(1-2p_i)a_{ci}]u_{ci}$$

are the total or a fraction of an additive genetic variance or covariance or an analog of an additive genetic variance or covariance. Second, $\sigma(\hat{X})$ is a component of both 8.28 and 8.29 and $\sigma^2(\hat{X})$ is a function of genetic, genotype-environment and nongenetic variances (see Chapters X and XII).

These expressions (8.28 and 8.29) are the crux of quantitative genetic theory relative to breeding. Are they valid? Are their predictions approximately correct? The consideration of these questions in section 16.1 is encouraging but not conclusive. Evidence from well designed work with agricultural species is required to (a) provide answers satisfactory to the breeding profession as a whole and (b) to determine whether an understanding grasp of current quantitative genetic theory should be included in the qualifications of a professional breeder, i.e., whether quantitative genetics should be a component of advanced level training programs.

What kind of experiments would be useful? The obvious crucial question is whether 8.28 and 8.29 provide good predictions--predictions accurate enough to be employed routinely as the basis for breeding program decisions. There is plenty of evidence indicating that selection supposed to be effective when additive genetic variance is present is actually effective in that circumstance. There is far less, if any, showing responses to be close to the amounts predicted. The probable reason is the cost of producing such evidence. As an example, the cost of checking whether response to one cycle of full-sib family selection for grain yield of maize conforms with prediction will be considered.[20] In this case, the prediction equation (8.28) becomes

$$E(\Delta\ddot{Y}_x) = (1/2)\sigma^2_{gx}[k/\sigma(\hat{X})] \qquad 15.42$$

in which X symbolizes grain yield, the selection criterion. To test its validity, estimates of σ^2_{gx} (the additive genetic variance of X), $\Delta\ddot{Y}_x$ (the actual response to one cycle of selection), and $\sigma^2(\hat{X})$, the phenotypic variance of among full-sib families in the field comparison basis for selection. The question to be examined is the cost (in data required) of making the standard error of $[\Delta\ddot{Y}_x - E(\Delta\ddot{Y}_x)]$ reasonably small relative to $E(\Delta\ddot{Y}_x)$ itself.

It appears (see Appendix 15.6) that if σ^2_{gx} is to be estimated in terms of the genetic variance among half-sib families, grown in more than one location in each of more than one year to avoid bias from genotype-environment interaction variance, data from as many as (or more than) 15,000 plots would be required if reasonably narrow confidence limits of an estimate of $[\Delta\ddot{Y}_x - E(\Delta\ddot{Y}_x)]$ are to be achieved. In the specific example examined non-validity of theory would not be demonstrated conclusively by absolute magnitudes of the estimate as great as 30 percent of the actual (and unknown) $E(\Delta\ddot{Y}_x)$. Can support for tests of theory of that magnitude be found?

There are alternatives (listed below) to the context in which the "cost" issue is examined in Appendix 15.6--alternatives by which the variance of estimates of σ^2_{gx} would be reduced-- but in each instance there is bias to be dealt with.

(1) Estimates of the genetic variance among (a) full-sib families or (b) S_1 families (see section 15.1.2.2) could be used as the source of information regarding the magnitude of additive genetic variance. In each of these alternatives the variance of the biased estimate of σ^2_{gx} would be smaller than when the variance among half-sib families is used as the source of information--because the quantity estimated is larger relative to σ^2_{gx} than in the case of half-sib family genetic variance.

(2) The difference between (a) the average of $\Delta\ddot{Y}_x$ over two or more cycles of selection and (b) the expectation of $(\Delta\ddot{Y}_x)$ per cycle instead of the single cycle value of $[\Delta\ddot{Y}_x - E(\Delta\ddot{Y}_x)]$ could be made the test of theory. In this case it is $E(\Delta\ddot{Y}_x)$ that may be biased--as a consequence of finite effective population size (see Table 14.19 and related text).

The author thinks that both of these alternatives deserve consideration but warns that in either case inferences from estimates will necessarily be made with care.

Another potential source of evidence on validity of theory are the changes in genetic variances and covariances during recurrent selection programs. Consider additive genetic variance in successive cycles of a single population program.

$$\sigma^2_{gx} = \Sigma_i 2q_i(1-q_i)[1+(1-2q_i)a_{xi}]^2 u^2_{xi}$$

$$+ \Sigma_i\Sigma_{j\neq i}\, 2(pt-rs)_{ij}[1+(1-2q_j)a_{xj}][1+(1-2q_i)a_{xi}]u_{xi}u_{xj} \qquad 6.19$$

which reduces to

$$\sigma^2_{gx} = \Sigma_i 2q_i(1-q_i)[1+(1-2q_i)a_{xi}]^2 u^2_{xi} \qquad 6.8$$

in the absence of linkage disequilibrium. It is most obviously a function of allele frequencies, each of which is expected to be increased by selection based on X. If, in general, they have been increased is of course indicated by whether $\ddot{Y}_X$, the population genotypic mean of X, has increased. Given estimates of both $\ddot{X}$ and σ^2_{gx}, at successive cycles of recurrent selection, significant questions are (a) are the observed magnitudes of $\ddot{X}$ in reasonable accord with the theoretical expectations of $\Delta\ddot{Y}_X$ as a function of additive genetic variance estimates and (b) are observed successive estimates of σ^2_{gx} explicable in terms of 6.8 or 6.19? Clearly, the data providing estimates of $\ddot{X}$ and σ^2_{gx} must be adequate.

It is appropriate to assume that the goals of breeders are always to increase $\ddot{Y}_W$, the average value of genotypes with respect to total value. As emphasized in section 12.4, in the long run that goal can only be insured by making total value the selection criterion. However, it must be recognized (see section 6.3.7) that there often are compelling reasons for use of a different measure as selection criterion. In those cases, $E(\Delta\ddot{Y}_W)$ depends on the genetic covariance between W and X, the selection criterion actually employed. The issue is considered in the context of mass selection in section 6.3.7 where the ratio of expectations of correlated and direct responses to selection is shown in 6.36. Substitute w, as a subscript, for the subscript c and the equation shows the ratio of the expected correlated response per unit of time of W to selection based on X to the direct response of W per unit of time to selection based on W itself. The magnitude of σ_{gxw}, the additive genetic covariance between W and X is shown to be critical. Very little information concerning the magnitudes of this particularly important genetic parameter is available--much more is needed.

16 Looking Ahead

The central focus of this book is on the genetic improvement of our agricultural species by selection. Background facts of particular importance are these.

(1) The great impact that recurrent selection has had in the past is obvious--consider the productivity of our major agricultural species relative to that of their distant ancestors. Though the contrast suggests that the genetic variation that enabled so much change may have been largely exhausted in the process, the weight of recent and numerous estimates of genetic variances indicates that the potential for further response to selection that remains in the best of random mating populations is not minimal and may be great.

(2) Alternatives in the application of selection are numerous but the inherent cost of an adequate experimental base for choices among selection systems and variations of them is formidable. In particular, the total cost would probably include decades of time whereas bases for decision are needed now. The value of an adequate theory base for comparison of alternative selection systems and variations of them is apparent (review of section 1.5 is recommended).

(3) The quality and utility of theory pertaining to the consequences of various forms of selection have been, and no doubt are being, questioned [see, for example, Kempthorne (1988)]. The argument is that the connections between deductions and the "real world" are seriously diminished by the simplifying assumptions that are, and must be, made in the theory-producing process. Simplifying assumptions are always made. The amount of damage done in particular instances is the real issue.

(4) Recent developments in molecular genetics have added to the available procedures for achieving genetic improvements via selection. In turn these have confronted Quantitative Genetics with new challenges.

• 16.1. Molecular Genetics and Breeding Methodology

The genetic potential of a population hinges on (1) content of its gene pool and (2) the effectiveness of recurrent selection for optimizing the frequencies of available alleles. Two promising applications of molecular genetics technology are identified and discussed in the following subsections. The challenge of food production for a growing world population (see sections 1.1 and 16.2) dictates that the frame of reference for discussion of these applications be long term recurrent selection aimed at maximum utilization of the genetic potential of food production species.

• 16.1.1. Marker-Assisted Selection (MAS)

For discussions of this supplement (or alternative) to conventional systems in which selection is based directly on measures of phenotype see Dentine (1992) and Dudley (1993). Both pointed out that the idea involved has been around for a long time but as a practcal procedure awaited molecular technolgy to supply markers associated with a sufficient portion of the entire genome and the technics for identifying their presence and distinguishing between alleles of individual markers. A type that appears promising is known as "restriction fragment length polymorphisms" (referred to by the acronym RFLP). These are formed by DNA-cutting enzymes that act at sites characterized by a specific sequence of nucleotides. Such a sequence may be present at many sites in the total DNA of a species. The result is that a single restriction enzyme can produce many RFLP's of which there are variations in length, sequence of nucleotides and location within the genome. Individual RFLPs can be distinguished by length and nucleotide sequence. Technologies by which this is accomplished need not be described but it is germane to note that the cost involved may not be prohibitive.

Because each specific RFLP is a fragment of the DNA of the organism, its nucleotide sequence may not be exactly the same in all members of that RFLP family, i.e., there may be "alleles" of the fragment. In fact, its utility as a marker hinges on polymorphism, the existence of alleles. Because many of these fragments may be long enough to encompass a gene the phenotype differences associated with alleles of a specific fragment may arise from effects of the allele itself or from alleles at QTL (quantitative trait loci) linked with the marker alleles. However, Dudley states, "Because markers usually are not QTL, but are linked to QTL, linkage between the marker allele and the QTL is important, (and)··· without linkage disequilibrium, associations between markers and QTL cannot be identified."

• 16.1.1.1. THE LIMIT OF INCREASED RESPONSE VIA MAS

The objective of MAS is, to the extent possible, substitution of genotype value estimates for measurements of phenotype as the basis for selection. It is reasonable to inquire how much can be gained thereby. An upper limit can be visualized in the context of mass selection. Assume

1. A large random mating population with H-W frequencies of single locus genotypes in linkage equilibrium,
2. Recognition of all single locus genotypes with non-zero effects on X (a measurable quantitative trait),

3. Knowledge of the magnitudes of effects on X of all single locus genotypes, and
4. Constancy through time of those effects.

From 6.10 we know that in the case of mass selection based on phenotype of the selection criterion the expected direct response is

$$E(\Delta\ddot{Y}_{xp}) = (\bar{X}_s - \bar{X})h_X^2 = k\sigma(X)\,\sigma_{gX}^2/\sigma^2(X) \qquad 6.10$$

Alternatively, if selection could be based on $Y = \Sigma_i y_i$ = the sum of single locus genotype effects in individuals,

$$E(\Delta\ddot{Y}_{XY}) = k\sigma(Y)\,\sigma_{gX}^2/\sigma^2(Y) \qquad 16.1$$

because then Y, genotype value, would have been substituted for the approximation (probably crude) provided by phenotype. It follows that

$$E(\Delta\ddot{Y}_{XY})/E(\Delta\ddot{Y}_{xp}) = \sigma(X)/\sigma(Y)$$

If all genes were additive, $\sigma^2(Y)$ would be equal to σ_{gX}^2, the additive genetic variance of X, with the result that $\sigma(X)/\sigma(Y) = 1/h_X$. Thus, in the case of a trait with heritability (h_X^2) equal to 0.16 $E(\Delta\ddot{Y}_{XY})/E(\Delta\ddot{Y}_{xp})$would be 2.5. Given some dominance, $\sigma^2(Y)$ would be somewhat larger than σ_{gX}^2 and $E(\Delta\ddot{Y}_{XY})/E(\Delta\ddot{Y}_{xp})$ somewhat smaller than $1/h_X$. For example, assuming two alleles of genes with effects on X, complete dominance of the more favorable allele, and $h_X^2 = 0.16$ the reduction of $E(\Delta\ddot{Y}_{XY})/E(\Delta\ddot{Y}_{xp})$ from its value if all genes were additive would be in the range from 0.09 to 0.32 if favorable allele frequencies of genes with effects on X were in the 0.3 to 0.7 range. The advantage over selection based on phenotype would be substantial (70-127 percent) instead of the 150 percent assuming no dominance.

A further advantage of MAS would accrue if marker genotypes could be determined early enough to enable selection of both males and females in animal species and both pollen and seed parents in plant species before instead of after mating or fertilization. Consider mass selection for grain yield in the case of maize. By identifing marker genotypes using DNA from coleoptile tissue values of $\Sigma_i y_i$ could be obtained as the basis for equally intense selection of both pollen and seed paraents. Thus the k of 16.1 could be twice as great as when selection is delayed until grain yield has been determined, i.e., precluding selection on performance in the case of pollen parents. As another example, consider the selection in dairy cattle of males to be progeny tested. In the past that selection has perforce been based on performance records of female relatives (dam, full sisters, and half sisters). The advantage of knowing $\Sigma_i y_i$ (on the basis of examination of DNA from amniotic cells) is not as obvious as in the preceding example but would be substantial. Though standard progeny testing of males can conceivably provide estimates of "breeding value" as good as inherent in $\Sigma_i y_i$, comparative costs would in practice determine size of the "genetic sample" in selection programs and thereby the magnitude, if any, of the "k-advantage". However, the ideal MAS procedure would enable a shorter selection cycle and thereby greater genetic improvement per unit of time.

The foregoing discussion of limits to greater responses to selection by use of MAS should serve to explain and justify the "gleam in the eye" of those engaged in perfecting the necessary technologies. However, problems having to do with the extent to which limits sketched can be approached need to be considered. They differ between MAS by (a) knowledge of approximate effects of single locus genotypes and (b) by knowledge for numerous markers of alleles at QTL linked with the marker plus the effect on X, if any, of the marker itself.

- 16.1.1.2. **WHAT PORTION OF THE CONCEPTUAL PROMISE OF MAS CAN BE REALIZED?**

Method A -- Estimate of $Y = \Sigma_i y_i$ as the selection criterion The requirement is an estimate of the effect of each single locus genotype. Each value of Y would , even if all genes were additive, be the sum of as many separate y-values as the number of genes with effects on X. Thus

$$Y = y_1 + y_2 + y_3 + \cdots + y_n.$$

No y-value can be known exactly--the available estimates of Y would necessarily be the sum of estimates of individual y's and the error variance of Y would be the sum of the estimation error variances of the y's. The cost of achieving the theoretical limit of increase in heritabilitly of Y relative to that of X would obviously depend on the number of genes with effects on X, the cost of identifying genotypes of those n genes, and the cost of estimating the individual y-values with reasonable precision. The latter cost would be greater for small effect genes. The number of genes and the distribution of magnitudes of their effects are unknowns for significant traits of food production species. It follows that the cost of a substantial increase in the heritability of Y from that of X may be great at best, i.e. even if all genes were additive, there were no gene x environment interaction effects (see section 2.13, Appendix 5.1, Chapter X, and section 15.1) and the individual effects were constant through time so that repeat estimates would never be needed. However, substantial temporal changes in the y's of concern are possible, if not inevitable, as a consequence of epistasis (see section 7.3) and time trends in TPE's for production of the species.

Whatever total cost in practice may be, it is obvious that the realized benefits of Form A of MAS may be relative to what they are advertised to be will be strongly influenced by cost.

Method B--selection based on in part on estimates of genetic effects associated with "marker" alleles The foregoing discussion of the impact of costs apply to Form B as to Form A. But there is more to be added when B is considered.

A significant aspect is the cost of linkage disequilibrium (LD). LD is obtained by limitation of effective population size, N_e. The perfect example (Edwards et. al., 1987 and Dudley, 1993) is the F_2 generation from a cross of pure lines. The parent generation is the F_1 in which there is only one genotype, hence N_e=1. The impact of small N_e is to reduce the probability of fixation of favorable alleles as an eventual result of recurrent selection (see section 2.8). That this should be true in an F_1 generated population is obvious since the probability of any favorable allele being absent in the F_1 is $(1-q)^2$. Also obvious is that the linkage association measured in an F_2 generation is specific to that population and would be different in any other population-- would have no value for selection in any other population.

Another cost arises because in the case of markers polymorphic in the population from which purelines are extracted only a portion are expected to be heterozygous in any specific F_1. If not, that marker would not be segregating in the F_2 and hence would not have utility in the generations beyond the F_1. The portion expected to be heterozygous in a random F_1 is $1 - \Sigma_j(q^2)_j$ in which q is the frequency of the j-th allele of the marker in the population from which pure lines are extracted. Thus only a fraction of the potential markers will have utility with reference to selection in any population initiated from an F_1 of pure lines.

The two costs identified above would exist, though probably at different levels, whatever the source of LD.

A third cost recognized by Dudley (1993) is the fact that marker-associated effects will dissipate by virtue of recombination in successive generations. This adds to the reasons why estimates for use in selection would have to be repeated at intervals. In this connection, note that the cost of LD, cited above would be recurrent.

Summary Enthusiasm for MAS is justified. However, in practice benefits will be tempered by costs. In the area of long term recurrent selection aimed at capitalizing as fully as possible on the potential of the gene pool of populations, Form A must be considered superior to Form B.

• 16.1.2. Transgenes for Improvement of Gene Pools

Genes transferred between inter-sterile species or populations using molecular genetics technology are identified as transgenes. In the case of genes having unique and valuable effects such transfers can be particulary beneficial. An example cited by Weaver and Hedrick (1992) was the transfer to tobacco of a gene for resistance to glyphosate. The gene was a mutant that had occurred and survived in *Salmonella* in the presence of glyphosate. By a series of sophisticated procedures the gene was transferred to tobacco plants. The transfer was significant because glyphosate is the active ingredient in the widely used herbicide, Roundup. The herbicide can be used to kill weeds around the mutant carrying tobacco plants without harming the tobacco.

This success encourages attempts to use transgenes to supplement the gene pools of important species in profitable ways. Would it not be significant if additions to the genomes of maize, wheat, or rice created the potential for nitrogen fixation in one or more of those species.

• 16.2. Limits of Response

The question posed is this--what is the maximum of $\ddot{Y}_W$ (the average of genotype values with respect to total value) in the case of a single agricultural species in a particular target population of environments? To date the question has not been answered with respect to any species-TPE combination.[1] The obvious next question is how information regarding such a maximum of $\ddot{Y}_W$ can be obtained. By definition it would be the value (with respect to total value) of an "ideal genotype"--the genotype composed throughout of the best of the possible single locus genotypes. Barring overdominance, this would mean homozygosity of the superior allele in the case of all genes with effects on W. We have learned that in the case of most alleles $E(\Delta q)$ is a small, possibly very small, quantity unless s' is larger than it can be when the number of genes with effects on the trait in question is large as it must be in the case of total value. It follows that homozygosity (or near homozygosity) can only be achieved via recurrent selection. And because Δq, the actual change of frequency of a superior allele is variable around $E(\Delta q)$ as a consequence of "drift" with variance equal to $q(1-q)/2N_e$ the probability of eventual fixation of the superior allele is a function of both s' and N_e, shown by Kimura (1957) to be a function of $N_e s'$, i.e., the probability that the allele will be fixed is larger when $N_e s'$ is larger and the probability that it will be lost instead of fixed is greater when $N_e s'$ is smaller. It appears, therefore, that recurrent selection, in which W is the selection criterion, continued long enough with $N_e s'$ large enough is the procedure by which maximum $\ddot{Y}_w$ can be approximated. However, the end result of such selection cannot be predicted. The potential for response is, locus by locus, a function of allele effects on total values (u_w's), allele frequencies, and levels of dominance--all unknowns. What can be accomplished can only be learned by experiense.

The basic element of theory that points to long term recurrent selection as the way to information (the only way) concerning maximum values of $\ddot{Y}_w$ cannot be questioned. The probable directions of changes in allele frequencies in closed populations result only from selection. And the qualitative logic of Kimura's finding that the probability of fixation of an allele is a function of its selection coefficient and effective size of the selection population is beyond question. Cycle by cycle, Δq is the sum of $E(\Delta q)$ and a random deviation of which the expectation is zero and the variance as noted above is $q(1-q)/2N_e$. Of course, the probabilities of fixation of an allele as stated by 2.33 and 2.34 are not "real world" quantities because of the unrealistic assumptions on which they rest. Remember those assumptions (section 2.8)--that both s and N_e are constant through the entire recurrent selection process until the allele in question is either "fixed" or lost. The practical value of those probability statements lies in the information they provide concerning the range of possibilities of the fixation probability, i.e., the guidance they provide with respect to design of long term recurrent selection programs.

• 16.3. Large Recurrent Selection Programs in the Context of World Population Growth

As background, thoughtful rereading of section 1.1 is recommended. We know that human beings, young and old, are starving in some parts of the world and in many other areas food is plentiful but for economic reasons is not be available in adequate amounts or quality to members of underclass families. We know that the reasons for hunger and malnutrition among unfortunates of our world are complex--distribution problems and costs aggravated by ethnic strife, national boundaries, and reluctance of the rich to share with the less fortunate. But when all is taken into account a basic question remains--is food production potential at this time sufficient to feed the present world population as needed for good health? I don't know but have the impression that distribution, not production capacity is the current over-riding problem. However, at some point, distribution barrier problems will be compounded by actual food shortages if population growth continues too long.

Brown et al (1991) report "The most recent World Bank projections show the 1991 world population of 5.4 billion growing to at least 12 billion late in the next century," and then react as follows:

> "If the world remains on the 12-billion-plus trajectory for much longer, population growth will be checked in more and more countries by famine and disease as living conditions fall below survival levels. .... We believe human numbers can go no higher than about 8 billion--half again as large as today's population--if the world is to avoid massive deaths from starvation and disease. Considering that one third of all people in the world today are under 15 years of age, and thus have yet to enter their reproductive years, holding world population at this number will take nothing less than a revolution in social values and reproductive behavior. This, in turn, depends on fundamental reforms in many developing countries in education, health care, and the status of women, along with a massive global reordering of priorities".

Their opinion is based in part on the fact that, for various reasons including the population increase, the amount of cropland per person has been decreasing for at least three decades and will inevitably continue to decrease as population increases.

In contrast, Lasher (1991), appearing less pessimistic than Brown et al (op. cit.) concerning population growth constraints in the more immediate future, reports that the "world population is now predicted to exceed the 10.2 billion level for eventual population stabilization calculated in 1984, and is likely to be closer to 11 billion." His projection that ZPG (zero population growth) can be reached at the 11 billion level by expansion of family planning practices assumes obviously that suffient food would be available as the world population doubles. Whether ZPG is brought about by malnutrition and starvation or by family planning, the role of food production is obvious.

Brown and associates outlined some of the ways in which food per hectare can be increased without mention of the possibilities of genetic improvements via breeding. Of the substantial increase in the grain yield of maize hybrids in the 1930-80 period it appears that over 50 percent has been the consequence of genetic improvement (Duvick, 1984 and 1992). He predicted further genetic improvement but refrained (wisely, it must be supposed) from quantifying that prediction. Roughly summarized, possibilities and prospects appear to be these: (1) ZPG will be reached (a) by malnutrition or (b) more humanely, by education, family planning, and world-wide sharing. Population growth limits have been predicted to be 8 billion and 11 billion or more, respectively. (2) Food per hectare can be increased by the combination of (a) Improved husbandry practices (a la Brown and associates) and (b) genotype improvements in selected agricultural species. <u>But</u> crop land per person will decrease as population increases and for one reason or another land productivity may suffer in the years ahead. (3) To enable a healthy population of 11 billion present food production capacity would <u>at minimum</u> need to be doubled.

Whether our world's capacity for the production of food can be doubled is problematical. To maximize the probability of doing so coordination of all promising approaches may clearly be required.

• 16.3.1. **Potential for Genetic Improvements**

This potential consists of (a) the currently present useful alleles of agricultural species of which population frequencies have not yet been brought to optimum levels and (b) new useful alleles arising continuously by mutation and constitutes a very significant natural resource of our world.[2] It has been, and is being, exploited beneficially by breeding programs but theory described herein indicates that a larger portion of genetic potential relative to total value can be realized by increasing the effective population size of recurrent selection programs and using total value itself as the selection criterion whenever feasible.[3]

The accumulated evidence provided by numerous estimates of genetic variances within populations together with observed responses to recurrent selection leaves little reason to doubt the further potential of selection programs focused on total value. However, actual potential measured in terms of world food production can never be approximated except by trial.

The concept of physiological limits (section 15.5.2.2) to productive performance cannot be doubted--who would argue that grain yield of wheat can be raised genetically to a level such that 100 hectares of the crop would suffice to feed the world? On the other hand, given the allele resource of the species, it is no more resonable to contend that limits have been reached or to assert with confidence that limits of further improvement will be, for example, ten, twenty, or forty percent.

• 16.3.2. **Considerations—Practical and Philosophical**

The reality of extensive starvation now and in the future poses a challenge to the conscience of the world. Individually, we recoil from the prospect of death by starvation for ourselves, our immediate families and our descendants. Collectively, it is reasonable to question whether hunger and peace can coexist. Maximizing the potential of recurrent selection as a component of an overall program aimed at providing adequate food supplies for a stable world population would require a major commitment by the world community and its leadership. Organizing and insuring the coordinated and continuing efforts entailed would be a task for a dedicated and very well informed personnel team.

This can not be visualized as a free-enterprise project. Its pay-off would be in health of the entire world society, not in monetary profit of any private company. And its full potential hinges on long time continuity that would or could not be guaranteed by private enterprise. By its nature it would necessarily be a public program of a multinational institution involving as many nations as possible. For a variety of reasons the United Nations appears a natural candidate. However, some portion of the professional expertise required might best be obtained by contracting with one or more private companies for (a) counsel regarding structure and (b) portions of the operation, of the overall program. And it is well worth emphasizing that the design of the long term public program would be simplified by recognition that near term improvements would be engineered, as now, by the private breeding companies.

Major favorable arguments are (1) a well conceived battery of large recurrent selection programs would constitute a significant component of a campaign to increase the world's food production capacity as needed during the realization of ZPG and (2) it could be a project that would engage the cooperation of all countries, in the interest of each and all of them, and thereby advance the cause of world peace.

The primary argument against the proposal would probably be cost--the logical dimensions of the project, outlined below, are such that annual cost would be considerable and the time span indefinite. Considered out of context, cost could be a major argument against such a project but in relation, for example, to the amounts spent around the world for military equipment and personnel, its cost would be well within reason.

16.3.3. **Program Dimensions**

An individual component would be focused on maximum genetic improvement of any one of the agricultural species with respect to performance in a particular TPE and each component, by itself, would be an open-ended recurrent selection project. Obviously, a total program aimed at increasing the world food production capacity could be very large--and its total operating cost very great. The critical decisions having to do with costs and program effectiveness are those concerning (1) the species-TPE focus of the separate components of the overall program, (2) foundation populations employed, and (3) size of genetic sample in the selection populations. Readers will recognize that each of these issues is complex and that the quality of decisions concerning them would depend on more than genetic rationale. Comments offered below are restricted to the most salient things that flow from quantitative genetic theory already presented plus a few others that are obvious and serve to demonstrate my own recognition that more than quantitative genetics would be involved.

The following should clearly be considered in the choice of species: (1) whether selection based on total value is likey to be both possible and reasonable, (2) ease with which crosses can be made--numerous crosses between selections are required to enable desirable effective population size of selection populations, (3) nutritional quality as human food, and (4) current productivity. We note that maize qualifies by all of these except nutritional quality as human food--it is not as widely used as a human food as, for example, rice and wheat and its protein is an incomplete source of essential amino acids. By contrast the soybean is an excellent source of protein and a nitrogen fixing plant but a species in which production of seeds using cross pollinations requires more time and is therefore more expensive.[4]

The argument for separate recurrent selection programs in different TPE's is that the rank order values of the alleles of some genes are affected by the environment in which the individuals of a species are grown, i.e., that different alleles of some genes are "best" in TPE's that differ in significant respects. That this argument is correct is supported by performance data having to do with the relative performance of cultivars in different environment populations. Day length, season length, and rainfall and temperature during critical portions of the growing season are examples of variables with significant effects relative to genotype values. Two things are obvious and worthy of comment here. First, judgements concerning TPE's in which the allele content of ideal genotypes may be different enough to justify separate recurrent selection programs should logicaly be made by persons with maximum background concerning interrelations of a specific species with environments and maximum bases for predicting husbandry changes during future years. Second, by virtue of the "greenhouse" effects of changes in atmosphere composition, the environment populations of geographical areas may change significantly in the decades ahead. It may be, therefore, that the delineations of TPE's should emphasize aspects most likely to remain constant and significant, e.g., latitude, probable rainfall and ground water differentials, etc. With respect to unpredictable aspects there can and must be dependence on accommodation by the selection process.to whatever changes are caused in rank order values of alleles.

An ideal foundation population would (a) possess, in some frequency, each of the potentially useful alleles of the species and (b) have high average value of genotypes with respect to the TPE for which maximum $\ddot{Y}_w$ is to be sought. Unfortunately, these are conflicting attributes. To maximize content of potentially useful alleles would call for compositing materials from diverse sub-populations of the species at the expense of the initial value of $\ddot{Y}_w$. Eventual potential with respect to $\ddot{Y}_w$ would theoretically be increased but a level useful in relation to an increasing world population might not come soon enough. Alternatively, a higher initial $\ddot{Y}_w$ could be achieved, with possibly little sacrifice in content of useful alleles, in a foundation population formed by intermating individuals from sub-populations both diverse in origin and having better than average levels of total value--achieved by prior breeding programs.

The utility of a recurrent selection program is a function of the magnitudes of N_es' and $E(\Delta q)$, the expected rate by which the frequency of a favorable allele in the S population is increased. According to theory sketched in Appendix 11.2, the ultimate potential gain in $\ddot{Y}_w$ inherent in advantageous alleles from mutation in one gene and one cycle of selection is proportional to N_es' and the probability of fixation of existing superior alleles is greater when N_es' is greater. Both N_es' and $E(\Delta q)$ are variable among alleles and both are functions of the details of the selection program. The general ways in which they are affected by T (the size of the genetic sample) and the trade-off between them are apparent from the expressions for N_es' and $E(\Delta q)/q(1-q)L$ in Table 14.5.[5] These are easily illustrated in terms of any of the selection systems for which formulas are shown in Table 14.5. Consider system 2 and suppose that the structure of criterion unit comparisons has been decided. Then relative to any single allele $\sigma(\hat{W})$, $(1+h)$, and u_w are constants leaving k (which depends on selection intensity, p) and M as the quantities affected by T. For illustrative purposes P(f) = probability of fixation, $E(\Delta q)/q(1-q)$, and N_es' are shown in Table 16.1 for a few combinations of T and p. The following formulas were employed.

$P(f) \approx 1 - e^{-2N_es'q_0}$ in which q_0 = the initial frequency of the allele.

$E(\Delta q)/q(1-q) = k(1+h)u_w/\sigma(\hat{W})L$ and

$N_es' = 2(kpT)u_w/\sigma(\hat{W})$ because M= pT where p = selection intensity.

The quantities tabulated were obtained assuming $\sigma(\hat{W}) = 10.8$, the value of $\sigma(\hat{X})$ from part b of Table 14.15 and $u_w = 0.1$ so that $u_w/\sigma(\hat{W}) = 0.0093$.

Table 16.1. Values of $N_e s'$, probabilities of fixation, and $E(\Delta q)/q(1-q)L$ as functions of T and p when $u_w = 0.1$, $\sigma(\hat{W})=10.8$, and selection is among S_4 families

T	p	k	kpT	$N_e s'$	P(f) $q_0 = .01$	P(f) $q_0 = .03$	$E(\Delta q)/q(1-q)L$
250	.01	2.66	6.65	0.124	.0025	.0074	$.46u_w/L$
	.05	2.06	25.75	.479	.010	.028	$.36u_w/L$
	.20	1.40	70.00	1.30	.026	.075	$.24u_w/L$
	.50	.80	100.00	1.86	.036	.105	$.14u_w/L$
2000	.01	2.66	53.20	0.99	.020	.058	$.46u_w/L$
	.05	2.06	206.00	3.83	.074	.205	$.36u_w/L$
	.20	1.40	560.00	10.42	.188	.465	$.24u_w/L$
	.50	.80	800.00	14.88	.257	.590	$.14u_w/L$
10000	.01	2.66	266.00	4.95	.094	.257	$.46u_w/L$
	.05	2.06	1030.00	19.16	.318	.683	$.36u_w/L$
	.20	1.40	2800.00	52.08	.647	.956	$.24u_w/L$
	.50	.80	4000.00	74.40	.774	.988	$.14u_w/L$

Note the following.

(1) When T is constant (at any level) k increases and kp decreases if p is decreased (selection intensity increased). Thus $E(\Delta q)$ is increased and $N_e s'$ is decreased.

(2) When p is constant (at any level) $N_e s'$ increases in exact proportion to any increase in T. Thus if selection is to be intense (p = 0.01 for example) to maximize $E(\Delta q)$, an x-fold increase in T would mean an x-fold increase in $N_e s'$.

Clearly, if high values of both $N_e s'$ and $E(\Delta q)$ are to be attained the size of project could be very large with cost of operation in proportion. It should be equally apparent that to optimize the coincident values of p and T given a program of specific size would require information not available--the improvement potential of favorable alleles with frequences large enough so that for them P(f) would be high as a consequence of only modest values of $N_e s'$, the improvement potential of favorable alleles with lower initial frequencies, and the probabilities (all genes considered) of new beneficial alleles via mutation. This suggests that in a program composed of large recurrent selection components, the numbers of species and of TPE's could very well be minimized to allow contrasting (T,p) combinations to be compared.

• 16.4. Quantitative Genetics in Relation to Genetic Improvement of Agricultural Species—Looking Ahead

These are my views--errors and inadequacies are mine--the redundancies are inevitable but seem to me justified for the sake of emphases. The imperatives for quantitative genetics are posed by interrelations between world population growth, food production capacity, and the dominant social attitudes of those who shape national and world policies.[6] The long term future of world population is uncertain but one of the scenarios that must be considered is that ZPG at about

11 billion will be reached within a century. Considering that crop land per person would inevitably be reduced by at least 50 percent if population is doubled from the present 5.5 billion, it is apparent that at minimum food production per hectare would have to be tripled. That may or may not be possible but the increment possible via recurrent selection deserves top-priority attention by quantitative genetics.

In the context of maximizing genetic improvement by recurrent selection within 90-100 years, major questions are (1) is our theory base for comparing selection systems and their operational details adequate and (2) what are the magnitudes and compositions of genetic potential?

Theory outlined in section 14.1 and appendix 11.2 provides a basis for long term recurrent selection decisions. As illustrated by the example presented in the preceding section (Table 16.1 and associated discussion) comparisons of systems and variations of systems are necessarily based on three quantities:

(1) the probabilities of fixation of favorable alleles that exist in low frequencies in the S population at the outset,

(2) the possible magnitude of the sum of beneficial effects from new favorable alleles arising by mutation, and

(3) the rates of allele frequency responses to selection practiced.

In section 14.1.2 the comparative evaluation of systems for use in long term recurrent selection is described as a two step process in which first the magnitude of $N_e s'$ to be achieved is decided and then systems are ranked in terms of $E(\Delta q)/q(1-q)$ per unit of time.[7] However, even though the potential of useful alleles from mutation is not considered in that discussion, the specification of an optimum $N_e s'$ requirement is recognized as impossible because distributions of the frequencies and u-values of alleles existing at the outset are always unknown.

The problem posed by the race between genetic improvement and world population is different in two ways. (1) The potential inherent in useful alleles that arise by mutation is considered to be significant and is directly proportional to $N_e s'$ and (2) The time span for genetic improvement is finite, something like a century or less, instead of infinite and hence $E(\Delta q)$ per unit of time is not secondary. It follows that the trade-off between $N_e s'$ and $E(\Delta q)$ is critical in this instance and optimum decisions concerning selection intensity (p) and genetic populataion size (T) can only be made given knowledge of the total increment of genetic gain inherent in (a) new alleles from mutation and (b) the existing favorable alleles that by virtue of their frequencies and u-values would have much higher fixation probabilities given higher than usual $N_e s'$ to capitalize on good alleles from mutation. To exemplify in terms of quantities in Table 16.1, by increasing selection intensity from p = 0.20 to p = 0.05 with T constant would increase k from 1.40 to 2.06--thus an increase of 47 percent in $E(\Delta q)$--while decreasing $N_e s'$ by 73 percent. The better choice depends, obviously, on information not now available. Another view is obtained if we suppose that a decision is made first concerning selection intensity. Suppose it is decided that selection intensity should be at the level of p = .05 Note that to increase it to p = .01, and gain 29 percent in k and $E(\Delta q)$, would require a five-fold increase in T (five times as many plots in the criterion unit comparison) if $N_e s'$ is not to be decreased. Then we observe that by increasing T a corresponding increase in $N_e s'$ would be obtained. These facts suggest that a logical procedure may be to make decisions in the following sequence--(1) the selection system and structure of criterion unit comparisons, (2) intensity of selection, and finally (3) size of the genetic sample.

16.4.1. High Priority Research

16.4.1.1. TESTS OF THEORY FOR COMPARISON OF SELECTION SYSTEMS

Some may question whether they are needed. However, if good data indicate that presently available theory does a good job, the confidence engendered would mean (a) more extensive use of that theory and (b) extended provisions for inclusion of relevant quantitative genetics in the advanced training of plant and animal breeding professionals. There are two kinds of meaningful research that occur to me. Direct comparisons of two or more selection systems applied to parallel foundation populations is an obvious approach. To my knowledge, very little work of this kind has been done or is in progress. However, the replicated experiment reported by Odiambo and Compton (1989) is a noteworthy example. That work had progressed through five cycles of selection and I have learned through correspondence that it is intended to be continued through at least two more cycles. In addition to its evidence concerning validity of theory for comparison of systems it will probably throw light on some aspects of quantitative genetics in maize. Unfortunately, such direct replicated comparisons are costly in terms of time in addition to operational requirements. Because of the time span involved, the required unflagging devotion to purpose has probably mitigated against the launching of such experiments. It should be noted, however, that in comparisons of selection systems no more than one or two cycles are required if sufficient replication is a part of the experimental design.

Another promising source of information concerning theory is computer simulation. As an example, the more general utility of a theory that assumed absence of multiple alleles can be tested by simulation comparisons using programs in which multiple alleles are modelled. The hypothesis tested would be whether prediction equations that assume no more than two alleles make essentially the same comparative evaluations of selection systems when multiple alleles are present as when they are absent. Finding that hypothesis acceptable would remove a significant reason for doubt concerning utility of present theory. Another simplifying assumption employed herein --that when $N_e s'$ is large enough for the purpose of maximizing genetic improvement in recurrent selection programs a close approach to continuing linkage equilibrium is assured--can be adequately tested by computer simulation. A meaningful argument for computer produced evidence is that it can be obtained quickly.

16.4.1.2. ASSAYS OF THE GENETIC IMPROVEMENT POTENTIAL OF NEW ALLELES ARISING BY MUTATION

Because this potential depends on actual rates of mutation the desired information must come from biological organisms, i.e., it cannot be produced using computers. Enfield and Darmo (1992)report that in random mating poulations of Tribolium from highly inbred lines realized heritability of pupa weight increased from .02, .04, and .03 in the initial 10 generations of selection to .04, .05. and .10 in the period from 20 t0 30 generations. The standard errors of the estimates were all .01. Population sizes in terms of number of mating pairs in each generation were 10, 30, and 50, respectively. Numbers of populations of the three sizes were 18, 6, and 6, respectively. In accord with theory the greatest increase in genetic variance (indicated by heritabilities) was in the populations with the largest effective population size. Heritabilities of pupa weight were said to range from .30 to .40 in populations sampled from the wild. Their reported results indicate that in the largest populations ($N_e \approx 100$) the heritability after about 25 generations had reached something like 25 percent of that in wild populations. Note that in accord with theory (Appendix 11.2) the increase observed was greatest in the largest populations. Clearly, more evidence from work of this general nature would be useful. However, being proportional to mutation probabilities per generation and number of genes, the "mutation potential" is probably variable among species. While short generation pilot organisms are attractive, as well as useful, as evidence sources, there is an obvious reason to be interested in information derived from agricultural species likely to be employed in the sort of large recurrent selection programs being discussed. In the case of maize, three generations per year, 30 generations and 10 cycles of S_1 family selection or 5 generations of S_4 family selection in 10 years, are possible and for duplication of the Enfield-Darmo investigation, many highly inbred lines from which to initiate experimental populations are available.

- 16.4.1.3. **PLATEAUS**

Though Dudley (1977) reported continuous responses to 76 generations of selection for high and low protein and oil contents in the grain of maize, a not infrequent consequence of long term recurrent selection is apparent cessation of response to the selection being practiced. It is probable that continuous recurrent selection programs conducted with the purpose of maximizing the productivity of chosen agricultural species would encounter such events. Possible causes include the following.

(1) Complete exhaustion of genetic variation, i.e., homozygosity of one of the alleles of each gene having to do with the selection criterion.

(2) Pleiotropy--alleles favorable relative to the selection criterion but unfavorable relative to reproduction.

(3) Physiological limit as postulated by Jain (op cit).

(4) Deterioration of environments of the target population of environments.

Faced with no indication of forward progress by successive phenotypic averages in a series of selection cycles, the first question posed is whether there actually has been no forward change in the average value of genotypes. This emphasizes the need for an adequate control to enable recognition of any non-zero trend in the value environments in which performance has been measured. Gardner (1977) reported an decrease in grain yield performance of maize in generations after the 15th cycle of effective mass selection that he could not explain on the basis of environment variation indicated by his control. He was forced to conclude that his S population and his control had not responded in parallel to an environment change unfavorable to his S population. His report emphasize the need for a control that can be expected to react the same as the S population to changes in environment.

The obvious second question is whether the program should be, or might as well be, terminated. Let's consider the listed possible explanations further but first recognize that to terminate would be inconsistent with the postulate that on-going mutation will supply the potential for non-trivial forward progress in programs of which the genetic sample is sufficiently large. The high priority to be placed on research concerning the magnitude of the mutation potential is emphasized. Management should also take account of the fact that because of the inevitable variance of the difference between the single cycle performance averages of the S population and the control the data of a number (unknown in advance) of cycles would always be required before it can be concluded that actual genetic progress in the S population has been less than a worthwhile amount, i.e., that for practical purposes a real plateau has been experienced.

In the absence of convincing evidence to the effect (1) that the "mutation potential" is, for practical purposes, trivial and (2) that progress in recent cycles had been no more than trivial, it can reasonably be argued that continuation with increased genetic sample size would be a better decision than termination. A part of that argument would be that the cost of organization and the early, "shake down" portion of the total program would made cost per cycle prior to the established no pay-off period considerably greater than during the continuation period.

The physiological limit explanation as proposed by Jain (op cit) poses a special challenge to the idea that good things can result by mutation. He postulates that the photosynthetic mechanism is uniquely impervious to change in the alleles by which it is controlled. Yet he holds out the hope of changing the limits it imposes by manipulative molecular engineering. My own reaction is that if the latter is possible and hence worth attempting, as it may be, change by capitalizing on spontaneous mutation (the natural process that has supported evolutionary change) should also be given its chance.

Finally, the best possible controls to measure environment effects on the S population could very well be previous generations of the S population itself. Suppose that in every cycle a random sample of S population seed were placed in storage to be used in a field comparison for estimation of the difference between the average genetic values of the S populations of different cycles. Let $\ddot{Y}_k$ = the average genetic value of the S population of the k-th cycle. Then the field comparison of the S populations of the k-th and (k+1)-th cycles would provide an estimate of $\ddot{Y}_{k+1} - \ddot{Y}_k$, the difference between average genetic values of the S populations of successive cycles free of any environmental effect bias because the samples from the two cycles had been grown in the same macro-environments and would have been so little different in genotype composition that genotype x environment interaction (postulated by Gardner) could hardly have been significant and, in any event, would be expected to vary randomly among cycles.

16.4.2. Personnel, Training Programs, and Cooperation

Our theory base for decisions in the design of recurrent selection programs must be used effectively in the interest of genetic contributions to world food production capacity. At the same time there is high priority research to be be conducted, followed if indicated to be desirable, by refinements of theory and/or details of recurrent selection programs in progress. The need for an adequate world cadre of persons having special competence in quantitative genetics as it relates to plant and animal breeding is obvious. Their primary functions: (1) to participate as appropriate in the advanced training of breeding professionals to be employed by private companies and public agencies and (2) to cooperate in the planning and, as needed, in the conduct and management of research and genetic improvement programs. Cooperation is emphasized in view of the probable large scale and long term nature of a substantial portion of research programs that they as a group would identify as particularly important.

Appendixes

• Terminology, Symbols, and Elements of Statistics

Words are the basic elements of verbal and written communication. Imagine trying to evoke the picture of a tree in the absence of faith that the meaning of the word is known. Think about the words "house" or "home". Each conveys a meaning that, without the word and general knowledge of its meaning, would take a lot more time and effort to transmit from the mind of one person to the mind of another.

The terminology of any specialized area consists of words having specific meanings in the context of that area -- meanings that may not be the same if and when such words are used in another area of thought.

Symbols, like words, are short-hand vehicles for the transmission of information, i.e., once their definitions have been stated and accepted , they save repetitive use of sets of words and therefore considerable time for both writer and reader. In a real sense, word and symbol are synonymous. But neither a word nor a symbol has utility without definition and knowledge of the definition.

• A1.1. Terminology

Variable -- any quantity or quality that is not the same for all individuals of a population. Thus associated with a specific population of plants there are numerous variables, e.g., number of leaves, height, age at time of first flower, flower color, number of petals per flower, etc.

More generally, any specified function of two or more values of a single variable or of two or more variables is also a variable (see definition of statistic below).

When the variable is qualitative the number of possible manifestations is finite and its distribution is said to be discrete or discontinuous. Examples: sex in dioecious species, single-locus genotypes, and number of locules in oranges.

When the variable is quantitative and can take any real value within its range of possibilities, its distribution is said to be continuous. Examples: height of humans or maize plants at maturity and annual or life-time production of milk by Holstein cows. Any symbol can be employed as the "name" of a variable. If X is chosen then, by convention or explicit definition, X_i denotes the value of X in the case of the i-th member of the population.

Distribution -- a symbolic mathematical statement concerning the probabilities of the possible values or manifestations of a variable or variables. The sum of the probabilities must of course be one.

Distributions are identified as univariate, bivariate, or multivariate depending on whether one, two, or more than two variables are being considered. The mathematical probability statement concerning a distribution always reflects and depends on underlying assumptions.

The basic distributions most generally employed in this text are the "normal" and the binomial.

Parameter (of a distribution) -- is a mathematical function of all permissible values of the variable or variables of a population. Of course in the case of real infinite populations the exact values of parameters can not be known.

Sample -- a sub-set of the individuals of a population. Random samples, as the words imply, are those consisting of individuals drawn randomly from the population sampled.

Statistic -- is a mathematical function of the variable values of the individuals in a sample. In general statisics are valued as estimates of parameters. Conceptually, any statistic could be computed from the values of an infinite number of similar random samples from the same population. The resulting values of the statistic would then comprise an infinite population of which the statistic would be the variable.

Expectation -- Consider the individual values of any variable. The average (mean) of all values in the population to which they belong is defined to be the expectation of any single value drawn at random from the population. The large case letter E is employed in this text to symbolize expectation and is always followed by another symbol in parentheses that identifies the variable in question. Thus, for example, E(Z) is always to be read, " the expectation of Z" where Z may symbolize any variable or any function of two or more variables.

Bias -- Consider a statistic designed to serve as an estimate of a populataion parameter. If the expectation of that statistic is equal to the parameter of interest it is said to be unbiased. The values of unbiased statistics vary around that of the parameter and approach the parameter value more closely as the amount of data from which they are computed becomes greater. On the other hand, the variance of a biased statistic also becomes smaller as the amount of data on which it is based becomes greater but its expectation, the quantity around which individual values of the statistic vary is not the value of the parameter. Consider a statistic (S) employed as an estimate of a parameter (P). If $E(S) = P$, the statistic is an unbiased estimate of P. But if $E(S) = P+\alpha$ (and $\alpha \neq 0.0$), the statistic is biased and α is the magnitude of the bias.

• A1.2. Distributions

Descriptive parameters: The most important in the context of this text are the mean and the second and third moments about the mean. Letting X symbolize individual values of the variable and $\ddot{X}$ the population mean of X, the r-th moment about the mean is defined to be the population mean of $(X-\ddot{X})^r$. Thus the population mean of $(X-\ddot{X})^2$ (known as the variance of X) is the second moment of X about its mean and the population mean of $(X-\ddot{X})^3$ is the third moment about the mean.

Because the squares of all numbers (positive or negative) are positive, variance must always be a positive number. The positive square root of the variance is called the standard deviation of X and is conventially symbolized as σ.

If a distribution is symmetrical, it is because, no matter the magnitude of X, positive and negative values of $(X-\ddot{X})$ are equally frequent. Then because $(X-\ddot{X})^3$ is positive when $(X-\ddot{X})$ is positive and negative when $(X-\ddot{X})$ is negative, the third moment about the mean is zero in the case of symmetrical distributions and near to zero in the case of distributions that are approximately symmetrical.

The "normal" distribution is continuous and symmetrical. The mathematical statement that defines it identifies the range of a normally distributed variable as being from minus infinity to plus infinity and is a function of two parameters, the mean and variance of the population. Because such a variable as the grain yield of maize per acre can never be infinitely large and is bounded by zero on the down side, it cannot be absolutely "normal" in its distribution. However, from tables for this distribution we find that the probability of values more than 2.5 or 3.0 standard deviations from the mean are only .0062 and .0013, respectively. Clearly, unless the standard deviation is quite large as a fraction of the population mean (larger, say, than $0.3\ddot{X}$) biological limits do not of necessity exclude close approximation to "normal" distribution.

The binomial distribution deals with the probabilities of alternative events. Let A and B be such alternatives of which one or the other is the only possibility and let the probabilities of A and B be p and q = 1-p, respectively. Then, given n independent opportunities and X = the number in which the outcome is A, the distribution statement provides the probabilities of the n+1 possible values of X. For any admissible value of X, the probability in question is

$$P(X) = [n!/X!(n-X)!]p^X q^{n-X}$$

in which, for example,

$n! = n$ factorial $= n(n-1)(n-2)(n-3)\cdots\cdots(2)(1)$ and, by convention, $(0)! = 1$. Considered collectively, these probabilities are the separate terms of the expansion of $(p+q)^n$ and must therefore sum to 1.0. This makes sense in view of the fact that in each case X must take one, but only one, of its n+1 possible values. The primary parameter of the binomial distribution is obviously the probability of one of the two alternative events (which defines the probability of the other since p+q = 1).

Because the distribution statement provides the probability of each of the possible values of X, the population mean and variance of X can be determined, as functions of p, to be np and npq, respectively.

The classical text-book example of the binomial distribution deals with coin-tosses and the number of "heads" or "tails" to be observed when a coin is tossed n times or when n coins are tossed in such a way that the n outcomes are all independent of each other.

In quantitative genetics an application of major importance has to do with the random change in the frequencies of alleles from one generation to the next in finite populations.

A1.3. Symbols and Statistics—Univariate Distributions

Throughout it is assumed that all samples are composed of random individuals and that symbols employed have meanings as follows when individual values of the variable are symbolized by the letter, X.

X_i = the value of the variable in the i-th individual,

$E(X) = \ddot{X}$ = the population mean of the variable X,

$x_i = X_i - E(X) = X_i - \ddot{X}$ and $E(x)$ = zero,

$E(X-\ddot{X})^2 = E(x^2) = \sigma^2(X)$ = the population variance of the variable X,

$\sigma(X) = [\sigma^2(X)]^{1/2}$ = the population standard deviation of X,

n = sample size -- number of individuals in the sample,

Σ = sum of, and $\Sigma_i X_i$ indicates the sum of X_i over all individuals of a population.

The statistics of primary importance in the case of univariate distributions are the mean and variance of random samples. Symbolically, they and their variances are as follows.

<u>Sample mean</u> $\bar{X} = \Sigma_i X_i/n$ and $E(\bar{X}) = \ddot{X}$

$$\sigma^2(\bar{X}) = \sigma^2(X)/n$$

<u>Sample variance</u> $s_X^2 = \Sigma_i(X_i-\bar{X})^2/(n-1) = [\Sigma_i X_i^2 - (\Sigma_i X_i)^2/n]/(n-1)$

$$\sigma^2(s_X^2) = 2[\sigma^2(X)]^2/(n-1) \quad \text{(when X is normally distributed)} \qquad \text{A1.1}$$

Note that the above expressions have to do with three univariate distributions, i.e., the distributions of X, of $\bar{X}$, and of s_X^2.

A1.4. Symbols and Statistics—Bivariate Distributions

The following are extra symbols required for reference to the statistics to be encountered when bivariate distributions are being considered. Any two letters can be used as "names" and numerical values of the two variables of a bivariate distribution. The letters X and Y (used here) are frequently employed in the statistical literature.

$\sigma(X,Y) = E(X-\ddot{X})(Y-\ddot{Y}) = E(xy)$ the <u>covariance</u> between the two variables in a bivariate distribution,

$\rho(X,Y) = \sigma(X,Y)/[\sigma(X)\sigma(Y)]$ = the product moment correlation between X and Y, and

$\beta(Y,X) = \sigma(X,Y)/\sigma^2(X) = \rho(X,Y)[\sigma(Y)/\sigma(X)]$ = the coeffient of linear regression of Y on X.

A1.5. Expectations

The following are expectations that will be employed on one or more occasions in this text. Some will be derived herein, others will not.

$$E(X^2) = \ddot{X}^2 + \sigma^2(X) \qquad \text{A1.4}$$

$$E(X^3) = \ddot{X}^3 + 3\ddot{X}\sigma^2(X) \qquad \text{A1.5}$$

(if the distribution of X is symmetrical)

$$E(X^4) = \ddot{X}^4 + 6\ddot{X}^2\sigma^2(X) + 3[\sigma^2(X)]^2 \qquad \text{A1.6}$$

(if the distribution is "normal")

$$E(XY) = \ddot{X}\ddot{Y} + \sigma(XY) \qquad \text{A1.7}$$

Variables are often modified (and will often be in this text) by the addition or subtraction of a constant or by multiplication or division by a constant. Doing so is referred to as <u>coding</u>. The basic relationships between expectations of coded and uncoded variables are intuitively obvious but are well worth listing. Letting c symbolize any constant,

$$E(X + c) = E(X) + c$$

The rationale, of course, is that if each value of the variable is increased by the amount c, the average of the modified variable is certainly larger than that of the original variable by the amount c. Of course, if c is positive but subtracted instead of added, the above becomes

$$E(X - c) = E(X) - c.$$

In the case of multiplication,

$$F(cX) = cE(X) \qquad \text{A1.8}$$

The obvious logic is that if each value of a variable is multiplied by c, the average of the modified variable must as a consequence be the average of the original variable multiplied by c. Because division by c is the equivalent of multiplying by 1/c,

$$E(X/c) = (1/c)E(X) = E(X)/c.$$

Finally, if each of two variables (say X and Y) is multiplied by a constant, e.g., by a and c, respectively, the products (XY and ac) may be viewed as a single variable and a single constant. It follows that

$$E(aX)(cY) = E(acXY) = acE(XY)$$

A1.5.1 Linear Functions

A <u>linear function</u> is one in which its variable components are combined by addition or subtraction. Consider the following example in which large and small case letters are understood to be variables and constants, respectively.

$$F = aX - bW + Z/c + d \qquad \text{A1.9}$$

Note that the definition stated above places no restriction on the nature of variables involved. For example, X could be the product of X_1 and X_2 (where X_1 and X_2 are the "names" of different variables) in which case that product (X_1X_2) could be the variable symbolized by X in A1.9.

Intuitively obvious is the fact that <u>the expectation of any linear function is the same function of the expectations of its separate terms</u>. Thus, for example, the expectation of A1.9 is

$$E(F) = \ddot{F} = aE(X) - bE(W) + (1/c)E(Z) + d \qquad \text{A1.10}$$

Linear functions are utilized repeatedly in this text. Examples: the expressions numbered 2.48, 2.49, 5.4, 10.9, and 10.18.

A1.5.2 Variance and Covariance of Linear Functions

Routine procedure for writing variances of linear functions can be established by considering the function given by A1.10 and its variance. By definition

$$\sigma^2(F) = E(F - \ddot{F})^2$$

and by A1.10

$$\ddot{F} = a\ddot{X} - b\ddot{W} + (1/c)\ddot{Z} + d$$

It follows that

$$F - \ddot{F} = a(X-\ddot{X}) - b(W-\ddot{W}) + (1/c)(Z-\ddot{Z}) \text{ and}$$

$$\sigma^2(F) = E(F-\ddot{F})^2 = E[a(X-\ddot{X}) - b(W-\ddot{W}) + (1/c)(Z-\ddot{Z})]^2$$

$$= E[a^2(X-\ddot{X})^2 + b^2(W-\ddot{W})^2 + (Z-\ddot{Z})^2/c^2$$

$$-2ab(X-\ddot{X})(W-\ddot{W}) + 2(a/c)(X-\ddot{X})(Z-\ddot{Z}) - 2(b/c)(W-\ddot{W})(Z-\ddot{Z})].$$

Then, by the definitions of variance and covariance,

$$\sigma^2(F) = a^2\sigma^2(X) + b^2\sigma^2(W) + \sigma^2(Z)/c^2 - 2ab\sigma(X,W) + 2(a/c)\sigma(X,Z)$$

$$- 2(b/c)\sigma(W,Z)$$

The rule-of-thumb for the variance of a linear function that emerges from this example consists of these steps.

(1) Drop any constant term (e.g. the d of A1.9) from the function and let R symbolize the remainder.

(2) Square the remainder (R) to obtain R^2 as a linear function derived from F.

(3) Write the variance of F as the expectation of R^2 using the fact that the expectation of a linear function is the same function of the expectaions of its separate terms.

By analogy, the covariance between any two linear functions (say, F and F') is obtained by

(1) dropping constant terms from both functions,

(2) Writing the product of the remainders, R and R', and

(3) Writing the covariance between F and F' as the expectation of the RR' linear function.

A1.5.2.1. APPLICATIONS

The expectation of X^2

$$X = (X-\ddot{X}) + \ddot{X}$$

$$X^2 = (X-\ddot{X})^2 + 2(X-\ddot{X})\ddot{X} + (\ddot{X})^2$$

$$E(X)^2 = E(X-\ddot{X})^2 + 2\ddot{X}E(X-\ddot{X}) + (\ddot{X})^2$$

$$= \sigma^2(X) + (\ddot{X})^2 \quad \text{because } E(X-\ddot{X}) = \text{zero} \qquad \text{A1.11}$$

The expectation of XY

$$XY = [(X-\ddot{X})+\ddot{X}][(Y-\ddot{Y})+\ddot{Y}]$$

$$E(XY) = E(X-\ddot{X})(Y-\ddot{Y}) + \ddot{X}E(Y-\ddot{Y}) + \ddot{Y}E(X-\ddot{X}) + \ddot{X}\ddot{Y}$$

$$= \sigma(X,Y) + \ddot{X}\ddot{Y} \qquad \text{A1.12}$$

The expectation of $\Sigma_i(X_i - \bar{X})^2$ when a sample of n individuals is drawn randomly from an infinite population: We know that

$$\Sigma_i(X_i - \bar{X})^2 = \Sigma_i X_i^2 - [\Sigma_i X_i]^2 = \Sigma_i X_i^2 - n\bar{X}^2 \text{ and hence that}$$

$$E[\Sigma_i(X_i - \bar{X})^2] = E(\Sigma_i X_i^2) - nE\bar{X}^2 \qquad \text{A1.13}$$

Any summation is a linear function written in condensed form. Thus

$$\Sigma_i X_i^2 = X_1^2 + X_2^2 + \cdots\cdots\cdots X_n^2$$

and by A1.11 and the fact that the expectation of a linear function is the same function of the expectations of its separate terms

$$E(\Sigma_i X_i^2) = n[\ddot{X}^2 + \sigma^2(X)].$$

The expectation and variance of the mean of a random sample of size n are $\ddot{X}$ and $\sigma^2(X)/n$, respectively. It follows by A1.11 that

$$E(\bar{X})^2 = \ddot{X}^2 + \sigma^2(X)/n$$

Then, substituting for $E(\Sigma_i X_i^2)$ and $E(\bar{X})^2$ in A1.13, yields

$$E\Sigma_i(X_i - \bar{X})^2 = n[\ddot{X}^2 + \sigma^2(X)] - n[\ddot{X}^2 - \sigma^2(X)/n] = (n-1)\sigma^2(X) \qquad \text{A1.14}$$

and $E(s_X^2) = E\Sigma_i(X_i - \bar{X})^2/(n-1) = \sigma^2(X)$.

The expectation of the variance of means of samples drawn randomly and without replacement from larger samples from the same infinite population. Assume that the larger samples (finite populations) are drawn randomly from the infinite population, that each of them is composed of T individuals, and that a sample composed of n individuals is drawn from each finite population. Let

$\bar{X}_n$ = the mean of a sample drawn from one of the finite populations,

$\bar{X}_T$ = the mean of that finite population, and

$\sigma^2(X)$ and $\ddot{X}$ symbolize the variance and the mean of the infinite population from which all finite populations are drawn.

Then

$$\sigma^2(\bar{X}_n - \bar{X}_T) = E(\bar{X}_n - \bar{X}_T)^2 = E(\bar{X}_n^2) - 2E(\bar{X}_n\bar{X}_T) + E(\bar{X}_T^2) \qquad \text{A1.15}$$

is the quantity to be established. By A1.11, A1.12, and the fact that the n members drawn without replacement from one of the finite populations are, nevertheless, random members of the infinite source population because all members of each finite population were random members of the single infinite source population

$$E(\bar{X}_n^2) = \ddot{X}^2 + \sigma^2(X)/n,$$

$$E(\bar{X}_T^2) = \ddot{X}^2 + \sigma^2(X)/T,$$

and $E(\bar{X}_n\bar{X}_T) = \ddot{X}^2 + \sigma(\bar{X}_n, \bar{X}_T) = \ddot{X}^2 + \sigma^2(X)/T$.

That $\sigma(\bar{X}_n, \bar{X}_T) = \sigma^2(X)/T$ is shown below. Substituting in A1.15 yields

$$\sigma^2(\bar{X}_n - \bar{X}_T) = [(T-n)/Tn]\sigma^2(X). \qquad \text{A1.16}$$

That $\sigma(\bar{X}_n, \bar{X}_T) = \sigma^2(X)/T$ becomes apparent when the compositions of $\bar{X}_n$ and $\bar{X}_T$ are considered.

$$\bar{X}_n = (X_1 + X_2 + \cdots\cdots X_n)/n = (x_1 + x_2 + \cdots\cdots x_n)/n + \ddot{X}$$

and $\bar{X}_T = (x_1 + x_2 + \cdots\cdots x_n + x_{n+1} + \cdots\cdots\cdots x_T)/T + \ddot{X}$

where $x = X - \ddot{X}$.

Because all X- values are assumed to be random members of the infinite source population so that the expectation is zero for all products between pairs of x's, the portion of the product between $\bar{X}_n$ and $\bar{X}_T$ that is relevant to their covariance is $(x_1{}^2 + x_2{}^2 + \cdots\cdots\cdots x_n{}^2)/nT$, from which

$$\sigma(\bar{X}_n,\bar{X}_T) = [Ex_1^2 + Ex_2^2 + \cdots\cdots Ex_n^2]/nT = \sigma^2(X)/T$$

The expectation of a mean square -- Consider data from a randomized block comparison of n varieties of a crop assuming r replications. This means that one plot of each variety was present in each of r "blocks" of land and that within each block the n plots were arranged randomly. The standard linear model (function) for measured plot values would be of the form

$$X_{ij} = \ddot{X} + v_i + B_j + e_{ij} \qquad \text{A1.17}$$

in which

$\ddot{X}$ = a population mean -- therefore, a constant,

v_i = the "effect" of the i-th variety in the sample compared,

B_j = the "effect" of the the j-th of the r blocks in the trial, and

e_{ij} = a random "effect" responsible for the difference between X_{ij} and $(\ddot{X} + v_i + B_j)$.

Both v_i and e_{ij} are defined to be random members from populations of which the means are zero. By virtue of this definition

$E(v_i) = E(v_i) =$ zero,

$E(v_i^2) = \sigma^2(v)$,

$E(e_{ij}^2) = \sigma^2(e)$,

and $E(v_i e_{ij}) =$ zero.

Consider the variety mean square and let the variety sum of squares and mean square be symbolized by SS(v) and MS(v), respectively. Then in this instance,

$$MS(v) = SS(v)/(n-1)$$

and by A1.8,

$$E[MS(v)] = E[SS(v)]/(n-1) \qquad \text{A1.18}$$

Let V_i symbolize the sum over the r replications of plot values for the i-th variety. Then by standard computation procedure in analyses of variance

$$SS(v) = \Sigma_i V_i^2/r - (\Sigma_i V_i)^2/rn = \Sigma_i(V_i - \bar{V})^2/r$$

from which, by A1.14

$$E[SS(v)] = \sigma^2(V)(n-1)/r$$

and, by A1.18

$$E[MS(v)] = \sigma^2(V)/r \qquad \text{A1.19}$$

The content of $\sigma^2(V)$ is easily determined using the rules-of-thumb for finding the variance of a linear function. Noting that $V_i = \Sigma_j(X_{ij}) = r\ddot{X} + rv_i + \Sigma_j B_j + \Sigma_j e_{ij}$ and that $r\ddot{X} + \Sigma_j B_j$ is a constant component of each V our steps are to square R (the non constant portion of V) and write $\sigma^2(V)$ as the expectation of R^2.

$$R = rv + \Sigma_j e_{ij} = rv + e_1 + e_2 + \cdots\cdots e_r.$$

$$R^2 = r^2v^2 + e_1^2 + e_2^2 + \cdots\cdots e_r^2$$ (cross products omitted -- their expectations are all zero because e's are random relative to each other and and to v).

$$E(R^2) = \sigma^2(V) = r^2\sigma^2(v) + r\sigma^2(e).$$

Finally, substituting in A1.19, we obtain

$$E[MS(v)] = \sigma^2(e) + r\sigma^2(v).$$

The example treated above is simpler than most (much simpler than some) but suffices to illustrate logic and procedure. Principles are the same in derivation of expectations of the mean products of analyses of covariance.

• Appendix 2.1. Wright's Coefficient of Inbreeding

This coefficient pertains to the impact of inbreeding on the probability of single-locus genotypes in diploid individuals (as indicated in equations 2.2). A plant or animal is said to be inbred if it derived from the mating of related individuals. To be related that pair of individuals must have one or more ancestors in common and if they do it is intuitively obvious that their offspring are more likely to be homozygous at every gene locus. We will consider one such locus, identify it as the locus of the B gene, and symbolize the frequency of its j-th allele (in individuals chosen to be parents) as q_j. Given random mating, the probability of homozygosity of the j-th allele (B_j) is

$$P(\text{homo/r.m.}) = q_j^2,$$

the obvious upper limit of P(homo) is

$$P(\text{homo})_{max} = q_i,$$

and the difference between q and q^2 is $q - q^2 = q(1-q)$.

Because the essential effect of inbreeding is to increase the amount of homozygosity, the most appropriate measure of inbreeding would take values, zero when matings are random and 1.0 when the probability of homozygosity of the j-th allele is q_j. In addition, its relation to the increase in the expected homozygosity in inbred individuals should be linear between those limits. That Wright's coefficient has those attributes is stated by equations 2.2 (section 2.5).

$$P(B_jB_j) = q_j^2 + fq_j(1-q_j)$$

It remains to demonstrate that this is correct. The formula derived by Wright (1922) is

$$f_X = \Sigma(1/2)^{n+m+1}(1+f_A) \qquad \text{A2.1.1}$$

in which

f_X = the coefficient of inbreeding of the individual, X,

n = the number of generations separating a common ancestor from the male parent of X, m = the number of generations separating the same common ancestor from the female parent of X,

f_A = the coefficient of inbreeding of that common ancestor, and

Σ indicates summation over all <u>independent</u> pathways of relationship (defined below) between the immediate parents of X.

Any individual that is the male parent of X or one of his ancestors and also the female parent or one of her ancestors is defined to be a "common ancestor". Let such a common ancestor be named A and let the homologous genes at the B locus in A be identified as B^1 and B^2. Here it is to be understood that the superscripts do not identify alleles and that each is <u>specific</u> for one of the two genes in A -- no gene in any individual other than A is to be identified as B^1 or B^2. It is possible for both B locus genes of X to be replicas of B^1 (or of B^2) in which case they would be "identical by descent" and X would be homozygous at the B locus. Note that this outcome does not depend on whether A itself is, or is not, homozygous. On the other hand, "identity by descent" requires that the two B genes of X be replicas of a common ancestral gene which depends in turn on the existence of a common ancestor in the pedigree of X.

Crow and Kimura (1970) state that the inbreeding coefficient, f, "is the probability that a pair of alleles in the two gametes that unite to form the individual are identical by descent." A proof of this is obtained by deriving a formula for the probability of "identity of descent" to check whether it is, in fact, the formula, A2.1.1, for f_X.

Given such a common ancestor as defined above, the probability that the B locus genes of X will be identical by descent from A hinges on the distance, counted in generations, between A and the parents of X. Assuming absence of both gametic selection and meiotic drive, the probability that a replica of the B^1 gene would be transmitted to any offspring of A would obviously be (1/2). This is so because gametes produced by A would, with equal proability, carry either B^1 or B^2. It follows that the probability of transmission of a replica of the B^1 gene to a second generation descendant would be $(1/2)(1/2) = (1/2)^2$ or to an n-th generation offspring would be $(1/2)^n$. Thus

P(transmission of B^1 to X <u>via</u> the male parent of X) = $(1/2)^{n+1}$,

P(transmission of B^1 to X <u>via</u> the female parent of X) = $(1/2)^{m+1}$, and

P_1 = P(that both B locus genes in X will be replicas of B^1) = $(1/2)^{n+m+2}$

By parallel arguments

P_2 = P(that both B locus genes in X will be replicas of B^2) = $(1/2)^{n+m+2}$.

It follows that, if B^1 and B^2 are random members of the population of B genes as they would be unless A is inbred, the total probability that the B locus genes of X would be "identical by descent" is

$$P' = P_1 + P_2 = (1/2)^{n+m+1}$$

On the other hand, if B^1 and B^2 are identical by descent, the B locus genes of X would be identical by descent if each was a replica of one of the B genes of A -- they would not have to be replicas of the same one because of the identity of B^1 and B^2. Because the gene transmitted by A to each of its offspring would <u>of necessity</u> be either B^1 or B^2

P_3 = P(transmission of B^1 or B^2 to X *via* the male parent of X) = $(1/2)^n$ and

P_4 = P(transmission of B^1 or B^2 to X *via* the female parent of X) = $(1/2)^m$.

Obviously then the probability that the B locus genes of X would be identical by descent from A given identity of B^1 and B^2 is

$$P'' = P_3P_4 = (1/2)^{n+m}$$

Now let

P_A be the probability that the B locus genes of A are identical by descent and

P_X be the probability that the B locus genes of X are identical by descent from A.

Then because the B genes of A either are, or are not, identical by descent

$$P_X = (1-P_A)P' + P_AP'' = (1-P_A)(1/2)^{n+m+1} + P_A(1/2)^{n+m}$$

$$= (1/2)^{n+m+1}(1 + P_A) \qquad \text{A2.1.2}$$

Note now that while all ancestors of A are also, by the defintion given above, common ancestors of X, they do not contribute to P_X if they only appear in the pedigree of X by virtue of being ancestors of A. That is because replicas of their B genes can only be transmitted to X *via* A and hence are accounted for by A2.1.2.

Finally, if A2.1.2 is summed over all "independent pathways of relationship between the immediate parents of X", the result is exactly parallel to A2.1.1. It is apparent that, as stated by Crow and Kimura, f_X = the total probability that the two B locus genes of X are "identical by descent".

It remains to show how f_X is computed using its formula (A2.1.1) and in the process to explain the meaning of "independent pathways of relationship". Consider the pedigree shown below.

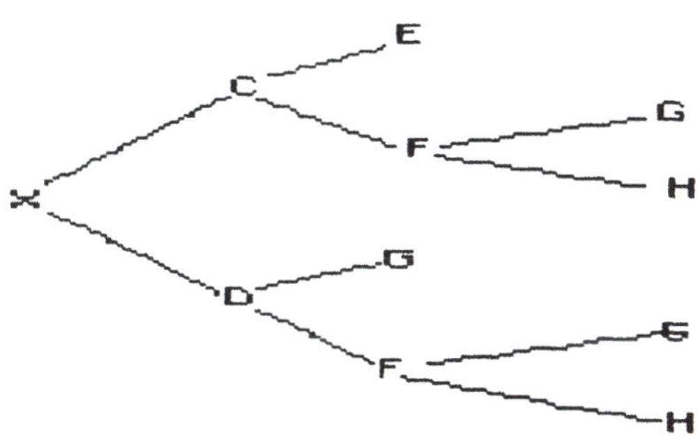

The pathway of relationship between the parents of X that runs through the common ancestor F can be indicated by the series of individuals, CFD. The various pathways from C to D that are evident from the pedigree are CFD, CFGD, CFGFD, and CFHFD. The independent pathways can be recognized by the fact that no individual appears twice in the series. By this test the first two in our list (CFD and CFGD) are independent pathways. While F provides the tie between C and D in the CFD series and appears in the second, it is G that provides the tie in the second. F receives a replica of one of the B genes of G but may not pass it along to D. However, a replica of that B gene may be received by D from G directly, i.e., the CFGD and CFD pathways are independent and mutually exclusive -- if one is operative the other cannot be. The last two of the list of four pathways are not independent of the CFD pathway in accord with the argument presented earlier. The two independent pathways along with the associated values of n and m and their contibutions to f_X are shown below.

Independent pathway	n	m	Contribution to f_X
CFD	1	1	.125
CFGD	2	1	.0625

$$f_X = 0.125 + 0.0625 = 0.1875$$

As a second example consider an individual produced by two generations of self-fertilization, the first individual self-fertilized being a non-inbred individual. The pedigree of X can be written as follows.

Focus first on the individual C. In its pedigree the single effective common ancestor D is both the male and female parent of C. Thus n = m = 0 and, by A2.1.1, $f_C = 1/2$. Going on to X the single effective common parent is C. Again n = m = 0 making, by A2.1.1,

$$f_X = (1/2)(1 + f_C) = 0.75$$

Note that f_C and f_X are the first two of the values shown for successive generations of self-fertilization shown in section 2.5.1.

• Appendix 2.2. Δ_s in the Case of Autosomal Genes in Dioecious Species

More symbols are required than in the treatment of Δ_s in monoecious organisms. Let M be the number of males and F the number of females in the population at the single cell zygote stage of the parent generation and let relative frequencies ond reproductive rates of genotypes be symbolized as follows.

	Males		Females	
Genotype	Rel. freq.	Reprod. rate	Rel. freq.	Reprod. rate
BB	P_m	W_{2m}	P_f	W_{2f}
Bb	R_m	W_{1m}	R_f	W_{1f}
bb	S_m	W_{0m}	S_f	W_{0f}

Frequencies of the B allele in the parent generation at that point are these,

$q_m = P_m + R_m$,

$q_f = P_f + R_f$, and

$$q_p = (q_m+q_f)/2 = (P_m + R_m + P_f + R_f)/2 \quad \text{A2.2.1}$$

Because each offspring gets one gamete from its male parent and one from its female parent, the number of individuals at the single-cell zygote stage of the offspring generation is

$$M(\bar{W}_m) = F(\bar{W}_f) \quad \text{A2.2.2}$$

in which

$$\bar{W}_m = P_m W_{2m} + 2R_m W_{1m} + S_m W_{0m} \quad \text{and}$$
$$\bar{W}_f = P_f W_{2f} + 2R_f W_{1f} + S_f W_{0f}$$

Assuming that B containing gametes and b containing gametes produced by heterozygous parents have equal probabilities of being involved in the formation of zygotes,

q_{gm} = the frequency of B in effective gametes from male parents

$$= M(P_m W_{2m} + R_m W_{1m})/M(\bar{W}_m),$$

q_{gf} = the frequency of B in effective gametes from female parents

$$= F(P_f W_{2f} + R_f W_{1f})/F(\bar{W}_f).$$

and the frequency of B in offspring at the single cell zygote stage is

$$q_0 = (q_{gm} + q_{gf})/2$$
$$= (P_m W_{2m} + R_m W_{1m})/2(\bar{W}_m) + (P_f W_{2f} + R_f W_{1f})/2(\bar{W}_f) \qquad \text{A2.2.3}$$

Because

$\Delta_s = q_0 - q_p$, substitution in terms of A2.2.3 and A2.2.1 yields

$$\Delta_s = (P_m W_{2m} + R_m W_{1m})/2(P_m W_{2m} + 2R_m W_{1m} + S_m W_{0m}) - q_m/s$$
$$+ (P_f W_{2f} + R_f W_{1f})/2(P_f W_{2f} + 2R_f W_{1f} + S_f W_{0f}) - q_f/2$$
$$= [(1-q_m)(P_m W_{2m} + R_m W_{1m}) - q_m(R_m W_{1m} + S_m W_{0m})]/2(\bar{W}_m)$$
$$+ [(1-q_f)(P_f W_{2f} + R_f W_{1f}) - q_f(R_f W_{1f} + S_f W_{0f})]/2(\bar{W}_f) \qquad \text{A2.2.4}$$

Assuming that the probabilities of participation in fertilizations by B and b carrying germ cells are the same whether or not the cell from the heterogametic parent contains the X-chromosome allows setting $P_m = P_f$, $R_m = R_f$, $S_m = S_f$, and $q_m = q_f = q_p = q$. Then A2.2.4 can be rewritten as follows.

$$\Delta_s = [(1-q)(PW_{2m} + RW_{1m}) - q(RW_{1m} + SW_{0m})]/2(\bar{W}_m)$$
$$+ [(1-q)(PW_{2f} + RW_{1f}) - q(RW_{1f} + SW_{0f})]/2(\bar{W}_f) \qquad \text{A2.2.5}$$

Finally, assuming that the frequency of the X-chromosome in effective germ cells from the heterogametic parent is 0.5 so that M = F, we have from A2.2.2 that $\bar{W}_m = \bar{W}_f = \bar{W}$. Then, substituting in A2.2.5 yields

$$\Delta_s = [(1-q)(PW_2 + RW_1) - q(RW_1 + SW_0)]/\bar{W}$$

in which

$$W_2 = (W_{2m} + W_{2f})/2,$$
$$W_1 = (W_{1m} + W_{1f})/2, \quad \text{and}$$
$$W_0 = (W_{0m} + W_{0f})/2.$$

• Appendix 5.1. The Effect of Test Environment on E(Δq), the Expectation of Allele Frequency Change in Response to Selection

The interaction known to exist between environment and the total genotype (section 2.12) implies interaction as well between environment and single-locus genotypes; in fact, the former could not be explained without assuming the latter. Following 5.5, let the value of genotype with respect to the selection criteron (X) and the TPE be

$$Y_X = \alpha_X + \Sigma_i y_{Xi} \quad \text{A5.1.1}$$

Because $E(\Delta q)$ is proportional to the covariance between $\hat{X}$ and $\hat{q}$, the selection criterion (X) is the only phenotypic variable (trait) that is relevant herein. Thus all Y's and y's will pertain to X and the trait-identifying subscript can be ommitted without loss of specificity. Rewriting A5.1.1 without the subscript

$$Y = \alpha + \Sigma_i y_i \quad \text{A5.1.2}$$

Because of genotype x environment interaction the value of genotype with respect to any population of environments other than the entire TPE will differ from the Y specified by A5.1.2. Let the value of genotype with respect to X and <u>the k-th macro-environment</u> be

$$Y_k = \alpha_k + \Sigma_i y_{ik} \quad \text{A5.1.3}$$

There is, of course, a specific value of y_{ik} (and of y_i) for each genotype possible at the i-th locus. Let

y_{imk} be the difference, in effect on X, between the m-th genotype at the i-th locus and the b_ib_i genotype, averaged over all micro-environments of the k-th macro-environment (assuming only two alleles, there will be just three of these quantities for each macro-environment of which one must be zero) and

y_{im} be the same difference averaged over all micro-environments of the TPE.

By virtue of these definitions,

$$y_{im} = \Sigma_k P_k y_{imk} \quad \text{A5.1.4}$$

where P_k is the fraction of the TPE represented by the k-th macro-environment.

Adding a nongenetic effect e_k to both sides of A5.1.3 we obtain

$$X_k = \alpha_k + \Sigma_i y_{ik} + e_k \quad \text{A.5.1.5}$$

so that when criterion units are compared in the k-th macro-environment

$$\hat{X}_k = \alpha_k + \Sigma_i \hat{y}_{ik} + \hat{e}_k$$

Note that the value of y_{ik} in A.5.1.3 and A.5.1.5 is determined by the actual i-th locus genotype of the individual, i.e., it is the y_{ik} of that specific i-th locus genotype.

Since criterion units can be distributed randomly relative to environments within a macro-environment (though <u>not</u> relative to all environments of the TPE), the covariance between $\hat{q}_i$ and $\hat{X}_k$ is

$$\sigma(\hat{q}_i, \hat{X}_k) = \sigma(\hat{q}_i, \hat{y}_{ik}) + \Sigma_{j \neq i} \sigma(\hat{q}_i, \hat{y}_{jk})$$

Because the y_{imk}'s are variable among macro-environments this covariance is variable from one macro-environment (or set of macro-environments) to another. Note, however, that in accord with A5.1.4

$$\hat{y}_i = \Sigma_k P_k \hat{y}_{ik}$$

and hence

$$\sigma(\hat{q}_i, \hat{y}_i) = \Sigma_k P_k \sigma(\hat{q}_i, \hat{y}_{ik})$$

Thus $\sigma(\hat{q}_i, \hat{y}_{ik})$ which varies among macro-environments, is equal on average to $\sigma(\hat{q}_i, \hat{y}_i)$, that is,

$$\sigma(\hat{q}_i, \hat{y}_{ik}) \neq \sigma(\hat{q}_i, \hat{y}_i) \quad \text{but}$$

$$E[\sigma(\hat{q}_i, \hat{y}_{ik})] = \sigma(\hat{q}_i, \hat{y}_i)$$

in the case of random macro-environments.

• Appendix 6.1. Algebra in Derivation of 6.14

The final form of equation 6.14 shown in the text rests on these two equalities:

$$[p^2 - r^2 + ps - rt - (p+r)(p+s-r-t)] = (pt - rs)$$

and $$[2pr + pt + rs - 2(p+r)(p+s)(r+t)] = (pt-rs)(1-2q_j).$$ The purpose here is to show the steps that lead from the left-hand member to the right-hand member of each of these two equality statements. They will be treated in turn. First

$$[p^2 - r^2 + ps - rt - (p+r)(p+s-r-t)] = (p+r)(p-r-p-s+r+t) + ps - rt$$

$$=(p+r)(t-s) + ps - rt = pt - ps + rt - rs + ps - rt = (pt - rs)$$

Second

$$[2pr + pt + rs - 2(p+r)(p+s)(r+t)]$$

$$= [pr + pt - (p+r)(p+s)(r+t)] + [pr + rs - (p+r)(p+s)(r+t)]$$

$$= [(r+t)(p-p^2-ps-pr-rs)] + [(p+s)(r-pr-pt-r^2-rt)]$$

$$= (r+t)[p-p(1-t)-rs] + (p+s)[r-r(1-s)-pt]$$

$$= (r+t)(pt - rs) + (p+s)(rs - pt) = (pt - rs)(r+t-p-s) = (pt - rs)(1-2q_j)$$

because

$$p+r+s+t = 1.0 \quad \text{so that}$$

$$p^2+ps+pr = p(p+r+s) = p(1-t),$$

$$pr+r^2+rt = r(p+r+t) = r(1-s), \text{ and}$$

$$(r+t-p-s) = [p+s+r+t - 2(p+s)] = 1 - 2(p+s) = 1 - 2q_j \text{ because } p+s = q_j$$

• Appendix 6.2 The Rate at which Linkage Disequilibrium Decreases in Large Random Breeding Populations

Consider any series of consecutive generations in an infinitely large random mating population unaffected by either natural or artificial selection and let the successive generations be numbered 0,1,2,3, ... Using the symbolism of section 6.2 the frequencies of two-locus genotypes in the gametes produced by the individuals of generation 0 would be as follows

Gamete genotype	Frequency
B_iB_j	p_0
B_ib_j	r_0
b_iB_j	s_0
b_ib_j	t_0

and, with (1) no selection and (2) random mating, frequencies of genotypes in the individuals of generation 1 and of genotypes in the gametes produced by them would be as shown below.

Genotypes	Frequency	Frequency of gamete genotypes[1] B_iB_j	B_ib_j	b_iB_j	b_ib_j
B_iB_j/B_iB_j	p_0^2	1	0	0	0
B_iB_j/B_ib_j	$2p_0r_0$	1/2	1/2	0	0
B_ib_j/B_ib_j	r_0^2	0	1	0	0
B_iB_j/b_iB_j	$2p_0s_0$	1/2	0	1/2	0
B_iB_j/b_ib_i	$2p_0t_0$	$(1-r)/2$	$r/2$	$r/2$	$(1-r)/2$
B_ib_j/b_iB_j	$2r_0s_0$	$r/2$	$(1-r)/2$	$(1-r)/2$	$r/2$
B_ib_j/b_ib_j	$2r_0t_0$	0	1/2	0	1/2
b_iB_j/b_iB_j	s_0^2	0	0	1	0
b_iB_j/b_ib_j	$2s_0t_0$	0	0	1/2	1/2
b_ib_j/b_ib_j	t_0^2	0	0	0	1

Absence of selection means that numbers of gametes produced are the same for all ten genotype classes of individuals. It follows that the frequency of B_iB_j gametes produced by generation 1 individuals is

$$p_1 = p_0^2 + p_0r_0 + p_0s_0 + (1 - r)p_0t_0 + rr_0s_0$$

$$= p_0(p_0 + r_0 + s_0 + t_0) - r(p_0t_0 - r_0s_0) = p_0 - r\Delta_0$$

where $\Delta_0 = p_0t_0 - r_0s_0$. The corresponding frequencies of the other genotypic classes of gametes are easily obtained in the same way. To summarize, the frequencies of gamete genotypes produced by generation 0 individuals are as follows.

Genotypes of gametes produced by generation 1 individuals	Frequency
B_iB_j	$p_0 - r\Delta_0 = p_1$
B_ib_j	$r_0 + r\Delta_0 = r_1$
b_iB_j	$s_0 + r\Delta_0 = s_1$
b_ib_j	$t_0 - r\Delta_0 = t_1$

Employing these frequencies, we find

$$\Delta_1 = (p_1t_1 - r_1s_1) = (p_0t_0 - r_0s_0)(1 - r) = \Delta_0(1 - r) \qquad \text{A6.3.1}$$

By analogy, it is appparent that

$$\Delta_2 = (p_2t_2 - r_2s_2)(1 - r) = \Delta_1(1 - r)$$

and, substituting for Δ_1 in terms of A6.3.1,

$$\Delta_2 = \Delta_0(1 - r)^2.$$

It follows that

$$\Delta_3 = \Delta_2(1 - r) = \Delta_0(1 - r)^3$$

and, by extension, that

$$\Delta_n = \Delta_0(1 - r)^n$$

This result was first obtained by Robbins(1918).

Appendix 8.1. The Covariance between $\hat{q}$ and $\hat{y}_{xj}$ (A) in Progeny Test Selection when Selection Unit Individuals Have Been Produced by One or More Generations of Self-Fertilization and (B) in Selection Among S_n Families

Consider an infinite population of individuals produced by n successive generations of self-fertilization and let it be identified as an S_n population. Assume that the first individuals to be self-fertilized were an infinite random sample from an S population formed by the union of random pairs of gametes from a gamete population in which the frequencies of two-locus gamete genotypes were as follows.

Gamete genotype	Frequency
B_iB_j	p
B_ib_j	r
b_iB_j	s
b_ib_j	t

Asume further that, in each of the n generations, the individuals self-fertilized were also those of an infinite random sample. The frequencies of two locus genotypes in the S_n generation are shown below along with (a) values of q_i for each genotype, (b) coded values of the expectation of u_{xj} in offspring from matings to random members of the non-inbred S generation, $*(y_{xj})'$, and (c) coded values of the expectation of y_{xj} in progenies obtained from the S_n individuals by self-fertilization, $*(\hat{y}_{xj})'' = (\hat{y}_{xj}) - u_{xj}$. The genotype frequencies are from Gates et al (1957) with some modification of symbols.

S_n Genotype	Frequency (F)	$\hat{q}_i$	$*(\hat{y}_{xj})'$	$*(\hat{y}_{xj})''$
B_iB_j/B_iB_j	$p^2 + (A+pr+ps)f + (pt-rs)\pi/2 + A(B-1)/2$	1	Z_{xj}	u_{xj}
B_iB_j/B_ib_j	$(A + 2pr)(1-f) - AB$	1	0.0	$a_{xj}u_{xj}/2$
B_ib_j/B_ib_j	$r^2 + (A+pr+rt)f - (pt-rs)\pi/2 + A(B-1)/2$	1	$-Z_{xj}$	$-u_{xj}$
B_iB_j/b_iB_j	$(A+2ps)(1-f) - AB$	1/2	Z_{xj}	u_{xj}
B_iB_j/b_ib_j	AB	1/2	0.0	$a_{xj}u_{xj}/2$
B_ib_j/b_iB_j	AB	1/2	0.0	$a_{xj}u_{xj}/2$
B_ib_j/b_ib_j	$(A+2rt)(1-f) - AB$	1/2	$-Z_{xj}$	$-u_{xj}$
b_iB_j/b_iB_j	$s^2 + (A+ps+st)f - (pt-rs)\pi/2 + A(B-1)/2$	0	Z_{xj}	u_{xj}
b_iB_j/b_ib_j	$(A + 2st)(1-f) - AB$	0	0.0	$a_{xj}u_{xj}/2$
b_ib_j/b_ib_j	$t^2 + (A+st+rt)f + (pt-rs)\pi/2 + A(B-1)/2$	0	$-Z_{xj}$	$-u_{xj}$

$A = pt + rs$ $\quad B = [(1-2\gamma+2\gamma^2)/2]^n$ $\quad \pi=[(1-2\gamma)/(1+2\gamma)]\{1-[(1-2\gamma)/2]^n\}$

γ = the probability of recombination between the i-th and j-th loci

$f = 1-(1/2)^n$ = the coefficient of inbreeding in the S_n population

$Z_{xj}= [1+(1-2v_j)a_{xj}]u_{xj}/2$ as derived in section 8.1.2

A8.1.1. Progeny Test Selection

By virtue of the assumptions of (a) random selection[1] of individuals and (b) infinite population size in each generation, there can be no change in allele frequencies from one generation to the next. Hence allele frequencies in the S_n population are

$$\ddot{q}_i = q_i = p + r \quad \text{and} \quad q_j = p + s$$

as they were in the S population. From the distribution data the mean of $*(\hat{y}_{xj})'$ is

$$E*(\hat{y}_{xj})' = Z_{xj}[p^2 - r^2 + (ps - rt)f + (pt-rs)\pi + 2(ps-rt)(1-f)$$
$$+ s^2 - t^2 + (ps-rt)f - (pt-rs)\pi]$$
$$= [(p+s)^2 - (r+t)^2]Z_{xj} = (p+s-r-t)Z_{xj}$$

and $\sigma(\hat{q}_i,\hat{y}_{xj}) = \Sigma F(\hat{q}_i)*(\hat{y}_{xj})' - q_i E*(\hat{y}_{xj})'$

$$= Z_{xj}[p^2 - r^2 + (ps-rt)f + (pt-rs)\pi$$
$$+ (ps-rt)(1-f) - (p+r)(p+s-r-t)]$$
$$= (pt-rs)(1+\pi)Z_{xj} = (pt-rs)_{ij}K_{ij}[1 + (1-2q_j)a_{xj}]u_{xj}/2 \qquad \text{A8.1.1}$$

where $K_{ij} = 1+\pi = [2 - (1-2\gamma)^{n+1}/2^n]/(1+2\gamma)$

• A8.1.2. Selection among S_n Families

Averages of both variables are required. As noted in the preceeding subsection, $\ddot{q}_i = p+r$. Then, working from the bivariate distribution, we obtain

$$E*(\hat{y}_{xj})'' = \Sigma F[*(\hat{y}_{xj})'']$$
$$= u_{xj}\{[p^2+(A+pr+ps)f + (pt-rs)\pi/2 + A(B-1)/2]$$
$$-[r^2 + (A+pr+rt)f - (pt-rs)\pi/2 + A(B-1)/2] + [(A+2ps)(1-f) - AB]$$
$$+[(A+2rt)(1-f) - AB] + [s^2 + (A+ps+st)f - (pt-rs)\pi/2 + A(B-1)/2]$$
$$-[t^2 + (A+st+rt)f + (pr-rs)\pi/2 + A(B-1)/2]\}$$
$$+ [a_{xj}u_{xj}/2][(A+2pr)(1-f) - AB + AB + AB + (A+2st)(1-f) - AB]$$
$$= (p+s-r-t)u_{xj} + (p+s)(r+t)(1-f)a_{xj}u_{xj}$$

and $\sigma[\hat{q}_i, E(\hat{y}_{xj})''] = \Sigma\{F(\hat{q}_i)E*(\hat{y}_{xj})''\}$

$$= (pt-rs)(1+\pi)u_{xj} + \{[(A+2pr)(1-f) - AB] + AB/2$$
$$+AB/2 - (p+r)(p+s)(r+t)(1-f)\}[a_{xj}u_{xj}/2]$$
$$= (pt-rs)(1+\pi)u_{xj} - (pt-rs)(p+s-r-t)(1-f)a_{xj}u_{xj}/2$$
$$= (pt-rs)[(1+\pi) + (1-2q_j)(1-f)a_{xj}/2]u_{xj} \qquad \text{A8.1.2}$$

• Appendix 9.1. Probabilities of Two-Locus Genotypes in Pure Lines Derived by Successive Self-Fertilizations and without Selection from a Random Mating, Linkage Equilibrium Source Population

Probabilities of the ten possible two-locus genotypes in individuals obtained from a random mating source population by self-fertilization of random individuals in n successive generations are shown in Appendix 8.1. Probabilities of the four homozygous genotypes (from Appendix 8.1) are as follows.

S_n Genotype	Frequency (F)
B_iB_j/B_iB_j	$p^2 + (A+pr+ps)f + (pt-rs)\pi/2 + A(B-1)/2$
B_ib_j/B_ib_j	$r^2 + (A+pr+rt)f - (pt-rs)\pi/2 + A(B-1)/2$
b_iB_j/b_iB_j	$s^2 + (A+ps+st)f - (pt-rs)\pi/2 + A(B-1)/2$
b_ib_j/b_ib_j	$t^2 + (A+st+rt)f + (pt-rs)\pi/2 + A(B-1)/2$

Here (as in Appendix 8.1) p, r, s, and t symbolize the frequencies of the B_iB_j, B_ib_j, b_iB_j, and b_ib_j genotypes in gametes produced by individuals of the source population; A = pt + rs, and $B = [(1-2\gamma+2\gamma^2)/2]^n$ in which γ = the frequency of recombination between the i-th and j-th loci.

Assuming n large enough so that only homozygous individuals (pure lines) are present in the population and remembering that the source population was in linkage equilibrium, we can set f = 1.0 and pt-rs = 0.0 in the above frequencies. Then in the case of the B_iB_j/B_iB_j genotype

$$F = p^2 + (A+pr+ps) + A(B-1)/2. \qquad \text{A9.1.1}$$

Because $0.0 \leq \gamma \leq 1/2$, the value of B can be no greater than $(1/2)^n$ so that

$$A(B-1)/2 = -A[1-(1/2)^n] \approx -A/2 = -(pt+rs)/2 \text{ when n is large.}$$

Making this substitution in A9.1.1 we obain

$$F = \text{frequency of } B_iB_j/B_iB_j = p^2 + pt + rs + pr + ps - (pt+rs)/2$$

$$= p(p+r+s) + pt/2 + rs/2 = p(1-t) + pt/2 + rs/2 = p-(pt-rs) = p$$

because by virtue of linkage equilibrium in the source population (pt-rs) = 0.

In the same way the frequencies of the B_ib_j/B_ib_j, b_iB_j/b_iB_j, and b_ib_j/b_ib_j genotypes can be shown to be r, s, and t, respectively. It follows that the frequencies of genotypes in gametes produced by the population of pure lines would be the same as in gametes produced by the source population and, hence, that a population of single crosses between random members of the pure line population would, like the source population, be in linkage equilibrium.

Appendix 9.2. Linkage Disequilibrium within S_n Families

The probabilities of two-locus (any two loci) genotypes in individuals of an S_n generation are shown in Appendix 8.1. The assumptions on which those probabilities were based were (1) that the S_1 generation was formed by self-fertilization of random individuals from a random mating source population and (2) that each later generation was formed by self-fertilization of random individuals from the immediately preceding generation. Note (1) that linkage equilibrium in the source population was not assumed and (2) that the specification that parents of each generation were random members of their own generation equates to assuming no selection of parental individuals.

The probabilities of of the B_iB_j/b_ib_j and B_ib_j/b_iB_j genotypes are the ones of interest here because their families (by self-fertilization) are the only ones in which segregation at both the i-th and the j-th locus occurs, i.e., the only ones in which pt-rs can be non-zero. The significant fact is that the probabilities of those two genotypes are shown to be the same. Letting γ = the rate of recombination between the i-th and j-th loci, the probabilities of two-locus genotypes in gametes produced by the parental types of interest are as follows.

Gamete genotype		Genotype of parent: B_iB_j/b_ib_j	Genotype of parent: B_ib_j/b_iB_j
B_iB_j	p =	$(1-\gamma)/2$	$\gamma/2$
B_ib_j	r =	$\gamma/2$	$(1-\gamma)/2$
b_iB_j	s=	$\gamma/2$	$(1-\gamma)/2$
b_ib_j	t=	$(1-\gamma)/2$	$\gamma/2$

From these quantities it is apparent that $(pt-rs)_{ij} = [(1-\gamma)^2 - \gamma^2]/4 = (1-2\gamma)/4$ in the case of families of B_iB_j/b_ib_j parents and $(2\gamma-1)/4$ in the case of families of B_ib_j/b_iB_j parents. Because the probabilities of the two parent types are the same [and the fact that $(pt-rs)_{ij} = 0.0$ in all other families], it follows that the expectation of the average of $(pt-rs)_{ij}$ for the entire generation is zero.

The genotype probabilities shown in Appendix 8.1 were derived assuming that parents in each generation were random individuals. On the other hand, some selection of plants to be self-fertilized is probably standard in actual practice. Will this selection alter the expected equality of probabilities of the two double-heterozygote genotypes among parents? Probably not. The genotyes are the same; barring position effects, the joint contributions of the two single-locus genotypes to value of the total genotype of individuals would be the same and hence selection should not favor one of the double heterozygotes over the other.

• Appendix 11.1. Variations of Effective Population Size (N_e)

The definition of an ideal population is stated in section 2.6.3 and the sampling variance of Δq in a random mating monoecious population is shown there to be

$$\sigma^2(d_c) = q_{st}(1-q_{st})/2N \qquad 2.33$$

where q_{st} = the expected allele frequency in selections of the t-th generation and N = number of selections when the selection units are individuals. In the case of an ideal population, this is ordinarily written

$$\sigma^2(\Delta q) = q(1-q)/2N$$

in which Δq denotes the difference between the frequencies of any particular allele in two successive generations, and q is the expectation of the frequency of that allele in the second of those generations.

Formulas for computation of N_e depend on the details of selection programs. The procedure for deriving them is described in section 2.7. It consists of the following two steps

(1) derivation of $\sigma^2(\Delta q)$ in the case of the program in question and

(2) solving for N_e after equating $\sigma^2(\Delta q)$ for the specific program to $q(1-q)/2N_e$.

As an example, the effective population size of a random mating dioecious population is considered in section 2.7.

In the context of selection programs, the genetic sample has been defined as a random sample of selection units from an infinite genetic population of such units. Given systems in which the selection units are individuals, the genetic population is the S population. When the selection unit is some sort of family the genetic population consists of such families derived randomly from the S population. Selections are a subset of selection units from the genetic sample but for present purpose could be its entirety because effective population size as defined derives from random variation in allele frequencies, i.e., variation independent of selection.

Consider any allele (say B) and its frequencies in two successive generations of an S population and in the selections from the first of those generations. If

q = the frequency of B in the first of the two generations of the S population and hence in the genetic population and

q_s = the frequency of B in the selections (or members of selected families) used as parents of the second generation of the S population, then

$$\Delta q = q_s - q \qquad \text{A.11.1.1}$$

because, by virtue of being an infinite population, allele frequency in every generation of the S population will be equal to frequency of the allele in the parents employed to produce it. The crucial assumptions are random mating and equal probabilities of involvement in fertilizations by B and non-B gametes. Then, from A11.1.1 and the fact that q, being the frequency of B in the infinite genetic population, is a constant

$$\sigma^2(\Delta q) = \sigma^2(q_s)$$

When parent stock is composed of (a) male parents and (b) female parents as it must be in dioecious species and may be in monoecious species,

$$q_s = (q_{sc} + q_{sr})/2$$

where q_{sc} and q_{sr} are the frequencies of the B allele in male and female parents (or common and random parents in the case of half-sib families), respectively. It follows, in such cases, that

$$\sigma^2(\Delta q) = \sigma^2(q_s) - [\sigma^2(q_{sc}) + 2\sigma(q_{sc}, q_{sr}) + \sigma^2(q_{sr})]/4 \qquad \text{A11.1.2}$$

in which $\sigma^2(q_{sc})$ and $\sigma^2(q_{sr})$ may or may not be equal.

• A11.1.1. Variance of Allele Frequency (Random Mating Diploid Populations)

Let q and Q symbolize the frequency of any allele (say B) in a genetic population and in individuals of that population, respectively. Then the well known distribution of Q is

Genotype	Frequency	Q
BB	q^2	1
Bb	$2q(1-q)$	1/2
bb	$(1-q)^2$	0

from which, as shown in section 2.5, $\sigma^2(Q) = q(1-q)/2$. We know that the value of Q in any individual depends in part on the values of Q in its parents and therefore it is reasonable to partition values of Q as follows.

$$Q_{ijk} = q + S_i + D_j + w_{ijk}$$

in which

S_i = the effect of the i-th male or pollen parent on allele frequency in its offspring,

D_j = the effect of the j-th female or seed parent on allele frequency in its offspring, and

w_{ijk} = the deviation of Q_{ijk} from $q + S_i + D_j$.

Because each parent contributes only one of the two genes that constitute the B locus genotype of any offspting, parental effects must be one-half of the Q of the parent. Thus S (or D) = Q/2 and

$$\sigma^2(S) = \sigma^2(D) = \sigma^2(Q)/4 = q(1-q)/8 \text{ and} \qquad \text{A11.1.3}$$

$$\sigma^2(w) = \sigma^2(Q) - \sigma^2(S) - \sigma^2(D) = q(1-q)/4 \qquad \text{A11.1.4}$$

- A11.1.2. **Special Cases**
- A11.1.2.1. SELECTION AMONG HALF-SIB FAMILIES (TWO-GENERATION CYCLES)

The structure of half-sib families may of course vary from one instance to another. By definition, all members of the same family have one parent in common but the numbers of random parents and of offspring per random parent are variable from one selection program to another and from family to family within programs. The latter variation can rarely be completely controlled. It is the variation in the distribution among selected families (the selections) of individuals employed as parents of the suceeding generation of the S population that affects effective population size.

It is apparent from A11.1.2 that the variances of q_{sc} and q_{sr} are requisite to obtaining expressions for N_e. Let

C_i be the effect of the common parent on allele frequencies of members of the i-th family,

R_{ij} be the effect of the j-th random parent of members of the i-th family on allele frequencies of its offspring,

w_{ijk} be the deviation, $Q_{ijk} - C_i - R_{ij}$, in the case of the k-th member of the i-th family,

c_{ij} be the number taken from the j-th full-sib family of the i-th selection unit to be used as the common parent of families in the following generation, and

r_{ij} be the corresponding number of individuals to be used as random parents.

Then

$$q_{sc} = q + \Sigma_i\{C_i\Sigma_j c_{ij} + \Sigma_j c_{ij}R_{ij} + \Sigma_j\Sigma_k w_{ijk}\}/\Sigma_i\Sigma_j c_{ij}.$$

Remembering that there are no covariances among C's, among R's, or between C's and R's because all matings are assumed to be random and that q is a constant within any one generation,

$$\sigma^2(q_{sc}) = \{\Sigma_i[\Sigma_j c_{ij}]^2\sigma^2(C) + \Sigma_i\Sigma_j c_{ij}^2\sigma^2(R) + \sigma^2(w)\Sigma_i\Sigma_j c_{ij}\}/(\Sigma_i\Sigma_j c_{ij})^2 \quad \text{A11.1.5}$$

By analogy,

$$q_{sr} = q + \Sigma_i\{C_i\Sigma_j r_{ij} + \Sigma_j r_{ij}R_{ij} + \Sigma_j\Sigma_k w_{ijk}\}/\Sigma_i\Sigma_j r_{ij} \text{ and}$$

$$\sigma^2(q_{sr}) = \{\Sigma_i[\Sigma_j r_{ij}]^2\sigma^2(C) + \Sigma_i\Sigma_j r_{ij}^2\sigma^2(R) + \sigma^2(w)\Sigma_i\Sigma_j r_{ij}\}/(\Sigma_i\Sigma_j r_{ij})^2 \quad \text{A11.1.6}$$

Now note (a) that the C's and R's of q_{sc} and q_{sr} are identical and (b) that the w's of q_{sc} are all different from any w of the expression for q_{sr}. Consequently,

$$\sigma(q_{sc}, q_{sr}) = \{\Sigma_i[\Sigma_j c_{ij}][\Sigma_j r_{ij}]\sigma^2(C) + \Sigma_i\Sigma_j c_{ij}r_{ij}\sigma^2(R)\}/(\Sigma_i\Sigma_j c_{ij})(\Sigma_i\Sigma_j r_{ij}) \quad \text{A11.1.7}$$

Then substitution in A11.1.2 in terms of A11.1.5, A11.1.6, and A11.1.7 yields

$$\sigma^2(\Delta q) = [K_c\sigma^2(C) + K_r\sigma^2(R) + K_w\sigma^2(w)]/4 \quad \text{A11.1.8}$$

in which

$$K_c = \Sigma_i[\Sigma_j c_{ij}/\Sigma_i\Sigma_j c_{ij} + \Sigma_j r_{ij}/\Sigma_i\Sigma_j r_{ij}]^2,$$

$$K_r = \Sigma_i\Sigma_j[c_{ij}/\Sigma_i\Sigma_j c_{ij} + r_{ij}/\Sigma_i\Sigma_j r_{ij}]^2, \text{ and}$$

$$K_w = 1/\Sigma_i\Sigma_j c_{ij} + 1/\Sigma_i\Sigma_j r_{ij}.$$

This expression, A11.1.8, is not attractive for practical application because individual values of c_{ij} and r_{ij} can rarely, if ever, be known in advance. However, reasonable substitutions for the various quantities called for in the K's are possible. Assuming that the genetic sample will consist of T half-sib families in each cycle of the program, the number of common parents required is T. Thus it is reasonable to set $\Sigma_i\Sigma_j c_{ij} = T$. Then, if the intent is to use F random parents per half-sib family, TF random parents are required and to set $\Sigma_i\Sigma_j r_{ij} = TF$ is also reasonable. Assuming that equal numbers of individuals to be used as common and random parents are taken from each selection, $\Sigma_j c_{ij}$ and $\Sigma_j r_{ij}$ are logically set equal to T/M and TF/M, respectively. On the other hand, it is not reasonable to expect that either c_{ij} or r_{ij} will in practice be constant from one full-sib family to another. It follows that while $E(c_{ij}) = T/FM$, if things go according to plan, $E(c_{ij})^2 = (T/FM)^2 + \sigma^2(c)$. In the same way, $E(r_{ij}) = T/M$ and $E(r_{ij})^2 = (T/M)^2 + \sigma^2(r)$. Making the following substitutions: $\Sigma_i\Sigma_j c_{ij} = T$, $\Sigma_i\Sigma_j r_{ij} = TF$, $\Sigma_j c_{ij} = T/M$, $\Sigma_j r_{ij} = TF/M$, $\Sigma_j (c_{ij})^2 = F(T/FM)^2$, and $\Sigma_j (r_{ij})^2 = F(T/M)^2$ in the K's of A11.1.10, we obtain $K_c = 4/M$, $K_r = 4/FM$, and $K_w = (F+1)/FT$. Then from A11.1.8

$$\sigma^2(\Delta q) = \{\sigma^2(C) + \sigma^2(R)/F + [p(F+1)/4F]\sigma^2(w)\}/M \qquad \text{A11.1.9}$$

in which $p = M/T$ = the fraction of units selected.

Finally, before substituting for $\sigma^2(C)$ and $\sigma^2(R)$ in terms of A11.1.3, remember that because in cycles after the first each genetic population is generated from a finite number of parents and hence that the parents of individuals in selected families are samples drawn without replacement from a finite population. Therefore, in accord with A1 16 and A11.1.3,

$$\sigma^2(C)/M = [q(1-q)/8][(T-M)/TM\} \qquad \text{A11.1.10}$$

instead of $q(1-q)/8M$ and

$$\sigma^2(R)/FM = [q(1-q)/8F][(T-M)/TM]. \qquad \text{A11.1.11}$$

However, the members of each full-sib family of which the half-sib families are composed are random members of the conceptual infinity of offspring that could be produced by each pair of parents. Thus no finite population correction of the value of $\sigma^2(w)$ given by A11.1.4 is required. Substituting in A11.1.9 in terms of A11.1.10, A11.1.11, and A11.1.4 yields

$$\sigma^2(q) = [q(1-q)/8M][F+1)/F]\{[(T-M)/T] + p/2\}$$

$$=[q(1-q)/8M][(F+1)/F][(2-p)/2] \qquad \text{A11.1.12}$$

Finally, setting $\sigma^2(\Delta q) = q(1-q)/2N_e$ and solving for N_e yields

$$N_e = 8MF/(F+1)(2-p) \qquad \text{A11.1.13}$$

- A11.1.2.2. SELECTION AMONG FULL-SIB FAMILIES (TWO GENERATION CYCLES)

Full-sib families may be viewed as the extreme of half-sib families in which F=1. This is the one aspect that sets them apart. Setting F=1 in A11.1.12 and A11.1.13 yields

$$\sigma^2(\Delta q) = [q(1-q)/4M][(2-p)/2] \quad \text{and}$$

$$N_e = 4M/(2-p)$$

- A11.1.2.3. SELECTION WITHIN FULL-SIB FAMILIES

Consider the special case where T full-sib families are produced by as many pair matings involving 2T individuals (T males and T females in a dioecious organism) and two individuals are taken from each family to be used as parents of the families of the next generation. The absence of selection among families means that M=T so that p=1. Making these substitutions and setting F= 1 in A11.1.12, we obtain

$$\sigma^2(q) = q(1-q)/8T$$

from which, by standard procedure,

$$N_e = 4T = 2N \quad (N = \text{total number of parents})$$

the answer given by Falconer(1981 and 1960).

A11.1.2.4. SELECTION AMONG FAMILIES PRODUCED BY SELF-FERTILIZATION

Following standard symbolism, an S_n family will be understood to be one obtained by n successive generations of self-fertilization -- the first of the self-fertilized ancestors being a member of a non-inbred random mating population. Now consider a population of S_n families derived, without either gametic or zygotic selection, from the same source population and let

Q_{ij} symbolize the frequency of any allele (to be symbolized here by the letter B) in the j-th member of the i-th family,

C_i symbolize the frequency of the allele in the immediate parent of the i-th family, and

w_{ij} symbolize the deviation of the frequency of the allele in the j-th member of the i-th family from the expected average of allele frequencies in the family.

Then because the expected average of Q in any family is obviously the frequency of the allele in the immediate parent of the family,

$$Q_{ij} = C_i + w_{ij}$$

and because the intra-family variation of w's is random relative to parental allele frequency,

$$\sigma^2(Q) = \sigma^2(C) + \sigma^2(w) \qquad \text{A11.1.14}$$

We know (section 2.5) that in the absence of selection the probabilities of single locus genotypes and their Q values in a population of individuals produced by n successive generations of self-fertilization are as follows.

Genotypes	Frequency	Q
BB	$q^2 + f_n q(1-q)$	1
Bb	$2q(1-q)(1-f_n)$	1/2
bb	$(1-q)^2 + f_n q(1-q)$	0

From this distribution the variance of Q is found to be

$$\sigma^2(Q) = q(1-q)(1+f_n)/2$$

where (from section 2.5.1) $f_n = 1-(1/2)^n$

is the coefficient of inbreeding of individuals resulting from n generations of self-fertilization. It follows that the variance of frequency of the allele in the parents of a population of S_n families

$$\sigma^2(C) = q(1-q)(1+f_{n-1})/2 \qquad \text{A11.1.15}$$

Then from A11.1.14

$$\sigma^2(w) = \sigma^2(Q) - \sigma^2(C) = q(1-q)[f_n - f_{n-1}]/2 \qquad \text{A11.1.16}$$

and, if m individuals from each of M selected families are employed as parents for regeneration of the S population

$$q_s = [m\Sigma_i C_i + \Sigma_i \Sigma_j w_{ij}]/Mm$$

from which

$$\sigma^2(\Delta q) = \sigma^2(q_s) = \sigma^2(C)/M + \sigma^2(w)/Mm$$

Finally, substitution for $\sigma^2(C)$ and $\sigma^2(w)$ in terms of A11.1.15 and A11.1.16 yields

$$\sigma^2(\Delta q) = q(1-q)[1+f_{n-1} + (1/2)^n/m]/2M$$

from which, by standard procedure,

$$N_e = M/[1 + f_{n-1} + (1/2)^n/m] \qquad \text{A11.1.17}$$

Whenever n and/or m are large enough to make $(1/2)^n/m$ small relative to $(1+f_{n-1})$

$$N_e \approx M/(1+f_{n-1}) \qquad \text{A11.1.18}$$

However, in the special case of S_1 families in which A11.1.18 yields $N_e = M$, the result from A11.1.17 is

$$N_e = M/[(2m+1)/2m]$$

which would be substantially smaller than M if m were as small as 1, 2, or 3.

• A11.1.3. Cycles Extended by One or More Extra Generations of Random Mating in the Population

The effective population size for the cycle as a whole is then obtained as $N_e = 1/\Sigma_k(1/N_{ek})$ in which N_{ek} = effective population size of the k-th plant generation of the cycle (see section 11.1.1). N_e in the one or more extra generations will be the total number of parents when the numbers of male or pollen parents and of female or seed parents are the same. On the other hand if m, the number of male parents, is not the same as f, the number of female parents, N_e in the extra generation(s) would be 4mf/(m+f) as shown in section 2.7.

• Appendix 11.2. Recent Mutations and Superior Alleles at Low Frequencies

Consider a large random mating population in which recurrent selection for total value, practiced for many generations, is to be continued in the future. Assume that neither the population of environments with respect to which genotype values are defined nor the definition of total value has changed during those generations of selection. Optimum selective coefficients (s'-values) of alleles would then be as before except for effects due to allele frequency changes or the introduction of new alleles by mutation.

Next consider a gene of which two alleles (B and b) were present when the recurrent selection began. Though multiple alletes may be more realistic, it will suffice to assume only two. Given our assumptions (s' constant and large size of the S population) the frequency of the better allele (B) might well be close to 1.0 unless its original value of $s'q_o$ had been low. However, if a third allele (B'), superior to B in its effect relative to total value, were to arise by mutation and not be lost by drift, its frequency could eventually replace the original best allele of the i-th gene. Actually there may be numerous superior alleles at relatively low frequencies because of recent mutant origin. <u>Can there be enough of these to be important</u>?

Among the gametes received by individuals of the S population in a single cycle of the selection program, let

π = the probability that any one of those gametes will possess an allele superior to B that arose by mutation within the most recent selection cycle.

If there are N individuals per cycle, there would be 2N opportunities for that event to occur and the expected number of new alleles superior to B would be $2N\pi$. The approximate probability of fixation of any one such allele would be, from 11.1

$$P_f \approx 1 - e^{-2N_e s' q_0} \quad \text{A11.2.1}$$

The indicated approximation recognizes that (a) the effects of superior mutant alleles are not likely to be exactly additive and (b) the denominator of 11.1 must be slightly less than 1.0 even though N_e is very large. Because the mutation event in question would be very rare, the chance of it occurring twice in the same generation would be infinitesimal; hence we can set $q_0 = 1/2N$. Substituting in A11.2.1, we obtain

$$P_f \approx 1 - e^{-2N_e s'/2N} \approx N_e s'/N$$

Multiplying $2N\pi$ (the number of mutant alleles superior to B expected per selection cycle) by $N_e s'/N$ (the probability of fixation of such an allele), we obtain $2\pi N_e s'$ as the probability that such an allele will occur in one selection cycle and subsequently be fixed in the population. Given that event, the genetic improvement of the S population would be $Z = 2\pi N_e s'(2u)$ and the expectation ($\ddot{Z}$) of total improvement from fixation of superior alleles arising by mutation in the same single cycle of the selection program is

$$\ddot{Z} = 4\Sigma_i \pi_i N_e (s')_i u_i \quad \text{A11.2.2}$$

in which

π_i = the probability of mutation of the i-th gene to a superior allele not previously present,

$(s')_i$ = the weighted average of the effective selective values of the possible new superior alleles of the i-th gene,

$2u_i$ = the weighted average of the difference in value between B'B' and BB genotypes, and

Σ_i indicates summation over the n genes of the entire genome of the organism.

A11.2.2 is not a usable formula because the many π's and u's cannot be quantitized. There is no reason to suppose that either u_i or π_i is the same from one gene to another but, with no basis for knowing the the individual values of either of them the available alternative is to act as if both are constants. Then, noting that s' is a linear function of u_i when no dominance and linkage equilibrium are assumed (see Table 11.3) and hence is a constant for any specific selection system when u_i is constant, A11.2.2 will be rewritten for our purpose as

$$\ddot{Z} \approx 4n\pi N_e s' u. \quad \text{A11.2.3}$$

What is the possible magnitude of $\ddot{Z}$? Can it be large enough to merit attention by breeders? The utility of A11.2.3 as a source of information hinges on our ability to supply plausible values for n, π, and u.

The number of genes (n) varies among species -- the number of "structural" genes in

Drosophila melanogaster is believed to be about 5000 and in higher organisms considerably higher, e.g., probably 50,000 or more in humans. In addition there is a large amount of "redundant" DNA which may have regulation functions and in which mutation could therefore be the source of useful heritable changes.

The probabilities of mutation to specific alleles are variable; many have been found to be in the range 10^{-4} to 10^{-6}. However, a majority of the alleles for which information is available are deleterious and there are reasons to believe that mutant alleles are more likely to be deleterious than beneficial. This suggests that probabilities of specific beneficial mutants may not be higher than 10^{-7}. On the other hand, remember that the DNA of single structural genes may be made up of as many as 1000 base pairs and, hence, that the probability (π) of a point mutation at any one of the many sites within the gene has to be much greater than the probability of mutation to a specific allele that depends on a single base pair change. Thus values of π much greater than 10^{-7} deserve to be considered.

If the selection criterion (X) is measured in terms of a unit such that $\ddot{X} = 100$ it can be shown that average values of u for the genes responsible for the major portion of observed genetic variance can reasonabley be assumed to be as large as 0.5. The u values of new superior alleles will be be > 0.0 and it is reasonable to suppose a skew distribution with frequencies greatest very close to zero and progressively less as u becomes greater. On that basis the value appropriate for use in A11.2.3 may well be considerably less than 0.5.

Values of $\ddot{Z}$ provided by A11.2.3 when values of parameters are as indicated are shown below.

n	$N_e s'$	u	π	$\ddot{Z}$
20,000	50	.5	10^{-4}	200
			10^{-5}	20
			10^{-6}	2
	20	.2	10^{-4}	80
			10^{-5}	8
			10^{-6}	0.8
	10	.1	10^{-4}	40
			10^{-5}	4
			10^{-6}	0.4

Relative to $\ddot{X} = 100$, $\ddot{Z} = 1$ would amount to a 1 percent improvement per cycle or, more generally, m percent as a consequence of mutation in m successive cycles. However, because the expected increase per cycle in the frequency of an allele is small, of the order of $kq(1-q)u/\sigma(\hat{X})$ [see 8.27 and Table 8.2], a large fraction of the full potential of a favorable allele arising by mutation would only be realized after many cycles of selection. Thus the realized genetic gain, from increased frequency of new favorable alleles, at the end of m cyles of recurrent selection would be only a fraction of $m\ddot{Z}$, a fraction that would be small if m were small but progressively larger as the recurrent slection program continued.

There is much that is speculative in the above presentation. Would the average of u be less (or greater) than 0.1? Not all new superior alleles would be additive in relation to the previously best allele of the gene in question. If dominant,their fixation probabilities would be greater than if additive as assumed; if recessive, the reverse would be true (see section 11.2.1). Have appropriate values of π and n been employed? Despite these questions, it appears proper to conclude that beneficial mutations may add significantly to the potential of long-term recurrent selection programs.

Appendix 11.3. Derivations of E(Δq) and E(Δÿ) as Functions of N_e

The problems addressed herein arise from the fact that in each cycle of any recurrent selection program q_i, the frequency of the B_i allele varies around its expectation and,hence, every function of q_i is also variable so that the expectation of the function requires attention.

Meanings of symbols employed herein are as stated in section 11.3.

A11.3.1. Generalizations from Appendix 1

First of all, remember that the expectation of any linear function of variables is the same function of its separate terms, i.e., if

$X = aY + bW - cZ$ where a, b, and c are constants

$E(X) = aE(Y) + bE(W) - cE(Z)$

Such variables as Y, W, and Z in the above may of course be powers or products of underlying variables.

Other general forms from Appendix 1 that are relevant here are the following in which X and Y symbolize any variables, $\ddot{X}$ and $\ddot{Y}$ are the expectations (the population means) of those variables, $\sigma^2(X)$ is the variance of X, and $\sigma(X,Y)$ is the covariance between X and Y.

$$E(X^2) = \ddot{X}^2 + \sigma^2(X) \qquad \text{(A1.6) A11.3.1}$$

$$E(X^3) = \ddot{X}^3 + 3\ddot{X}\sigma^2(X) \qquad \text{(A1.7) A11.3.2}$$

(if the distribution of X is symmetrical)

$$E(X^4) = \ddot{X}^4 + 6\ddot{X}^2[\sigma^2(X)] + 3[\sigma^2(X)]^2 \qquad \text{(A1.8) A11.3.3}$$

(if the distribution of X is normal as well as symmetrical)

$$E(XY) = \ddot{X}\ddot{Y} + \sigma(X,Y) \qquad \text{(A1.9) A11.3.4}$$

These provide the bases for obtaining the expectations of relevant functions of allele frequencies subject only to the assumption that whatever a-normality there may be in the distributions of q's is small enough to be of little significance. Those distributions are binomial and approach the normal unless E(q) is near 0.0 or 1.0 and N_e is small. The expectaion of q_1 is $q_0 + E(\Delta q)_0$. It and the variance of q_1 around it will be symbolized here by $\bar{q}_1$ and $\sigma^2(\bar{d}_0)$, respectively[1].

Employing A11.3.1, A11.3.2, A11.3.3, and 11.21a the following are obtained.

$$E[q_1(1-q_1)] = E(q_1) - E(q_1^2) = \bar{q}_1 - \bar{q}_1^2 - \sigma^2(\bar{d}_0) = \bar{q}_1(1-\bar{q}_1)(1-1/2N_e) \qquad \text{A11.3.5}$$

$$E[q_1^2(1-q_1)] = E(q_1^2) - E(q_1^3) = \bar{q}_1^2 + \sigma^2(\bar{d}_0) - \bar{q}_1^3 - 3\bar{q}_1\sigma^2(\bar{d}_0)$$

$$= \bar{q}_1^2(1-\bar{q}_1) + [1-3\bar{q}_1]\bar{q}_1(1-\bar{q}_1)/2N_e \qquad \text{A11.3.6}$$

$$E[q_1^2(1-q_1)^2] = E(q_1^2) - 2E(q_1^3) + E(q_1^4)$$

$$= \bar{q}_1^2[1-\bar{q}_1]^2 + [1-6\bar{q}_1+6\bar{q}_1^2]\bar{q}_1(1-\bar{q}_1)/2N_e$$

$$+3\bar{q}_1^2(1-\bar{q}_1)^2/(2N_e)^2 \qquad \text{A11.3.7}$$

- A11.3.2. **Expected Change in Frequency of the B Allele**

The content of Table 8.2 reminds us that $E(\Delta q) = E(\Delta q/q)$ varies with the nature of criterion units employed in the selection program. However, with respect to the selection systems considered, only two functions of q are involved.

- A11.3.2.1. WHEN SELECTION UNITS ARE S POPULATION INDIVIDUALS OR FAMILIES COMPOSED OF NON-INBRED S POPULATION INDIVIDUALS

Among such systems the quantity, c, is variable but is not a function of q. And among selection programs of the same sort both k, the standardized selection differential, and $\sigma(\hat{X})$ vary from program to program and cycle to cycle. However k is not a function of q and $\sigma(\hat{X})$ is so little affected by variation in allele frequency at a single locus that for the present purpose it can be considered independent of q. In these cases, assuming linkage equilibrium (see 5.24 and Table 8.2),

$$E(\Delta q/q) = q(1-q)[1+(1-2q)a]uck/\sigma(\hat{X})\backslash=[q(1-q)(1+a) - q^2(1-q)2a]ucK \qquad \text{A11.3.8}$$

in which

$$K = k/\sigma(X_u)$$

and c = the coefficient shown in Table 8.2 that is variable among selection systems.

From A11.3.8 it follows that

$$E(\Delta q/q{=}q_0)_1 = \{(1+a)Eq(1-q) - 2aEq^2(1-q)\}ucK.$$

This, by A11.3.5 and A11.3.6 followed by some rearrangement, yields

$$E(\Delta q/q{=}q_0)_1 = \bar{q}_1(1-\bar{q}_1)\{[1+(1-2\bar{q}_1)a][1-1/2N_e]-(1-2\bar{q}_1)a/N_e\}ucK \qquad \text{A11.3.9}$$

- A11.3.2.2. WHEN SELECTION IS AMONG S_n FAMILIES ON THE BASIS OF THEIR OWN PERFORMANCE

In these cases

$$E(\Delta q/q) = q(1-q)[(1+h) + (1-h)(1-2q)a/2]uK$$

from which

$$E(\Delta q/q{=}q_0)_1 = \{[(1+h) + (1-h)a/2]Eq(1-q) - (1-h)aEq^2(1-q)\}uK.$$

Again, by A11.3.5, A11.3.6 and rearrangement, this yields

$$E(\Delta q/q{=}q_0)_1 = \bar{q}_1(1-\bar{q}_1)\{[(1+h)+(1-h)(1-2\bar{q}_1)a/2][1-1/2N_e]$$

$$- [(1-h)(1-2\bar{q}_1)a/2]/N_e\}uK \qquad \text{A11.3.10}$$

- A11.3.4. **Expected Change in the Average Value of Single-Locus Genotypes when the T Population is the Cross between Two S Populations**

The S population is, by definition, always composed of individuals resulting from random matings. For that reason and because of infinite size

$$\ddot{y} = 2qu + 2q(1-q)au \qquad 5.10$$

It follows that

$$(\Delta\ddot{y})_1 = \ddot{y}_2 - \ddot{y}_1 = 2(q_2 - q_1)u + 2[q_2(1-q_2) - q_1(1-q_1)]au$$
$$= 2(q_2 - q_1)u + 2(q_2 - q_1)(1-2q_1)au - 2(q_2 - q_1)^2 au \qquad \text{A11.3.11}$$

Remembering now that

$$(q_2 - q_1) = (q_{1s} - q_1) = E(\Delta q/q{=}q_1) + \bar{d}_1$$

let $E(\Delta q/q{=}q_1) = z_1$ so that $(q_2 - q_1) = z_1 + \bar{d}_1$. Then by substitution in A11.3.11 followed by rearrangement,

$$(\Delta\ddot{y})_1 = 2z_1[1+(1-2q_1)a]u + 2\bar{d}_1[1+(1-2q_1)a]u$$
$$-2z_1^2 au - 4z_1\bar{d}_1 au - 2\bar{d}_1^2 au. \qquad \text{A11.3.12}$$

Because

$$2[1+(1-2q)a]u = \partial\ddot{y}/\partial q \qquad 5.11$$

when T = S, the first term on the right side of A11.3.12 is $E(\Delta q/q)[\partial y/\partial q]$. In previous chapters the approximation

$$E(\Delta\ddot{y}) \approx E(\Delta q/q)[\partial y/\partial q]$$

has been employed, i.e., the term $2z^2au = 2[E(\Delta q/q)]^2au$ and the impact of finite population size (terms involving $\bar{d}$) have been ignored.

As the basis for writing $(\Delta\ddot{y})_1$ the expectations of the separate terms of the right-hand member of A11.3.12 are required. Remember as background that d has been defined (section 5.5) as a random deviation from the regression (assumed to be linear) of q on X and that the variation of d_1 is random relative to q_1. Consequences are that[2]

$$E(\bar{d}_1) = E(d_1) = 0.0$$

$$\sigma(d_1,q_1) = \sigma(\bar{d}_1,q_1) = 0.0 \qquad \text{(by A11.3.4)}$$

and $\sigma(d_1,z_1) = 0.0$

It follows that

$$E(\bar{d}_1)[1+(1-2q_1)a]u = uE(\bar{d}_1) + auE(\bar{d}_1) - 2auE(\bar{d}_1,q_1) = 0.0$$

and, from A11.3.12, that

$$E(\Delta\ddot{y})_1 = 2Ez_1[1+(1-2q_1)a]u - 2E(\bar{d}_1^2)au \qquad \text{A11.3.13}$$

Because $2z_1^2au$ must ordinarily be trivial in magnitude (see section 7.1) it is ignored in A11.3.13 as it has been earlier. Finally, by 11.20 and A11.3.1

$$E(\bar{d}_1^2)au = [E(\bar{d}_1^2) + \sigma^2(\bar{d}_1)]au \approx \bar{q}_1(1-\bar{q}_1)au/2N_e$$

in which the indicated approximation reflects the equating of $\bar{q}_1$ with $q_1 + E(\Delta q/q)_1$. Substitution in All.3.13 yields

$$E(\Delta\ddot{y})_1 = 2Ez_1[1+(1-2q_1)a]u - 2\bar{q}_1(1-\bar{q}_1)au/2N_e \qquad \text{A11.3.14}$$

The expectation of $(\Delta\ddot{y})_0$ is obtained by substituting z_0 for z_1 and q_0 for $\bar{q}_1$ in A11.3.14. Then because (a) $z_0 = E(\Delta q/q{=}q_0)_0$, (b) $2[1+(1-2q)a]u = \partial\ddot{y}/\partial q$, and (c) q_0 is a constant once the S population of cycle 0 has been chosen,

$$E(\Delta\ddot{y})_0 \approx E(\Delta q/q=q_0)_0[\partial\ddot{y}/\partial q] + q_0(1-q_0)au/N_e. \qquad A11.3.15$$

in which q_0 is employed in place of $q_0+E(\Delta q/q=q_0)_0$ in the final term. Note also that the first term of the right-hand member of A11.3.15 is the basis of all expressions for $E(\Delta\ddot{Y})$ shown in preceding chapters -- expressions that did not take account of effective population size.

- A11.3.3.1. WHEN SELECTION UNITS ARE NON-INBRED INDIVIDUALS OR FAMILIES OF NON-INBRED INDIVIDUALS

The portion of $E(\Delta q/q)$ that is a function of q is the same for all selection systems in this class. Variation or potential variation occurs only in c, k, or $\sigma(\hat{X})$. Thus, in all cases

$$z = E(\Delta q/q)_1 = q_1(1-q_1)[1+(1-2q_1)a]ucK \qquad A11.3.16$$

Substitution in A11.3.14 yields

$$E(\Delta\ddot{y})_1 = 2Eq_1(1-q_1)[1+(1-2q_1)a]^2u^2cK - 2\bar{q}_1(1-\bar{q}_1)au/2N_e$$
$$= 2u^2\{Eq_1(1-q_1)(1+2a+a^2) - 4aE(q_1^2)(1-q_1) - 4a^2Eq^2(1-q)^2\}cK$$
$$- 2\bar{q}_1(1-\bar{q}_1)au/2N_e$$

Substitutions in terms of A11.3.5, A11.3.6, and A11.3.7, followed by rearrangements yields

$$E(\Delta\ddot{y})_1 = 2\bar{q}_1(1-\bar{q}_1)\{[1+(1-2\bar{q}_1)a]^2(1-1/2N_e) - Z/2N_e\}u^2cK$$
$$- \bar{q}_1(1-\bar{q}_1)au/N_e \qquad A11.3.17$$

in which

$$Z = 4a[(1-2\bar{q}_1) + a - \bar{q}_1(1-\bar{q}_1)(5-3/2N_e)a]$$
$$\approx 4a\{(1-2\bar{q}_1) + [1 - 5\bar{q}_1(1-\bar{q}_1)]a\}$$

because $(5-3/2N_e)$ is, for practical purposes, equal to 5.0 unless N_e is very small.

- A11.3.3.2. WHEN SELECTION IS AMONG S_n FAMILIES ON THE BASIS OF THEIR OWN PERFORMANCE

For this system

$$z = E(\Delta q/q) = q(1-q)[(1+h) + (1-h)(1-2q)a/2]K \qquad A11.3.18$$

where

h = the inbreeding of the parents of the S_n families.

Substituting in A11.3.14, expanding, taking expectations of terms in q, and rearranging, the following is obtained.

$$E(\Delta\ddot{y})_1 = 2\,\bar{q}_1(1-\bar{q}_1)\{[(1+h) + (1-h)(1-2\bar{q}_1)a/2][1+(1-2\bar{q}_1)a](1-1/2N_e)$$
$$- Z/2N_e\}u^2K - \bar{q}_1(1-\bar{q}_1)au/N_e \qquad A11.3.19$$

in which

$$Z = a\{(3+h)(1-2\bar{q}_1) + 2(1-h)[1-\bar{q}_1(1-\bar{q}_1)(5 - 3/2N_e)]a\}$$
$$\approx a\{(3+h)(1-2\bar{q}_1) + 2(1-h)[1-5\bar{q}_1(1-\bar{q}_1)]a\}$$

because $(5 - 3/N_e) \approx 5$ unless N_e is very small.

- A11.3.4. **Expected Change in the Average Value of Single-Locus Genotypes when the T Population is the Cross between Two S Populations**

Let p be the frequency of B in the S_1 population and q be the frequency of B in the S_2 population. Then

$$\ddot{y} = (p+q)u + (p+q - 2pq)au \qquad 5.13$$

from which, because $p_2 = p_1 + (\Delta p)_1$ and $q_2 = q_1 + (\Delta q)_1$,

$$(\Delta\ddot{y})_1 = (\Delta p)_1[1+(1-2q_1)a]u + (\Delta q)_1[1+(1-2p_1)a]u - 2(\Delta p)_1(\Delta q)_1 au \qquad A11.3.20$$

Consider the second term of the right-hand member of A11.3.20 and let $z_q = E(\Delta q/q=q_1)$ and $(\Delta q)_1 = q_2 - q_1 = z_q + \bar{d}_1$ as before. Then

$$E(\Delta q)_1[1+(1-2p_1)a]u = Ez_q[1+(1-2p_1)a]u + E(\bar{d}_1)[1+(1-2p_1)a]u$$
$$= Ez_q[1+(1-2p_1)a]u \qquad A11.3.21$$

because (1) $E(\bar{d}_1) = 0.0$ and (2) the deviations, d_1, around $q_1 + E(\Delta q/q)_1$ in S_2 are random relative to the frequency (p_1) of B in S_1 and hence

$$E(\bar{d}_1)[1+(1-2p_1)a]u = 0.0$$

Because the structures of the first first two terms of the right hand member of A11.3.20 are the same (they differ only because, in each, p and q take the places of q and p in the other), it is intuitively obvious that

$$E(\Delta p)_1[1+(1-2q_1)a]u = Ez_p[1+(1-2q_1)a]u \qquad A11.3.22$$

in which $z_p = E(\Delta p/p=p_1)$.

Finally, from A11.3.20, A11.3.21, A11.3.22 it is clear that

$$E(\Delta\ddot{y})_1 = Ez_q[1+(1-2p_1)a]u + Ez_p[1+(1-2q_1)a]u - 2E(\Delta p)_1(\Delta q)_1 au$$

Ignoring the final term, because $2(\Delta p)_1(\Delta q)_1 au$ can never be more than a trivial fraction of $\Delta\ddot{y}$,

$$E(\Delta\ddot{y})_1 \approx Ez_q[1+(1-2p_1)a]u + Ez_p[1+(1-2q_1)a]u \qquad A11.3.23$$

in which

$$z_q = E(\Delta q)_1 \text{ and } z_p = E(\Delta p)_1.$$

Note the absence in this expression of terms similar to the $2\bar{q}_1(1-\bar{q}_1)/2N_e$ that is a part of A11.3.14. This reflects the fact that the random components of allele frequency changes in S_1 and S_2 are independent.

- A11.3.4.1. WHEN SELECTION UNITS ARE NON-INBRED INDIVIDUALS OR FAMILIES OF NON-INBRED INDIVIDUALS

Given that p and q symbolize the frequencies of B in S_1 and S_2, respectively, the required z's for insertion in A11.3.23 are, in accord with A11.3.16,

$$z_p = E(\Delta p) = p(1-p)[1+(1-2p)a]uc_1K_1 \qquad \text{in which } K_1 = k_1/\sigma(X_u)_1$$

and $$z_q = E(\Delta q) = q(1-q)[1+(1-2q)a]u\,c_2K_2 \qquad \text{in which } K_2 = k_2/\sigma(X_u)_2$$

The subscripts to the c's and to the K's and their components reflect the possibility of different selection systems in S_1 and S_2. Then by substitution in A11.3.23,

$$E(\Delta\ddot{y})_1 = Ep_1(1-p_1)[1+(1-2p_1)a][1+(1-2q_1)a]u^2c_1K_1$$
$$+ Eq_1(1-q_1)[1+(1-2q_1)a][1+(1-2p_1)a]u^2\,c_2K_2$$

from which

$$E(\Delta\ddot{y})_1 = \bar{p}_1(1-\bar{p}_1)[1+(1-2q)a]\{[1+(1-2\bar{p}_1)a](1-1/2N_{e1}) - (1-2\bar{p}_1)a/N_{e1}\}u^2c_1K_1 + \bar{q}_1(1-\bar{q}_1)[1+(1-2p)a]\{[1+(1-2\bar{q}_1)a](1-1/2N_{e2}) - (1-2\bar{q}_1)a/N_{e2}\}u^2c_2K_2 \quad \text{A11.3.24}$$

Note that this section deals with the system, RSP, considered in section 9.4 and that A11.3.24 (when summed over all loci) converges, as it should, to 9.24 as N_e becomes large (and the compositions of K_1 and K_2 are taken into account). Note also that when gene effects are additive ($a=0.0$) A11.3.24 reduces to

$$E(\Delta\ddot{y})_1 = \bar{p}_1(1-\bar{p}_1)u^2c_1K_1(1-1/2N_{e1}) + \bar{q}_1(1-\bar{q}_1)u^2c_2K_2(1-1/2N_{e2})$$

- A11.3.4.2. **WHEN SELECTION IS AMONG S_n FAMILIES ON THE BASIS OF THEIR OWN PERFORMANCE**

For this system, in accord with A11.3.18

$$z = E(\Delta q/q) = q(1-q)[(1+h) + (1-h)(1-2q)a/2]K$$

where h = the inbreeding of the parents of the S_n families.

Substituting in A11.3.23, expanding, taking expectations of terms in q, and rearranging, the following is obtained.

$$E(\Delta\ddot{y})_1 = \bar{q}_1(1-\bar{q}_1)[1+(1-2p_1)a]\{(1+h)+(1-h)(1-2q_1)a/2](1-1/2N_{e2}) - [(1-h)(1-2\bar{q}_1)a/2]/N_{e2}\}u^2K_2 + \bar{p}_1(1-\bar{p}_1)[1+(1-2q_1)a]\{(1+h)+(1-h)(1-2\bar{p}_1)a/2](1-1/2N_{e1}) - [(1-h)(1-2\bar{p}_1)a/2]/2N_{e1}\}u^2K_1 \quad \text{A11.3.25}$$

- A11.3.4.3. **WHEN $T = S_1 \times S_2$ AND SELECTION IS BASED ON PERFORMANCE OF TEST CROSS PROGENIES**

Given that v_1 and v_2 symbolize the frequencies of B in the testers for selection in S_1 and S_2, respectively, and again that p and q symbolize frequencies of B in the two S populations, the required z's are

$$z_p = p(1-p)[1+(1-2v_1)a]uc_1K_1$$

and $$z_q = q(1-q)[1+(1-2v_2)a]uc_2K_2$$

Substitution in A11.3.23 yields

$$E(\Delta\ddot{y})_1 = Ep(1-p)[1+(1-2v_1)a][1+(1-2q)a]u^2c_1K_1 + Eq(1-q)[1+(1-2v_2)a][1+(1+2p)a]u^2c_2K_2$$
$$= \bar{p}_1(1-\bar{p}_1)[1+(1-2v_1)a][1+(1-2q)a](1-1/2N_{e1})u^2c_1K_1 + \bar{q}_1(1-\bar{q}_1)[1+(1-2v_2)a][1+(1-2p)a](1-1/2N_{e2})u^2c_2K_2 \quad \text{A11.3.26}$$

As in A11.3.24 the numerical subscripts recognize that the testers and elements of the K's may be different for S_1 and S_2. This general expression takes more specific form when the selection system is specified in greater detail.

In RSSC, if S_2 is visualized as the S population and S_1 as the tester, $K_1 = 0.0$, $K_2 = K = k(1+f)/\sigma(\ddot{X})$, $N_{e2} = N_e$, and $p = v_2 = v$. Making these substitutions in A11.3.26 and summing over all loci yields the following that converges, as it should, to 9.6 as N_e becomes large and $\bar{q}_1$ is replaced by q.

$$E(\Delta\ddot{y})_1 = \bar{q}_1(1-\bar{q}_1)[1+(1-2v)a]^2u^2cK(1-1/2N_e)$$

- A11.3.5. **Reciprocal Recurrent Selection**

In RRS, $v_2 = p$ (= frequency of B in S_1), $v_1 = q$ (= frequency of B in S_2), $K_1 = k_1(1+f_1)/\sigma(\hat{X})_1$, $K_2 = k_2(1+f_2)/\sigma(\hat{X})_2$, $z_p = p(1-p)[1+(1-2q)a]uc_1K_1$ and $z_q = q(1-q)[1+(1-2p)a]uc_2K_2$. Making these substitutions in A11.3.23 yields

$$\begin{aligned} E(\Delta\ddot{y})_1 &= Eq_1(1-q_1)[1+(1-2p_1)a][1+(1-2p_1)a]u^2c_2K_2 \\ &\quad +Ep_1(1-p_1)[1+(1-2q_1)a][1+(1-2q_1)a]u^2c_1K_1 \\ &= \bar{q}_1(1-\bar{q}_1)\{[1+(1-2\bar{p}_1)a]^2 + 2\bar{p}_1(1-\bar{p}_1)a^2/N_{e2}\}(1-1/2N_{e2})u^2c_2K_2 \\ &\quad + \bar{p}_1(1-\bar{p}_1)\{[1+(1-2\bar{q}_1)a]^2 \\ &\quad + 2\bar{q}_1(1-\bar{q}_1)a^2/N_{e1}\}(1-1/2N_{e1})u^2c_1K_1 \end{aligned} \quad \text{A11.3.27}$$

In A11.3.27 the numerical subscripts identify K, c, and N_e with the S_1 or S_2 population, respectively. On the other hand, as subscripts to p or q they identify allele frequencies with cycles of the selection program.

When A11.3.27 is summed over all loci and subscripts to the p's and q's are ignored, it converges as it should, to 9.13 as the N_e's becomes large. The partial analogy between A11.3.26 and A11.3.27 is not surprising since both relate to selection on the basis of test-cross progenies. The terms in A11.3.27 for which there are no counterparts in A11.3.26 are present because the "testers" for S_1 and S_2 are the opposite members of the pair of S populations.

• Appendix 12.1. Selection Criterion Weighting Coefficients That Maximize Genetic Improvement per Cycle

It is shown in section 12.1 (see eq. 12.5) that when the selection criterion is $\hat{X} = \Sigma_m b_m \hat{X}_m$ (with m = 1,2,3, ····n) the expectation of $\Delta\ddot{Y}_w$ per cycle of selection is[1]

$$E(\Delta\ddot{Y}_w) = [k/\sigma(\hat{X})]\,[b_1G_{1w} + b_2G_{2w} \cdots\cdots b_nG_{nw}]$$

where all symbols are employed as in Chapter XII. Standard calculus procedure is employed to find the b-values that maximize $E(\Delta\ddot{Y}_w)$. Its partial derivative is obtained with respect to each of the b's. Setting each partial derivative equal to zero yields n equations which, solved simulataneously, provide the required set of weighting coefficients . Consider the partial derivative with respect to b_1.

$$\partial E(\Delta\ddot{Y}_w)/\partial b_1 = [k/\sigma^2(\hat{X})]\{\sigma(\hat{X})G_{1w} - (b_1G_{1w} + \cdots\cdots b_nG_{nw})[\partial\sigma(\hat{X})/\partial b_1]\} \quad \text{A12.1.1}$$

Now note that

$$\begin{aligned} \partial\sigma(\hat{X})/\partial b_1 &= \partial[\sigma^2(\hat{X})]^{1/2}/\partial b_1 = (1/2)[\partial\sigma^2(\hat{X})/\partial b_1][\sigma^2(\hat{X})]^{-1/2} \\ &= [\partial\sigma^2(\hat{X})/\partial b_1]/2\sigma(\hat{X}) \end{aligned}$$

and that since $\hat{X} = \Sigma_m b_m \hat{X}_m$

$$\sigma^2(\hat{X}) = \Sigma_m(b_m)^2\sigma_{mm} + \Sigma_m\Sigma_{m'\neq m}b_mb_{m'}[\sigma_{mm'}] \quad \text{A12.1.2}$$

Remembering that the range of m is from 1 to n and (from section 12.2) that

σ_{mm} = the variance of $\hat{X}_m$ and

$\sigma_{mm'}$ = the covariance of $\hat{X}_m$ with $\hat{X}_{m'}$

it is apparent (from A12.1.2) that the variance of $\hat{X}$ can be written as

$$\sigma^2(\hat{X}) = (b_1)^2\sigma_{11} + 2b_1[b_2\sigma_{12} + b_3\sigma_{13} + \cdots b_n\sigma_{1n}] + R$$

where R is the sum of all terms in $\sigma^2(\hat{X})$ that do not involve b_1 and, hence that

$$\partial\sigma(\hat{X})/\partial b_1 = [b_1\sigma_{11} + b_2\sigma_{12} + \cdots\cdots b_n\sigma_{1n}]/\sigma(\hat{X})$$

Substituting this value of $\partial\sigma(\hat{X})/\partial b_1$ in A12.1.1 and setting the derived value of $\partial E(\Delta\ddot{Y}_w)/\partial b_1$ equal to zero, we obtain

$$b_1\sigma_{11} + b_2\sigma_{12} + \cdots\cdots b_n\sigma_{1n} = G_{1w}K \qquad \text{A12.1.3}$$

where

$$K = \sigma^2(\hat{X})/(b_1G_{1w} + b_2G_{2w} + \cdots\cdots b_nG_{nw}).$$

The expressions obtained when the derivatives of $E(\Delta\ddot{Y}_w)$ with respect to b_2, b_3, etc. are equated to zero are all analogous to A12.1.3. In general, $\partial E(\Delta\ddot{Y}_w)/\partial b_m = 0.0$ yields

$$b_m\sigma_{mm} + \Sigma_{m'\neq m}[b_{m'}\sigma_{mm'}] = G_{mw}K$$

For example, when n = 3, i.e., when the selection criterion is a linear function of three measures of phenotype, the set of equations is

$$\begin{aligned} b_1\sigma_{11} + b_2\sigma_{12} + b_3\sigma_{13} &= G_{1w}K \\ b_1\sigma_{12} + b_2\sigma_{22} + b_3\sigma_{23} &= G_{2w}K \qquad \text{A12.1.4} \\ b_1\sigma_{13} + b_2\sigma_{23} + b_3\sigma_{33} &= G_{3w}K \end{aligned}$$

where

$$K = \sigma^2(\hat{X})/(b_1G_{1w} + b_2G_{2w} + b_3G_{3w}).$$

Because K is a non-linear function of the whole set of b's, equations A12.1.4 or similar ones are non-linear, which suggests that they can only be solved by iteration. In fact, however, K can be assigned any arbitrary numerical value. That is because only the relative values of the b's affect the rank order of the $\hat{X}$ values of any series of criterion units. For that reason, there is an infinity of b-value sets (that vary with respect to absolute but not relative values of the b's) that maximize $E(\Delta\ddot{Y}_w)$. When K is given a specific numerical value, the equations become linear and their simultaneous solution yields a specific set of optimum b-values. It is convenient, and the general practice, to set K = 1.0. Then the b's provided by the equations would be such that

$$\sigma^2(\hat{X}) = \Sigma_m b_m G_{mw}$$

if the true values, rather than estimates, of the σ_{mm}, $\sigma_{mm'}$, and G_{mw} parameters were available for use.

Finally, the foregoing is written as if the procedure (from calculus) that is described and employed always maximizes the function investigated. Of course, depending on the function, it might minimize instead. In this instance, however, it is clear that it maximizes because it provides a set of b-values that makes $E(\Delta\ddot{Y}_w) > 0.0$. Such a value must be a maximum because other b-values that would make $E(\Delta\ddot{Y}_w)$ negative are obviously possible, e.g., those obtained by reversing the algebraic signs of the b's provided by the equations.

• Appendix 12.2. Total Value Discussed in Terms of Maize and Swine

Maize: In the U.S., maize is utiliized in two principal ways. Most frequently, the grain is allowed to mature before being harvested for use, primarily as feed for animals though a smaller portion is processed for human consumption (corn meal, corn oil, corn syrup, etc.). The remainder of the plant is left in the field to worked into the soil. Alternatively, the whole plant (excep roots and stubble) is harvested, while green and before the grain is mature, to be fed to animals in the form of silage.

Consider the first practice, in which the focus is on the grain. In this case the grain (cob-free) can justifiably be viewed as the product and, as a first approximation, grain yield can be viewed as value of the product.[1] Questions arise, however, concerning (1) how yield should be measured and (2) attributes of the grain that may affect its value.

What about the grain from ears that would be missed by a good mechanical picker (ears on or near the ground as the result of plant weaknesses)? When plots are hand-harvested, it is tempting to gather and include the grain from them. However, such ears do not contribute to the yield obtained when the crop is harvested using mechanical equipment, as now common in developed countries. "Harvestable yield" (the grain obtained by equipment of the type employed in general agriculture) is the marketable yield and has the desirable attribute that genotypes susceptible to lodging or ear droppage are penalized in proportion to the practical effects of those weaknesses.

What about variation in the composition of harvested grain (moisture, protein, oil, amino acids)? Because the purpose here is to illustrate principles, not to develop a refined definition of total value, only moisture and protein content will be discussed further. Other things being equal, feeding value is proportional to dry matter content of the grain. Hence yield adjusted to a standard moisture content is a better measure of feeding value than is gross "harvestable yield". Because large amounts of maize are fed to non-ruminant animals its feeding value for average purposes is doubtless enhanced by higher protein content. If so, product value would be measured more precisely with protein content taken into account. These arguments suggest the following as a reasonable definition.

$$\text{Product value} = KX_1 = [d_1 + d_2(X_2 - \ddot{X}_2]X_1 \qquad \text{A12.2.1}$$

where X_1 = "harvestable yield" per unit of land area adjusted to a standard moisture content (the moisture content used as the standard for X_1 could be any but, for convenience, let it be the level required to insure against spoilage of grain in storage),

X_2 = protein content in percent,

$\ddot{X}_2$ = average protein content,

K = value of grain per unit of weight,

d_2 = increase in value (per unit of weight) per one percent increase in protein, and

d_1 = value of grain per unit of weight when protein content is equal to its average.

Some production costs are the same for all genotypes, others may not be. In this case (maize for grain) it appears reasonable to assume that costs per hectare for such items as land, labor, fertilizer and herbicides are the same for all genotypes. Certainly, in the field comparisons of a selection program, all criterion units would be managed in the same way. Hence most production costs would be the same for all units. The costs that are affected by genotype are the ones that affect value and should, therefore, be appropriately recognized in the definition of total value. An example is the cost of reducing the moisture in grain to the level required for safe storage. Late maturing genotypes tend to have higher yield potentials but their grain will ordinarily have higher moisture contents at standard harvesting dates. Let

d_0 = the total of fixed production costs,

d_3 = the cost of moisture removal per unit of moisture,

X_3 = the percent moisture in harvested grain, and

X_{3a} = the moisture level to which "harvestable yield" is adjusted to obtain X_1.

Then, assuming $X_3 > X_{3a}$, the moisture that must be removed to achieve a safe level for storage is $X_1(X_3 - X_{3a})/(1 - X_3)$ and where cost for grain drying is included

$$\text{Cost of production} = d_0 + d_3X_1(X_3 - X_{3a})/(1 - X_3) \qquad \text{A12.2.2}$$

On the basis of A12.2.1 and A12.2.2 total value would be defined as

$$W = (\text{Product value}) - (\text{Cost of production})$$
$$= [d_1 + d_2(X_2 - \ddot{X}_2) - d_3(X_3 - X_{3a})/(1 - X_3)]\,X_1 - d_0 \qquad \text{A12.2.3}$$

Whatever its flaws, A12.2.3 serves to illustrate what the words, "definition of total value", have been intended to convey. Amount of grain produced, moisture content, and protein content are identified as "value" traits; disease resistances, prolificacy (ears per plant), earliness, lodging resistance, etc. are not. These latter traits are all important; appropriate changes in them achieved by selection will increase the productivity of maize, but measurements of them are not required for quantification of total value. <u>Their effects on total value are inherent in the measures of "value" traits.</u>

Obviously, proper definition of total value hinges on the uses contemplated for the products of the plant. For example, in the case of maize for use as silage, moisture content of the grain at harvest would not affect value as it does when storage of shelled grain in bins is a part of the use process. Just as obviously, if the object of a breeding program were development of a high-oil variety for the production of grain for the oil-processing industry, oil content of the harvested grain would properly appear in the definition of total value.

<u>Swine</u>: Because almost all production costs could be assumed proportional to land area, it was advantageous to define total value of maize in terms of the grain produced on a unit of land. In the case of swine, there is no unit for which costs are constant and none for which value is constant. Lacking such a unit, total value will be discussed in terms of animals assumed to weigh 100 kilgrams, the approximate live weight at which the majority of swine are marketed and slaughtered.

Assuming constant weight at slaughter, the value of the principal product (pork) per animal depends on carcass weight (dressing percent) and value per unit of carcass weight. The latter depends primarily on lean tissue as a fraction of the total and, to a lesser extent, on the quality of the lean tissue and its distribution among cuts (ham, loin, shoulder, etc.).

Production costs divide rather naturally between those encountered before and after weaning. The principal post-weaning costs that are variable among genotypes are those for feed, housing, equipment and labor. Of these, the last three are roughly in inverse proportion to growth rate, at least in intensive swine production operations. Pre-weaning costs consist largely of the overhead costs for parent animals. These reflect primarily, the numbers of offspring and maintainence costs per female parent. The first of these depends on conceptions per female and litter size, maintainence cost is inversely proportional to body size. Reduced conception rate may, of course, reflect a deficiency in numbers or quality of sperm in the semen of the male or a defect in the reproductive physiology of the female.

In the context of large scale intensive swine production, let

X_1 = number of offspring weaned per female per month,[2]

X_2 = average body weight of brood sows,

X_3 = feed required per unit of body weight increase during growth to 100 kilograms,

X_4 = days to reach 100 kilograms of body weight,

X_5 = lean tissue as a fraction of carcass weight,

X_6 = dressing percent = (carcass weight)/(slaughter weight),

d_0 = the average cost per month of brood sows that is independent of weight and assumed independent of genotype,

d_1 = value per kilogram of the whole carcass when X_5 is equal to its mean value ($\ddot{X}_5$) and

d_2 to d_5 be numerical coefficients of which meanings are indicated below.

Then cost of production, product value and total value can be approximated as follows.

Cost of production of a 100 kg. market hog

$$= [d_2(X_2)^{0.75} + d_0]/X_1 + d_3X_3 + d_4X_4 \qquad \text{A12.2.4}$$

Value of carcass from a 100 kg. market hog

$$= [d_1 + d_5(X_5 - \ddot{X}_5]100X_6 \qquad \text{A12.2.5}$$

W = product value - production cost

$$=[d_1 + d_5(X_5 - \ddot{X}_5)]100X_6 - [d_2X_2)^{0.75} + d_0]/x_1 - d_3X_3 - d_4X_4 \qquad \text{A12.2.6}$$

The quantity, $(X_2)^{0.75}$, is employed because the total digestible nutrient requirement for weight maintainence of animals is known to be proportional to a quantity closely approximated by (body weight)$^{0.75}$ and it appears reasonable to assume space requirements proportional to that same quantity. The quantities symbolized by d_2 through d_5 should be apparent from the way they enter the above expressions. For example, it is obvious that d_3 = the cost of a unit of feed for pre-market pigs and that d_4 = cost per day per pig, exclusive of cost for feed.

As in the case of maize, there are traits that affect total value that do not appear in its definition, A12.2.6. Both semen quality of males and number of functional nipples of females may affect number of weanlings produced but both are appropriately taken into account by X_1, the number of offspring per brood sow per month.

Comments; The proposals embodied in A12.2.3 and A12.2.6 are offered (1) for illustrative purposes and (2) as first approximations to work from in the development of completely satisfying definitions. It would be presumptious of this author to venture beyond statements concerning the principles involved in definition of total value.

The examples offered demonstrate that total value, appropriately defined, will ordinarily be a function of more than one variable affected by genotype and a number of coefficients for which numerical values are required, e.g., d_1 through d_5 in A12.2.6. Provision of those values is an exercise in economics (not genetics) that calls for knowledge of the organism, the production process and the relations between attributes and value of the product or products. Thus the breeder must be an economist of sorts as well as a geneticist or must have effective communication with one or more economists willing and able to provide the help needed.

Unfortunately, the proper numerical values for coefficients in total value definitions cannot be expected to be the same 10 or 20 years from now as they are now. The cost (d_3) of moisture removal in the case of maize depends on the cost of fuel for dryers which is obviously not static and the premium (d_2) for increased protein content depends on costs of protein supplements such as soybean meal. Thus forecasting is called for because the most appropriate values for use during the course of a recurrent selection program are those that will apply when the program has reached the pay-off stage and its product is being used in commercial agriculture.

Other important issues encountered in the case of swine are (1) that in dioecious organisms some of the traits relevant to total value are expressed in only one of the sexes and (2) the age at which data collection is possible varies among value traits (contrast, for example, growth rate to market weight and litter size). The result, to be discussed more explicitly in Chapter XIV, is that depending on the criterion unit, or combination of criterion units, employed in a selection program, it may not be possible to include all value traits in the selection criterion.

• Appendix 14.1. Allele Frequency (q) and the Magnitude of q(1−q)

The facts to be recognized are (1) that allele frequency responses to one cycle of selection are small unless the effect (u) of the allele is unusually large and (2) that $q(1-q)$ changes much less than q itself. Let

$\Delta = E(\Delta q)$, the expectation of one cycle change in allele frequency,

q = the frequency of an allele in the foundation population of a selection program,

$\bar{q}_1 = q + \Delta$ = the expected frequency of the allele after the initial cycle of selection.

Then $D = \bar{q}_1(1-\bar{q}_1) - q(1-q) = (q+\Delta)(1-q-\Delta) - q(1-q)$

$$= \Delta(1-2q) - \Delta^2 = \Delta(1-2q-\Delta) \qquad \text{A14.1.1}$$

To assess how large this quantity can be and how it may be changed by greater selection intensity we obviously need information concerning possible magnitudes of Δ. The size of $E(\Delta q)$ in the case of mass selection is discussed in section 6.3.3 and values as functions of q and a (= level of dominance) assuming k=1 and $u/\sigma(\hat{X}) = 0.1$ are shown in Table 6.3. In the special situation ($a=0.0$) considered in section 14.3.1.1, the largest value of Δ is 0.025 when $q = 0.5$. Substitution in A14.1.1 yields $D = -\Delta^2 = -0.000625$, a trivial change in $q(1-q)$.

However, exploration of values of $\Delta = E(\Delta q)$ that are possible in the case of selection systems that might be employed in T = S and RSP programs is required.

$$E(\Delta q_i) = \sigma(\hat{q}_i, \hat{X})k/\sigma(\hat{X}) \qquad 5.18$$

Substituting for $\sigma(\hat{q}_i, \hat{X})$ in terms of expressions shown in Table 8.2 in the event of linkage equilibrium in the selection population yields

$$E(\Delta q) = q(1-q)[1+(1-2q)a]uck/\sigma(\hat{X})$$

when the selection and criterion units are both composed of non-inbred S population individuals and

$$E(\Delta q) = q(1-q)[(1+h)+(1-h)a/2]uk/\sigma(\hat{X})$$

when selection is among S_n families on the basis of their own performance. These expressions reduce to

$$E(\Delta q) = q(1-q)uck/\sigma(\hat{X}) \qquad A14.1.2$$

and

$$E(\Delta q) = q(1-q)(1+h)uk/\sigma(\hat{X}) \qquad A14.1.3$$

when $a = 0.0$. In these expressions

c = a constant (shown in Table 8.2) that varies among selection systems, and

h = the inbreeding coefficient of the parents of S_n families.

Because (1) c< 1.0 except in the case of mass selection and (2) $\sigma(\hat{X})$ is larger in mass selection than in any variation of family selection, it is clear from A14.1.2 and A14.1.3 that $E(\Delta q)$ is larger when selection is among S_n families than when other systems are employed. Consider selection among pure lines so that h =1.0. Assuming u = 0.5, k = 2.06 or 1.76 (on the basis of selection intensities of 0.05 or 0.10), $\sigma(\hat{X}) = 11.04$ in accord with the value shown in Table14.13, and q_0 =0.2, 0.5, or 0.8 we obtain the following answers from A14.1.3.

	$E(\Delta q)$	
q_0	k = 2.06	k'= 1.76
0.2	.0299	.0255
0.5	.0466	.0399
0.8	.0299	.0255

Remembering that $\bar{q}_1 = q_0 + E(\Delta q)$, we obtain the following

	$\bar{q}_1$		$\bar{q}_1(1-\bar{q}_1)$	
q_0	k = 2.06	k'= 1.76	k = 2.06	k'= 1.76
0.2	.230	.226	.1771	.1749
0.5	.547	.540	.2478	.2484
0.8	.830	.826	.1411	.1437

Because the differences between values of $E(\Delta q)$ are small when the selection intensities .05 and .10 are compared, the differencies between consequent values of $\bar{q}_1(1-\bar{q}_1)$ are trivial when compared to the difference between k and k'. Given other pairs of selection intensities in which one member is twice the magnitude of the other, values of $E(\Delta q)$ would of course be different but the reader can quite easily verify that the difference between values of $\bar{q}_1(1-\bar{q}_1)$ would in any case be very small relative to that between the values of k and k'. It follows that when $r = p = q$ and the selection intensities in the T = S and RSP programs are such that N_e is the same in each of the two populations of the RSP program as in the single population of the T = S program, the approximation stated by 14.14 is very close.

Appendix 14.2. Algebra

Let $a = \bar{p}_1$ and $b = \bar{q}_1$. Then

$$2[(\bar{p}_1+\bar{q}_1)/2][1-(\bar{p}_1+\bar{q}_1)/2] = 2[(a+b)/2][1-(a+b)/2]$$

$$= [a+b][2-(a+b)]/2 = [2a+2b-a^2-2ab-b^2]/2$$

and when both a^2 and b^2 are subtracted and added within the brackets

$$= [2a-2a^2+2b-2b^2+a^2-2ab+b^2]/2 = [a(1-a)+b(1-b)+(a-b)^2/2]$$

$$= [\bar{p}_1(1-\bar{p}_1)+\bar{q}_1(1-\bar{q}_1)+(\bar{p}_1-\bar{q}_1)^2/2].$$

Appendix 15.1. Total Genetic Variance

It is apparent from 15.5 that derivation of σ^2_Y requires the forms of (1) the variances of single locus genotype effects and (2) the covariances of single locus genotype effects at pairs of loci. Once σ^2_Y has been obtained, subtraction of additive genetic variance (given by 6.19) leaves σ^2_d, the dominance variance. The bivariate distribution of y_i-u_i and y_j-u_j is shown below.

Genotype	Frequency	y_i-u_i	y_j-u_j
B_iB_j/B_iB	p^2	u_i	u_j
B_iB_j/B_ib_j	$2pr$	u_i	a_ju_j
B_ib_j/B_ib_j	r^2	u_i	$-u_j$
B_iB_j/b_iB_j	$2ps$	a_iu_i	u_j
B_iB_j/b_ib_j	$2pt$	a_iu_i	a_ju_j
B_ib_j/b_iB_j	$2rs$	a_iu_i	a_ju_j
B_ib_j/b_ib_j	$2rt$	a_iu_i	$-u_j$
b_iB_j/b_iB_j	s^2	$-u_i$	u_j
b_iB_j/b_ib_j	$2st$	$-u_i$	a_ju_j
b_ib_j/b_ib_j	t^2	$-u_i$	$-u_j$

Remember that subtraction of the constants, u_i and u_j, does not change the variance of either variable or the covariance between them. The averages of the two coded variables are easily shown to be

$$E(y_i-u_i) = (p+r-s-t)u_i + 2(p+r)(s+t)a_iu_i \text{ and}$$

$$E(y_j-u_j) = (p+s-r-t)u_j + 2(p+s)(r+t)a_ju_j$$

Proceeding in the usual way we find that

$$\sigma^2(y_i) = [p^2+2pr+r^2+s^2+2st+t^2-(p+r-s-t)^2]u_i^2$$
$$-4(p+r-s-t)(p+r)(s+t)a_iu_i^2$$
$$+[2(p+r)(s+t)-4(p+r)^2(s+t)^2]a_i^2u_i^2. \qquad A15.1.1$$

Remembering that $p+r = q_i$, $s+t = 1-q_i$, $p+s = q_j$, and $r+t = 1-q_j$, it is not very difficult to show from A15.1.1 that

$$\sigma^2(y_i) = 2q_i(1-q_i)[1+2(1-2q_i)a_i+(1-2q_i+2q_i^2)a_i^2]u_i^2 \qquad A15.1.2$$

Going on to the covariance between y_i and y_j we find that

$$\sigma(y_i,y_j) = [p^2-r^2-s^2+t^2-(p+r-s-t)(p+s-r-t)]u_iu_j$$
$$+[2ps-2rt-2(p+r)(s+t)(p+s-r-t)]a_iu_iu_j$$
$$+[2pr-2st-2(p+s)(r+t)(p+r-s-t)]u_ia_ju_j$$
$$+2[pt+rs-2(p+r)(s+t)(p+s)(r+t)]a_iu_ia_ju_j \qquad A15.1.3$$

Using the defined relations between two locus gamete frequencies (p, r, s, and t) and allele frequencies (q_i and q_j) and proceeding with reasonable care, we find the following equivalent form of A15.1.3.

$$\sigma(y_i,y_j) = 2(pt-rs)_{ij}u_iu_j + 2(pt-rs)_{ij}(1-2q_i)a_iu_iu_j + 2(pt-rs)_{ij}(1-2q_j)u_ia_ju_j$$
$$+2[pt+rs-2(p+r)(s+t)(p+s)(r+t)]a_iu_ia_ju_j \qquad A15.1.4$$

Returning to 15.5, we note that

$$\sigma_Y^2 = \Sigma_i\sigma^2(y_i) + \Sigma_i\Sigma_{j\neq i}\sigma(y_i,y_j) = \sigma_g^2 + \sigma_d^2$$

Thus using A15.1.2 and A15.1.4, it is apparent that

$$\sigma_g^2 + \sigma_d^2 = 2\Sigma_iq_i(1-q_i)\{1+2(1-2q_i)a_i+[1-2q_i+2q_i^2]a_i^2\}u_i^2$$
$$+2\Sigma_i\Sigma_{j\neq i}\{(pt-rs)_{ij}u_iu_j+(pt-rs)_{ij}(1-2q_i)a_iu_iu_j+(pt-rs)_{ij}(1-2q_j)u_ia_ju_j$$
$$+[pt+rs-2(p+r)(s+t)(p+s)(r+t)]a_iu_ia_ju_j\} \qquad A15.1.5$$

Additive genetic variance, σ_g^2, was identified in Chapter VI as

$$\sigma_g^2 = 2\Sigma_iq_i(1-q_i)\{1+2(1-2q_i)a_i+[1-4q_i+4q_i^2]a_i^2\}u_i^2$$
$$+2\Sigma_i\Sigma_{j\neq i}\{(pt-rs)_{ij}u_iu_j+(pt-rs)_{ij}(1-2q_i)a_iu_iu_j+(pt-rs)_{ij}(1-2q_j)u_ia_ju_j$$
$$+(pt-rs)_{ij}(p+s-r-t)(p+r-s-t)a_iu_ia_ju_j\} \qquad (6.19)\ A15.1.6$$

In this expression 6.19 has been modified by expansion of $[1+(1-2q_i)a_i]^2$ and $[1+(1-2q_i)a_i][1+(1-2q_j)a_j]$ and by recognition that $(1-2q_j)$ and $(1-2q_i)$ are equal to $(p+s-r-t)$ and $(p+r-s-t)$, respectively. σ_d^2 is obtained by subtracting A15.1.6 from A15.1.5; the result is

$$\sigma_d^2 = 4\Sigma_i q_i^2(1-q_i)^2 a_i^2 u_i^2 + 2\Sigma_i\Sigma_{j\neq i}[pt+rs-2(p+r)(s+t)(p+s)(r+t)$$

$$-(pt-rs)(p+s-r-t)(p+r-s-t)]a_i u_i a_j u_j$$

$$=4\Sigma_i q_i^2(1-q_i)^2 a_i^2 u_i^2 + 4\Sigma_i\Sigma_{j\neq i}(pt-rs)^2 a_i u_i a_j u_j \qquad \text{A15.1.7}$$

That the second term of σ_d^2 collapses to $4\Sigma_i\Sigma_{j\neq i}(pt-rs)^2 a_i u_i a_j u_j$ is not at all obvious and needs now to be demonstrated. Consider the quantity in A15.1.7 that is enclosed in square brackets. With minor rearrangements it can be rewritten as follows

$$pt[1-(p+s-r-t)(p+r-s-t)]+rs[1+(p+s-r-t)(p+r-s-t)]-2(p+r)(s+t)(p+s)(r+t)$$

The three major components will now be considered in order.

$$C_1 = pt[1-p^2-pr+ps+pt-ps-rs+s^2+st-ps+pr+r^2-rs-rt+pt+rt-st-t^2]$$

$$= pt[1-(p^2+pr+ps)+(s^2+ps+st)+(r^2+pr+rt)-(rt+st+t^2)+2pt-2rs]$$

$$=pt[1-p(1-t)+s(1-r)+r(1-s)-t(1-p)+2pt-2rs]$$

$$=pt[1-p+pt+s-rs+r-rs-t+pt+2pt-2rs]= pt[4(pt-rs)+2r+2s]$$

The second of the three components is very similar to the first. Taking their two differences into account, it is obvious that

$$C_2 = rs[1+(p+s-r-t)(p+r-s-t)] = rs[1+p(1-t)-s(1-r)-r(1-s)+t(1-p)-2pt+2rs]$$

$$=rs[1+p-pt-s+rs-r+rs+t-pt-2pt+2rs = rs[2p+2t-4(pt-rs)]$$

When the third component is expanded into sixteen terms there are eight that contain pt and eight that contain rs. Factoring pt and rs out of the two sets, we obtain

$$C_3 = 2pt(ps+s^2+pr+pt+rs+st+r^2+rt) + 2rs(p^2+ps+pr+pt+rs+st+rt+t^2)$$

$$=2pt[s(p+s+r+t)+r(p+r+t+s)+pt-rs] + 2rs[p(p+s+r+t)+t(s+r+t+p)+rs-pt]$$

$$=2pt[s+r+(pt-rs)] + 2rs[p+t-(pt-rs)]$$

Combining the three components, we obtain

$$C_1 + C_2 + C_3 = 4pt(pt-rs) - 4rs(pt-rs) - 2pt(pt-rs) + 2rs(pt-rs)$$

$$+2prt+2pst+2prs+2rst-2pst-2prt-2prs-2rst = 2(pt-rs)^2$$

It follows that

$$2\Sigma_i\Sigma_{j\neq i}[pt+rs-2(p+r)(s+t)(p+s)(r+t)-(pt-rs)(p+s-r-t)(p+r-s-t)]a_i u_i a_j u_j$$

$$= 4\Sigma_i\Sigma_{j\neq i}(pt-rs)^2 a_i u_i a_j u_j$$

• Appendix 15.2. Effects of Genotypes at Marked and Linked Loci—Theory

Consider an F_2 population of which the parents, once removed, were two homozygous lines, a "marked" locus of which the three possible genotypes (MM, Mm, and mm) in that F_2 can be distinguished by their phenotypes, and any other segregating gene (the i-th) with its two alleles identified as B_i and b_i. The associated two locus genotypes of the ancestral homozygous lines would then be either B_i M/B_i M and b_i m/b_i m or B_i m/B_i m and b_i M/b_i M, identified in what follows as the coupling or repulsion alternatives. Letting γ sybolize the probability of recombination between the two loci, the two locus genotypes of functional F_1 gametes along with their frequencies would be as follows (assuming absence of gametic selection and/or meiotic drive).

Genotype	Frequencies: Coupling	Repulsion
$B_i M$	$(1-r)/2$	$r/2$
$B_i m$	$r/2$	$(1-r)/2$
$b_i M$	$r/2$	$(1-r)/2$
$b_i m$	$(1-r)/2$	$r/2$

The distributions of genotypes and their coded values in the F_2 population would be as shown below.

Genotype	Genotype frequencies: Coupling	Repulsion	Coded values of genotypes
B_iB_iMM	$(1-r)^2/4$	$r^2/4$	$u_i + u_m$
B_ib_iMM	$r(1-r)/2$	$r(1-r)/2$	$a_iu_i + u_m$
b_ib_iMM	$r^2/4$	$(1-r)^2/4$	$-u_i + u_m$
B_iB_iMm	$r(1-r)/2$	$r(1-r)/2$	$u_i + a_mu_m$
B_ib_iMm	$[(1-r)^2 + r^2]/2$	$[(1-r)^2 + r^2]/2$	$a_iu_i + a_mu_m$
b_ib_iMm	$r(1-r)/2$	$r(1-r)/2$	$-u_i + a_mu_m$
B_iB_imm	$r^2/4$	$(1-r)^2/4$	$u_i - u_m$
B_ib_imm	$r(1-r)/2$	$r(1-r)/2$	$a_iu_i - u_m$
b_ib_imm	$(1-r)^2/4$	$r^2/4$	$-u_i - u_m$

Working from these distributions it is easily shown that averages of the two locus genotype values for the three marked locus groups of individuals are as follows.

Marker genotype	Average values of genotypes: Coupling	Repulsion
MM	$(1-2r)u_i + 2r(1-r)a_iu_i + u_m$	$(2r-1)u_i + 2r(1-r)a_iu_i + u_m$
Mm	$(1-2r+2r^2)a_iu_i + a_mu_m$	$(1-2r+2r^2)a_iu_i + a_mu_m$
mm	$(2r-1)u_i + 2r(1-r)a_iu_i - u_m$	$(1-2r)u_i + 2r(1-r)a_iu_i - u_m$

Next, letting MM, Mm and mm identify the genotype value averages, as well as the marked locus genotypes, of the three groups, we find in the coupling case that

$$(MM - mm)/2 = (1-2r)u_i + u_m \qquad \text{A15.2.1}$$

and

$$(2Mm - MM - mm)/2 = (1-2r)^2 a_iu_i + a_mu_m \qquad \text{A15.2.2}$$

and in the repulsion case that

$$(MM - mm)/2 = (2r-1)u_i + u_m \qquad \text{A15.2.3}$$

$$(2Mm - MM - mm)/2 = (1-2r)^2 a_iu_i + a_mu_m \qquad \text{A15.2.4}$$

Edwards et al (1987) termed (2Mm - MM - mm)/(MM - mm) the Dominance/Additive (D/A) ratio and discussed its attributes as a measure of level of dominance.

Note that A15.2.2 and A15.2.4 are identical but that A15.2.1 and A15.2.3 are not. Note also that each of those four expressions applies when there is only one gene that affects the trait being considered and that does not segregate independently from the marker gene ($r < 1/2$). To encompass all possibilities we recognize that

$$(MM - mm)/2 = A = \Sigma_i^c(1-2r_i)u + \Sigma_i^r(2r_i-1)u_i + u_m \qquad A15.2.5$$

and that

$$(2Mm - M - m)/2 = D = \Sigma_i(1-2r_i)^2 a_i u_i + a_m u_m \qquad A15.2.6$$

In A15.2.5, Σ_i^c and Σ_i^r indicate summation over genes in the "coupling" and "repulsion" relations

to the marked locus. Uncertainties inherent in interpretation of D/A ratios are clarified by considering special cases.

(a) Suppose there are no genes that affect the trait and are not linked ($r_i < 1/2$) with the marked gene. Then from A15.2.5 and A15.2.6

$$E(D)/E(A) = a_m u_m / u_m = a_m,$$

the level of dominance of the marker gene (if it has any effect on the trait).

(b) Suppose there is one gene that affects the trait and is linked with the marked locus. Then from A15.2.5 and A15.2.6

$$E(D)/E(A) = [(1-2r_i)^2 a_i u_i + a_m u_m]/[(1-2r_i)u_i + u_m]$$

or $$E(D)/E(A) = [(1-2r_i)^2 a_i u_i + a_m u_m]/[(2r_i-1)u_i + u_m]$$

depending on whether the initial linkage relation was coupling or repulsion. Note that by virtue of linkage $(1-2r_i)$ is a positive and $(2r_i-1)$ a negative quantity. Thus the observed D/A ratio could be strongly affected by whether the linkage relation between B and M was coupling or repulsion.

(c) In general the number of genes (other than the marker gene) affecting the trait and the proportions for which the initial linkage relations were coupling or repulsion are unknowns. Thus realistically, it must be recognized that

$$E(D)/E(A) = [\Sigma_i(1-2r_i)^2 a_i u_i + a_m u_m]/[\Sigma_i^c(1-2r_i)u + \Sigma_i^r(2r_i-1)u_i + u_m]$$

Because (a) $(2r_i-1)$ is a negative quantity whenever $r_i < 1/2$ and (b) the initial linkage relation between B and M of a portion of the genes that affect the trait will inevitably be repulsion, there is a strong and always unknown upward bias in E(D)/E(A) relative to the average of a for all genes affecting the trait.

• Appendix 15.3. The Genetic Variances of Design III Data

The genetic units of a design III investigation are the progenies obtained by backcrossing random individuals from an F_n population to each of the pair of highly inbred lines (assumed to be homozygous) that were crossed to yield the F_1 from which subsequent F_n populations could be derived. Thus F_2 (or F_n) individuals employed were each the parent of a pair of backcross progenies. Let z_{1k} and z_{2k} be the expected genotypic mean values of the progenies of the k-th F_n individual mated to members of inbred lines 1 and 2, respectively;

$\bar{z}_k = (z_{1k}+z_{2k})/2$ symbolize the expected average genotypic value of the k-th pair of backcross progenies; and

$d_k = (z_{1k}-z_{2k})$.

Then, in accord with 15.10 and Table15.11,

$\sigma^2(mG) = \sigma^2(\bar{z})$

and $\sigma^2(mIG) = \sigma^2(d)/2$.

The first of these equalities is obvious; the second reflects the fact (easily verified) that in the analysis of variance of any 2 x n table of quantities the interaction variance is one-half the variance of the pair differences.

By analogy with 5.7

$z_{1k} = \alpha + \Sigma_i \bar{y}_{1ki}$

in which $\bar{y}_{1ki}$ is the average value of genotye at the i-th locus in the progeny identified by subscripts 1 and k. It follows that

$\bar{z}_k = \alpha + \Sigma_i(\bar{y}_{1ki} + \bar{y}_{2ki})/2 = \alpha + \Sigma_i(\bar{y}_{ki})$

in which

$\bar{y}_{ki} = (\bar{y}_{1ki} + \bar{y}_{2ki})/2$

and

$d_k = \Sigma_i(\bar{y}_{1ki} - \bar{y}_{2ki}) = \Sigma_i(d_i)$.

To reduce the symbols required let each value of y averaged in $\bar{y}_{1ki}$ or $\bar{y}_{2ki}$ be coded by subtraction of u_i (or u_j in the case of values averaged in $\bar{y}_{1kj}+\bar{y}_{2kj}$) and let coded quantities be identified as such by a preceding asterisk as before. Then

$*\bar{z}_k = \bar{z}_k - \Sigma_i u_i = \alpha + \Sigma_i *\bar{y}_{ki}$,

$\sigma^2(*\bar{z}) = \sigma^2(\bar{z}) = \sigma^2(mG) = \Sigma_i\sigma^2(*\bar{y}_i) + \Sigma_i\Sigma_{j\neq i}\sigma(*\bar{y}_i, *\bar{y}_j)$, A15.3.1

and $\sigma^2(d) = 2\sigma^2(mIG) = \Sigma_i\sigma^2(d_i) + \Sigma_i\Sigma_{j\neq i}\sigma(d_i, d_j)$ A15.3.2

We now require the various values of $*\bar{y}_k$ and d_k (listed in Tables A15.3.1 and A15.3.2).[1] As the first of two examples of how these quantities were derived, consider the progenies obtained when an F_n individual having the B_iB_j/B_iB_j genotype is mated with inbred lines of which the two-locus genotypes of lines 1 and 2 are B_iB_j/B_iB_j and b_ib_j/b_ib_j, respectively. All gametes from the F_n individual carry the B allele at the i-th locus, those from line 1 carry the B allele, and those from line 2 the b allele. Thus the genotypes of all individuals in the F_n x line 1 progeny are BB at the i-th locus and those in the F_n x line 2 progeny are Bb. It follows that

$*\bar{y}_{1i} = 2u_i - u_i = u_i$,

$*\bar{y}_{2i} = (u_i + a_iu_i) - u_i = a_iu_i$,

and, as shown in Table A15.3.1, that

$*\bar{y}_i = (*\bar{y}_{1i} + *\bar{y}_{2i})/2 = (u_i + a_iu_i)/2$

and $d = *\bar{y}_{1i} - *\bar{y}_{2i} = u_i - a_iu_i$.

As our second example, consider the progenies obtained when an F_n individual having the B_iB_j/b_ib_j genotype is mated with inbred lines of which the two locus genotypes of lines 1 and 2 are

Table A15.3.1. Values of $*\bar{y}$ and d for the i-th and j-th genes when the genotypes of inbred lines 1 and 2 are B_iB_j/B_iB_j and b_ib_j/b_ib_j, respectively

F_n genotype	freq.	$*\bar{y}_i$	$*\bar{y}_j$	d_i	d_j
B_iB_j/B_iB_j	p^2	$(u_i+a_iu_i)/2$	$(u_j+a_ju_j)/2$	$u_i-a_iu_i$	$u_j-a_ju_j$
B_iB_j/B_ib_j	2pr	$(u_i+a_iu_i)/2$	$a_ju_j/2$	$u_i-a_iu_i$	u_j
B_ib_j/B_ib_j	r^2	$(u_i+a_iu_i)/2$	$(a_ju_j-u_j)/2$	$u_i-a_iu_i$	$u_j+a_ju_j$
B_iB_j/b_iB_j	2ps	$a_iu_i/2$	$(u_j+a_ju_j)/2$	u_i	$u_j-a_ju_j$
B_iB_j/b_ib_j	2pt	$a_iu_i/2$	$a_ju_j/2$	u_i	u_j
B_ib_j/b_iB_j	2rs	$a_iu_i/2$	$a_ju_j/2$	u_i	u_j
B_ib_j/b_ib_j	2rt	$a_iu_i/2$	$(a_ju_j-u_j)/2$	u_i	$u_j+a_ju_j$
b_iB_j/b_iB_j	s^2	$(a_iu_i-u_i)/2$	$(u_j+a_ju_j)/2$	$u_i+a_iu_i$	$u_j-a_ju_j$
b_iB_j/b_ib_j	2st	$(a_iu_i-u_i)/2$	$a_ju_j/2$	$u_i+a_iu_i$	u_j
b_ib_j/b_ib_j	t^2	$(a_iu_i-u_i)/2$	$(a_ju_j-u_j)/2$	$u_i+a_iu_i$	$u_j+a_ju_j$

Table A15.3.2. Values of $*\bar{y}$ and d for the i-th and j-th genes when the genotypes of inbred lines 1 and 2 are B_ib_j/B_ib_j and b_iB_j/b_iB_j, respectively

F_n genotype	freq.	$*\bar{y}_i$	$*\bar{y}_j$	d_i	d_j
B_iB_j/B_iB_j	p^2	$(u_i+a_iu_i)/2$	$(u_j+a_ju_j)/2$	$u_i-a_iu_i$	$a_ju_j-u_j$
B_iB_j/B_ib_j	2pr	$(u_i+a_iu_i)/2$	$a_ju_j/2$	$u_i-a_iu_i$	$-u_j$
B_ib_j/B_ib_j	r^2	$(u_i+a_iu_i)/2$	$(a_ju_j-u_j)/2$	$u_i-a_iu_i$	$-a_ju_j-u_j$
B_iB_j/b_iB_j	2ps	$a_iu_i/2$	$(u_j+a_ju_j)/2$	u_i	$a_ju_j-u_j$
B_iB_j/b_ib_j	2pt	$a_iu_i/2$	$a_ju_j/2$	u_i	$-u_j$
B_ib_j/b_iB_j	2rs	$a_iu_i/2$	$a_ju_j/2$	u_i	$-u_j$
B_ib_j/b_ib_j	2rt	$a_iu_i/2$	$(a_ju_j-u_j)/2$	u_i	$-a_ju_j-u_j$
b_iB_j/b_iB_j	s^2	$(a_iu_i-u_i)/2$	$(u_j+a_ju_j)/2$	$u_i+a_iu_i$	$a_ju_j-u_j$
b_iB_j/b_ib_j	2st	$(a_iu_i-u_i)/2$	$a_ju_j/2$	$u_i+a_iu_i$	$-u_j$
b_ib_j/b_ib_j	t^2	$(a_iu_i-u_i)/2$	$(a_ju_j-u_j)/2$	$u_i+a_iu_i$	$-a_ju_j-u_j$

B_ib_j/B_ib_j and b_iB_j/b_iB_j. The expected frequencies of the two locus genotypes produced by the F_n individual are as follows in which γ symbolizes the probability of recombination between the i-th and j-th genes.

Gamete genotype	Expected frequency
B_iB_j	$p = (1-\gamma)/2$
B_ib_j	$r = \gamma/2$
b_iB_j	$s = \gamma/2$
b_ib_j	$t = (1-\gamma)/2$

We see that the frequencies of the B_i and b_i alleles in the F_n gametes are both 1/2--(p+r) and (s+t). Then It follows that

$$E(*\bar{y}_{1i}) = [(2u_i-u_i)+(u_i+a_iu_i-u_i)]/2 = (u_i+a_iu_i)/2$$

Next, because all gametes from line 2 must carry the b_i allele, genotypes at the i-th locus in the F_n x line 1 progeny are expected to be half Bb and half bb. Consequently

$$E(*\bar{y}_{2i}) = [(u_i+a_iu_i-u_i)+(0-u_i)]/2 = (a_iu_i-u_i)/2$$

$$E(*\bar{y}_i) = [E(*\bar{y}_{1i})+E(*\bar{y}_{2i})]/2 = a_iu_i/2$$

and $E(d_i) = [E(*\bar{y}_{1i})-E(*\bar{y}_{2i})] = u_i$

The last two of these quantities are shown in both Table A15.3.2 and Table A15.3.1. The redundancy in these two tables may be justified as an aid to understanding. Actually the only values that are different in the two tables are those of d_i in the final column and they differ only in algebraic sign.

Given the distributions of Tables A15.3.1 and A15.3.2, the derivations of the component variances and covariances of $\sigma^2(*\bar{z})$ and $\sigma^2(d)$ are straightforward. However, they are eased by noting that in each instance there is a constant that can be subtracted from or added to each value of the variable without effect on variance or covariance. Consider, for example, the variance of d. If u is subtracted from each of the ten values listed, the distribution collapes to the following.

d	frequency
-au	$p^2+2pr+r^2 = q_i^2$
0	$2ps+2pt+2rs+2rt = 2q_i(1-q_i)$
au	$s^2+2st+t^2 = (1-q_i)^2$

mean of d = (s+t-p-r) = (1-2q)au

$$\sigma^2(d_i) = [q_i^2+(1-q_i)-(1-2q_i)]a_i^2u_i^2 = 2q_i(1-q_i)a_i^2u_i^2 \quad \text{A15.3.3}$$

By similar procedure

$$\sigma^2(*\bar{y}_i) = 2q_i(1-q_i)u_i^2/4 \quad \text{A15.3.4}$$

$$\sigma(*\bar{y}_i,*\bar{y}_j) = 2(pt-rs)_{ij}u_iu_j/4 \quad \text{A15.3.5}$$

$$\sigma(d_i,d_j) = 2(pt-rs)_{ij}a_iu_ia_ju_j \quad \text{A15.3.6}$$

when the F_1 linkage phase was "coupling", and

$$\sigma(d_i,d_j) = -2(pt-rs)_{ij}a_iu_ia_ju_j \quad \text{A15.3.7}$$

when the F_1 linkage phase was "repulsion".

Finally, substituting in A15.3.1 in terms of A15.3.4 and A15.3.5, we obtain

$$\sigma^2(mG) = \Sigma_i q_i(1-q)_i u_i^2/2 + \Sigma_{ij}(pt-rs)_{ij}u_iu_j$$

and substituting in A15.3.2 in terms of A15.3.3, A15.3.6, and A15.3.7, we obtain

$$2\sigma^2(mIG) = 2\Sigma_i q_i(1-q)_i a_i^2 u_i^2 + 4\Sigma_{ij}^{c}(pt-rs)_{ij}a_iu_ia_ju_j - 4\Sigma_{ij}^{r}(pt-rs)_{ij}a_iu_ia_ju_j$$

• Appendix 15.4. Biased vs. Unbiased Estimates of Genetic Variances from Design III Data

When the same n backcross progeniy pairs are grown in r replications in each of e macro-environments, the analysis of variance of data obtained on any one variable (trait) would logically be that shown in Table 15.4.1.[1] Note that when data have been obtained in only one macro-environment (e = 1) this analysis collapses to the form shown in Table 15.11 in which

$$\sigma^2(m) = \sigma^2(mG) + \sigma^2(mE)$$

and $\sigma^2(mI) = \sigma^2(mIG) + \sigma^2(mIE)$

Table A15.4.1. ANOVA, Design III data, n progeny pairs grown in e macro-environments

Variance source	d.f.	Mean Square	Expectation of mean square
Macro-environments(E)	e-1		
Repls. in E's	e(r-1)		
Lines in E's	e		
Progeny pairs (P)	(n-1)	M_1	$\sigma^2(e) + 2r\sigma^2(mE) + 2re\sigma^2(mG)$
P x E	(n-1)(e-1)	M_2	$\sigma^2(e) + 2r\sigma^2(mE)$
Lines x P	(n-1)	M_3	$\sigma^2(e) + r\sigma^2(mIE) + re\sigma^2(mIG)$
Lines x P x E	(n-1)(e-1)	M_4	$\sigma^2(e) + r\sigma^2(mIE)$
Remainder among plots in E's	e(2n-1)(r-1)	M_5	$\sigma^2(e)$

e = number of macro-environments

It is apparent from the table that an unbiased estimate of $\sigma^2(mG)$ is provided by $E(M_1-M_2)/2re$ and that a biased estimate is provided by $(M_1-M_5)/2re$. In the case of $\sigma^2(mIG)$, an unbiased estimate is provided by $(M_3-M_4)/re$ and a biased estimate by $(M_3-M_5)/re$. Of course, when data have been obtained in only one macro-environment neither unbiased estimate is available.

As explained in section 15.1.1.3, the variance of any linear function of mean squares is a function of (a) the expectations of the mean squares and (b) the degrees of freedom associated with the mean squares employed. Thus given numerical values of those parameters approximations to variances of interest are easily obtained.[2] Our purpose here will be served by the following assumptions.

1. That the total number of field plots to be employed is T=1800 and hence that, if n backcross progeny pairs are to be grown in r replications in each of e macro-environments, n=T/2re.

2. That $\sigma^2(mG) = \sigma^2(mE) = 8$, $\sigma^2(mIG) = \sigma^2(mIE) = 16$, $\sigma^2(e) = 100$, and r = 3. Rationale for numerical assumptions of these kinds is discussed 14.1.1 and 14.1.2. Note that in both cases the genetic variance and the genotype x environment interaction variances are assumed to be equal and that the magnitudes assumed make both $\sigma^2(mIG)/2\sigma^2(mG)$ and $[\sigma^2(mIG) + \sigma^2(mIE)]/2[\sigma^2(mG) + \sigma^2(mE)]$ equal to 1.0.

The following four possibilities with respect to r,e, and n are considered: (a) r=3, e = 1, and n = 300; (b) r= 3, e = 2, and n = 150; (c) r = 3, e = 3, and n = 100; and (d) r=3, e =6, and n = 50. Note the trade-off between number of macro-environments and number of backcross progeny pairs.

Example (a)

Mean Square	d.f.	E(Mean Square)
M_1	299	196
M_2	0	--
M_3	299	196
M_4	0	--
M_5	1198	100

$E[(M_1-M_5)/2r = \sigma^2(mG) + \sigma^2(mE) = 16$

$\sigma^2[(M_1-M_5)/2r] = [2(196)^2/299 + 2(100)^2/1198]/36 = 7.6016$

$\sigma[(M_1-M_5)/2r]= 2.76$ Coeffient of variation = (100)(2.76)/16 = 17.2%

$E[(M_3-M_5)/r = \sigma^2(mIG) + \sigma^2(mIE) = 32$

$\sigma^2[(M_3-M_5)/r] = [2(196)^2/299 + 2(100)^2/1198]/9 = 30.4064$

$\sigma[(M_1-M_5)/r]= 5.51$ Coeffient of variation = (100)(5.51)/32 = 17.2%

Example (b)

Mean Square	d.f.	E(Mean Square)
M_1	149	244
M_2	149	148
M_3	149	244
M_4	149	148
M_5	1196	100

Biased estimates

$E[(M_1-M_5)/2re = \sigma^2(mG) + \sigma^2(mE)/2 = 12$

$\sigma^2[(M_1-M_5)/2re] = 5.6657$ C.V. = 19.8%

$E[(M_3-M_5)/re = \sigma^2(mIG) + \sigma^2(mIE)/2 = 24$

$\sigma^2[(M_3-M_5)/re] = 22.6628$ C.V. = 19.8%

Unbiased estimates

$E[(M_1-M_2)/2re = \sigma^2(mG) = 8$

$\sigma^2[(M_1-M_2)/2re] = 7.5913$ $\quad$ $C.V. = 34.4\%$

$E[(M_3-M_4)/re = \sigma^2(mIG) = 16$

$\sigma^2[(M_3-M_4)/re] = 30.3654$ $\quad$ $C.V. = 34.4\%$

Example (c)

Mean Square	d.f.	E(Mean Square)
M_1	99	292
M_2	198	148
M_3	99	292
M_4	198	148
M_5	1194	100

Biased estimates

$E[(M_1-M_5)/2re = \sigma^2(mG) + \sigma^2(mE)/3 = 10.67$

$\sigma^2[(M_1-M_5)/2re] = 5.3681$ $\quad$ $C.V. = 21.7\%$

$E[(M_3-M_5)/re = \sigma^2(mIG) + \sigma^2(mIE)/3 = 21.33$

$\sigma^2[(M_3-M_5)/re] = 21.4728$ $\quad$ $C.V. = 21.7\%$

Unbiased estimates

$E[(M_1-M_2)/2re = \sigma^2(mG) = 8$

$\sigma^2[(M_1-M_2)/2re] = 5.9992$ $\quad$ $C.V. = 30.6\%$

$E[(M_3-M_4)/re = \sigma^2(mIG) = 16$

$\sigma^2[(M_3-M_4)/re] = 23.9970$ $\quad$ $C.V. = 30.6\%$

Example (d)

Mean Square	d.f.	E(Mean Square)
M_1	49	436
M_2	245	148
M_3	49	436
M_4	245	148
M_5	1188	100

Biased estimates

$E[(M_1-M_5)/2re = \sigma^2(mG) + \sigma^2(mE)/6 = 9.33$

$\sigma^2[(M_1-M_5)/2re] = 5.9999$ $\quad$ $C.V. = 26.2\%$

$E[(M_3-M_5)/re = \sigma^2(mIG) + \sigma^2(mIE)/6 = 18.67$

$\sigma^2[(M_3-M_5)/re] = 23.9996$ $\quad$ $C.V. = 26.2\%$

Unbiased estimates

$E[(M_1-M_2)/2re = \sigma^2(mG) = 8$

$\sigma^2[(M_1-M_2)/2re] = 6.1249$ $\quad$ $C.V. = 30.9\%$

$E[(M_3-M_4)/re = \sigma^2(mIG) = 16$

$\sigma^2[(M_3-M_4)/re] = 24.4995$ $\quad$ $C.V. = 30.9\%$

Eye-catching is the fact that coefficients of variation (C.V.'s) are the same throughout for estimates (biased or unbiased) of $\sigma^2(mG)$ and $\sigma^2(mlG)$. This resulted from the assumptions, for the purpose of assigning values of mean square expectations, (a) that $\sigma^2(mG) = \sigma^2(mlG)/2$ and (b) that $\sigma^2(mE)/\sigma^2(mG)$ and $\sigma^2(mlE)/\sigma^2(mlG)$ were equal. With other numerical value assumptions those coefficients of variation would not have been equal.

Significant Facts demonstrated

(1) Coefficients of variation were in every instance smaller for biased estimates. Though variances of unbiased estimates are slightly larger, the C.V. differences were almost entirely due to the fact that the expectation of the estimate was larger in the case of biased estimates. The smaller C.V.'s mean that the biased estimates are to be preferred if there are good reasons to believe that $\sigma^2(mE)/\sigma^2(mG)$ and $\sigma^2(mlE)/\sigma^2(mlG)$ are in fact very close to equal.

(2) As n, the number of backcross progeny pairs, is decreased to allow collection of data in more macro-environments without increase in total number of plots employed the gap between the C.V.'s of biased and unbiased estimates narrowed but remained smaller for the biased estimate. The principal reason was that the expectation of the biased estimate was still greater than that of the unbiased estimate. However, the potential bias was much less with e=6 than with e = 1. Note that as number of macro-environments was increased from 1 to 3 the genotype x environment variance included in the biased estimates was diminished from $\sigma^2(mE)$ to $\sigma^2(mE)/3$-- from $\sigma^2(mE)$ to $\sigma^2(mE)/3$.

• Appendix 15.5. Average Genotype Value Difference between the F_1 and F_2 Generations from a Cross of Pure Lines

Each gene that is homozygous (BB or bb) in an F_1 geneation must be homozygous for the same allele in the F_2 generation derived by selfing or random mating of F_1 individuals. Thus only the genes heterozygous (Bb) in the F_1 can contribute to the difference in average genotype values of the two generations and, in the absence of epistasis,

$$\ddot{Y}_{F1} - \ddot{Y}_{F2} = \Sigma_i(\ddot{y}_{i\,F1} - \ddot{y}_{i\,F2}) \qquad \text{A15.5.1}$$

where summation is over loci heterozygous in the F_1. The distribution of single-locus genotype values in the F_2 will obviously be as follows.

Genotype	Probability	y
B_iB_i	q^2	2u
B_ib_i	$2q(1-q)$	u+au
b_ib_i	$(1-q)^2$	0

Here the effect of mutation is considered trivial with respect to probabilities of genotypes and survival of both gametes and zygotes is assumed to be independent of genotypes. Then by 5.10 and by virtue of q = 1/2 when the genotype is Bb in the F_1

$$\ddot{y}_{i\,F2} = 2qu_i + 2q(1-q)a_iu_i = u_i + a_iu_i/2$$

and

$$\ddot{y}_{i\,F1} - \ddot{y}_{i\,F2} = (u_i + a_iu_i) - (u_i + a_iu_i/2) = a_iu_i/2$$

because the value of the B_ib_i genotye of the F_1 is $u_i + a_iu_i$. Finally, by A15.5.1

$$\ddot{Y}_{F1} - \ddot{Y}_{F2} = \Sigma_i\ddot{y}_{i\,F1} - \Sigma_i\ddot{y}_{i\,F2} = \Sigma_i a_iu_i/2$$

Appendix 15.6. The Standard Error of $[\Delta\ddot{Y}_x - E(\Delta\ddot{Y}_x)]$

The variances of estimates of $\Delta\ddot{Y}_X$ and $E(\Delta\ddot{Y}_X)$ are required. When 15.42 is being used for prediction in advance, the variance of the estimate of $\sigma(\hat{X})$ being employed must be considered. On the other hand, when the agreement between an estimate of $\Delta\ddot{Y}_X$ and its expectation in the case of selection already practiced is being questioned, the observed value of $\sigma(\hat{X})$ would be appropriate for use in the computation of $E(\Delta\ddot{Y}_X)$ and the variance of $E(\Delta\ddot{Y}_X)$, given the value of $\sigma(\hat{X})$ actually involved becomes $[k/2\sigma(\hat{X})]^2\sigma^2(\bullet\sigma^2_{gx})$.

A15.6.1. Additive Genetic Variance

The genetic variance among half-sib families in the S population is an obvious source of information concerning magnitude of the additive genetic variance (see Table 15.8). Of course, the data for estimation would require comparison of families in more than one location and more than one year to enable an estimate not biased by genotype-environment interaction variance. The common parents of families could be either females (seed parents) or males (pollen parents)-- females will be assumed for the purpose of this discussion. Near equivalence of the family genetic variance to σ^2_{gx} would require that the number of random (male) parents be quite large in the case of each family. This can be accomplished by (1) using bulked pollen from numerous random plants in pollination of the female (common parent) plants or (2) wind pollination if sufficient isolation of the parent population is possible. When the common parents are seed parents the number of kernels available would limit the number of plots per half sib family and variance due to cytoplasmic effects (if any) would be confounded with the variance arising from genetic differences among the seed parents. It will become apparent that data on a large number of families will be required if the estimate of genetic variance among families is to be as accurate as required for the purpose being considered. Therefore, to avoid excessive block size, it will be assumed that F families would be divided into sets of size n and the families of each set would be compared on a within set basis in r replicates at each of v locations in each of t years. The portion of the analysis of variance of data from the field comparison of families (from Table 15.5) that would be used in estimation of $\sigma^2(\bar{z}) = (1/4)\sigma^2_{gx}$ is the following,

Source of Variance	d.f.	Mean square	Expectation of Mean Square
Families w. sets	(F/n)(n-1)	M_1	$\sigma^2(e) + r\sigma^2(\bar{z}ab) + rt\sigma^2(\bar{z}a) + rv\sigma^2(\bar{z}b) + rvt\sigma^2(\bar{z})$
F x Y within sets	(F/n)(n-1)(t-1)	M_2	$\sigma^2(e) + r\sigma^2(\bar{z}ab) + rv\sigma^2(\bar{z}b)$
F x L within sets	(F/n)(n-1)(v-1)	M_3	$\sigma^2(e) + r\sigma^2(\bar{z}ab) + rt\sigma^2(\bar{z}a)$
F x Y x L w. sets	(F/n)(n-1)(t-1)(v-1)	M_4	$\sigma^2(e) + r\sigma^2(\bar{z}ab)$

from which the estimate of $\sigma^2(\bar{z})$ is

$$\bullet[\sigma^2(\bar{z}) \approx (1/4)\sigma^2_{gx}] = [M_1 - M_2 - M_3 + M_4]/rvt$$

and, from 15. 3, the variance of the estimate is

$$\sigma^2[\bullet\sigma^2(\bar{z})] = \{2[E(M_1)]^2/f_1 + 2[E(M_2)]^2/f_2 + 2[E(M_3)]^2/f_3 + 2[E(M_4)]^2/f_4\}/(rvt)^2 \qquad \text{A15.6.1}$$

To obtain values of the E(M)'s assumptions concerning the magnitudes of variance components are required. The following from Table 14.4 are employed here.

$\sigma^2(\bar{z}) = 6.25 \quad \sigma^2(\bar{z}a) = \sigma^2(\bar{z}b) = \sigma^2(\bar{z})/4 \quad \sigma^2(\bar{z}ab) = \sigma^2(\bar{z})/2 \quad \sigma^2(e) = 100$

Degrees of freedom and mean square expectations are listed below in the case of two sets of assumptions concerning r, v, and t. In both examples n = 51 and F = 204 were employed.

Example 1 (r = 1, v = 8, t=3)			Example 2 (r = 2, v = 6, t = 2)		
M.S.	d.f.	E(M)	M.S.	d.f.	E(M)
1	200	270.3	1	200	281.2
2	400	115.6	2	200	125.0
3	1400	107.8	3	1000	112.5
4	2800	103.1	4	1000	106.3
$\sigma[\bullet\sigma^2(\bar{z})] = 1.4264$			$\sigma^2[\bullet\sigma^2(\bar{z})] = 1.7277$		

We see that decreasing the numbers of locations and years (decreasing vt from 24 to 12) in order to increase number of replications from one to two would be counterproductive. Example 2 will not be considered further. Accepting 1.4264 as the variance of an estimate of $(1/4)\,\sigma^2_{gx}$ we obtain $(2)^2(1.4264) = 5.7056$ as the associated variance of $(1/2)\,\sigma^2_{gx}$. Note (1) that the number of degrees of freedom of each of the mean squares of which $\bullet[\sigma^2(\bar{z}) \approx (1/4)\,\sigma^2_{gx}]$ is a linear function of F and (2) that the variance of the estimate is therefore inversely proportional to F (see A15.6.1). Thus, in the context of Example 1, the variance of the estimate of one-half the additive genetic variance of grain yield, X, would vary as shown below with F, the number of families from which data are obtained.

F	204	510	1020	2040
$\sigma^2[\bullet(1/2)\,\sigma^2_{gx}]$	5.71	2.28	1.14	0.57

• A15.6.2. **The Estimate of $\Delta\ddot{Y}_x$**

The obvious basis for this estimate would be a field comparison of random members of the unselected S population with random members of the population generated by random matings among selections. Let these populations be symbolized as the S_0 and S_1 populations, respectively. If the genetic sample from the S_0 population is large enough so that the effective population size of the S_1 population is also large, it will be reasonable to assume that genetic variance among members of the two populations will be nearly equal.[1] Consequently, it will also be reasonable to assume that $\sigma^2_Y + \sigma^2(Ya) + \sigma^2(Yb) + \sigma^2(Yab)$, the total of genotype variance plus genotype-environment interaction variance among individuals, will be close to equal in the two populations and that the average of genotype-environment effects will be essentially the same in the two populations if random members of the two are compared in a randomized block trial in a single macro-environment. Suppose that block size is two (one plot of each population), that yield is measured on 20 plants per plot, that the number of blocks is m, and that the total of genotype plus genotype-environment interaction variances among plants is 90.[2] It follows that the within block variance of the population difference would be $2(100 + 90/20) = 209$ and that of the average difference $[\bar{d} = \bullet(\Delta\ddot{Y}_X)]$ in the entire comparison would be 209/m. Magnitudes associated with several values of m are shown below.

m	100	200	400	800	1600
$\sigma^2(\bar{d})$	2.09	1.04	0.52	0.26	0.13

- A15.6.3. **The Variance of $[\Delta\ddot{Y}_x - E(\Delta\ddot{Y}_x)]$**

Because, as indicated earlier, $E(\Delta\ddot{Y}_X)$ is a multiple of $[k/\sigma(\hat{X})]$ values of k and $\sigma^2(\hat{X})$ are now required. For this example it will be assumed (a) that selection intensity is 0.05 so that $k = 2.06$, (b) that the parameters of the field comparison are $r = t = 1$ and $v = 3$ and (c) that $\sigma^2(\hat{X}) = 59.58$.[2] Using these quantities we obtain $[k/\sigma(\hat{X})]^2 = 0.0712$ and

$$\sigma^2[E(\Delta\ddot{Y}_X)] = (0.0712)\,\sigma^2[\bullet(1/2)\,\sigma^2_{gx}]. \qquad \text{A15.6.2}$$

Finally, standard errors of the difference between the estimate, $\bullet(\Delta\ddot{Y}_X)$, and the expectation, $E(\Delta\ddot{Y}_X)$, of direct response to a single cycle of a specific variation of selection among full-sib families depend on the numbers, F and m (defined above), employed in estimation procedures. A set of such standard errors is shown below. Total number of plots to be grown are shown in parentheses.

	F		
m	204	510	1020
100	1.58 (5096)	1.50 (12440)	1.47 (24680)
200	1.20 (5296)	1.10 (12640)	1.06 (24880)
400	0.93 (5696)	0.83 (13040)	0.78 (25280)
800	0.82 (6496)	0.65 (13840)	0.58 (26080)
1600	0.73 (8096)	0.54 (15440)	0.46 (27680)

- A15.6.4. **Discussion**

The standard errors shown above are functions of all the assumptions employed in their computation (magnitudes of variance components, the details of procedure in estimation of σ^2_{gx}, the selection system, etc.). Nevertheless, they illustrate the potential cost of assessing the relative validity of a basic type of prediction theory in quantitative genetics. What exactly do they tell us?

Remember that in the actual situation the real magnitude of σ^2_{gx} is not known; an estimate of it must be used in computation of $E(\Delta\ddot{Y}_X)$. Our exercise tells us that standard errors will be approximately those shown if variance components, procedures followed in estimating σ^2_{gx} and $\Delta\ddot{Y}_X$, and details of the selection system considered are all as assumed. In practice the crucial data items would be (a) the prediction of response to selection practiced that is a function of an estimate of σ^2_{gx} and (b) an estimate of the actual response to selection. Consider the possibility that (1) the real (but unknown) magnitudes of variance components are those assumed above, (2) $E(\Delta\ddot{Y}_X)$ is computed from the estimate of σ^2_{gx} obtained as described, and (3) $\Delta\ddot{Y}_X$ is estimated as described. The hypothesis to be tested is that $D = [\Delta\ddot{Y}_X - E(\Delta\ddot{Y}_X)] = 0.0$. Let us suppose that F and m were 510 and 400, respectively, and hence that the standard error of •D = 0.83. Then if the theory is absolutely correct and the expectation of $D = [\Delta\ddot{Y}_X - E(\Delta\ddot{Y}_X)]$ is, therefore, zero; observed values of •D not greater than 1.6 in absolute value would all be less than two standard errors less than or greater than zero and hence consistent with either validity or non-validity of theory. Given the real situation considered,

$$E(\Delta\ddot{Y}_X) = [k/\sigma(\hat{X})][(1/2)\sigma^2_{gx}] = (2.06/7.72)(12.5) = 3.34$$

and non-validity of theory would not be demonstrated conclusively by absolute magnitudes of •D as great as 48 percent of $E(\Delta\ddot{Y}_X)$.

Notes

CHAPTER 1

1. Marker assisted selection, MAS, greatly facilitated by advances in molecular genetic technology, is an indirect mode of selection based on phenotype. For thoughtful discussions of its potential see Dentine (1992), Stuber (1992), and Dudley (1993)).

2. Clearly, value to the agricultural industry and to society as a whole are likely to be different except, perhaps, in some presently non-existent utopian society.

CHAPTER 2

1.The word "relatively" recognizes the mild inconstancy resulting from mutation.

2. In finite populations there is genetic variation from one generation to another that results from sampling from the infinite population of genotypes possible. This variation is referred to as random genetic drift, random drift, or simply drift. See Section 2.6.3 for further discussion.

3. Departures from this rule may be required when two or more aspects of phenotype are considered at the same time. The allele of a pleiotropic gene that is best with respect to one of the traits may not be best with respect to another.

4. Segregation would be maintained in populations of infinite size but despite the effect of selection would eventually be lost in finite populations (see section 2.8).

5. The quantity defined here is referred to by Crow and Kimura (1970) as variance effective population number (size) as distinguished from what they identify as inbreeding effective population number.

6. Here "in the same way" implies that all of the closed populations are identical with respect to effective population size, number of generations since being established from the source population, and the absence of selection assumed by equations 2.2.

7. The demonstration in this section assumes only two alleles per gene but can be extended to the multiple allele situation.

CHAPTER 3

1.For convenience, I have ignored the possibility that two or more alleles of the same gene may be equally good.

CHAPTER 5

1. This does not imply that each genotype has the same value relative to each environment in the specified population of environments. On the contrary, assuming genotype x environment interaction the value of every particular genotype is expected to be variable from one environment to another. It is the average value of the genotype over the entire population of environments that we assume to be constant.

2. The single exception is encountered, as noted by Hill (1970), when the y's for BB and bb genotypes are equal but not equal to the y for the Bb genotype. In that event u would be zero, making au = 0.0 despite the real deviation of the heterozygote y from the mean of the homozygote y's. However, given a gene with a bearing on the trait in question, evidenced by the difference between the impacts of the heterozygous and homozygous genotypes, the probability of exact equality of the homzygous effects must be infinitesimal. Thus the theoretical inadequacy seems to me trivial for practical purposes.

3. In the case of truncation selection the most extreme units are those selected, i.e., the $\hat{X}$ value of each selection is greater (or less) than that of every unit not among those selected.

4. For definitions of terms relating to evironments see sections 2.13 and 3.1.

CHAPTER 6

1. These are frequencies, in the case of gametes, in the infinite population that conceptually might be produced by the parents of a single generation and, in the case of genotypes, in the infinite population associated with random pairs from that gamete population.

2. For details of algebra see Appendix 6.1.

3. The definition of additive genetic variance provided by 6.19 agrees with those derived by Fisher (1918 and 1930), Wright (1935), Kempthorne (1957), Falconer (1960), and Crow and Kimura (1970) when two alleles per segregating locus, no epistasis, and H-W frequencies of single locus genotypes are assumed. Fisher employed "genetic" in place of "additive genetic".

4. For a detailed demonstration, see Crow and Kimura (1970).

APPENDIX 6.2

1. r = probability of recombination between the i-th and j-th loci.

CHAPTER 7

1. F(x) symbolizes the fraction of a distribution in which values of the variable are less than x.

2. The fact that for practical purposes u can be considered equal to $u/\sigma(\hat{X})$ deserves emphasis. Strictly, $\sigma^2(\hat{X}) = 1 + \sigma^2(y)$ as noted earlier. However, among the combinations of variables considered the largest value of $\sigma^2(y)$ encountered was .0827 (given a = 2.0, u = 0.2, and q = 0.3). In that case, $\sigma(\hat{X})$ was $(1.0827)^{1/2} = 1.04$ and $u/\sigma(\hat{X})$ was 0.192.

3. Obviously the final outcome of recurrent selection is not certain in any case when the genetic population is finite. For example, when original values of both p and q are greater than one-third there is a finite probability that by "drift" both frequencies could become smaller than one-third and then, in response to selection, fixation of both b_1 and b_2 would be favored.

CHAPTER 8

1. $\sigma(\hat{X})$ varies with the criterion unit, not with the selection unit.

2. $E^*(\hat{y})$ will be used consistently to symbolize $E(\hat{y}) - K$ where K is an identified constant.

3. This relation between the inbreeding of selection unit individuals and the response expectations was demonstrated by Comstock et al (1949). They, however, did not recognize the possible effects of inbreeding on $\sigma(\hat{q}_i,\hat{X})$ by virtue of effects on $\sigma(\hat{q}_i,\hat{y}_{xj})$.

4. However, they implied erroneously that effectiveness in progeny test selection for "general combining ability" should be proportional to genetic variance among the test progenies. This point will receive further attention in Chapter 9.

APPENDIX 8.1

1. Here "random selection" is meant to exclude both natural and artificial selection so that the probability of survival of individuals is not a function of their genotypes.

CHAPTER 9

1. As originally described, RSSC involved a homozygous tester but the procedure can of course be followed using other kinds of testers.

2. Because seed by self-fertilization is required from both parents and test cross seed must be produced by at least one of them, a reasonable frequency of two-eared plants is required in at least one of the RRS poulations and there would be some direct selection for ear number in the S population used as the seed parent unless all plants in that population produced two or more ears.

3. However, the rates at which they would be achieved would ordinarily differ. Reasons include the facts that the k's and $\sigma(\hat{X})$'s of 9.18 and 9.19 would ordinarily vary among selection systems, that the numerical coefficient of 9.18 and 9.19 is not the same for all systems, and that cycle length is not the same for all systems.

4. For other reasons RRS might not be the best choice. For example, in animal species number of offspring that can be obtained per individual often places serious limits on the effectiveness of progeny test selection applied to females.

5. For the purposes of this section "inbred line" and "pure line" will be used interchangeably and will mean one of which all members are homozygous for the same allele at most (say, 97 percent or more) of the loci that were segregating in the non-inbred population from which the line was derived. As shown in section 2.5, this level of inbreeding can be achieved by five generations of self-fertilization.

6. Estimated from the performance of progenies obtained by crossing the lines with a genetically heterogeneous tester.

7. Such lines, whether or not from a linkage equilibrium source population, are now commonly referred to as being derived by "single seed descent".

8. Failure to treat selection among double-cross or three-way hybrids in detail appears justified by the fact that they have now been largely replaced in maize production by single

cross hybrids. In addition, theory for single crosses provides a basis for inference concerning effects of selection among the other two types of hybrids.

9. Assuming, if there is more than one non-lethal allele, that the homozygote of each is equivalent in all effects to the lethal heterozygote involving the same non-lethal allele.

10. Note that here the subscript v signifies that $\ddot{Y}_v$ pertains to viable genotypes; it does not identify a trait. The average value of viable genotypes with respect to trait C would be symbolized as $\ddot{Y}_{vC}$.

11. In this connection $\pi_i K_i$ is to be interpreted as the product of K's over only the loci heterozygous for a lethal allele. Note also that effects of single-locus genotypes are being assumed to be multiplicative.

12. Equation 9.29 was obtained from the $\sigma(\hat{q}_i,\hat{X})$ shown in Table 8.1 by setting h = 1.0. It would be 1.0 in a pure line and would be nearly 1.0 in what are referred to herein as pure lines.

13. Actually, given equality in plants per plot, replications, etc., $\sigma(\hat{X})_p/\sigma(\hat{X})_1 < 1.0$ is probable because the single genotype of a pure line would exhibit more genotype x environment interaction than the average for an S_1 family. Still the ratio of the two standard deviations would hardly be less than 0.8.

14. In the case of maize harvestable yield of grain (measured at an appropriate calendar date and adjusted to a constant moisture content) constitutes a practical approximation to total value. Harvestable yield denotes the grain that would be harvested by a standard commercial picker. It does not include grain from "dropped" ears or ears "on the ground" as the result of lodging.

CHAPTER 10

1. Allen et al (1978) followed this up and pointed out that, if they could be identified, macro-environments for which the covariance of z and (fz) is positive and large would be superior ones in which to collect the data to be used as the basis for selection.

2. This would be the case for single parent families (those produced by self-fertilization or by asexual reproduction from a single individual or clone) or bi-parental (full-sib) families. It is not so in the case of half-sib families within which full-sibs, if present, are more closely related than members of different full-sib families within the half-sib family.

3. Obviously, the year effect reflects the uniqueness of climate in that year but may also reflect a difference from other years with respect to disease incidence or pest populations. Location effects will reflect variation in soil and the history of crops grown and cultural practices in preceding years.

4. Definitions of year and location effects are similar to the definition of macro-environment effects, set out in section 2.12. Thus, for example, the effect of a specific year is defined to be the average of the effects of all micro-environments possible within that year and is therefore not variable among locations.

5. Considering the two parents of any single member of a half-sib family, the one that is also a parent of every other member of the family is called the "common parent", the other is the "random parent".

6 The assignment of each full-sib family to a different pen and the assignment of each individual to its own pen are versions of this that have already been discussed.

7. In the denominator of the last term of 10.31, p = the number of replications in field comparisons of criterion units. The usual "r" symbolizes "random parent" in this expression.

8. It is clear from the context of their paper that they meant selection differential in standard units (k).

9. This assumes that the genotype x environment interaction variances would always be non-zero.

CHAPTER 11

1. For this to be true when multiple peak epistasis is present Y_{wi} must be defined as the value of the best genotype at the "peak" toward which the population is brought given the allele frequencies of the foundation population (see sections 3.6.4 and 7.3).

2. Kimura's result, equation 2.52, was not intended to yield probabilities of fixation in the case of overdominant genes in single population selection systems (see Robertson, 1962).

3. These deductions are supported by the data of Carr and Nassar (1970).

4. Unpublished work by Comstock and Ankelesaria (1982).

5. Some consequences in the case of self-fertilizing crop species were demonstrated in the study reported by Bailey and Comstock (1976).

6. If, as assumed throughout, there is no meiotic drive, no gametic selection and no covariance between the genotypes of parents and the numbers of their offspring in the genetic population.

7. Exceptions, of course, are functions of an allele frequency that is known or that has been assigned an arbitrary value.

8. This assumes equal contributions from each of the three parent lines.

9. But intuition suggests that, given $q_0 = 0.3$ instead of 0.7, the reverse would be true in early cycles.

APPENDIX 11.3:

1. Note the distinction between $\bar{q}_1$ and q_1. The latter is the B allele frequency in the infinite S_1 population and the expectation of allele frequency in samples from that population. The former is the a priori expectation of q_1 itself, i.e., the expectation of q_1 before it has been determined by the selection phase of cycle 0.

2. Because when regression is linear the expectation of each deviation from regression is zero no matter the value of the independent variable.

CHAPTER 12

1. For an extensive treatment of this question and variations of it, see Baker (1986).

2. For discussions of "breeding value" see Appendix 6.1 or Falconer (1960).

3. Of course some species are used for more than one purpose. For example, some cattle are grown primarily for milk production, others for beef and the appropriate definition of total value is obviously, not the same for dairy as for beef cattle.

4. If no statement to the contrary, $k > 0.0$ is always assumed.

5. The base index defined and treated by Williams (1962) would make total value the selection criterion if measurements of all value traits were available and included in the index.

6. Recall that measurement units are arbitrary. Thus, for example, if the unit were made twice as large for the first trait, the effects would be (1) to make each measurement half as large numerically, (2) to halve each value of G_{12} and σ_{12} and make each value of G_{11} and σ_{11} one-quarter as large, and (3) to double the value of c_1. Neither the expectations of $\Delta\ddot{Y}_w$ nor the values of E_p/E_m would be changed. It follows that the decision to let $G_{11} = G_{22}$ and $\sigma_{11} = \sigma_{22}$ in these examples was arbitrary and not critical relative to E_p/E_m.

7. The context of their discussion justified their speaking in terms of heritability. Here G_{mm}/σ_{mm} is not referred to as heritability because unless individuals are both the selection and criterion units G_{mm}/σ_{mm} is not heritability as defined for the purpose of this text.

8. Estimates can be made in terms of certain live animal measurements. However, the precision of such estimates may be low and, at best, is uncertain. Moreover, estimates of carcass value would require, at minimum, the combination of estimates of (1) dressing percent and (2) fraction of lean tissue in the carcass.

9. It is worth noting that the practical problem of parameter estimation is eased in this special case by the fact that all the G's are ascertainable multiples of one parameter, the additive genetic variance of the trait.

10. Unless the gene in question exhibits overdominance relative to total value and $q_i > (1+a_{wi})/2a_{wi}$

11. Note the following: (1) given linkage disequilibrium, $E(\Delta q) > 0.0$ for all useful alleles is not assured whatever the weighting coefficients, (2) a sufficient approach to linkage equilibrium can be expected when effective population size is adequate by the probability of fixation criterion (Chapter 11), and (3) given more than one criterion unit, there are weighting coefficients, determined by the c-values and the relationships between the selection unit and the criterion units that insure $E(\Delta q) > 0.0$ (Section 12.6).

12. For example, it was possible to forecast the decline in the value of lard relative to other components of the swine carcass that occurred during the first half of this century.

APPENDIX 12.2:

1. Indentifying grain as the product writes off stalks, husks, and cobs as waste of which variation in value is of no consequence.

2. Obviously, the time unit employed should be meaningful in terms of record keeping and the management plan for the program.

CHAPTER 13

1. This hinges on whether overdominance is significant in the genetics of total value.

2. As in selection for improvement of any single breed of dairy cattle or selection within two or more populations of swine for improvemnt of the rotation cross of those populations.

3. Remember that, in the absence of epistasis, $E(\Delta\ddot{Y}) = \Sigma_i(\Delta\ddot{y}_i)$.

4. A "category of genes" includes, in the context of this sentence, all genes for which the

probabilities of fixataion of the desired allele are the same when the applicable values of N_es' are the same.

5. Whether the ranking of sytems would be different depending on procedure employed has not been investigated by the author.

CHAPTER 14

1. Obviously, the silks in question must have been shielded from pollen transported by natural vectors.

2. In the case of system 8, the 2 in the denominator reflects selection based only on seed parent performance.

3. Values were assigned assuming plot error coefficients of variation equal to 10 and 12 percent for non-inbred materials and near homozygous lines, respectively, and linearly related to inbreeding coefficients in other instances.

4. Because the speed with which the population average of genotype values approaches the value of the ideal genotype is directly proportional to Δq.

5. Subject to the core assumptions of this text.

6. At first glance, readers may be puzzled by the small size of quantities shown in Table 14.13. Remember, however, that these are expected changes in the average value of genotypes at a single locus as opposed to the sum of such quantities for all genes that have an impact on the trait in question. The values shown under L(3) for system 5 range from .0016 to .0096. Thinking of all the genes that affect the trait let us imagine that variation among them in q, u, and a is such that the average of $E(\Delta\ddot{y})$ for all genes is .005 and that there are 300 effective genes. Assume further that selection intensity is such that k = 2.0. The value of $E(\Delta Y) = \Sigma_i E(\Delta\ddot{y})$ would then be 2(300)(.005) = 3.0, i.e., 3% of our assumed population average, $\ddot{X} = 100$.

7. The reader is reminded that the numerical subscript in 14.10 and 14.11 identifies cycle of the recurrent selection program, that the initial cycle is cycle 0, and hence that the subcript 1 designates the second cycle in the series (0,1,2,·········).

8. Negative or positive depending on whether a is greater or less than zero.

9. This is tedious and time consuming if done with no more powerful equipment than a non-programmed, hand operated electronic computer, particularly when effects of variation of all parameters involved are to be quantitized. On the other hand, if done using a modern high speed computer, the task can be accomplished with relative ease.

10. The effects of u being larger or smaller than 1/2 are not shown in Tables 14.15 - 14.18 but are considered later in this section.

11. Because the $\sigma(\hat{X})$'s of part (b) assumed v = 3 and t = r = 1 as did those ot table 14.13.

12. $E(\ddot{y}_1) = E(\ddot{y}_0) + E(\Delta\ddot{y})_0$ and $E(\ddot{y}_2) = E(\ddot{y}_1) + E(\Delta\ddot{y})_1$.

13. Of course for reasons not taken into account in the comparisons shown in tables of this chapter, this system might not be employed by commercial breeders.

14. Longevity and lifetime production of blueberries and dairy cows are obvious trailing components of total value in those species.

15. Here π denotes the probability of the new superior allele in each effective gamete.

16. Remember that values of $E(\Delta q)/q(1-q)L$ shown pertain to alleles with additive effects ($a = 0$) and that in the case of mass selection in an annual crop, $L = 1$.

17. Not exactly in the case of selection systems 6, 7, and 8 but very nearly so unless p is large.

CHAPTER 15

1. In practice, the first of these can be roughly correct. Years cannot be random except to the extent that the physical attributes (such things as temperature, rainfall, incidence of pests, etc. and the distribution of these during the growing season of the plant) that constitute the environment of a year are themselves random from one year to the next. We are aware, however, that the variation of these things is partly cyclic, i.e., not entirely random. The locations employed can, on the other hand, be a resonably random sample from those of the population about which information is sought. Of course, if family x environment interaction variances were uniform among macro-environments the accuracy with which they would be estimated would not depend on randomness of the years and locations in which data are collected.

2. This expression (15.2) is the generalization that covers the variance of $s_X^2 = \Sigma_i(X_i-\bar{X})^2/(n-1)$ given by A1.1. The structure of s_X^2 is that of a mean square and the expectation of s_X^2 is $\sigma^2(X)$.

3. However, in the case of full-sib families or half-sib families of which the constant parents are females, some portion of $\sigma^2(\bar{z})$ may arise from variation in maternal or cytoplasmic effects.

4. Note, however, that the variance estimated from data on such families may not relate to a population of the type for which kinds of variances have been defined. For example, the genetic variance estimated from data on clonal families would not be the total variance of a random mating population unless the parental genotypes were a random sample from such a population.

5. To circumvent the problem posed by environment covariance between relatives in estimates of the genetic variance of IQ, investigators have employed (a) covariance between identical twins separated at birth by adoption (see, for example, Bouchard et al, 1990) and (b) covariance between biological parents and offspring adopted at birth. The hope in each of these approaches is that the adoption event resulted in the relatives being exposed to random members of the environment population.

6. If the premises on which the derivation of genetic covariance between parent and offspring had allowed epistasis but not linkage disequilibrium the result (when the constant and random parents belong to the same population) would have been, according to Cockerham (op. cit.),

$$\sigma(Y_{xp},Y_{xo}) = \Sigma_{AO}^{n}(1/2)^{A}\sigma_{AD}^{2}$$

in which

$$\sigma_{10}^{2} = \Sigma_i 2q_i(1-q_i)[1+(1-2q_i)a_i]^2 u_i^2 \quad \text{as given by 6.8.}$$

7. The slow growing animals selected should not include individuals that have grown slowly because of injuries or obvious illnesses.

8. Because the preponderance of positive a's responsible for significant magnitudes of heterosis and inbreeding depression means that $a_iu_ia_ju_j$ would more often than not be a positive quantity.

9. Unfortunately, the heading of the third column of Table 7 in their publication is in error. It should have been (1/4)Vd instead of (1/2)Vd.

10. Whether maternal effects are non-trivial in the case of particular traits is not always obvious. Consider weaning weight and litter size in swine. There can be no doubt that maternal effects contribute to phenotypic variation of weaning weight. On the other hand, it is far from obvious that there is a non-genetic effect of mother on size of litter produced by daughters. In species in which it is possible to compare the variance among paternal half-sib families with the variance among maternal half-sib families, experimental evidence can be obtained. Maternal effects would be indicated by greater variance among maternal than among paternal half-sib families.

11. Of course if data had been collected on the progenies of two or more sets, the analyses of data from the separate sets would logically be pooled as discussed in section 15.1.1 and illustrated by comparison to Tables 15.1 and 15.5.

12. Unless at a substantial majority of loci, at which the inbred line progenitors of the F_2 differed in genotype, the better allele was present in the same line. If the phenotypic means of the two lines were not distinctly different a reasonably equal division of the favorable alleles and a consequent approach to equality in numbers of positive and negative values of (pt-rs) would be indicated.

13. The evidence from those studies support the same conclusions in the case of such other traits as plant height, ear height, and number of ears.

14. Of course non-significance does not establish that the gene does not affect the trait--the null hypothesis cannot be proven.

15. In accord with the description of the binomial distribution in Appendix 1, the probability that x genes in a random sample of n will be found to be genes with effects on a specific trait in

$$[n!/x!(n-x)!]P^x(1-P)^{n-x}$$

where P = the fraction of all genes that affect the trait. Thus if P = 0.04 the probability that two of 20 would be effective genes is $[20!/(2)!(18)!](0.04)^2(0.96)^{18} = 0.1452$. The probabilities of all values of x less than three (out of 20) are as follows:

x	Prob. of x
0	0.4420
1	0.3683
2	0.1452
Sum	0.9555

This leaves 1-0.0445 as the probability of finding x=3 or more effective genes among those in a random sample of 20. Binomial distribution probabilities are included in various sources of mathematical tables. Those shown above are included in Standard Mathematical Tables, 15th edition (1967), The Chemical Rubber Company.

16. Readers should be aware that this example is presented for illustrative purposes, not as real evidence relative to number of genes with effects on quantitative traits.

17. They must be 0.5 in the F_2 though not exactly 0.5 in samples from that infinite population.

18. However, if the two pure line ancestors of the F_2 were phenotypically very different in the trait being investigated a corresponding difference in genotypic value would be indicated. If such a difference were real it would be because at loci where different alleles were present in the lines there were more alleles favorable to the trait in one line than in the other. In that case, there would be more coupling than repulsion of favorable alleles and, considering all pairs of genes affecting the trait, $(pt-rs) > 0$ would be more frequent than $(pt-rs) < 0$. The result could then be a positive value of $\Sigma_i\Sigma_{j\neq i}(pt-rs)_{ij}u_iu_j$ large enough to introduce an upward bias

19. All quantitative genetic theory rests on simplifying assumptions. With respect to theory presented and developed herein, the author's intent has been to identify the most important of such assumptions. Section 16.1 discusses impacts and their possible and probable significance.

20. Moll (1991) found that full-sib family selection for grain yield of maize was effective through sixteen cycles of recurrent selection but did not examine how well response agreed with prediction.

<u>Appendix 15.3</u>

1. The subtraction of u from both $\bar{y}_{1ki}$ and $\bar{y}_{2ki}$ leaves d_{ki} unchanged; hence the d_k's are not designated as coded.

<u>Appendix 15.4</u>

1. Again, if similar data were obtained on more than one set of progeny pairs, the analyses of data from the several sets would logically be pooled as exemplified by the content of Table 15.5 as a modification of the analysis shown in Table 15.1.

2. Approximation because the basic equality, $\sigma^2(M) = 2[E(M)]/f$ in which f = degrees of freedom and M symbolizes a mean square, is correct when the variable in question is normally distributed.

<u>Appendix 15.6</u>

1. Remember that $E(\Delta q)$ in response to one cycle of selection is small unless the u-value of the allele is unusually large.

2. Underlying assumptions are that (a) the measurement unit is such that $\ddot{X} = 100$, (b) $\sigma_g^2 = 25$, (c) ratios of of genotype-environment interaction variances to the genetic variance of the unit (individual or family) in question are as indicated in Table 14.4 and section A15.6.1, and (d) σ_Y^2, the total genetic variance among individuals of a random mating population, consists of the additive genetic variance plus σ_d^2, the dominance variance. From Table 5.1 of Hallauer and Miranda (1981) it appears reasonable to assume $\sigma_d^2 = 0.8\sigma_g^2$. Using these quantifications, we obtain $25 + 20 + (45)/4 + (45)/4 + (45)/2 = 90$ as a reasonable approximation to σ_Y^2.

Using the above assumptions with $\sigma^2(e) = 100$ and $\sigma^2(\bar{z}) = \sigma_g^2/2 + \sigma_d^2/4$ (from Table 15.8) added, $\sigma^2(\hat{X})$ in full-sib family selection when v=3 and r = t =1 becomes (by 14.4)
$25/2 + 20/4 + 17.5/4v + 17.5/4t + 17.5/2vt + 100/rvt = 59.58$.

CHAPTER 16

1. The results of 76 generations of selection for high oil and high protein in the grain of maize reported by Dudley (1977) did not indicate that maximums had been reached despite the fact that increases of 279 and 133 percent, respectively, of initial oil and protein levels had been achieved. Moreover, the foundation populations, derived from a single open-pollinated variety, could hardly have possessed all plus alleles for either oil or protein.

2. The genetic variance that enabled the changes in species during their evolutionary history together with the genetic variation now present attest to mutation as a continuing souce of new favorable, as well as unfavorable, alleles.

3. Implied by the definition of total value is that it is a logical function of value traits measured in random members of the target population of environments. Thus, for example, when resistance to particular pathogens is judged so important that criterion units must be compared in terms of their performance when they have been "diseased" means that total value cannot be the selection criterion because the macro-environment created is not a random member of the TPE.

4. The author's opinion, for what it is worth is that animal species would not be high priority candidates for inclusion in a program of large recurrent selection projects. Except when animals are employed primarily to harvest forage from land not useful for production of food from plant species, the caloric value of their products is likely to be more expensive in terms of land use than food from plants.

5. Remember that these expressions are functions of X, the selection criterion and hinge on simplifying assumptions that include both linkage equilibrium in the S population and no dominance in the interaction of the "best" allele with other alleles of the same gene. Because $E(\Delta q) > 0.0$ for alleles that favor total value is only guaranteed by using total value as the selection criterion, that special case will be the focus here. Second, it is reasonable to assume little linkage disequilibrium in the S populations of programs in which N_e would be large. And finally dominance of the better allele would not cause a qualitative change in general conclusions.

6. It is possible that prevailing opinion will dictate that each nation, using assistance already in place, be responsible for its own solution to its population-food problems. As justification it could be emphasized that first Japan, then China, using incentives and education, reduced their population growth rates by about 50 percent within 7 year periods (Brown et al 1991, 103).

7. That procedural proposal reflected the fact that primary emphasis was being placed on ultimate fixation of favorable alleles as opposed to the time required to achieve the associated genetic improvement.

References

Allard, R.W. 1960. Principles of plant breeding. John Wiley and Sons. New York.

Allen, F.L., R.E. Comstock, and D.C. Rasmusson. 1978. Optimal environments for yield testing. Crop Science 18: 747-751.

Bailey, Theodore B.,Jr., and R.E. Comstock. 1976. Linkage and the synthesis of better genotypes in self-fertilizing species. Crop Science 16:363-370.

Baker, Marvel L., L.N. Hazel, and C.F. Reinmiller. 1943. The relative importance of heredity and environment in the growth of pigs at different ages. J. Anim. Sci. 2: 3-13.

Baker, R. J. 1986. Selection indices in plant breeding. CRC Press Inc., Boca Raton, FL

Becker, Walter A. 1984. Manual of quantitative genetics (4th ed). Academic Enterprises, Pullman WA.

Bell, A.E., C.H. Moore, and D.C. Warren. 1955. The evaluation of new methods for the improvement of quantitative characteristics. Cold Spr. Harb. Symp. Quant. Biol. 20:197-211.

Bernstein, F. 1925. Zusammenfassende betrachtungen uber die erblichen blutstrukturen des menschen. Zeit. ind. Abst. Vererb. 37:237-269.

Borlaug, N.E. 1953. New approach to the breeding of wheat varieties resistant to Puccinia graminis tritici. Phytopathology 43:467.

Bouchard, Thomas J., Jr., David T. Lykken, Matthew McGue, Nancy L. Segal, and Auke Tellegen. 1990. Sources of human psychological differences: The Minnesota study of twins reared apart. Science 250:223-228.

Brieger F.G. 1950. The Genetic basis of heterosis in maize. Genetics 35(4):420-445.

Brown, Lester R. 1990. The illusion of progress. State of the world 1990. W.W. Norton & Company, New York & London. pp. 3 -16.

Brown, Lester R., Christopher Flavin, and Sandra Postel. 1991. Saving the Planet, Chapter7. W.W. Norton & Company, New York & London.

Bruce A.B. 1910. The Mendelian theory of heredity and the augmentation of vigor. Science 32: 627-628.

Carr, R.N., and R.F. Nassar. 1970 Effects of selection and drift on the dynamics of finite populations. I. Ultimate probability of fixation of a favorable allele. Biometrics 26: 41-49.

Castle, W. E. 1921 An improved method for estimating the number of genetic factors concerned in cases of blending inheritance. Science 54 (no.1393): 223.

Castleberry, R.M., C.W. Crum, and C.F. Krull. 1984. Genetic yield improvement of U.S. maize cultivars under varying fertility and climatic environments. Crop. Science 24: 33-36.

Cockerham, C.C. 1954. An extension of the concept of partitioning hereditary variance for analysis of covariances among relatives when epistasis Is present. Genetics 39: 859-882.

Cockerham, C.C., and B.S. Weir. 1973. Descent measures for two loci with some applications. Theor. Pop. Biol. 4:300-330.

Cockerham, C.C., and B.S. Weir. 1977. Digenic descent measures for finite populations. Genet. Res. 30: 121-147.

Collins, G.N. 1921. Dominance and the vigor of first generation hybrids. Amer. Natr. 55:116-133.

Compton, W.A., and R.E. Comstock. 1976. More on modified ear-to-row selection in corn. Crop Science 16: 122.

Comstock, R.E. 1955. Theory of quantitative genetics: Synthesis. Cold Spring Harbor Symposia On Quantitative Biology 20: 93-102.

___________. 1964. Selection procedures in corn improvement. Proceedings of the Nineteenth Annual Hybrid Corn Industry-Research Conference, pp 1-8.

___________. 1969. Number of genes affecting growth in mice. Genetic Lectures. Oregon State University Press 1:137-148.

___________. 1977. Quantitative genetics and the design of breeding programs. Proceedings of the International Conference on Quantitative Genetics. Edward Pollak, Oscar Kempthorne, and Theodore B. Bailey, Jr. eds., pp. 705-718. Iowa State University Press, Ames, Iowa.

___________. 1979. Inbred lines vs. the populations as testers in reciprocal recurrent selection. Crop Science 19: 881-886.

Comstock, R.E., and F.D. Enfield. 1981. Gene number estimation when multiplicative genetic effects are assumed - growth in flour beetles and mice. Theor. Appl. Genet. 59: 373-379.

Comstock, R.E., and R.H. Moll. 1963. Genotype-environment interactions. Statistical Genetics and Plant Breeding, NAS-NRC 982, pp. 164 - 196.

Comstock R.E, and H.F. Robinson. 1948. The components of genetic variance in populations of biparental progenies and their use in estimating the average degree of dominance. Biometrics 4: 254-266.

Comstock, R.E., and H.F. Robinson. 1952. Estimation of average dominance of genes. In Heterosis. J. W. Gowen, ed., Iowa State College Press. Ames, Iowa. pp 494-516.

Comstock, R.E., H.F. Robinson, and P.H. Harvey. 1949. A breeding procedure designed to make maximum use of both general and specific combining ability. Agron. J. 41:360-367.

Crow, James F. 1948. Alternative hypotheses of hybrid vigor. Genetics 33: 477-487.

__________. 1952. Dominance and overdominance. In Heterosis, J.W. Gowen, ed., Iowa State College Press. Ames, Iowa. pp 282-297.

Crow, James F., and Motoo Kimura. 1970. An intrdoduction to population genetics theory. Harper & Rowe. New York, Evanston, and London.

Darrah, L.L., S.A. Eberhart, and L.H. Penney. 1972. A maize breeding study in Kenya. Crop Science 12: 605-608.

Dentine, M.R. 1992. Marker-assisted selection in cattle. Animal Biotechnology 3(1): 81-93.

Dudley, J.W. 1977. 76 Generations of selection for oil and protein percentage in maize. Proceedings of the International Conference on Quantitative Genetics. Edward Pollak, Oscar Kempthorne, and Theodore B. Bailey, Jr. eds., pp. 459-473. Iowa State University Press, Ames, Iowa.

__________. 1993. Molecular markers in plant improvement: Manipulation of genes affecting quantitative traits. Crop Science 33: 660-668.

Duvick, D.N. 1977. Genetic rates of gain in hybrid maize yields during the past 40 years. Maydica 22:187-196.

__________. 1992. Genetic contributions to advances in yield of U.S. maize. Maydica 37: 69-79.

Duvick, Donald N. 1984. Genetic contributions to yield gains of U.S. hybrid maize. Genetic Contribution to Yield Gains of Five Major Crop Plants. Chapter 2. pp15-47. ASA, CSSA, 677 South Segoe Road, Madison, WI.

East, E.M. 1908. Inbreeding in corn. 1907 Report Connecticut Agr. Expt. Sta., pp 419-428.

__________. 1910. A Mendelian interpretation of variation that is apparently continuous. Amer. Nat. 44: 65-82.

__________. 1936. Heterosis. Genetics 21:375-397.

Edwards, M.D., C.W. Stuber, and J.F. Wendel. 1987. Molecular-marker-facilitated investigations of quantitative-trait loci in maize. I. Numbers, genomic distribution, and types of gene action. Genetics 116:113-125.

Empig, L.T., C.O. Gardner, and W.A. Compton 1971. Theoretical gains for different population improvement procedures. Nebraska Agric. Exp. Sta. Bull. MP26.

Empig, L.T., C.O. Gardner, and W.A. Compton 1972. Theoretical gains for different population improvement procedures. Nebraska Agric. Exp. Sta. Bull. MP26 (revised).

Enfield, F.D. 1977. Selection experiments in Tribolium designed to look at gene action issues. Proceedings of the International Conference on Quantitative Genetics. Edward Pollak, Oscar Kempthorne, and Theodore B. Bailey, Jr. eds., pp.177-190. Iowa State University Press, Ames, Iowa.

__________. 1980. Long-term effects of selection; the limits to response. Proc. Symp. Selection Exp. Lab Domestic Animals. Commonwealth Agric Bureau, UK, pp 69-86.

Enfield, F.D., R.E. Comstock, Robert Goodwill, and O. Braskerud. 1969. Selection for pupa weight in Tribolium castaneum. II. Linkage and level of dominance. Genetics 62: 849-857.

Enfield, Franklin D., and Elizabeth Darmo. 1992. Quantitative genetic variation from new mutations: Response to directional selection for pupa weight in highly inbred lines of Tribolium. Personal communication from Prof. Enfield.

Falconer, D.S. 1953. Selection for large and small size in mice. J. Genetics 51: 470-501.

__________. 1960. Introduction to quantitative genetics. Oliver and Boyd The Ronald Press Co., New York.

__________. 1981. Introduction to quantitative genetics, 2nd ed., Longman, Inc., New York.

Fisher, R. A. 1918. The correlation between relatives on the supposition of Mendelian inheritancae. Trans. Roy. Soc. Edinb. 52: 399-433.

__________. 1930. The genetical theory of natural selection. Clarendon Press, Oxford.

__________. 1930. The Genetical Theory of Natural Selection, 2nd ed., Dover, New York.

__________.1949. The theory of inbreeding. 2nd Ed. Oliver and Boyd, London.

Fisher, R.A., and F. Yates. 1963. Statistical tables for biological, agricultural, and medical research. 6th ed. Hafner Publ. Co., New York.

Gardner, C.O. 1961. An evaluation of effects of mass selection and seed irradiation with thermal neutrons on yield of corn. Crop Science 1:241-245.

Gardner, C.O. 1963. Estimates of genetic parameters in cross-fertilizing plants and their implications in plant breeding. Statistical genetics and plant breeding. NAS-NRC 982. pp 225-252.

__________. 1977. Quantitative genetic studies and population improvement in maize and sorghum. Proceedings of the international conference on quantitative genetics. Edward Pollak, Oscar Kempthorne, and Theodore B. Bailey, Jr. eds., pp. 475-487. Iowa State University Press, Ames, Iowa.

Gardner, C.O., and J.H. Lonnquist. 1959. Linkage and the degree of dominance of genes controlling quantitative characters in maize. Agron. J. 51: 524-528.

Gates, C.E., R.E. Comstock, and H.F. Robinson. 1957. Generalized genetic variance and covariance formulae for self-fertilized crops assuming linkage. Genetics 42:749-763.

Haldane, J.B.S. 1924. A mathematical theory of natural selection. Part I. Trans. Camb. Phil. Soc. 23:19-41.

__________. 1924a. A mathematical theory of natural selection. Part II. Biol. Proc. Camb. Phil. Soc., Biol. Sci. 1:158-163.

__________.1924b. A mathematical theory of natural selection. Part III. Proc. Camb. Phil. Soc. 23:363-372.

__________. 1924c. A mathematical theory of natural selection. Part IV. Proc. Camb. Phil. Soc. 23:235-243.

__________.1927. A mathematical theory of natural selection. Part V. Selection and mutation. Proc. Camb. Phil. Soc. 28:838-844.

___________.1930. A mathematical theory of natural selection. Part VI. Isolation. Proc. Camb. Phil. Soc. 26:220-230.

___________.1930a. A mathematical theory of natural selection. Part VII. Selection intensity as a function of mortality rate. Proc. Camb. Phil. Soc. 27:131-136.

___________.1930b. A mathematical theory of natural selection. Part VIII. Metastable populations. Proc. Camb. Phil. Soc. 27:137-142.

___________.1932. A mathematical theory of natural selection. Part IX. Rapid selection. Proc. Camb. Phil. Soc. 28:244-248.

Hallauer, A.R. and S.A. Eberhart. 1970 Reciprocal full-sib selection. Crop Science 10: 315-316.

Hallauer, Arnel R., and J.B. Miranda. 1981. Quantitative genetics in maize breeding. Iowa State University Press, Ames.

Hardy, G.H. 1908. Mendelian proportions in a mixed population. Science 28: 49-50.

Harlan, Jack R. 1975. Our vanishing genetic resources. Science 188(4188): 618-621.

Harris, D.L. 1964. Expected and predicted progress from index selection involving estimates of population parameters. Biometrics 20: 46-72.

Hartl, D.L. 1977. Applications of meiotic drive in animal breeding and population control, Proceedings of the International Conference on Quantitative Genetics. Edward Pollak, Oscar Kempthorne, and Theodore B. Bailey, Jr. eds., pp 63-88, Iowa State University Press, Ames, Iowa.

Hayes, Herbert Kendall. 1963. A professor's story of hybrid corn. Burgess Publishing Company, Minneapolis, MN.

Hayes, Herbert Kendall, and Forrest Rhinehart Immer. 1942. Methods of plant breeding. McGraw-Hill Book Company, Inc. New York and London. pp 205-206.

Hazel, L.N. 1943. The genetic basis for constructing selection indexes. Genetics 28:476-490.

Hazel, L.N., and J.L. Lush. 1942. The efficiency of three methods of selection. J. Hered. 33:393-399.

Henderson, C. R. 1963. Selection index and expected genetic advance. Statistical Genetics and Plant Breeding, W.D. Hanson and H.F. Robinson, eds., NAS - NRC Publication No.982, pp.141-163.

Hill, W.G. 1970. Theory of limits to selection with line crossing. In Biomathematics, Vol. 1. Edited by K. Krickeberg, R.C.Lewontin, J. Neyman and M. Schreiber. Springer-Verlag. Berlin, Heidelberg, New York. pp 210-245.

________. 1982. Predictions of response to artificial selection from new mutations. Genet. Res. Camb. 40:255-278.

Hill, W.G., and Alan Robertson. 1966. The effect of linkage on limits to artificial selection. Genetical Res., Camb. 8: 269-294.

Hill, W.G., and Alan Robertson. 1968. Linkage disequilibrium in finite populations. Theor. Appl. Genet. 38:226-231.

Ho, Y.T., and R.E. Comstock. 1980. Combining superior alleles from two homozygous populations in a cross-fertilizing species. Genet. Res. Camb. 36: 277-287.

Horner, E.S., H.W. Lundy, M.C. Lutrick, and W. H. Chapman. 1973. Comparison of three methods of recurrent selection in maize. Crop Science 13: 485-489.

Hull, F.G. 1945. Recurrent selection for specific combining ability in corn. J. Amer. Soc. Agron. 37:134-145.

Hull, Fred H. 1952. Recurrent selection and overdominance. in Heterosis, John W. Gowen, Ed. Iowa State College Press. Ames.

Jain H.K. 1989. Biological and technical constraints on crop and animal productivity : Report on a dialogue (edited by Vernon W. Ruttan). Staff Paper P89-45, Revised December 1989. Department of Agricultural and Applied Economics, University of Minnesota. St. Paul. (pp 11-12).

Jones, Donald F. 1917. Dominance of linked factors as a means of accounting for heterosis. Genetics 2:466-479.

Jones, L.P., W.A. Compton, and C.O. Gardner. 1971. Comparison of full and half-sib reciprocal recurrent selection. Theoretical and applied genetics 41 : 36-39.

Kempthorne, O. 1954. The correlation between relatives in a random mating population. Proc. Roy. Soc. B 143:103-113.

____________. 1957. An introduction to genetic statistics. John Wiley and Sons. New York.

____________.1988. An overview of the field of quantitative genetics. Proceedings of the Second International Conference on Quantitative Genetics. Bruce S. Weir, Eugene J. Eisen, Major M. Goodman, and Gene Namkoong, Eds., pp 47-56, Sinauer Associates Inc., Sunderland, MA

Kimura, M. 1957. Some problems of stochasatic processes in genetics. Ann. Math. Stat. 28: 882-901.

Kimura, M., and T. Ohta. 1969. The average number of generations until fixation of a mutant gene in a finite population. Genetics 61 : 763-771.

Lasher, Craig. 1991. Issues Before The 45th General Assembly of the United Nations (Population section), An Annual Publication of the United Nations Association of the United States of America. John Tessitore and Susan Woolfson, Eds., pp 92-100. D.C. Heath and Company, Lexington, MA and Toronto.

Lerner, I.M. 1950. Population genetics and animal improvement. The University Press, Cambridge.

Lerner, I. Michael, and William J. Libby. 1976. Heredity, evolution, and society. W.H. Freeman and Company. San Fransisco.

Lewontin, R.C. 1974. The genetic basis of evolutionary change. Columbia University Press, New York.

Lewontin, R.C., and J.L. Hubby. 1966. A molecular approach to the study of genic heterozygosity in natural populations. II. Amount of variation and degree of heterozygosity in natural populations of Drosophila psuedoobscura. Genetics 54 : 595-609.

Lonnquist, J.H. 1964. A modification of the ear-to-row procedure for the improvement of maize populations. Crop Science 4 : 227-228.

Lush, Jay L. 1937. Animal breeding plans. Iowa State College Press. Ames.

__________.1943 Animal breeding plans., 2nd ed. Iowa State College Press. Ames.

Lush, J.L. 1947. Family merit and individual merit as bases for selection. Part I. The American Naturalist LXXXI: 241-261.

McNew, R.W., and A.E. Bell. 1976. Comparison of crossbred and purebred selection for a heterotic trait in highly selected populations of Tribolium. The Journal of Heredity 67: 275-283.

Moll, R.H., M.F. Lindsey, and H.F. Robinson. 1964. Estimates of genetic variances and level of dominance in maize. Genetics 49: 411-423.

Moll, R.H., C. Clark Cockerham, C.W. Stuber, and W.P. Williams. 1978. Selection responses, genetic-environmental interactions, and heterosis with recurrent selection for yield in maize. Crop Science 18: 641-645.

Odhiambo, M.O., and W.A.Compton. 1989. Five cycles of replicated S_1 vs. reciprocal full-sib index selection in maize. Crop Science 29:314-319.

Park, Y. C. 1977. Theory for the number of genes affecting quantitative characters I. Estimation of and variance of the estimation of gene number for quantitative traits controlled by additive genes having equal effects. Theoretical and Applied Genetics 50:153 -161.

Parlevliet, J.E. 1979. The multiline approach in cereals to rusts; aspects, problems, and possibilities. Proc. satellite symposium on use of multilines for reducing rust epidemics. Indian J. Genet. Plant Breed. 39:21-28. New Delhi.

Pederson, D.G., and D.F. Matzinger. 1972. Selection for fresh weight in Arabidopsis thaliana under two mating systems. Theor. Appl. Genet. 42:75-80.

Race, R.R., and R. Sanger. 1962. Blood groups in man. 4th ed. F.A. Davis Co. Philadelphia.

Rahnefeld, G.W., W.J. Boylan, R.E. Comstock, and Madho Singh. 1963. Mass selection for post-weaning growth in mice. Genetics 48:1567-1583.

Rawlings, J.O., and D.L.Thompson. 1962. Performance level as criterion for the choice of maize testers. Crop Science 2:217-220.

Robertson, A. 1955. Selection in animals: synthesis. Cold Spring Harbor Symp. Quant. Biol. 20:225-229.

__________. 1960. A theory of limits in artificial selection. Proc. Roy. Soc. London B 153:234-249.

__________. 1962. Selection for heterozygotes in small populations. Genetics 47:1291-1300.

__________.1970. A theory of limits in artificial selection with many linked loci. In Biomathematics, Vol. 1, Edited by K. Krickeberg, R.C.Lewontin, J. Neyman and M. Schreiber. Springer-Verlag. Berlin, Heidelberg, New York.

Robbins, Rainard B. 1918. Some applications of mathematics to breeding problems. III. Genetics 3:375 - 389.

Robinson, H.F., R.E.Comstock, and P. H. Harvey. 1949. Estimates of heritability and the degree of dominance in corn. Agron. J. 41:353-59.

Robinson, H.F., R.E.Comstock, and P.H. Harvey. 1955. Genetic variances in open pollinated varieties of corn. Genetics 40: 45-60.

Rojas, B.A., and G.F. Sprague. 1952. A comparison of variance components in corn yield trials. III. General and specific combining ability and their interaction with locations and years. Agron. J. 44: 462-466.

Russell, W.A. 1974. Comparative performance for maize hybrids representing different eras of maize breeding. Proceedings of the Twenty-Ninth Annual Corn and Sorghum Research Conference 29: 81-101.

Russell, W.A., and S.A. Eberhart. 1975. Hybrid performance of selected maize lines from reciprocal recurrent and testcross selection programs. Crop Science 15:1-4.

Sentz, J.C., H.F. Robinson, and R.E. Comstock. 1954. Relation between heterozygosis and performance in maize. Agron. J. 46:514-520.

Shull, G.H. 1908. The composition of a field of maize. Amer. Breeder's Assoc. Rept. 4:296-301.

________. 1911. The genotypes of maize. Amer. Nat. 45:234-252.

Smith, H.F. 1936. A discriminant function for plant selection. Ann. Eugen. London 7:240-250.

Sprague, G.F., and S.A. Eberhart. 1977. Corn Breeding. in Corn and Corn Improvement, G.F. Sprague ed., American Society of Agronomy, Inc. Madison, WI. pp 305-362.

Sprague, G.F., and W.T. Federer. 1951. A comparison ofvariance components in corn yield trials. II. Error, year x variety, location x variety, and variety components. Agron. J. 43: 535-41.

Standard Mathematical Tables., 15th Edition, Edited by Samuel M. Selby. The Chemical Rubber Co., CRC Publications, Boca Raton, FL.

Stonaker, H.H., and J.L.Lush. 1942. Heritability of conformation in poland-china swine as evaluated by scoring. J. Animal Sci. 1: 99-105.

Stuber, C.W. 1992. Biochemical and molecular markers in plant breeding. pp. 37-61. In J. Janick (ed.) Plant Breeding Reviews, John Wiley & Sons, Inc., NY.

Vencovsky, R., and C.R.M. Godoi. 1978. Immediate response and probability of fixation of favorable alleles in some selection schemes. Proc. Int. Biom. Conf. Boston, Mass. pp 292-297.

Weaver, Robert F., and Phillip E. Hedrick. 1992. Genetics. 2nd Ed. Wm. C. Brown Publishers, Dubuque, IA.

Webel, O.D., and J.H. Lonnquist. 1967. An evaluation of modified ear-to-row selection in a population of corn. Crop Science 7: 651-655.

Weinberg, W. 1908. Uber den nachweis der vererbung beim menschen. Jahresh. Verein f. vaterl. Naturk. Wurttemberg. 64: 369-382.

Weir, B.S., and C.C. Cockerham. 1977. Two locus theory in quantitative genetics. Proc. Internat. Conf. on Quant. Genetics. Edward Pollak, Oscar Kempthorne, and Theodore B. Bailey, Jr., eds., Iowa State Univ. Press. Ames. pp. 247-269.

Williams, J.S. 1962. The evaluation of a selection index. Biometrics 18:375-393.

Wright, S. 1921. Systems of mating II. The effects of inbreeding on the genetic composition of a population. Genetics 6:124-143.

_________.1922. Coefficients of inbreeding and relationship. Amer. Naturalist 56:330-338.

_________.1935. The analysis of variance and the correlations between relatives with respect to deviations from an optimum. J. Genet. 30: 243-256.

_________.1952. The genetics of quantitative variability. in quantitative inheritance, eds. Reeve, E.C.R. & Waddington, C.H.. London: H.M.S.O.

_________.1955. Classification of the factors of evolution. Cold Spring Harbor Symp. Quant. Biol. 20:16-24.

_________.1969. Evolution and the genetics of populations. Vol. 2. The theory of gene frequencies. Univ. of Chicago Press, Chicago.

_________.1977. Evolution and the genetics of populations. Vol. 3. Experimental results and evolutionary deductions. Univ. of Chicago Press, Chicago.

Young, S.S.Y. 1961. A further examination of the relative efficiency of three methods of selection for genetic gains under less-restricted conditions. Genet. Res. 2:106-121.

Printed and bound by CPI Group (UK) Ltd, Croydon, CR0 4YY
16/09/2026
14963449-0004